Jiaju Zhou

Handbook of Active Marine Natural Products

Handbook of Active Marine Natural Products

Jiaju Zhou

Volume 1: Terpenoids, Part 1
ISBN 978-3-11-065324-3, e-ISBN (PDF) 978-3-11-065513-1,
e-ISBN (EPUB) 978-3-11-065331-1

Volume 2: Terpenoids, Part 2
ISBN 978-3-11-065325-0, e-ISBN (PDF) 978-3-11-065515-5,
e-ISBN (EPUB) 978-3-11-065328-1

Volume 3: Alkaloids, Part 1
ISBN 978-3-11-065326-7, e-ISBN (PDF) 978-3-11-065519-3,
e-ISBN (EPUB) 978-3-11-065327-4

Volume 4: Alkaloids, Part 2
ISBN 978-3-11-057083-0, e-ISBN (PDF) 978-3-11-065390-8,
e-ISBN (EPUB) 978-3-11-065373-1

Volume 5: Polyketides and Steroids
ISBN 978-3-11-065363-2, e-ISBN (PDF) 978-3-11-065392-2,
e-ISBN (EPUB) 978-3-11-065371-7

Volume 6: Aliphatic Metabolites
ISBN 978-3-11-065400-4, e-ISBN (PDF) 978-3-11-065579-7,
e-ISBN (EPUB) 978-3-11-065414-1

Volume 8: Peptides and Others
ISBN 978-3-11-065402-8, e-ISBN (PDF) 978-3-11-065583-4,
e-ISBN (EPUB) 978-3-11-065413-4

Jiaju Zhou

Handbook of Active Marine Natural Products

Volume 7: *O*-Heterocycles and Aromatics

DE GRUYTER

Author
Prof. Jiaju Zhou
Chinese Academy of Sciences
1303 Department, 10 Building
31 Zhong Guan Cun Nan Dajie
100081 Beijing
China
jjzhou@mail.ipe.ac.cn

ISBN 978-3-11-065401-1
e-ISBN (PDF) 978-3-11-065581-0
e-ISBN (EPUB) 978-3-11-065416-5

Library of Congress Control Number: 2019941348

Bibliographic information published by the Deutsche Nationalbibliothek
The Deutsche Nationalbibliothek lists this publication in the Deutsche Nationalbibliografie; detailed bibliographic data are available on the Internet at http://dnb.dnb.de.

Typesetting: Integra Software Services Pvt. Ltd.
Printing and binding: CPI books GmbH, Leck
Cover image: Science Photo Library/Douwma, Georgette

www.degruyter.com

Preface

The English edition *Handbook of Active Marine Natural Products* (*HAMNP*) *with 8 Volumes* is a selective version of the Marine Natural Products Dataset. The whole dataset was collected and developed by the Molecular Design Group, Institute of Process Engineering, Chinese Academy of Sciences during 1998–2016. Totally, it covers 19,722 entries of secondary metabolites from marine living things, where 8,350 compound entries have pharmacological activity data. The 8,350 compound entries were arranged into eight volumes to form the set of handbooks as follows:

Volume 1: Terpenoids, Part 1
Volume 2: Terpenoids, Part 2
Volume 3: Alkaloids, Part 1
Volume 4: Alkaloids, Part 2
Volume 5: Polyketides and Steroids
Volume 6: Aliphatic Metabolites
Volume 7: O-Heterocycles and Aromatics
Volume 8: Peptides and Others

This set of eight HAMNP books gathers the structure, origin, and bioactivity, as well as other relevant information, of 8,350 active marine natural products from 3,025 marine organisms.

The HAMNP handbooks represent a largest collection of active secondary metabolites from marine organisms, and all kinds of scientific data have been reorganized as well-formatted data so that the books became helpful to researchers as a convenient reference. The materials covered in these books include those through systematic collection up to 2012, and further accompanied with the latest data published in several core journals until 2016.

The work covered in these HAMNP books was accomplished in two phases. The initial phase ranged from 1998 to 2001 and the main phase from 2011 to 2018. In the original version of the dataset, more than 22,000 compounds have been collected, including duplicated compounds from different authors. The comprehensive data compilation process include data specification definition, cross-validation, assessment confirmation, identification of duplicated structures, and merging of relevant information, leading to the final accomplishment of the current 19,722 datasets.

In brief, the main compilation process of the HAMNP books is given as follows. First, collect the name list, origin, and structure of chemical compounds from successive annual reviews (see Core References R01 and R02 in Introduction) and literature reviews. Second, double-check the documents to verify and complete other information. Third, confirm structural information and other types of data using orthogonal information from other sources with cross-validation methods. Fourth, the structures of more than 22,000 compounds are rechecked and the information is integrated by manual identification and computer programs. Finally, the comprehensive

https://doi.org/10.1515/9783110655810-201

information on the 19,722 compounds constitutes the dataset. Here, 8,350 active sets were picked up from the dataset to form the current HAMNP handbook.

Three problems need to be solved to compile a multidisciplinary reference book. First, every definition and concept should be explicit when expanding knowledge, connotation, and extension included, without any research details. Second, the reliability assessment is essential for all kinds of data, because the devil is in the detail. Third, it is essential to search, identify, and integrate data of duplicated chemical compounds. Fortunately, well-developed software packages can help us automatically identify the majority of duplicated chemical compounds. The remaining issues can be resolved along with manual processing.

It is the guiding principle of the author to make the book to be pithy, thorough, precise, and intelligible. In fact, we always view ourselves as HAMNP's readers, with the exclusive objective to let readers gain the most useful knowledge in the shortest possible time.

The core contents and highlights of the HAMNP books are the "three diversities," that is, the diversity of chemical structures, the diversity of biological resources, and the diversity of pharmacological activities. In terms of chemical structure diversity, we refer to the classification system from references, then further improve and expand it based on the latest research and development to define our classification framework of structures. Once readers browse the contents of the books, the classification system is straightforward. For the diversity of biological resources, it is recommended to refer to Index 3 in each volume – Compound Marine Organism Source Index, and Index 4 in each volume – Compound Marine Source Sampling Geographic Location Index. For the diversity of pharmacological activities, it is recommended to refer to Index 5 in each volume – Compound Pharmacological Activity Index.

These HAMNP handbooks are expected to help readers who are engaged in research, in teaching, and in the development of marine natural products. It should also benefit college students, postgraduates, marine resource managers, and those who are interested in the chemistry and pharmacology of marine natural products. We would feel fortunate if it works as expected.

Zhou jiaju

Jiaju Zhou
Institute of Process Engineering (IPE),
Chinese Academy of Sciences (CAS)
February 2019

Contents

About the Author

Prof. Jiaju Zhou was born in October 1939 in Tianjin, China. He graduated from Rare Earth Inorganic Chemistry Specialty, Chemistry Department, Peking University, in 1963 under a six-year program. Before he retired in 2008, Zhou was the leader of Molecule Design Group, IPE, CAS. Zhou's areas of research include rare earth chemistry, mineral analytical chemistry, chemical industry process simulation (in IPE, CAS and UBC, Canada), design of crystal structural database (in OSRD, NIST, Gaithersburg, MD, USA), scientific database R&D, and computer-aided and artificial intelligence drug design. Zhou developed the first TCM database (TCMDB) with 23,033 entries. Since 2008, he has worked on Marine Natural Products project and has developed the Marine Natural Products Database (MNPDB) with 19,722 entries.

https://doi.org/10.1515/9783110655810-202

Introduction

The *Handbook of Active Marine Natural Products* covers eight volumes. This book is *Volume 7: O-Heterocycles and Aromatic*, which includes 951 active compounds.

Format of Compound Entry. A compound entry starts with a title line, which has two items: the compound's unique code (from 1 to 951 for volume 7) and the main name. The following seven items form the title line as a body, and the graphic structure is placed at the end:

Title line (code number, main name)
A. Synonyms of the compound (if any)
B. Structural type
C. Formula (relative molecular mass)
D. Physicochemical properties
E. Marine source(s)
F. Pharmacological data (if any)
G. Reference(s)
Graphic structure

Chemical Names and A. Synonyms. Generally, a compound may have one scientific name and several trivial names. In the handbooks, based on original articles, we select one name as the "main name." The main name appeared at the title line of each compound entry. In most cases, a trivial name was selected as the main name, and in some cases, the main name is a scientific name. Any synonyms, if any, are presented after the title line as an item of the entry body.

B. Structural Type. Structural type is the second item, ordered by the contents order.

F. Normalization of Pharmacological Data. All of 951 MNP components in this book have pharmacological data, which are very valuable. Because different expressions are used for the same kind of data in different articles, we have to define and normalize thousands of pharmacological terms, so that the data could be expressed in a unified way, and be easily understood by readers.

Stereochemistry in Graphic Structure. We protracted all compound structures down to the atomic bond level including complicated glycosides, with stereochemical information based on the data in the original papers. For example, the structure with full stereochemistry of compound 331 (+)-Aeroplysinin 1 is

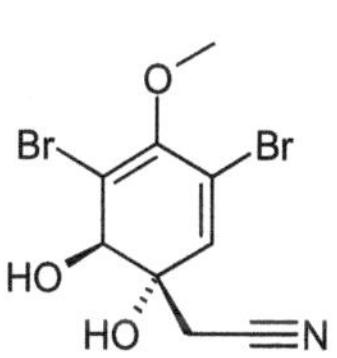

https://doi.org/10.1515/9783110655810-203

Let us further explain the data structure of source terms and pharmacological terms.

Source Terms

The source data of compound 331 (+)-Aeroplysinin 1 is

<u>**Source**</u>:
Sponge *Aplysina aerophoba* [Syn. *Verongia aerophoba*],
Sponge *Aplysina archeri*,
Sponge *Aiolochroia crassa*,
Sponge *Verongula rigida*,
Sponge *Psammaplysilla purpurea*,
Sponge *Ianthella* sp.

The format is as follows (banding the English-type name and the Latin name together):

<u>**Source**</u>:
English-type name + Latin name of source 1,
English-type name + Latin name of source 2,
English-type name + Latin name of source 3,
English-type name + Latin name of source 4,
English-type name + Latin name of source 5,
English-type name + Latin name of source 6.

Pharmacological Terms

The pharmacological terms in the handbooks are presented in a multilayered structure. In the top layer, there are more than 20 types of the most important pharmacological activity terms. They are cytotoxic (*in vitro* anticancer), antineoplastic (*in vivo* anticancer), antibacterial, antifungal, antiviral, anti-HIV, anti-inflammatory, antioxidant, antimalarial, NO production inhibitors, enzyme inhibitors, cardiovascular activity, smooth muscle relaxant and stimulant, toxin and medium lethal dose (LD_{50}), and so forth. Readers need to be familiar with these Tope lever pharmacological terms (see Table 1).

For each term, there is a regulation about how to describe related pharmacological data. The following is an example. Under the subtitle "<u>Pharm</u>:" of compound 331 (+)-Aeroplysinin 1, a set of multiple biodata is presented as follows:

<u>**Pharm:**</u>
Antibacterial (*Photobacterium phosphoreum*, IC_{50} = 3.5 μmol/L);

Table 1: Twenty-Four Main Pharmacological Terms in Tope Lever.

Order in Index 5	Pharmacological Terms in Tope Lever
1	Anti-AD
2	Antibacterial
3	Antifungal
4	Anti-HIV
5	Anti-inflammatory
6	Antileishmanial
7	Antimalarial
8	Antineoplastic (in vivo)
9	Antioxidant
10	Antiplasmodial
11	Antitrypanosomal
12	Antituberculosis
13	Antiviral
14	Cardiovascular activity
15	Cell cycle inhibitor
16	Cell division inhibitor
17	Cell growth inhibitor
18	Cell adhesion inhibitor
19	Cytotoxic (in vitro)
20	Enzyme inhibitors
21	NO production inhibitors
22	Smooth muscle relaxant and stimulant
23	Toxin
24	Medium lethal dose (LD_{50})

antimicroalgal (marine microalgae *Coscinodiscus wailesii,* IC_{50} = 5.6 µmol/L, *Prorocentrum minimum,* IC_{50} = 7.0 µmol/L);
antiviral (HIV-1, IC_{50} = 14.6 µmol/L);
anti-angiogenic (EVLC-2 cell line, 2.5 µmol/L, target MMP-2, effect decrease; HUVEC cell line, 2.5 µmol/L, target MMP-2, effect decrease; RF-24 cell line, 2.5 µmol/L, target MMP-2, effect decrease; BAEC cell line, 3 µmol/L, target MMP-2, effect decrease; BAEC cell line, 3 µmol/L, target PA, effect decrease; BAEC cell line, 3 µmol/L, target PAI, effect increase);
anti-inflammatory (HUVEC cell line, 10 µmol/L, target MCP-1, effect decrease; HUVEC cell line, 10 µmol/L, target TSP-1, effect decrease; HUVEC cell line, 10 µmol/L, target COX-2, effect decrease; HUVEC cell line, 20 µmol/L, target Il-1*α*, effect decrease; HUVEC cell line, 20 µmol/L, target MMP-1, effect decrease);

apoptogenic (BAEC cell line, 10 μmol/L, target Cleaved lamin-A, effect increase; BAEC cell line, 10 μmol/L, target Caspase-2, -3, -8, -9, effect increase; BAEC cell line, 10 μmol/L, target Cytochrome C, effect increase in cytoplasm; HUVEC cell line, 10 μmol/L, target p-Bad, effect increase);
anti-angiogenic (HUVEC (hmn umbilical vein endothelial cells), IC_{50} = 4.7 μmol/L; EVLC-2 (SV40 large T-antigen immortalized hmn umbilical vein cells), IC_{50} = 3.0 μmol/L; RF-24 (papillomavirus 16 E6/E7 immortalized hmn umbilical vein cells), IC_{50} = 2.8 μmol/L; HMEC (hmn microvascular endothelial cells), IC_{50} = 2.6 μmol/L; BAEC (bovine aortic endothelial cells,), IC_{50} = 2.1 μmol/L);
cytotoxic (HT1080, IC_{50} = 2.3 μmol/L, HTC116, IC_{50} = 4.7 μmol/L, HeLa, IC_{50} = 3.0 μmol/L, THP-1, IC_{50} = 10.0 μmol/L, NOMO-1, IC_{50} = 17.0 μmol/L, HL60, IC_{50} = 5.0 μmol/L);
antibiotic (Gram-positive bacteria and several dinoflagellate microalgae causing toxic blooms);
EGFR inhibitor (IC_{100} = 0.5 μmol/L);
antitumor (strong effect on EGFR tumor cell lines, blocking proliferation of EGFR-dependent hmn breast cancer cell lines MCF7 and ZR-75-1);
molluscacidal;
Note: (+)-Aeroplysinin 1 has a wide spectrum of bioactivities. In preclinical studies, it has been shown to have promising anti-inflammatory, antiangiogenic and antitumor effects. It might have a pharmaceutical interest for the treatment of different pathologies.

The format is as follows:

Pharm:
Term name 1 (formatted detail information)
Term name 2 (formatted detail information)
Term name 3 (formatted detail information)
Term name 4 (formatted detail information)
Term name 5 (formatted detail information)
Term name 6 (formatted detail information)
Term name 7 (formatted detail information)
Term name 8 (formatted detail information)
Term name 9 (formatted detail information)
Term name 10 (formatted detail information)
Term name 11 (formatted detail information)
Term name 12 (formatted detail information)
(**Note:** if any)

Under the *term name Cytotoxic*, a set of multiple cytotoxic biodata is presented as follows:

cytotoxic (
HT1080, IC_{50} = 2.3 μmol/L,
HTC116, IC_{50} = 4.7 μmol/L,
HeLa, IC_{50} = 3.0 μmol/L,
THP-1, IC_{50} = 10.0 μmol/L,
NOMO-1, IC_{50} = 17.0 μmol/L,
HL60, IC_{50} = 5.0 μmol/L);
EGFR inhibitor, IC_{100} = 0.5 μmol/L;
MCF7 and ZR-75-1 (blocking proliferation)

The format is as follows:
Term name (*in vitro/in vivo*,
target cancer cell 1, quantitative data,
target cancer cell 2, quantitative data,
target cancer cell 3, quantitative data,
target cancer cell 4, quantitative data,
target cancer cell 5, quantitative data,
target cancer cell 6, quantitative data,
brief description of related mechanism if any).

In order to standardize abbreviations of cancer cells, such as P_{388}, A549, HT29, MEL28, CCRF-CEM, and DLD-1, we defined and used 438 cancer cell codes (CCC) in the handbooks. For explanation of these codes, please see "List of Cancer Cell Codes."

By means of the formatted and structuralized methods, we normalized expressions of almost all the pharmacological data presented in the books. For complete information in volume 7, of all 959 normalized pharmacological activity terms, please see "Index 5 Compound Pharmacological Activity Index."

In summary, these handbooks with eight volumes provide an integrated collection of 8,350 marine natural products chemical components isolated from 3,025 marine organisms and a large amount of pharmacological activity data of these components. It might be used not only as a handbook to look for structures and bioactivities of marine natural products and marine organisms source information, but also as a fundamental platform for studying the marine natural products with a systematic and integrative approach.

Acknowledgments

First, as the author of those books, I would like to give my heartfelt thanks to Dr. David Lide and B.J. Lide, who were my directors 30 years ago when I worked in OSRD, NIST (former NBS), USA, in 1985–1986 for nine months. They gave me a rare opportunity to learn how to use a software platform and how to treat a complicated scientific information data system. It is my research experience in NBS that helped me to compile easily the current huge project on Marine Natural Products. At the same time, I also give my *sincere* thanks to my NBS's colleagues: Dr. John Rumble, Mrs. Geraldine Dalton, Mrs. Phoebe Fagan, and other OSRD members.

Then, I would like to give my genuine thanks to the following close friends. They gave my MNP project continual concerns and supports for years: Dr. Jun Xu, Professor and Director, Research Center for Drug Discovery, Sun Yet-Sen University, 132 East Circle, University City, Guangzhou, 510006, China, and Dr. Leming Shi, Professor and Director, Center for Pharmacogenomics, School of Life Sciences and Shanghai Cancer Center, Fudan University, Shanghai 200438, China (lemingshi@fudan.edu.cn).

Third, I like to give my honest thanks to my following group members. For many years, all of them gave various devices to me:

1 Dr. Jing Lei, Associate Professor, Educational Equipment Research and Development Centre, Ministry of Education of the People's Republic of China, Beijing 100080, China (early research in her doctor thesis)
2 Dr. Bing Liu, Lead Dev Prophix Software Inc. 350 Burnhamthorpe Road West, Suite 1000 Mississauga, Ontario L5B 3J1, Canada (data collection in the early stage)
3 Master Yingxin Qiao, Software Engineer, National Library of China, Beijing 100081, China (data source searching and original paper collection)
4 Dr. Haibo Liu, Associate Professor, The Institute of Medicinal Plant Development, Chinese Academy of Medical Sciences, Beijing 100193, China (special software development for automatic edition)
5 Dr. Tao Peng, Associate Professor, College of Robotics, Beijing Union University, Beijing 1001011, China (special software development for index generation)
6 Dr. Aihua Xie, Associate Professor, School of Pharmacy, Hebei Chinese Medical University, Shijiazhuang, Hebei 050200, China (part of data collection)
7 Dr. Chenzhong Liao, Professor, Dean of Department of Pharmacy, School of Biological and Medical Engineering, Hefei University of Technology, Hefei 230009, China (original paper collection)
8 Dr. Jianfeng Pei, Associate Professor, Center for Quantitative Biology, Academy for Advanced Interdisciplinary Studies, Peking University, Beijing 100871, China (data collection in the early stage)
9 Dr. Xianfeng He, Associate Professor, Scientific Researcher, EMMS Group, State Key Laboratory of Multiphase Complex Systems, Institute of Process Engineering, Chinese Academy of Sciences, Beijing 100190, China (data collection in the early stage)
10 Madam Guirong Xie, Associate Professor, Institute of Process Engineering, Chinese Academy of Sciences, Beijing 100190, China (part of data compilation)
11 Mr. Wucheng Tang, Engineer, Institute of Process Engineering, Chinese Academy of Sciences, Beijing 100190, China (part of original paper collection)

Finally, I thank my family members. Without their complete and never-ending support, this book would never have been possible.

Core References

(Guiding References 5)

R01 D. J. Faulkner, Marine Natural Products (review), Nat. Prod. Rep., 1986, 3, 1–33; 1987, 4, 539–576; 1988, 5, 613–663; 1990, 7, 269–309; 1991, 8, 97–147; 1992, 9, 323–364; 1993, 10, 497–539; 1994, 11, 355–394; 1995, 12, 223–269; 1996, 13, 75–125; 1997, 14, 259–302; 1998, 15, 113–158; 1999, 16, 155–198; 2000, 17, 7–55; 2001, 18, 1R–49R; 2002, 19, 1–49

R02 J. W. Blunt, et al, Marine Natural Products (review), Nat. Prod. Rep., 2003, 20, 1–48; 2004, 21, 1–49; 2005, 22, 15–61; 2006, 23, 26–78; 2007, 24, 31–86; 2008, 25, 35–94; 2009, 26, 170–244; 2010, 27, 165–237; 2011, 28, 196–268; 2012, 29, 144–222; 2013, 30, 237–323; 2014, 31, 160–258; 2015, 32, 116–211

R03 J. Buckingham (Executive Editor), Dictionary of Natural Products, Chapman & Hall, London, Vol. 1–Vol. 7 1994; Vol. 8, 1995; Vol. 9, 1996; Vol. 10, 1997; Vol. 11, 1998

R04 CRC Press, Dictionary of Natural Products on DVD, version 20.2, 2012

R05 Jean-Michel Kornprobst, Encyclopedia of Marine Natural Products, Vol. 1–Vol. 3, 2nd Edition, WILEY BLACKWELL, Germany, 2014

(Dictionaries 17)

R06 P.M. Kirk, P.F. Cannon, D.W. Minter and J.A. Stalpers, Dictionary of the Fungi, 10th Edition, CABI Europe-UK, 2011

R07 Miaoying Cai, et al., Names of Bacteria, 2nd Edition, Science Press, Beijing, 1996

R08 Rui-Fu Yang et al, Dictionary of Bacterial Names with English Explanation and Chinese Translation, Chemical Industry Press, Beijing, 2011

R09 Zongxun Wang et al. (Institute of Botany, Chinese Academy of Sciences), New Edited Plant Names in Latin-Chinese-English, Aerial Industry Press, Beijing, 1996

R10 Zhong-Yan Qi and Xi-Xing Liu, New Names of Invertebrate Animals in Latin-Chinese, Science Press, Beijing, 1999

R11 Ling-Ti Lu and Jia-Ran Zhu, Dictionarium Lantino-Sinicum de Scientia et technologia, The Commercial Press, Beijing, 2017

R12 Ji-Sheng Chen, et al., English-Chinese Dictionary of Life Science, Scientific and technological Literature Press, Beijing, 1992

R13 P. Singleton and D. Sainsbury (Qing-jun Ma and Cheng-hua Shi et al. translated), Dictionary of Microbiology and Molecular Biology, Chemical Industry Press, Beijing, 2008

R14 Scientific Terms Laboratory of Science Press, English-Chinese Dictionary of Chemistry and Chemical Engineering, 4th Edition, Science Press, Beijing, 2000

R15 Scientific Terms Laboratory of Science Press, English-Chinese Dictionary of Chemistry and Chemical Engineering, 5th Edition, Science Press, Beijing, 2016

R16 Jian Zhuge and Zheng-Xiang Wang, Modern English-Chinese Dictionary of Biotechnology, Science Press, Beijing, 2003

R17 Jing-Ying Tan, English-Chinese Biological Dictionary of Biochemistry and Molecular Biology, 2nd Edition, Science Press, Beijing, 2007

R18 Scientific Terms Laboratory of Science Press, English-Chinese Biological Dictionary, 2nd Edition, Science Press, Beijing, 1997

R19 Scientific Terms Laboratory of Science Press, Chinese-English Biological Dictionary, 2nd Edition, Science Press, Beijing, 1998

R20 Yu Hui, A New Century Chinese-English Dictionary, Foreign Language Teaching and Research Press, Beijing, 2003

R21 Zong-Guo Huang and Mei-Ling Jin, Dictionary of Marine Biology, Ocean Press, Beijing, 1994 (in Chinese)

R22 Wenbao Chang, et al., Dictionary of Chemistry, Science Press, Beijing, 2008 (in Chinese)

(Book References 11)

R23 Hua-Shi Guan and Shu-Guang Wang, Zhong-hua Hai-yang Ben-cao, Marine Natural Products, 3 Volumes, Chemical Industry Press and Shanghai Science and Technology Press, Beijing, 2009 (in Chinese)

R24 C. J. Alexopoulos, M. Blackwell and C. W. Mims, (Yijian Yao and Yu Li translated), Introductory Mycology, Fourth Edition, John Wiley & Sons, Inc., 1996, Chinese Agricultural Press, Beijing, 2002

R25 Janet S. Dodd, The ACS Style Guide, A Manual for Authors and Editors, 2nd Edition, American Chemical Society, Washington, DC, 1997

R26 Shu-Xian Ren, Invertebrates, 2 volumes, Peking University Press, Beijing, 1990 (in Chinese)

R27 R. Mcneill Alexander, (translated by Lan-zhi Du), The invertebrates, Chemical Industry Press, Beijing, 2013 (in Chinese)

R28 Yanghua Yi and Binghua Jiao, Modern Marine pharmacology, Science Press, Beijing, 2006 (in Chinese)

R29 Chang-Yun Wang and Chang-Lun Shao, Marine pharmacology, Science Press, Beijing, 2011 (in Chinese)

R30 Rensheng Xu, et al., Chemistry of Natural Products, 2nd Edition, Science Press, Beijing, 2004 (in Chinese)

R31 Yue-Zeng Chen, General Biology, Higher Education Press, Beijing, 1997 (in Chinese)

R32 Jiaju Zhou, Guirong Xie and Xinjian Yan, Encyclopedia of traditional Chinese Medicines, Molecular Structures, Pharmacological Activities, Natural Sources and Applications, Vol. 1–Vol. 6, Springer, Heidelberg Dordrecht London New York, 2011

R33 Jiaju Zhou, Guirong Xie and Xinjian Yan, TCM Series of Active Components, 10 books, Science Press, Beijing, 2012 (in Chinese)

How to Use the HAMNP Books

In essence, from data computerization point of view, scientific knowledge is the expression of interrelation between research objects in different types. During a long coastline without computer, people learn and spread scientific knowledge in traditional ways, including education, reading, and exchanging information with exchange each other. In today's world, using computer's powerful functions, we have a new way to learn systematical, complete knowledge. In short, a study process in the new way is to *search and learn some relationships*.

Next, we discuss concretely how to use the HAMNP books.

In these books, there are three kinds of data and three pairs of important relations. Three kinds of data are (1) marine living sources (source); (2) secondary metabolites (compounds); and (3) pharmacological activities (pharm-activity). The three pairs of important relations are (1) relationship between source and compounds; (2) relationship between compounds and pharm-activity; and (3) relationship between source and pharm-activity. In the case of asking questions, each relation has two directions; hence, together there are six types of questions:

Type 1: from known source to unknown compound
Type 2: from known compound to unknown source
Type 3: from known compound to unknown pharm-activity
Type 4: from known pharm-activity to unknown compound
Type 5: from known source to unknown pharm-activity
Type 6: from known pharm-activity to unknown source (Figure 1)

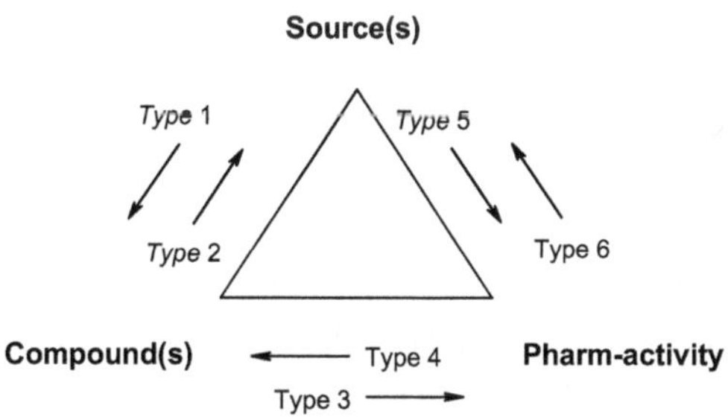

Figure 1: Kinds of Data and Six Types of Questions.

(1) An Illustration of Type 1 (and Type 3, Type 5) Question

Up to now, what *O*-heterocycles (1–259) and aromatic compounds (260–951) are isolated from ascidian of genus *Synoicum*?

From index 3 of volume 7, one will get the following related data in detail:
Synoicum blochmanni 591, 592, 593, 595, 599, 600, 601, 602.

https://doi.org/10.1515/9783110655810-204

Synoicum globosum 595, 596.
Synoicum prunum 248, 249, 250, 591.
Synoicum sp. 585, 586, 587, 588, 603.

The answer of the current question must be that 3 *O*-heterocycle compounds (248, 249, and 250) and 14 aromatic compounds (585–588, 591–593, 595, 596, 599, and 600–603) are isolated from ascidian of genus *Synoicum*.

Then, readers can enjoy to studying these 17 compounds by reading the book, including their pharm-activity (question of types 3 and 5). For example, with entry 248, a reader will know that the compound 248 Prunolide A had already been isolated from ascidian *Synoicum prunum* (Australia) and it has cytotoxic activity (inhibits growth of HeLa at 25μmol/L, weak).

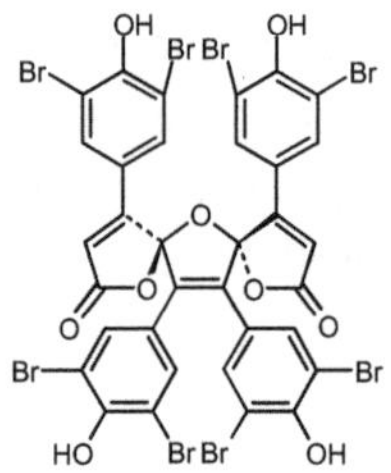

(2) An Illustration of Type 4 (and Type 2, Type 6) Question

"What are isolated aromatic compounds (260–951) with antioxidant activity? And what are their marine sources?"

To browse Index 5 of volume 7, searching "Antioxidant" and "Antioxidant inactive," the following results were obtained (The "Antioxidant inactive " means that antioxidant pharmacological assay has been done, but the result is not active or activity being very low):

Antioxidant 278, 280, 401, 559, 767, 768, 782, 889, 921, 941, 942.
Antioxidant, DPPH radical scavenger 267, 282, 284, 289, 290, 292, 293, 294, 295, 296, 297, 301, 303, 305, 306, 343, 361, 362, 381, 382, 383, 384, 460, 462, 546, 550, 639, 691, 698, 699, 701, 770, 771, 807, 812, 814, 819, 864, 865, 913.
Antioxidant, free radical scavenger 325.
Antioxidant, free radicals scavenger, DPPH radical 288, 292.
Antioxidant, free radicals scavenger, $NO^{\bullet}$ radical 288, 292.
Antioxidant, free radicals scavenger, $O_2^{\bullet -}$ radical 292.
Antioxidant, free radicals scavenger, $ONOO^{-}$ radical 288, 292.
Antioxidant, hydroxyl radical scavenger 657.
Antioxidant, increased antioxidative glutathione peroxidase, catalase, and superoxide dismutase activities 553.

Antioxidant, $O_2^{\bullet-}$ radical scavenger 288.
Antioxidant, radical scavenger 333, 334.
Antioxidant, reduction of ROS overproduction 547, 553.
Antioxidant, significant 656.
Antioxidant, thiobarbituric acid reactive substance TBARS assay 271, 272.
Antioxydant, DPPH scavenger 529.
Antioxidant inactive, DPPH radical scavenger 388, 433, 434, 435, 436.
Antioxidant inactive, superoxide anion radical scavenger 388, 433, 434, 435, 436.

Further, from the entry bodies of the 62 active aliphatic compounds (267, 271, 272, 278, 280, 282, 284, 288–290, 292–297, 301, 303, 305, 306, 325, 333, 334, 343, 361, 362, 381–384, 401, 460, 462, 529, 546, 547, 550, 553, 559, 639, 656, 657, 691, 698, 699, 701, 767, 768, 770, 771, 782, 807, 812, 814, 819, 864, 865, 889, 913, 921, 941, and 942), all sources can be obtained (See Table 2).

Table 2: Answer to the Above Type 4, 2, and 6 Questions.

Vol.	Code	Compound Name	Type	Related Sources
7	267	Citrinin H_2	*m*-Benzenediol derivatives	Marine-derived fungus *Penicillium* sp. MFA446 from green alga *Ulva pertusa* (Korea waters)
7	271	Acremonin A	*p*-Benzenediol derivatives	Marine-derived fungus *Acremonium* sp.
7	272	Acremonin A 5-*β*-glucopyranoside	*p*-Benzenediol derivatives	Marine-derived fungus *Acremonium* sp.
……	……	……	……	……
7	921	*δ*-Indomycinone	Pyrano[*b*] anthraquinones	Marine-derived streptomycete *Streptomyces* sp. B8300 (Lagoon, Gulf of Mexico), marine-derived streptomycete *Streptomyces* sp. (psychrophilic, cold water, depth of 4680 m sediment core, culture broth)
7	941	5,7-Dihydroxy-1-(hydroxymethyl)-6*H*-anthra[1,9-*bc*]thiophen-6-one	Miscellaneous polycyclic aromatics	Bryozoan *Dakaria subovoidea* (Japan waters)
7	942	5,7-Dihydroxy-6-oxo-6*H*-anthra[1,9-*bc*]thiophene-1-carboxylic acid, methyl ester	Miscellaneous polycyclic aromatics	Bryozoan *Dakaria subovoidea* (Japan waters)

In summary, by using three parts of the books – the contents (ordered by structural classifications), the text (8,350 compound entries in volumes 1–8), and the indexes, readers can easily gain well-formatted systematically related knowledge in multidisciplinary fields.

List of Abbreviations and Acronyms

$[^{3}H]$AMPA	$[^{3}H]$-1-amino-3-hydroxy-5-methyl-4-isoxazolepropionic acid
$[^{3}H]$CGS-19755	*N*-methyl-D-aspartic acid (NMDA) receptor antagonist
$[^{3}H]$CPDPX	$[^{3}H]$-1,3-dipropyl-8-cyclopentylxanthine
$[^{3}H]$DPDPE	opioid peptide
$[^{3}H]$KA	$[^{3}H]$-kainic acid
‡	homonym mark
3Y1	rat fibroblasts
5-FU	5-fluorouracil
6-MP	6-mercaptopurine
6-OHDA	6-hydroxydopamine
AAI	antioxidant activity index (final DPPH concentration/EC_{50})
ABRCA	amphotericin B-resistant *Candida albicans*
ABTS$^{\bullet+}$	2,2′-azino-bis-(3-ethyl benzthiazoline 6-sulfonic acid), radical
ACAT	Acyl-CoA: cholesterol acyl transferase
ACE	angiotensin-converting enzyme
AChE	acetylcholinesterase
ACTH	adrenocorticotropic hormone
ADAM9	ADAM9 protease
ADAM10	ADAM10 protease
ADM	adriamycin
AGE	advanced glycation end products
AIDS	acquired immune deficiency syndrome
AKT	ribosomal protein
AKT1	protein kinase
ALK	protein kinase
AMPB	amphotericin B
AP-1	transcription factor
APOBEC3G	hmn innate intracellular antiviral factor (recombinant protein)
aq	aqueous solution
ARCA	amphotericin-resistant *Candida albicans*
ARK5	protein kinase
ATCC	American Type Culture Collection
ATPase	adenosine triphosphatase
Aurora-B	protein kinase
AXL	protein kinase
AZT	3′-azido-3′-deoxythymidine
BACE	β-secretase
BACE1	β-secretase
BCG	Bacille Calmette-Guérin
Bcl-2	a cell survival promoting factor
BoMC	further abbreviation on *Bioorg. Med. Chem.*
BoMCL	further abbreviation on *Bioorg. Med. Chem. Lett.*
bp	boiling point
c	concentration
CaMKIII	protein kinase
cAMP	cyclic adenosine monophosphate
CAPE	caffeic acid phenethyl ester

https://doi.org/10.1515/9783110655810-205

caspase-3	caspase-3 protein
CB	cytochalasin B
CC_{50}	IC_{50} of cytotoxicity (concentration of the 50% cytotoxic effect)
CCR5	chemokine receptor 5
CD	concentration required to double the specific activity
Cdc2	cyclin-dependent kinase
Cdc25	protein Cdc25 phosphatase
Cdc25a	protein phosphatase
Cdc25b	recombinant hmn phosphatase
CDDP	*cis*-diaminedichloroplatinum (cisplatin)
CDK	cyclin-dependent kinase
CDK1	protein kinase
CDK2	protein kinase
CDK4	protein kinase
CDK4/cyclin D1	cyclin-dependent kinase 4 (CDK4) in complex with its activator cyclin D1
P25	protein kinase
p25	protein kinase
CDK7	protein kinase
c-erbB-2	protein kinase
CETP	cholesteryl ester transfer protein
cGMP	cyclic guanylic acid, cyclic guanosine monophosphate
CGRP	calcitonin gene-related peptide
ChAT	choline acetyltransferase
CMV	CMV protease
CNS	central nervous system
COMPARE	COMPARE is an algorithm to analyze data
ConA	concanavalin A
COX-1	cyclooxygenase-1
COX-2	cyclooxygenase-2
CPB	further abbreviation on *Chem. Pharm. Bull.*
$cPLA_2$	cytosolic 85 kDa phospholipase
CPT	camptothecin
CRPF	chloroquine-resistant *Plasmodium falciparum*
CRPF FcM29	chloroquine-resistant *Plasmodium falciparum* FcM29
CSPF	chloroquine-sensitive *Plasmodium falciparum*
Cyp1A	aromatase cytochrome P450 1A
CYP1A	cytochrome P450 1A
CYP450 1A	cytochrome P450 1A
d	day
D	diameter (mm)
Delta	difference in $\log_{10}$ GI_{50} (mol/L) value of the most sensitive cell line and MG-MID value
DGAT	diacylglycerol acyltransferase
DHFR	dihydrofolate reductase
DMSO	dimethyl sulfoxide
DNA	deoxyribonucleic acid
DOX	doxorubicin
DPI	diphenylene indonium
DPPH	1,1-diphenyl-2-picrylhydrazyl free radical

DRPF	drug-resistant *Plasmodium falciparum*
DRS	drug-resistant *Staphylococcus* sp.
DSPF	drug-sensitive *Plasmodium falciparum*
DYRK1A	protein kinase
EBV	Epstein–Barr virus
EC	effective concentration
EC_{50}	medium effective concentration
ED_{50}	effective dose for 50%
ED_{50}	medium effective dose (sometimes for the medium effective concentration)
EGF	epidermal growth factor
EGFR	epidermal growth factor receptor
EL-4	lymphoma cell line with resistance to natural killer cells
ELISA	enzyme-linked immunosorbent assay
EPI	epirubicin
ERK	extracellular signal-regulated protein kinase
ESBLs	extended spectrum β-lactamase
EurJOC	further abbreviation on *Eur. J. Org. Chem.*
FAK	protein kinase
FBS	fetal bovine serum
FLT3	a protein tyrosine kinase
Flu-A	influenza virus type A
Flu-B	influenza virus type B
fMLP/CB	*N*-formyl-L-methionyl-L-leucyl-L-phenylalanine/cytochalasin B
FOXO1a	downstream target of PTEN tumor suppressor
fp	freezing point
FPT	farnesyl protein transferase
FRCA	fluconazole-resistant *Candida albicans*
FtsZ	a structural homolog of eukaryotic tubulin, a GTPase
FXR	farnesoid X receptor
GABA	γ-aminobutyric acid
GI_{50}	the concentration of sample necessary to inhibit the growth to 50% of the control.
GlyR	glycine-gated chloride channel receptor
gp41	a transmembrane protein of HIV-1 (recombinant protein)
gpg	guinea pig
GPR12	G protein-coupled receptor 12; it can be a significant molecular target for treating a variety of neurological disorders
GRP78	molecular chaperone (chaperone)
GST	glutathione S-transferases
GTP	guanosine triphosphate
GU4	*Candida albicans*-sensitive GU4 strain
GU5	*Candida albicans*-resistant GU5 strain
h	hour
H1N1	influenza virus H1N1
H3N2	influenza virus H3N2
H5N1	influenza virus A H5N1
HBV	hepatitis B virus
HC_{50}	medium hemolytic concentration
HCMV	hmn cytomegalovirus

HCV	hepatitis C virus
HD	a positive control compound, no concrete explanation in original paper (J. Qin, et al., BoMCL, 2010, 20, 7152)
HER2	tyrosine kinase
HF	hypersensitivity factor
HIF-1	hypoxia-inducible factor-1
HIV	hmn immunodeficiency virus
HIV-1	hmn immunodeficiency virus type 1
HIV-1 IIIB	hmn immunodeficiency virus type 1 IIIB
HIV-1 in	hmn immunodeficiency virus type 1 integrase
$HIV\text{-}1_{RF}$	hmn immunodeficiency virus RF
HIV-1-rt	hmn immunodeficiency virus type 1 reverse transcriptase
HIV-2	hmn immunodeficiency virus type 2
HIV-rt	hmn immunodeficiency virus reverse transcriptase
HLE	hmn leukocyte elastase
HMG-CoA	3-hydroxy-3-methylglutaryl coenzyme A reductase
hmn	human
HNE	hmn neutrophil elastase
$HO^{\bullet}$	hydroxyl radical
hPPARd	hmn peroxisome proliferator-activated receptor delta
HSV	herpes simplex virus
HSV-1	herpes simplex virus 1
HSV-2	herpes simplex virus 2
hTopo l	hTopo l isomerase
HXB2	T-cell tropic viral strain
IC	inhibiting concentration
IC_{50}	median inhibiting concentration
IC_{90}	inhibiting concentration for 90%
IC_{100}	absolute inhibiting concentration
ICR	imprinting control region mouse
ID	inhibition diameter (mm)
ID_{50}	median inhibiting dose
IDE	insulin-degrading enzyme
IDO	indoleamine 2,3-dioxygenase
IFV	influenza virus
IgE	immunoglobulin E
IGF1-R	protein kinase
IgM	immunoglobulin M
IL	interleukin
IL-1	interleukin-1
IL-1α	interleukin-1α
IL-1β	interleukin-1β
IL-2	interleukin-2
IL-4	interleukin-4
IL-5	interleukin-5
IL-6	interleukin-6
IL-8	interleukin-8
IL 12	interleukin-12
IL-13	interleukin-13

IM	immunomodulator
IMPDH	inosine monophosphate dihydrogenase
IN	integrase
iNOS	inducible nitric oxide synthase
InRt	inhibitive rate
ip	intraperitoneal injection
iv	intravenous injection
IZ	inhibition zone (mm)
IZD	inhibition zone diameter (mm)
IZR	inhibition zone radii (mm)
JACS	further abbreviation on *J. Am. Chem. Soc.*
Jak2	Janus kinase 2
JCS Perkin I	further abbreviation on *J. Chem. Soc., Perkin Trans. I*
JMC	further abbreviation on *J. Med. Chem.*
JNK	c-Jun NH_2-terminal kinase
JNP	further abbreviation on *J. Nat. Prod.*
JOC	further abbreviation on *J. Org. Chem.*
KDR	a protein tyrosine kinase
KU-812	hmn basophilic granulocyte
LAV	T-cell tropic viral strain
LC_{50}	concentration at which only 50% of the cells are viable
LCV	lymphocyte viability
LD	lethal dose
LD_{100}	100% lethal dose
LD_{50}	medium lethal dose
LD_{99}	99% lethal dose
LDH	lactate dehydrogenase
LOX	lipoxygenase
LPS	lipopolysaccharide
LTB_4	leukotriene B_4
LTC_4	leukotriene C_4
LY294002	phosphatidylinositol-3-kinase inhibitor; used as a positive control in anti-inflammatory assay
MABA	microplate Alamar blue assay
MAGI test	also called single life cycle test, reflects only one round of infection
MAPKAPK-2	mitogen-activated protein kinase-activated protein kinase 2
MAPKK	mitogen-activated protein kinase kinase
MBC	minimum bactericidal concentration
MBC_{90}	minimum bactericidal concentration for 90%
$MBEC_{90}$	minimum biofilm eradication counts for 90%
MCV	poxvirus *Molluscum contagiosum* virus
MDR	multidrug resistance
MDR1	major facilitator superfamily 1; one type of efflux pump in *C. albicans*, which functions as an H^+-antiporter
MDRPF	multidrug-resistant *Plasmodium falciparum*
MDRSA	multidrug-resistant *Staphylococcus aureus*
MDRSP	multidrug-resistant *Streptococcus pneumoniae*
MEK1 wt	protein kinase
MET wt	protein kinase

MG-MID	mean value of $\log_{10}$ GI_{50} (mol/L) over all cell lines tested
MIA	minimal inhibitory amounts (μg/disk)
MIC	minimum inhibitory concentration
MIC_{50}	minimal inhibitive concentration for 50%
MIC_{80}	minimal inhibitive concentration for 80%
MIC_{90}	minimal inhibitive concentration for 90%
MID	minimum inhibitory dose
min	minute
MLD	minimum lethal dose
MLR	mixed lymphocyte reaction
MMOA	molecular mechanism of action
MMP	matrix metalloproteinases
MMP-2	matrix metalloproteinase-2
MoBY-ORF	molecular barcoded yeast open-reading frame library method
mp	melting point
MPtpA	mycobacterial protein tyrosine phosphatase A
MPtpB	mycobacterial protein tyrosine phosphatase B
mPTPB	*Mycobacterium tuberculosis* protein tyrosine phosphatase B
MREC	methicillin-resistant *Escherichia coli*
MRSA	methicillin-resistant *Staphylococcus aureus*
MRSE	methicillin-resistant *Staphylococcus epidermidis*
MSR	macrophage scavenger receptor
MSSA	methicillin-sensitive *Staphylococcus aureus*
MSSE	methicillin-sensitive *Staphylococcus epidermidis*
MT1-MMP	membrane type 1 matrix metalloproteinase
MT4	MT4 cells containing HIV-1 IIIB virus
MTT	3-(4,5-dimethylthiazole-2-yl)-2,5-diphenyltetrazolium bromide
MTT assay	a cytotoxicity measurement method, tetrazolium-based colorimetric assay, see L. V. Rubinstein, et al., Nat. Cancer Inst., 82, 1113–1118 (1990)
mus	mouse
n	number of parallel experiments
nACh	nicotinic acetylcholine
NADH	reduced nicotinamide adenine dinucleotide
NDM-1	New Delhi metallo-β-lactamase-1
NEK2	protein kinase
NEK6	protein kinase
NF-κB	NF-κB serves as a central regulator of hmn immune, inflammatory, and antiapoptotic responses (Ghosh et al., 1998, Ann. Rev. Immunol, 16, 225–260)
NFRD	NADH-fumarate reductase
NGF	nerve growth factor
NMDA	*N*-methyl-D-aspartate
$NO^{\bullet}$	nitric oxide free radical
NPR	further abbreviation on *Nat. Prod. Rep.*
$O_2^{\bullet -}$	superoxide free radical
$ONOO^-$	peroxy nitrite free radical
ORAC	oxygen radical absorbance capacity
orl	oral
p24	p24 protein

P2Y receptors	one type of purine receptors, which includes P1 (adenosine receptors) and P2 receptors [ionotropic P2X and metabotropic (G protein-coupled) P2Y]
$P2Y_{11}$ receptor	one of eight P2Y subtypes
P450	cytochrome P450
p56lck	tyrosine kinase
PAcF	platelet activating factor
PAF	platelet aggregation factor
PD	Parkinson's disease
pD_2 (=pEC_{50})	negative logarithm (–log *M*) of molar concentration required to produce 50% of the maximum response (EC_{50})
PDE5	phosphodiesterase 5
PDGF	platelet-derived growth factor
PfGSK-3	kinase
Pfnek-1	a NIMA-related protein kinase of *Plasmodium falciparum*
PfPK5	kinase
PfPK7	kinase
PGE_2	prostaglandin E2
PHK	primary hmn keratinocytes
PIM1	protein kinase
PK	protein kinase
PKA	protein kinase A
PKC	protein kinase C
PKC-ε	protein kinase C-ε
PKD	ribosomal protein
PKG	protein kinase G
PLA	phospholipase A
PLA_2	phospholipase A_2
PLCγ1	ribosomal protein
PLK1	protein kinase
PM	further abbreviation on *Planta Med.*
PMA (=TPA)	phorbol-12-myristate-13-acetate
PMNL	hmn polymorphonuclear leukocyte
PP	protein phosphatase
PP1	protein phosphatase PP1
PP2A	protein phosphatase PP2A
$pp60^{V\text{-}SRC}$	tyrosine kinase
PPAR	peroxisome proliferator-activated receptor
PPDK	pyruvate phosphate dikinase
PR	protease
PRK1	protein kinase
PRNG	penicillin-resistant *Neisseria gonorrhoeae*
PRSP	penicillin-resistant *Staphylococcus pneumoniae*
PTEN	tumor suppressor, an identified tumor suppressor gene located on hmn chromosome 10q23.3
PTK	protein tyrosine kinase
PTP1B	protein tyrosine phosphatase 1B, an important target for treatment of type II diabetes
PTPB	protein tyrosine phosphatase B
PTPS2	protein tyrosine phosphatase S2

PV-1	*Polio* virus
PXR	pregnane X receptor
QR	NAD(P)H: quinone reductase
Range	difference in $\log_{10}$ GI_{50} (mol/L) value of the most sensitive cell line and the least sensitive cell
rat	white rat
rbt	rabbit
RLAR	rat lens aldose reductase
RNA	ribonucleic acid
ROS	reactive oxygen species (involved in genesis of various cancers, arteriosclerosis, rheumatism, and aging)
RS321	code of a yeast
RSV	respiratory syncytial virus
RT	reverse transcriptase
RU	response unit of binding capacity to HIV-1 targets, 1 RU = 1 pg/mm^2
RyR1-FKBP12	RyR1-FKBP12 Ca^{2+} channel, a tetrameric heterodimeric channel protein (~2000 kDa) associated with smaller 12 kDa immunophilin FKBP12
S6	ribosomal protein
SAK	a protein kinase
SARS	severe acute respiratory syndrome
ScRt	scavenging rate
SF162	macrophage-tropic viral strain
SI	IC_{50} of testing cells/IC_{50} of HUVECs
SI	selective index = cytotoxic CC_{50}/target EC_{50}
SI	selective index = cytotoxic IC_{50}/target IC_{50}
SI	selective index = cytotoxic IC_{50}/target MIC
SI	selective index = cytotoxic TC_{50}/target IC_{50}
SIRT2	hmn sirtuin type 2 (a NAD^+-dependent cytoplasmic protein that is co-localized with HDAC6 on microtubules. SIRT2 has been shown to deacetylate α-tubulin and to control mitotic exit from the cell cycle)
sp.	species
spp.	species (plural)
SR	sarcoplasmic reticulum
SRB	sulforhodamine B assay
SRC	protein kinase
SV40	SV40 virus
Syn.	synonym
T/C	survival ratio [survival time of treated animal (*T*) was compared to that of control animal (*C*) expressed as a percent (*T/C*%)]
TACE	α-secretase (a serine protease)
Taq DNA polymerase	a DNA polymerase isolated from the thermophilic bacterium *Thermus aquaticus*
TBARS	thiobarbituric acid-reactive substance assay
TC_{50}	50% cytotoxic concentration
TEAC	Trolox equivalent antioxidant capacity
TGI	100% growth inhibition
TMV	tobacco mosaic virus
TNFα	tumor necrosis factor-α
TPA (=PMA)	12-*O*-tetradecanoyl phorbol 13-acetate

TPK	tyrosine protein kinase
TRP	transient receptor potential cationic channel
TRPA1	transient receptor potential cationic channel of subfamily A1
TRPV1	transient receptor potential cationic channel of subfamily V1
TRPV3	transient receptor potential cationic channel of subfamily V3
TXB_2	thromboxane B_2
TZM-bl	host cell in HIV-1 neutralization assay
USP7	a deubiquitylating enzyme hydrolyzing isopeptide bond at C-terminus of ubiquitin is an emerging cancer target
VCAM	vascular cell adhesion molecule
VCAM-1	vascular cell adhesion molecule-1
VCR	vincristine
VEGF	vascular endothelial growth factor
VEGF-A	vascular endothelial growth factor A
VEGFR2	tyrosine kinase VEGFR2
VE-PTP	protein phosphatase
VGSC	voltage-gated sodium channel
VHR	vaccinia open-reading frame H1-related protein phosphatase
Vif	viral infectivity factor of HIV-1
VP-16	etoposide (Sigma product), a positive control for cytotoxic assay
VRE	vancomycin-resistant *Enterococcus* sp.
VREF	vancomycin-resistant *Enterococcus faecium*
VSE	vancomycin-sensitive *Enterococcus* sp.
VSSC	voltage-sensitive sodium channel
VSV	*Vesicular stomatitis* virus
WST-8	2-(2-methoxy-4-nitrophenyl)-3-(4-nitrophenyl)-5-(2,4-disulfo-phenyl)-2*H*-tetrazolium, monosodium salt
XTT	sodium 3′-[1-(phenylaminocarbonyl)-3,4-tetrazolium] bis(4-methoxy-6-nitrobenzene)sulfonic acid
YU2-V3	viral strain

List of Cancer Cell Codes

This set of codes for 438 cancer cells, named as *CCC codes*, are defined and tried out in the books by the author. The codes of some normal cells are also listed below.

293T	kidney epithelial cells
3T3-L1	murine fibroblasts
5637	superficial bladder cancer (cell)
786-0	hmn renal cancer (cell)
9KB	hmn *epidermatoid* nasopharyngeal carcinoma (cell)
A-10	rat aorta cells
A2058	hmn (cell)
A278	hmn ovarian tumor (cell)
A2780	hmn ovarian tumor (cell)
A2780CisR	hmn ovarian tumor (cell)
A2780/DDP	hmn ovarian tumor (cell)
A2780/Tax	hmn ovarian tumor (cell)
A375	hmn melanoma (cell)
A375-S2	hmn melanoma (cell)
A431	hmn epidermic cancer (cell)
A498	hmn renal cancer (cell)
A549	hmn nonsmall cell lung cancer (cell)
A549 NSCL	hmn nonsmall cell lung cancer (cell)
A549/ATCC	hmn nonsmall cell lung cancer (cell)
ACC-MESO-1	hmn malignant pleural mesothelioma (cell)
ACHN	hmn renal cancer (cell)
AGS	gastric adenocarcinoma (cell)
AsPC-1	hmn pancreatic cancer (cell)
B16	mouse melanoma (cell)
B16F1	mouse melanoma (cell)
B16-F-10	mouse melanoma (cell)
BC	hmn breast cancer (cell)
BC-1	hmn breast cancer (cell)
BCA-1	hmn breast cancer (cell)
BEAS2B	normal hmn lung bronchial cells
Bel7402	hmn liver cancer (cell)
BG02	normal hmn embryonic stem cells
BGC823	hmn gastric cancer (cell)
BOWES	hmn cells
BR1	DNA repair competent Chinese hamster ovary (cell)
BSC	normal monkey kidney cells
BSC-1	normal African Green Monkey kidney cells
BSY1	breast cancer (cell)
BT-483	hmn breast carcinoma (cell)
BT549	hmn galactophore cancer (cell)
BT-549	hmn breast cancer (cell)
BXF-1218L	hmn bladder cancer (cell)
BXF-T24	hmn bladder cancer (cell)
BXPC	hmn pancreas cancer (cell)

https://doi.org/10.1515/9783110655810-206

BXPC3	hmn pancreas cancer (cell)
C6	rat glioma (cell)
C26	hmn colon carcinoma (cell)
C38	murine colon adenocarcinoma (cell)
CA46	hmn Burkitt's lymphoma (cell)
Ca9-22	hmn gingival carcinoma (cell)
CaCo-2	hmn epithelial colorectal adenocarcinoma (cell)
CAKI-1	hmn renal cancer (cell)
Calu	prostate carcinoma (cell)
Calu3	nonsmall cell lung cancer (cell)
CCRF-CEM	hmn T-cell acute lymphoblastic leukemia (cell)
CCRF-CEMT	leukemia (cell)
CEM	hmn leukemia (cell)
CEM-TART	T-cells that express both HIV-1 tat and rev
CFU-GM	hmn/murine hematopoietic progenitor cells
CHO	Chinese hamster ovary cells
CHO-K1	subclone of normal Chinese hamster ovary cells
CML K562	chronic myelogenous leukemia (cell)
CNE	hmn nasopharyngeal carcinoma (cell)
CNE2	hmn nasopharyngeal carcinoma (cell)
CNS SF295	hmn brain tumor (cell)
CNXF-498NL	hmn glioblastoma cancer (cell)
CNXF-SF268	hmn glioblastoma cancer (cell)
Colo320	hmn colorectal cancer (cell)
Colo357	hmn colorectal cancer (cell)
Colon26	colorectal cancer (cell)
Colon38	mus colorectal cancer (cell)
Colon205	colorectal cancer (cell)
Colon250	colorectal cancer (cell)
CV-1	monkey kidney fibroblasts
CXF-HCT116	hmn colon cancer (cell)
CXF-HT29	hmn colon cancer (cell)
DAMB	hmn mammary carcinoma (cell)
DG-75	hmn B lymphocyte (cell)
DLAT	Dalton's lymphoma ascites tumor (cell)
DLD-1	hmn colorectal adenocarcinoma (cell)
DLDH	hmn colorectal adenocarcinoma (cell)
DMS114	hmn lung cancer (cell)
DMS273	hmn small cell lung cancer (cell)
Doay	hmn medulloblastoma (cell)
Dox40	hmn myeloma (cell)
DU145	prostate cancer (cell)
DU4475	breast cancer (cell)
E39	hmn renal carcinoma (cell)
EAC	Ehrlich ascites carcinoma (cell)
EKVX	hmn nonsmall cell lung cancer (cell)
EM9	topoisomerase I-sensitive Chinese hamster ovary (cell)
EMT-6	mouse tumor cells
EPC	carp epithelium (cell)

EVLC-2	SV40 large T-antigen immortalized hmn umbilical vein cells
F1	hmn amniotic epithelial cells
FADU	pharynx-sq cancer (cell)
Farage	hmn lymphoma (cell)
Fem-X	melanoma (cell)
Fl	hmn amniotic epithelial cell line
FM3C	mus mammary tumor (cell)
G402	hmn renal leiomyoblastoma
GM7373	bovine endothelial (cell)
GR-III	adenocarcinoma (cell)
GXF-251L	hmn stomach cancer (cell)
H116	hmn colorectal cancer (cell)
H125	hmn colorectal cancer (cell)
H441	hmn lung adenocarcinoma (cell)
H460	hmn lung cancer (cell)
H522	hmn nonsmall cell lung cancer (cell)
H1299	hmn lung adenocarcinoma (cell)
H1325	hmn nonsmall cell lung cancer (cell)
H1975	hmn cancer (cell)
H2122	hmn nonsmall cell lung cancer (cell)
H2887	hmn nonsmall cell lung cancer (cell)
H69AR	multidrug-resistant small cell lung cancer (cell)
H929	hmn myeloma (cell)
H9c2	rat cardiac myoblasts
HBC4	breast cancer (cell)
HBC5	breast cancer (cell)
HBL100	breast cancer (cell)
HCC366	hmn nonsmall cell lung cancer (cell)
HCC2998	hmn colorectal cancer (cell)
HCC-S102	hepatocellular carcinoma (cell)
HCT	hmn colorectal cancer (cell)
HCT8	hmn colorectal cancer (cell)
HCT15	hmn colorectal cancer (cell)
HCT29	hmn colon adenocarcinoma (cell)
HCT116	hmn colorectal cancer (cell)
HCT116/mdr+	overexpress mdr+ hmn colorectal cancer (cell)
HCT116/topo	resistant to etoposide hmn colorectal cancer (cell)
HCT116/VM46	multidrug-resistant colorectal cancer (cell)
HEK-293	normal hmn epithelial kidney cells
HEL	hmn embryonic lung fibrocytes
HeLa	hmn cervical epithelial carcinoma (cell)
HeLa-APL	hmn cervical epithelial cancer (cell)
HeLa-S3	hmn cervical epithelial cancer (cell)
Hep2	hmn liver carcinoma (cell)
Hep3B	hmn liver cancer (cell)
HepA	hmn liver cancer ascites (cell)
Hepa1c1c7	mus liver cancer (cell)
HepG	hmn liver cancer (cell)
HepG2	hmn liver cancer (cell)

HepG3	hmn liver cancer (cell)
HepG3B	hmn liver cancer (cell)
HEY	hmn ovarian carcinoma (cell)
HFF	hmn foreskin fibroblasts
HL60	hmn promyelocytic leukemia (cell)
HL7702	hmn liver tumor (cell)
HLF	hmn lung fibroblasts
HM02	hmn gastric adenocarcinoma (cell)
HMEC	hmn microvascular endothelial cells
HMEC1	hmn microvascular endothelial cells
HNXF-536L	hmn head and neck cancer (cell)
HOP-18	hmn nonsmall cell lung cancer (cell)
HOP-62	hmn nonsmall cell lung cancer (cell)
HOP-92	hmn nonsmall cell lung cancer (cell)
Hs578T	hmn breast cancer (cell)
Hs683	hmn oligodendroglioma (black dots) (cell)
HSV-1	nonmalignant cell
HT	hmn lymphoma (cell)
HT29	hmn colorectal cancer (cell)
HL60	M. Daferner, et al., Z. Naturforsch., Teil C, 1999, 54, 474
HT115	hmn colorectal cancer (cell)
HT460	hmn tumor (cell)
HT1080	hmn fibrosarcoma (cell)
HTC116	hmn acute promyelocytic leukemia (cell)
HTCLs	hmn tumors (cells)
HuCCA-1	hmn cholangiocarcinoma cancer (cell)
Huh7	hmn hepatoma (cell)
HUVEC	hmn umbilical vein endothelial cell
HUVECs	hmn umbilical vein endothelial cell
IC-2^{WT}	murine cell line
IGR-1	hmn melanoma (cell)
IGROV	hmn ovarian cancer (cell)
IGROV1	hmn ovarian cancer (cell)
IGROV-ET	hmn ovarian cancer (cell)
IMR-32	hmn neuroblastoma (cell)
IMR-90	hmn diploid lung fibroblasts
J774	mus monocyte/macrophage (cell)
J774.1	mus monocyte/macrophage (cell)
J774.A1	mus monocyte/macrophage (cell)
JB6 Cl41	mouse epidermal cells
JB6 P^+Cl41	mouse epidermal cells
JurKat	hmn leukemia (cell)
JurKat-T	hmn T-cell leukemia (cell)
K462	hmn leukemia (cell)
K562	hmn chronic myelogenous leukemia (cell)
KB	hmn nasopharyngeal carcinoma (cell)
KB16	hmn nasopharyngeal carcinoma (cell)
KB-3	hmn *epidermoid* carcinoma (cell)
KB-3-1	hmn *epidermoid* carcinoma (cell)

KB-C2	hmn carcinoma (cell)
KB-CV60	hmn carcinoma (cell)
KBV200	MDR nasopharyngeal carcinoma (cell)
Ketr3	hmn renal cancer (cell)
KM12	hmn colorectal cancer (cell)
KM20L2	hmn colorectal cancer (cell)
KMS34	hmn myeloma (cell)
KU812F	hmn leukemia (cell)
KV/MDR	multidrug-resistant cancer (cell)
KYSE30	hmn esophageal cancer (cell)
KYSE70	hmn esophageal cancer (cell)
KYSE180	hmn esophageal cancer (cell)
KYSE520	hmn esophageal cancer (cell)
L_{1210}	mouse lymphocytic leukemia (cell)
L_{1210}/Dx	doxorubicin-resistant L_{1210} (cell)
L363	hmn myeloma (cell)
L-428	leukemia (cell)
L5178	mouse lymphosarcoma (cell)
L5178Y	mouse lymphosarcoma (cell)
L-6	rat skeletal myoblasts (cell)
L929	mouse fibroblasts
LLC-PK_1	pig kidney cells
LMM3	mouse mammary adenocarcinoma (cell)
LNCaP	hmn prostate cancer (cell)
LO2	hmn liver cells
LoVo	hmn colorectal cancer (cell)
LoVo-DOX	hmn colorectal cancer (cell)
LOX	hmn melanoma (cell)
LOX-IMVI	hmn melanoma (cell)
LX-1	hmn lung cancer (cell)
LXF-1121L	hmn lung cancer (cell)
LXF-289L	hmn lung cancer (cell)
LXF-526L	hmn lung cancer (cell)
LXF-529L	hmn lung cancer (cell)
LXF-629L	hmn lung cancer (cell)
LXFA-629L	lung adenocarcinoma (cell)
LXF-H460	hmn lung cancer (cell)
M14	melanoma (cell)
M16	murine colon adenocarcinoma (cell)
M17	adriamycin-resistant breast cancer (cell)
M17-Adr	adriamycin-resistant breast cancer (cell)
M21	melanoma (cell)
M5076	ovarian sarcoma (cell)
MAGI	Hela-CD4-LTR-β-gal (indicator) cells containing HIV-1 IIIB virus
MALME-3	melanoma (cell)
MALME-3M	melanoma (cell)
MAXF-401	hmn breast cancer (cell)
MAXF-401NL	hmn breast cancer (cell)
MAXF-MCF7	hmn breast cancer (cell)

MCF	hmn breast cancer (cell)
MCF-10A	hmn breast epithelial (cell)
MCF7	hmn breast cancer (cell)
MCF7 Adr	drug-resistant hmn breast MCF7 cancer (cell)
MCF7/Adr	drug-resistant hmn breast MCF7 cancer (cell)
MCF7/ADR-RES	drug-resistant hmn breast cancer MCF7 (cell)
MCF12	hmn esophageal cancer (cell)
MDA231	hmn breast cancer (cell)
MDA361	hmn breast cancer (cell)
MDA435	hmn breast cancer (cell)
MDA468	hmn breast cancer (cell)
MDA-MB	hmn breast cancer (cell)
MDA-MB-231	hmn breast cancer (cell)
MDA-MB-231/ATCC	hmn breast cancer (cell)
MDA-MB-435	hmn breast cancer (cell)
MDA-MB-435s	hmn breast cancer (cell)
MDA-MB-468	hmn breast cancer (cell)
MDA-N	hmn breast cancer (cell)
MDCK	Madin–Darby canine (cell)
ME180	cervical cancer (cell)
MEL28	hmn melanoma (cell)
MES-SA	hmn uterine (cell)
MES-SA/DX5	hmn uterine (cell)
MEXF-276L	hmn melanoma (cell)
MEXF-394NL	hmn melanoma (cell)
MEXF-462NL	hmn melanoma (cell)
MEXF-514L	hmn melanoma (cell)
MEXF-520L	hmn melanoma (cell)
MG63	hmn osteosarcoma (cell)
MGC-803	hmn cancer (cell)
MiaPaCa	hmn pancreas cancer (cell)
Mia-PaCa-2	hmn pancreas cancer (cell)
MKN1	hmn gastric cancer (cell)
MKN7	hmn gastric cancer (cell)
MKN28	hmn gastric cancer (cell)
MKN45	hmn gastric cancer (cell)
MKN74	hmn gastric cancer (cell)
MM1S	hmn myeloma (cell)
Molt3	leukemia (cell)
Molt4	hmn T lymphocyte leukemia (cell)
Mono-Mac-6	mononuclear cells
MPM ACC-MESO-1	hmn malignant pleural mesothelioma
MRC-5	normal hmn diploid embryonic cells
MRC5CV1	SV40-transformed hmn fibroblasts
MS-1	mice endothelial cells
MX-1	hmn mammary carcinoma xenografts
N18-RE-105	neuronal hybridoma (cell)
N18-T62	mus neuroblastoma (cell)
NAMALWA	leukemia (cell)

NBT-T2 (BRC-1370)	rat bladder epithelial cells
NCI-ADR	hmn ovarian sarcoma (cell)
NCI-ADR-Res	hmn ovarian sarcoma (cell)
NCI-H23	hmn nonsmall cell lung cancer (cell)
NCI-H69	hmn lung cancer (cell)
NCI-H82	hmn lung cancer (cell)
NCI-H187	hmn small cell lung cancer (cell)
NCI-H226	hmn nonsmall cell lung cancer (cell)
NCI-H322M	hmn nonsmall cell lung cancer (cell)
NCI-H446	hmn lung cancer (cell)
NCI-H460	hmn nonsmall cell lung cancer (cell)
NCI-H510	hmn lung cancer (cell)
NCI-H522	hmn nonsmall cell lung cancer (cell)
neuro-2a	mouse neuroblastoma (cell)
NFF	nonmalignant neonatal foreskin fibroblasts
NHDF	normal hmn dermal fibroblasts
NIH3T3	nontransformed fibroblasts
NIH3T3	normal fibroblasts
NMuMG	nontransformed epithelial cells
NOMO-1	hmn acute myeloid leukemia
NS-1	murine cells
NSCLC	hmn bronchopulmonary nonsmall cell lung cancer
NSCLC HOP-92	hmn nonsmall cell lung cancer (cell)
NSCLC-L16	hmn bronchopulmonary nonsmall cell lung carcinoma
NSCLC-N6	hmn bronchopulmonary nonsmall cell lung cancer (cell)
NSCLC-N6-L16	hmn bronchopulmonary nonsmall cell lung carcinoma
NUGC-3	hmn gastric cancer (cell)
OCILY17R	hmn lymphoma (cell)
OCIMY5	hmn myeloma (cell)
OPM2	hmn myeloma (cell)
OVCAR-3	ovarian adenocarcinoma (cell)
OVCAR-4	ovarian adenocarcinoma (cell)
OVCAR-5	ovarian adenocarcinoma (cell)
OVCAR-8	ovarian adenocarcinoma (cell)
OVXF-1619L	ovary cancer (cell)
OVXF-899L	ovary cancer (cell)
OVXF-OVCAR3	ovary cancer (cell)
P_{388}	mus lymphocytic leukemia (cell)
P_{388}/ADR	P_{388} adriamycin-resistant (cell)
P_{388}/Dox	mus leukemia cells expressing resistance toward doxorubicin
P_{388}D1	mus macrophage cells
PANC1	hmn pancreas cancer (cell)
panc89	pancreatic cancer (cell)
PAXF-1657L	hmn pancreas cancer (cell)
PAXF-PANC1	hmn pancreas cancer (cell)
PBMC	hmn normal peripheral blood mononuclear cells
PC12	hmn lung cancer (cell)
PC-12	rat pheochromocytoma (cell)
PC3	hmn prostate cancer (cell)

PC3M	hmn prostate cancer (cell)
PC3MM2	hmn prostate cancer (cell)
PC-9	hmn lung cancer (cell)
PRXF-22RV1	hmn prostate cancer (cell)
PRXF-DU145	hmn prostate cancer (cell)
PRXF-LNCAP	hmn prostate cancer (cell)
PRXF-PC3M	hmn prostate cancer (cell)
PS (=P_{388})	PS system, P_{388} mouse lymphocytic leukemia (cell)
PV1	nonmalignant cell
PXF-1752L	mesothelioma cancer (cell)
QG56	hmn lung carcinoma (cell)
QGY-7701	hmn hepatocellular carcinoma (cell)
QGY-7703	hmn liver cancer (cell)
Raji	hmn EBV-transformed Burkitt's lymphoma B cell
RAW264.7	mouse macrophages
RB	hmn prostate cancer (cell)
RBL-2H3	rat basophilic cells
RF-24	papillomavirus 16 E6/E7 immortalized hmn umbilical vein cells
RKO	hmn colon cancer (cell)
RKO-E6	hmn colon cancer (cell)
RPMI7951	hmn malignant melanoma (cell)
RPMI8226	hmn myeloma (cell)
RXF-1781L	renal cancer (cell)
RXF-393	renal cancer (cell)
RXF-393NL	renal cancer (cell)
RXF-486L	renal cancer (cell)
RXF-631L	renal cancer (cell)
RXF-944L	renal cancer (cell)
S_{180}	mouse sarcoma (cell)
S_{180}A	sarcoma 180 ascite cells
SAS	hmn oral cancer
SCHABEL	mouse lymphoma cancer (cell)
SF268	hmn brain tumor (cell)
SF295	hmn brain tumor (cell)
SF539	hmn brain tumor (cell)
SGC7901	hmn gastric cancer (cell)
SH-SY5Y	hmn neuroblastoma (cell)
SK5-MEL	hmn melanoma (cell)
SKBR3	hmn breast cancer (cell)
SK-Hep1	hmn liver carcinoma (cell)
SK-MEL-2	hmn melanoma (cell)
SK-MEL-5	hmn melanoma (cell)
SK-MEL-28	hmn melanoma (cell)
SK-MEL-S	hmn melanoma (cell)
SK-N-SH	neuroblastoma (cell)
SK-OV-3	ovarian adenocarcinoma (cell)
SMMC-7721	hmn liver cancer (cell)

SN12C	hmn renal cancer (cell)
SN12k1	hmn renal cancer (cell)
SNB19	hmn brain tumor (cell)
SNB75	hmn CNS cancer (cell)
SNB78	hmn brain tumor (cell)
SNU-C4	hmn cancer (cell)
SR	leukemia (cell)
St4	gastric cancer (cell)
stromal cell	bone marrow stromal cells
SUP-B15	leukemia (cell)
Sup-T1	T-cell lymphoma cancer cells
SW480	hmn colorectal adenocarcinoma (cell)
SW620	hmn colorectal adenocarcinoma (cell)
SW1573	hmn nonsmall cell lung cancer (cell)
SW1736	hmn thyroid cancer (cell)
SW1990	hmn pancreatic cancer (cell)
T24	hmn liver cancer (cell)
T-24	hmn transitional bladder carcinoma (cell)
T47D	hmn breast cancer (cell)
THP-1	hmn acute monocytic leukemia (cell)
TK10	hmn renal cancer (cell)
tMDA-MB-231	hmn breast cancer (cell)
tsFT210	mouse cancer (cell)
TSU-Pr1	invasive bladder cancer (cell)
TSU-Pr1-B1	invasive bladder cancer (cell)
TSU-Pr1-B2	invasive bladder cancer (cell)
U251	CNS tumor/glioma (cell)
U266	myeloma (cell)
U2OS	hmn osteosarcoma (cell)
U373	glioblastoma/astrocytoma (cell)
U373MG	hmn brain cancer (cell)
U-87-MG	caucasian glioblastoma (cell)
U937	hmn monocytic leukemia (cell)
UACC-257	melanoma (cell)
UACC62	melanoma (cell)
UO-31	hmn renal cancer (cell)
UT7	hmn leukemia (cell)
UV20	DNA cross-linking agent-sensitive Chinese hamster ovary (cell)
UXF-1138L	hmn uterus cancer (cell)
V79	Chinese hamster (cell)
Vero	green monkey kidney tumor (cell)
WEHI-164	mus fibrosarcoma (cell)
WHCO1	hmn esophageal cancer (cell)
WHCO5	hmn esophageal cancer (cell)
WHCO6	hmn esophageal cancer (cell)
WI26	hmn lung fibroblasts
WiDr	hmn colon adenocarcinoma (cell)

WMF	hmn prostate cancer (cell)
XF498	hmn CNS cancer (cell)
XRS-6	topoisomerase II-sensitive Chinese hamster ovary (cell)
XVS	topoisomerase II-sensitive CHO cell
ZR-75-1	hmn breast cancer (cell)

1 Oxygen Heterocycles

1.1 *β*-Lactones

1 Salinosporamide A

Antibiotic NPI 0052 Type: *β*-Lactones. $C_{15}H_{20}ClNO_4$ Needles (EtOAc/2,3,3-trimethylpentane), mp 169–171 °C, $[\alpha]_D^{25} = -72.9°$ (c = 0.55, MeOH). Source: Marine-derived actinomycete *Salinispora tropica* CNB-392. Pharm: Cytotoxic (NCI-H226, IC_{50} < 0.011 μmol/L; NCI-H522, IC_{50} = 0.043 μmol/L; HCC2998, IC_{50} = 0.018 μmol/L; CNS cancer SNB75, IC_{50} < 0.011 μmol/L; SK-MEL-28, IC_{50} < 0.011 μmol/L; melanoma SNB75, IC_{50} = 0.032 μmol/L; A498, IC_{50} = 0.011 μmol/L; RXF-393, IC_{50} = 0.023 μmol/L; MDA-MB-435, IC_{50} < 0.011 μmol/L; CCRF-CEM, IC_{50} > 100 μmol/L; DU145, IC_{50} > 100 μmol/L); cytotoxic (NCI's 60-cell-line assay mean GI_{50} < 0.011 μmol/L; HCT116, GI_{50} = 0.035 μmol/L, LOX-IMVI, GI_{50} < 0.011 μmol/L; HOP-92, GI_{50} < 0.011 μmol/L; OVCAR-3, GI_{50} < 0.011 μmol/L; PC3, GI_{50} < 0.011 μmol/L); proteasome inhibitor. Ref: P. G. Williams, et al, JOC, 2005, 70, 6196

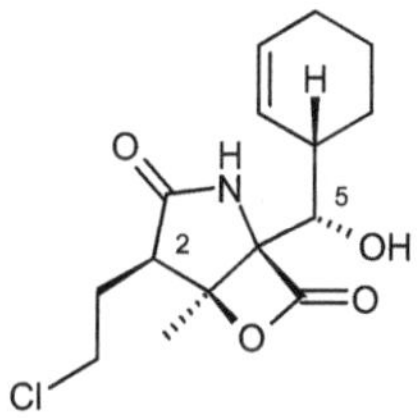

2 Salinosporamide B

Antibiotic NPI 0047 Type: *β*-Lactones. $C_{15}H_{21}NO_4$ Cryst. (EtOAc), mp 143–145 °C, $[\alpha]_D^{25} = -54.5°$ (c = 0.29, MeOH). Source: Marine-derived actinomycete *Salinispora tropica* CNB-392. Pharm: Cytotoxic (HCT116, GI_{50} = 20 μmol/L). Ref: P. G. Williams, et al, JOC, 2005, 70, 6196

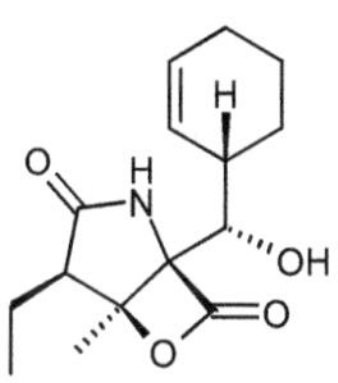

3 Salinosporamide K

Type: *β*-Lactones. $C_{13}H_{17}NO_4$ Source: Marine-derived actinomycete *Salinispora pacifica* (Fiji). Pharm: 20*S*-Proteasome inhibitor (Chymotrypsin-like). Ref: A. S. Eust´aquio, et al, Chem. Bio. Chem., 2011, 12, 61

https://doi.org/10.1515/9783110655810-001

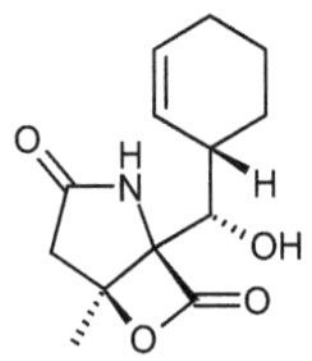

1.2 Furans

4 Acetylsumiki's acid

Type: Furans. $C_8H_8O_5$ Source: Marine-derived fungus *Cladosporium herbarum* from sponge *Callyspongia aerizusa, Epicoccum* sp. Pharm: Antibacterial (*Bacillus subtilis* and *Staphylococcus aureus*, 5 µg/disc). Ref: R. Jadulco, et al, JNP, 2001, 64, 527

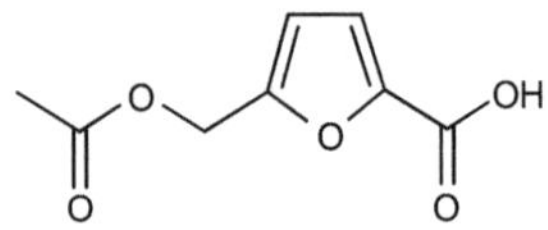

5 4,5-Dihydro-2,4-dimethyl-3-furancarboxaldehyde

Type: Furans. $C_7H_{10}O_2$ Oil, $[\alpha]_D^{20}$ = +38.9° (c = 0.61, $CHCl_3$). Source: Marine-derived fungus *Ulocladium botrytisfrom* from sponge *Callyspongia vaginalis.* Pharm: Antifungal. Ref: U. Höller, et al, EurJOC, 1999, 2949

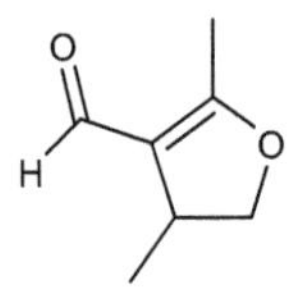

6 2,3-Dihydro-2-hydroxy-2,4-dimethyl-5-*trans*-propenylfuran-3-one

Type: Furans. $C_9H_{12}O_3$ Prisms (petrol), mp 99 °C, $[\alpha]_D^{20}$ = +1.8° (c = 0.29, EtOH). Source: Marine-derived fungus *Ascochyta salicorniae* from green alga *Ulva* sp. (North Sea coast, Germany), *Stemphylium radicinum* and *Ascochyta salicorniae.* Pharm: Tyrosine kinase inhibitor ($TKp56^{lck}$, 40 µg/mL, InRt = 44%, 200 µg/mL, InRt = 91%); antiplasmodial (*Plasmodium falciparum* K1, IC_{50} = 1.763 µg/mL; *Plasmodium falciparum* NF 54, IC_{50} = 1.760 µg/mL); antifungal (50 µg/disk, *Microbotryum violacea,* IZD = 1 mm; *Eurotium repens,* IZD = 2 mm). Ref: C. Osterhage, et al, JOC, 2000, 65, 6412

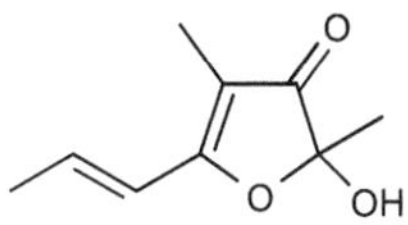

7 Echinofuran‡

8,12-Epoxy-1 (10),4 (15),7,11-guaiatetraene Type: Furans. $C_{15}H_{18}O$ Oil, $[\alpha]_D^{20} = -91°$ (c = 0.138, $CHCl_3$). Source: Gorgonian *Echinogorgia praelonga* (Okinawa). Pharm: Cell division inhibitor (fertilized sea urchin egg assay). Ref: J. Tanaka, et al, JNP, 1992, 55, 1522

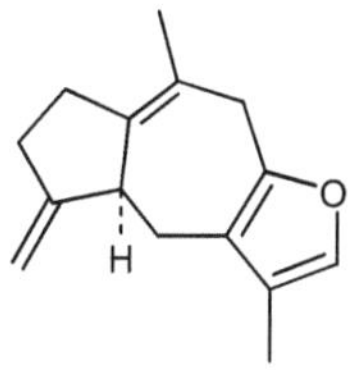

8 (2*S*,3*S*,5*R*)-5-[(1*R*)-1-Hydroxydec-9-enyl]-2-pentyltetrahydrofuran-3-ol

Type: Furans. $C_{19}H_{36}O_3$ Plates (hexane), mp 54.5–55.0 °C, $[\alpha]_D^{21} = +15.0°$ (c = 1, $CHCl_3$). Source: Brown alga *Notheia anomala*. Pharm: Nematocide (potent and selective inhibitor of larval development of parasitic nematodes). Ref: R. G. Warren, et al, Aust. J. Chem., 1980, 33, 891 | S. Hatakeyamu, et al, Tet. Lett., 1985, 26, 1333 | R. A. Barrow, et al, Aust. J. Chem., 1990, 43, 895 | L. M. Murray, et al, Aust. J. Chem., 1991, 44, 843 | H. Chikashita, et al, Chem. Lett., 1993, 477 | Z.M. Wang, et al, JOC, 1998, 63, 1414 | Y. Mori, et al, Tet. Lett., 1999, 40, 731 | C. G. García, et al, Tet. Lett., 2000, 41, 4127

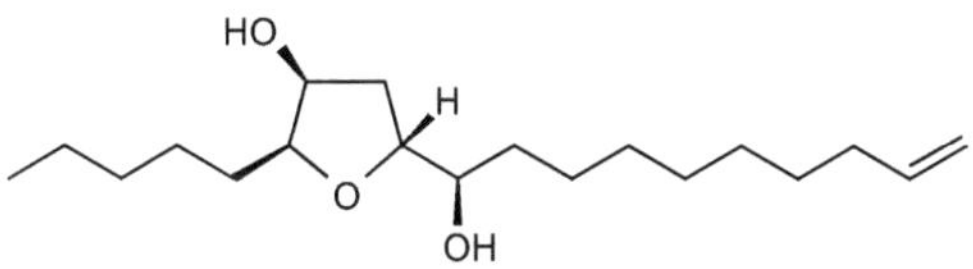

9 (2*S*,3*S*,5*S*)-5-[(1*S*)-1-Hydroxydec-9-enyl]-2-pentyltetrahydrofuran-3-ol

Type: Furans. $C_{19}H_{36}O_3$ Source: Brown alga *Notheia anomala* (Southern Australia). Pharm: Nematocide (potent and selective inhibitor of larval development of parasitic nematodes). Ref: R. G. Warren,et al, Aust. J. Chem., 1980, 33, 891 | R. A. Barrow, et al, Aust. J. Chem., 1990, 43, 895 | L. M. Murray, et al, Aust. J. Chem., 1991, 44, 843 | R. J. Capon, et al, Tetrahedron, 1998, 54, 2227 | C. G. García, et al, Tet. Lett., 2000, 41, 4127

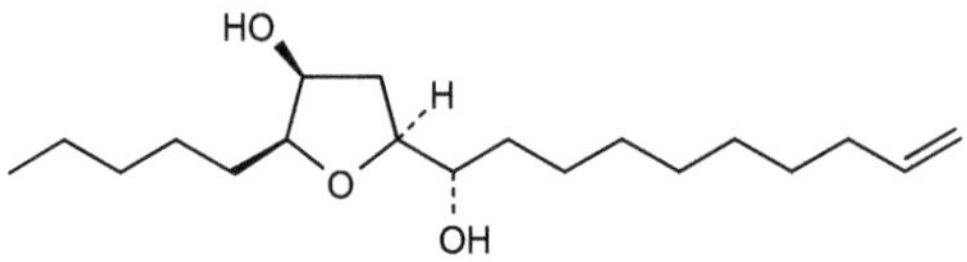

10 Laurencione (cyclic hemiacetal form)

Dihydro-3 (2*H*)-furanone Type: Furans. $C_5H_8O_3$ Light green oil. Source: Red alga *Laurencia spectabilis* (Oregon). Pharm: Toxic (brine shrimp). Ref: M. W. Bernart, et al, Phytochemistry, 1992, 31, 1273 | N. De Kimpe, et al, JOC, 1995, 60, 5162 | N. De Kimpe, et al, Synthesis, 1996, 1131

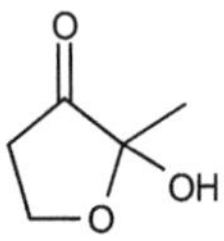

11 Mutafuran H

Type: Furans. $C_{16}H_{26}Br_2O_4$ Source: Sponge *Xestospongia testudinaria* (Sanya, Hainan, China). Pharm: AChE inhibitor (potent). Ref: X. Zhou, et al, Chem. Phys. Lipids, 2011, 164, 703

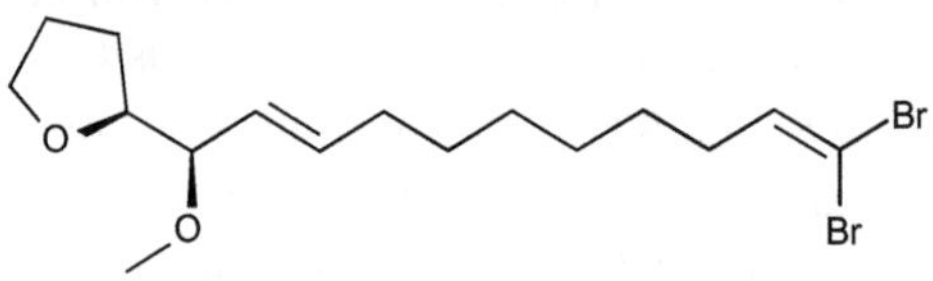

12 Plakorfuran A

Type: Furans. $C_{19}H_{30}O_4$ Source: Sponge *Plakortis simplex* (Yongxing I., South China Sea, China). Pharm: Antifungal (weak). Ref: X. -F. Liu, et al, Tetrahedron, 2012, 68, 4635

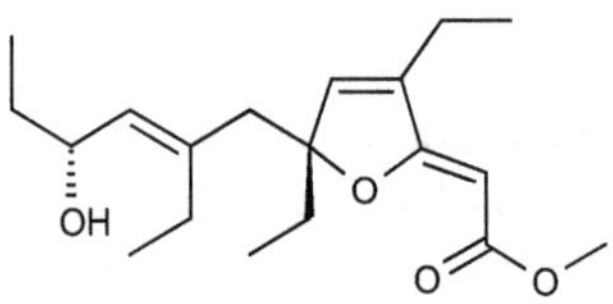

13 Plakorsin B

Type: Furans. $C_{22}H_{38}O_3$ Amorph. solid. Source: Sponge *Plakortis simplex* (Taiwan waters). Pharm: Cytotoxic (KB16, IC_{50} = 3.43 μg/mL; Colon250, IC_{50} = 0.28 μg/mL). Ref: Y. C. Shen, et al, JNP, 2001, 64, 324

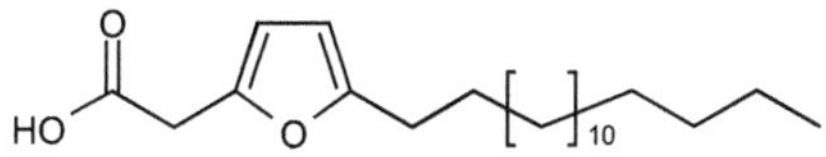

14 Sargafuran

Type: Furans. $C_{27}H_{36}O_4$ Oil, $[\alpha]_D = -1.3°$ (c = 0.9, $CHCl_3$). Source: Brown alga *Sargassum macrocarpum* (Japan waters). Pharm: Antibacterial; anti-acne (prevent or improve propionibacterium acne). Ref: Y. Kamei, et al, J. Antibiot., 2009, 62, 259

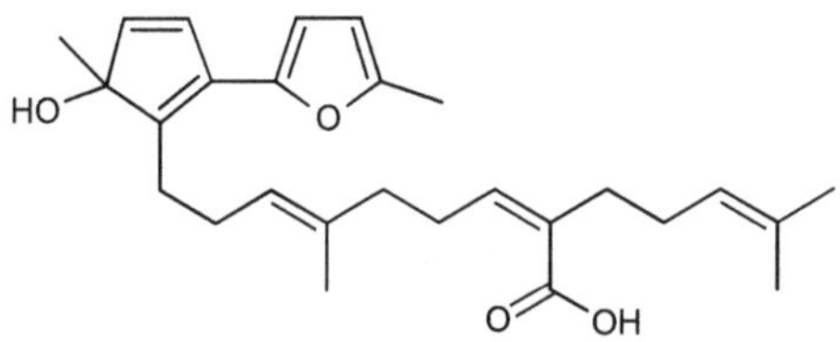

15 Testufuran A

Type: Furans. $C_{18}H_{25}BrO_3$ Yellow oil, $[\alpha]_D^{23} = -18°$ (c = 0.04, MeOH). Source: Sponge *Xestospongia testudinaria*. Pharm: Stimulator of adipogenesis (preadipocyte differentiation-inducing activity (30 μmol/L, 2+, 130 μmol/L, 3+)). Ref: T. Akiyama, et al, Tetrahedron, 2013, 69, 6560

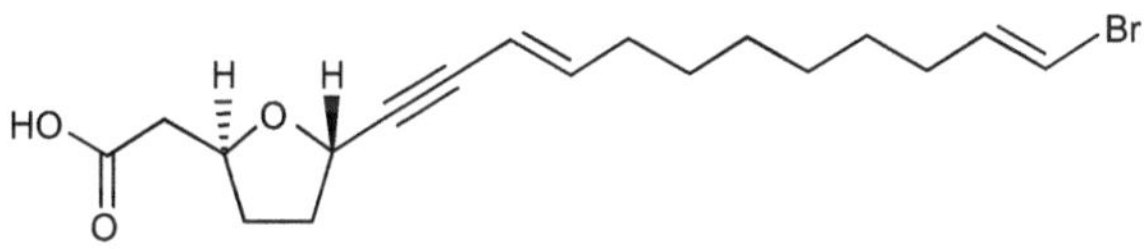

16 (*all-Z*)-5-(2,5,8,11-Tetradecatetraenyl)-2-furanacetic acid

Type: Furans. $C_{20}H_{26}O_3$ Yellow oil. Source: Sponges *Dictyonella incisa* (Mediterranean Sea) and *Hymeniacidon hauraki* (New Zealand). Pharm: Cytotoxic (P_{388}, IC_{50} = 13.4 μg/mL). Ref: P. Ciminiello, et al, Experientia, 1991, 47, 739 | M. R. Prinsep, et al, JNP, 1994, 57, 1557

17 Varitriol

Type: Furans. $C_{15}H_{20}O_5$ Oil, $[\alpha]_D^{25} = +18.5°$ (c = 2.3, MeOH). Source: Marine-derived fungus *Emericella variecolor* M75-2. Pharm: Cytotoxic (In NCI's 60-cell panel, for selected renal, CNS and breast cancer cell lines, varitriol displayed increased potency). Ref: J. Malmstrøm, et al, JNP, 2002, 65, 364 | M. Saleem, et al, NPR, 2007, 24, 1142 (rev)

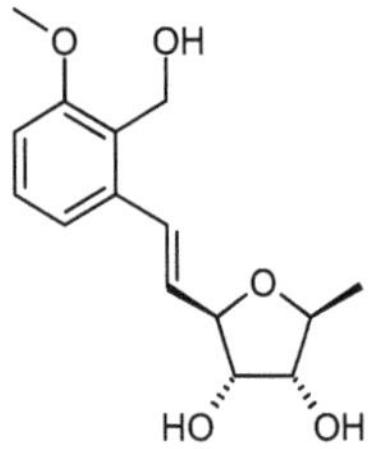

1.3 Butanolides

18 Acetoxyfimbrolide A

Type: Butanolides. $C_{11}H_{12}Br_2O_4$ $[\alpha]_D^{25} = -4.1°$ (c = 1.3, $CHCl_3$). Source: Red alga *Delisea fimbriata*. Pharm: Antibiotic; antifungal. Ref: R. Kazlauskas, et al, Tet. Lett., 1977, 37 | J. A. Pettus, et al, Tet. Lett., 1977, 41

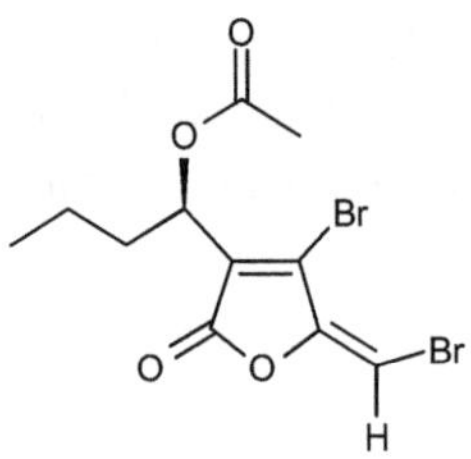

19 Acetoxyfimbrolide B

Type: Butanolides. $C_{11}H_{12}Br_2O_4$ $[\alpha]_D^{25} = +29°$ (c = 5.8, $CHCl_3$). Source: Red alga *Delisea fimbriata*. Pharm: Antibiotic; antifungal. Ref: R. Kazlauskas, et al, Tet. Lett., 1977, 37 | J. A. Pettus, et al, Tet. Lett., 1977, 41

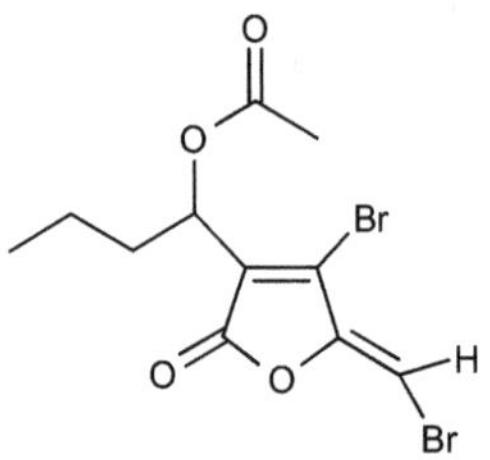

20 Acetoxyfimbrolide D

Type: Butanolides. $C_{11}H_{12}BrIO_4$ $[\alpha]_D^{25} = +26°$ (c = 1.4, $CHCl_3$). Source: Red alga *Delisea fimbriata*. Pharm: Antibiotic; antifungal. Ref: R. Kazlauskas, et al, Tet. Lett., 1977, 37 | J. A. Pettus, et al, Tet. Lett., 1977, 41

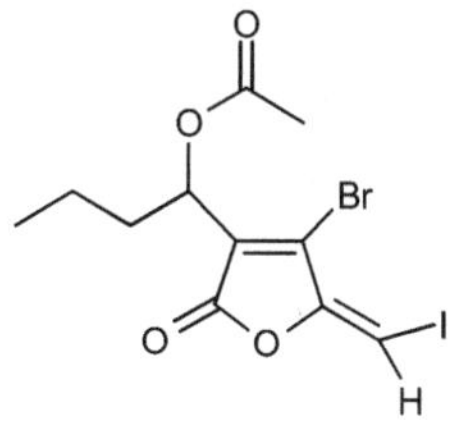

21 Antibiotic MKN 003C

Type: Butanolides. $C_{13}H_{22}O_3$ $[\alpha]_D^{25}$ = +20.3° (*c* = 0.16, CH_3OH). Source: Marine-derived streptomycete *Streptomyces* sp. M02750 (Korea watersn waters). Pharm: Antifoulant. Ref: K. W. Cho, et al, JNP, 2001, 64, 664

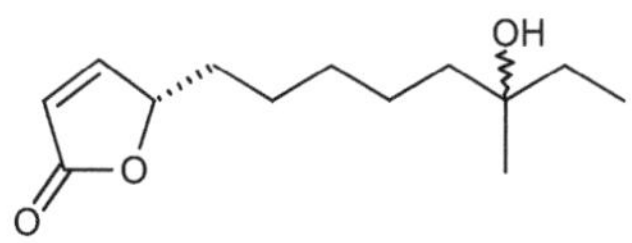

22 Asperlactone

Type: Butanolides. $C_9H_{12}O_4$ Yellow oil. Source: Marine-derived fungus *Exophiala* sp. Pharm: Antifungal; antibacterial. Ref: M. J. Garson, et al, JCS Perkin I, 1984, 1021

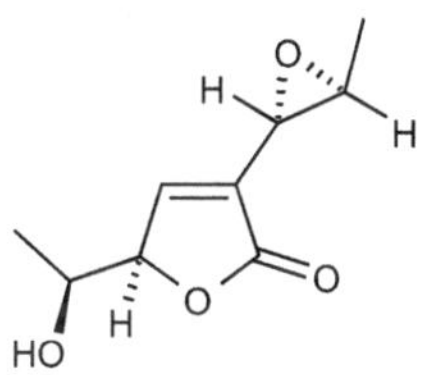

23 Aspernolide A

Type: Butanolides. $C_{24}H_{24}O_7$ Gummy solid, $[\alpha]_D^{28}$ = +88.7° (*c* = 0.58, $CHCl_3$). Source: Marine-derived fungus *Aspergillus terreus* PT06-2 (grown in high salt medium 10% salinity), marine-derived fungus *Aspergillus terreus* SCSGAF0162. Pharm: Cytotoxic (several cancer cell lines, mild); antiviral (HSV virus, IC_{50} = 68.16 μmol/L). Ref: R. R. Parvatkar, et al, Phytochemistry, 2009, 70, 128 | Y. Wang, et al, Mar. Drugs, 2011, 9, 1368 | X. -H. Nong, et al, Mar. Drugs, 2014, 12, 6113

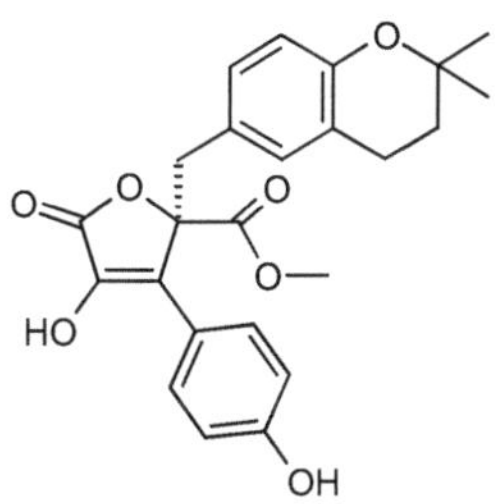

24 Aspernolide B

Terrelactone Type: Butanolides. $C_{24}H_{26}O_8$ Light brown syrup, $[\alpha]_D^{28}$ = +48.3° (*c* = 0.29, MeOH); pale yellow oil, $[\alpha]_D^{25}$ = +71° (*c* = 0.5, $CHCl_3$). Source: Marine-derived fungi *Aspergillus terreus* from soft coral *Sinularia kavarattiensis* (India waters) and *Aspergillus terreus* PT06-2 (Grown in high salt medium 10% salinity). Pharm: Antibacterial (*Enterobacter aerogenes, Staphylococcus aureus* and *Pseudomonas aeruginosa,* all MICs > 100 μmol/L); antifungal (*Candida albicans,* MIC > 100 μmol/L, control Ketoconazole, MIC = 5 μmol/L). Ref: R. R. Parvatkar, et al, Phytochemistry, 2009, 70, 128 | Y. Wang, et al, Mar. Drugs, 2011, 9, 1368

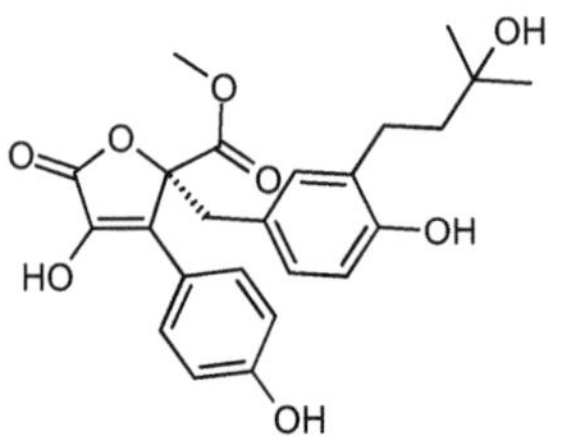

25 Aspiketolactonol

Type: Butanolides. $C_9H_{12}O_4$ Colorless oil (MeOH), $[\alpha]_D^{25}$ = +10.5° (*c* = 0.12, MeOH). Source: Deep-sea fungus *Aspergillus* sp. 16-02-1 from sediment (Lau Basin hydrothermal vent, depth of 2255 m, temperature 114 °C). Pharm: Cytotoxic (MTT method, 100 μg/mL: HL60, InRt = 11.7%, control Docetaxol, InRt = 49.9%). Ref: X. Chen, et al, Mar. Drugs, 2014, 12, 3116

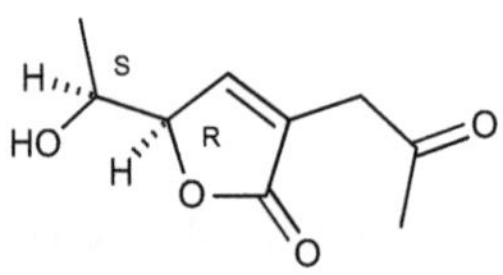

26 Aspilactonol A

Type: Butanolides. $C_9H_{14}O_3$ Colorless oil (MeOH), $[\alpha]_D^{25}$ = +24.6° (*c* = 0.23, MeOH). Source: Deep-sea fungus *Aspergillus* sp. 16-02-1 from sediment (Lau Basin hydrothermal vent, depth of 2255 m, temperature 114 °C). Pharm: Cytotoxic (MTT method, 100 μg/mL: HL60, InRt = 17.7%, control Docetaxol, InRt = 49.9%). Ref: X. Chen, et al, Mar. Drugs, 2014, 12, 3116

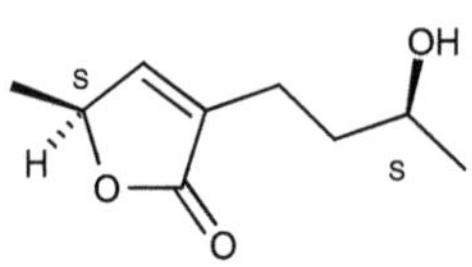

27 Aspilactonol B

Type: Butanolides. $C_9H_{14}O_5$ Colorless oil (MeOH), $[\alpha]_D^{25} = +34.5°$ (c = 0.41, MeOH). Source: Deep-sea fungus *Aspergillus* sp. 16-02-1 from sediment (Lau Basin hydrothermal vent, depth of 2255 m, temperature 114 °C). Pharm: Cytotoxic (MTT method, 100 µg/mL: HL60, InRt = 22.2%, control Docetaxol, InRt = 49.9%). Ref: X. Chen, et al, Mar. Drugs, 2014, 12, 3116

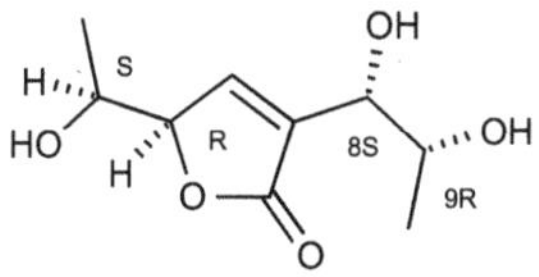

28 Aspilactonol C

Type: Butanolides. $C_9H_{14}O_5$ Colorless oil (MeOH), $[\alpha]_D^{25} = +27.2°$ (c = 0.32, MeOH). Source: Deep-sea fungus *Aspergillus* sp. 16-02-1 from sediment (Lau Basin hydrothermal vent, depth of 2255 m, temperature 114 °C). Pharm: Cytotoxic (MTT method, 100 µg/mL: HL60, InRt = 16.7%, control Docetaxol, InRt = 49.9%; K562, InRt = 20.0%, Docetaxol, InRt = 55.6%). Ref: X. Chen, et al, Mar. Drugs, 2014, 12, 3116

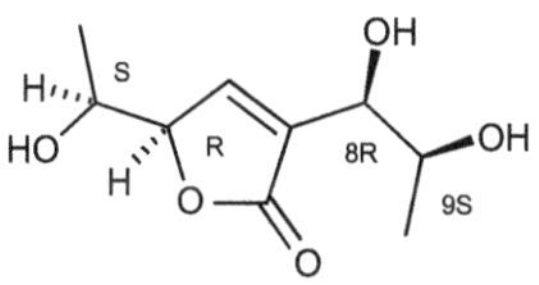

29 Aspilactonol D

Type: Butanolides. $C_{10}H_{16}O_5$ Colorless oil (MeOH), $[\alpha]_D^{25} = -1.7°$ (c = 0.38, MeOH). Source: Deep-sea fungus *Aspergillus* sp. 16-02-1 from sediment (Lau Basin hydrothermal vent, depth of 2255 m, temperature 114 °C). Pharm: Cytotoxic (MTT method, 100 µg/mL: HL60, InRt = 20.0%, control Docetaxol, InRt = 49.9%; BGC823, InRt = 13.2%, Docetaxol, InRt = 61.5%). Ref: X. Chen, et al, Mar. Drugs, 2014, 12, 3116

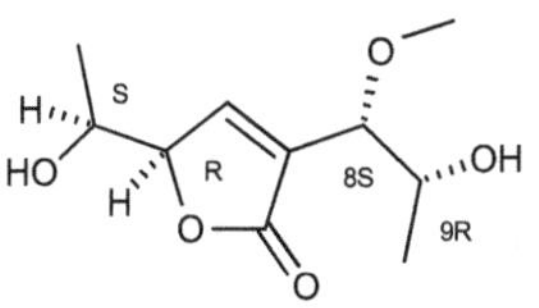

30 Aspilactonol E

Type: Butanolides. $C_9H_{14}O_4$ Mixture with Aspilactonol F, Colorless oil (MeOH), $[\alpha]_D^{25} = +78.1°$ (c = 1.00, MeOH). Source: Deep-sea fungus *Aspergillus* sp. 16-02-1 from sediment (Lau Basin hydrothermal vent, depth of 2255 m, temperature 114 °C). Pharm: Mixture with Aspilactonol F, Cytotoxic (MTT method, 100 μg/mL: HeLa, InRt = 13.4%, control Docetaxol, InRt = 45.1%; K562, InRt = 14.1%, Docetaxol, InRt = 55.6%). Ref: X. Chen, et al, Mar. Drugs, 2014, 12, 3116

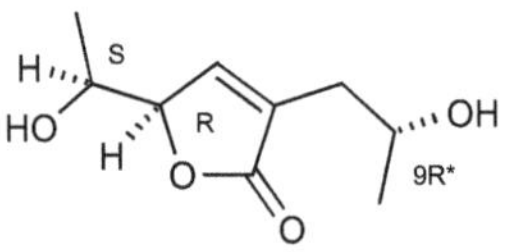

31 Aspilactonol F

Type: Butanolides. $C_9H_{14}O_4$ Mixture with Aspilactonol E, Colorless oil (MeOH), $[\alpha]_D^{25}$ = +78.1° (*c* = 1.00, MeOH). Source: Deep-sea fungus *Aspergillus* sp. 16-02-1 from sediment (Lau Basin hydrothermal vent, depth of 2255 m, temperature 114 °C). Pharm: Mixture with Aspilactonol E, Cytotoxic (MTT method, 100 μg/mL: HeLa, InRt = 13.4%, control Docetaxol, InRt = 45.1%; K562, InRt = 14.1%, Docetaxol, InRt = 55.6%). Ref: X. Chen, et al, Mar. Drugs, 2014, 12, 3116

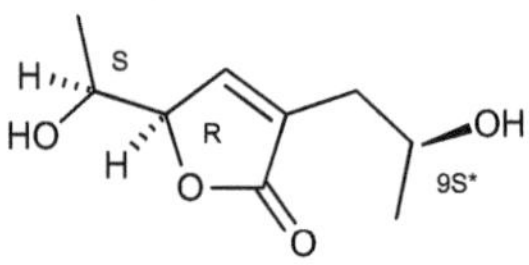

32 4-Bromobeckerelide

Type: Butanolides. $C_9H_{13}BrO_4$ Source: Red alga *Beckerella subcostatum*. Pharm: Antimicrobial. Ref: K. Ohta, Agrc. Biol. Chem., 1977, 41, 2105

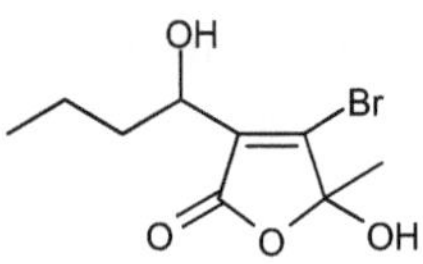

33 (+)-Butyrolactone I

Type: Butanolides. $C_{24}H_{24}O_7$ Cryst. (CH_2Cl_2), mp 95 °C (dec), $[\alpha]_D^{17}$ = +86° (*c* = 0.5, EtOH). Source: Marine-derived fungus *Aspergillus terreus* PT06-2 (grown in high salt medium 10% salinity), terrestrial and marine-derived *Aspergillus terreus* (strains IFO 8835, DRCC 152 and HKI0499). Pharm: Antiviral (influenza virus H1N1, lower cytotoxicity, IC_{50} = 143.1 μmol/L, CC_{50} = 976.4 μmol/L, control Ribavirin, IC_{50} = 100.8 μmol/L); cytotoxic (HL60, IC_{50} = 57.5 μmol/L, control VP-16 (Etoposide), IC_{50} = 0.042 μmol/L); kinase inhibitor (high selectivity towards CDK1 and CDK2). Ref. M. Kitagawa, et al, Oncogene, 1994, 9, 2549 | K. V. Rao, et al, CPB, 2000, 48, 559 | Y. Wang, et al, Mar. Drugs, 2011, 9, 1368

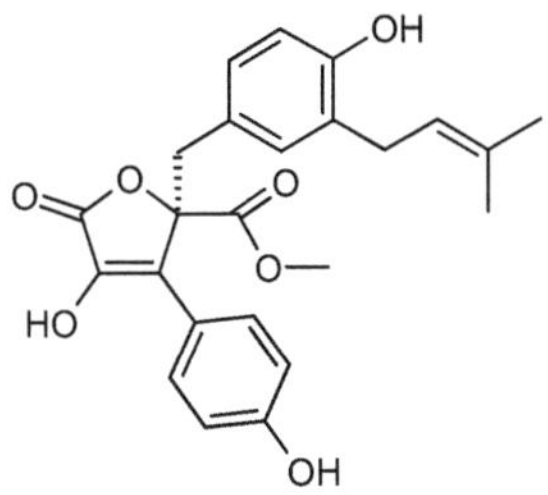

34 (+)-Butyrolactone II

Type: Butanolides. $C_{19}H_{16}O_7$ Plates (CH_2Cl_2), mp 94–96 °C, $[\alpha]_D^{25}$ = +85° (c = 1.0, EtOH). Source: Marine-derived fungus *Aspergillus terreus* PT06-2 (grown in high salt medium 10% salinity). Pharm: Cytotoxic. Ref: Y. Wang, et al, Mar. Drugs, 2011, 9, 1368

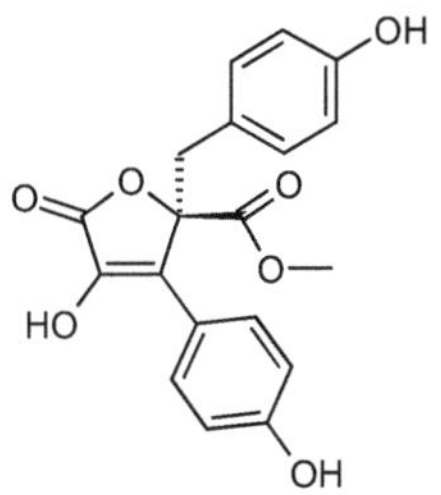

35 (+)-Butyrolactone III

Type: Butanolides. $C_{24}H_{24}O_8$ Yellowish mass (CH_2Cl_2), mp 70 °C, $[\alpha]_D^{25}$ = +80° (c = 0.7, EtOH). Source: Marine-derived fungus *Aspergillus terreus* PT06-2 (grown in high salt medium 10% salinity). Pharm: Cytotoxic. Ref: Y. Wang, et al, Mar. Drugs, 2011, 9, 1368

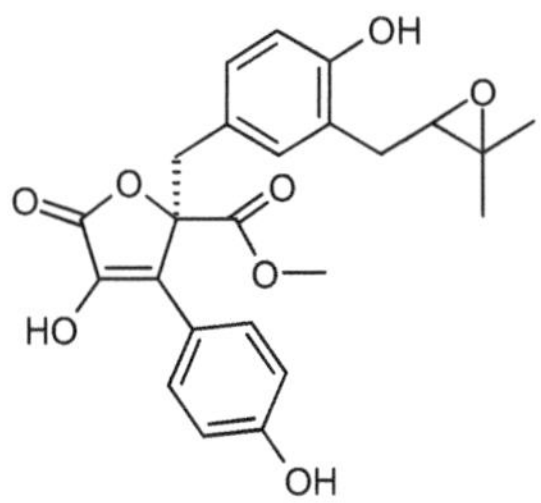

36 9-Chloro-8-hydroxy-8,9-deoxyasperlactone

Type: Butanolides. $C_9H_{13}ClO_4$ $[\alpha]_D^{26}$ = +76° (c = 0.01, $CHCl_3$). Source: Marine-derived fungus *Aspergillus ostianus* TUF 01F313 from an unidentified sponge (Pohnpei I., Federated States of Micronesia). Pharm: Antibacterial (25 µg/disk, inhibits growth of *Ruegeria atlantica*, IZD = 10.1 mm). Ref: M. Namikoshi, et al, J. Antibiot., 2003, 56, 755 | M. Saleem, et al, NPR, 2007, 24, 1142 (rev)

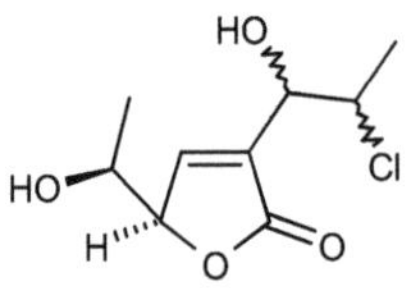

37 8-Chloro-9-hydroxy-8,9-deoxyasperlactone

Type: Butanolides. $C_9H_{13}ClO_4$ $[\alpha]_D^{24}$ = +22.5° (c = 0.17, $CHCl_3$). Source: Marine-derived fungus *Aspergillus ostianus* TUF 01F313 from an unidentified sponge (Pohnpei I., Federated States of Micronesia). Pharm: Antibacterial (5 μg/disk, inhibits growth of *Ruegeria atlantica*, IZD = 12.7 mm). Ref: M. Namikoshi, et al, J. Antibiot., 2003, 56, 755 | M. Saleem, et al, NPR, 2007, 24, 1142 (rev)

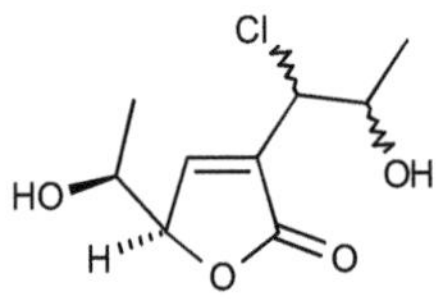

38 Cillifuranone

Type: Butanolides. $C_{10}H_{12}O_4$ Pale yellow needles. Source: Marine-derived fungus *Penicillium chrysogenum* from sponge *Tethya aurantium* (Limski Channel, Adriatic Sea, Croatia). Pharm: Antifungal (*Septoria tritici*, 100 μmol/L, 20% growth inhibition); antibacterial (*Xanthomonas campestris*, 100 μmol/L, 24% growth inhibition). Ref: J. Wiese, et al, Mar. Drugs, 2011, 9, 561

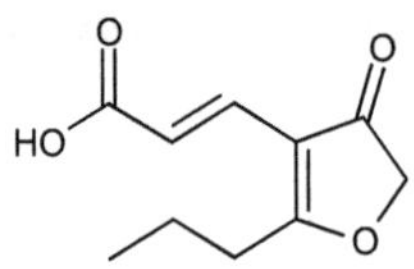

39 Eutypoid B

Type: Butanolides. $C_{17}H_{14}O_4$ Source: Marine-derived fungus *Penicillium* sp. (sea water sample, North Sea). Pharm: Glycogen synthase kinase-3β inhibitor. Ref: D. Schulz, et al, JNP, 2011, 74, 99

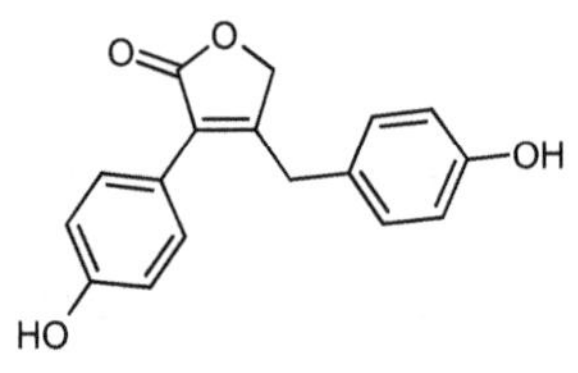

40 Eutypoid C

Type: Butanolides. $C_{17}H_{14}O_5$ Source: Marine-derived fungus *Penicillium* sp. (sea water sample, North Sea). Pharm: Glycogen synthase kinase-3β inhibitor. Ref: D. Schulz, et al, JNP, 2011, 74, 99

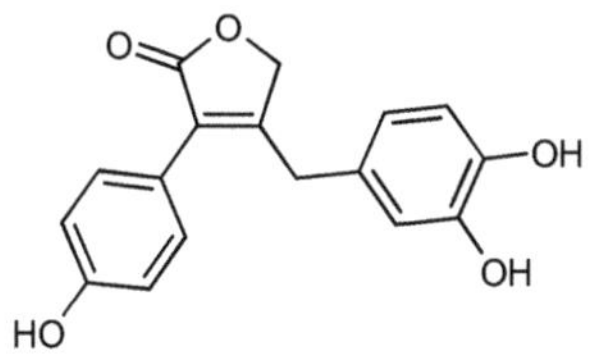

41 Eutypoid D

Type: Butanolides. $C_{17}H_{14}O_5$ Source: Marine-derived fungus *Penicillium* sp. (water sample, North Sea). Pharm: Glycogen synthase kinase-3β inhibitor. Ref: D. Schulz, et al, JNP, 2011, 74, 99

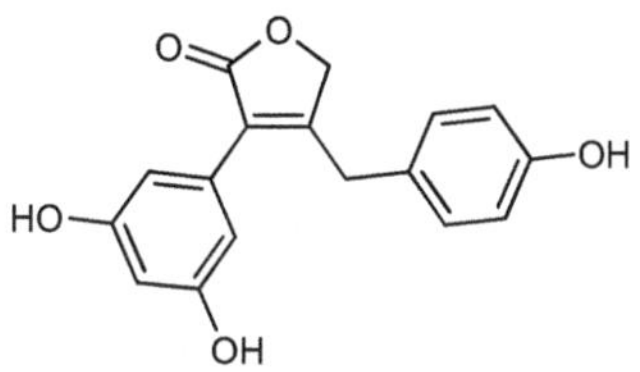

42 Eutypoid E

Type: Butanolides. $C_{17}H_{14}O_6$ Source: Marine-derived fungus *Penicillium* sp. (water sample, North Sea). Pharm: Glycogen synthase kinase-3β inhibitor. Ref: D. Schulz, et al, JNP, 2011, 74, 99

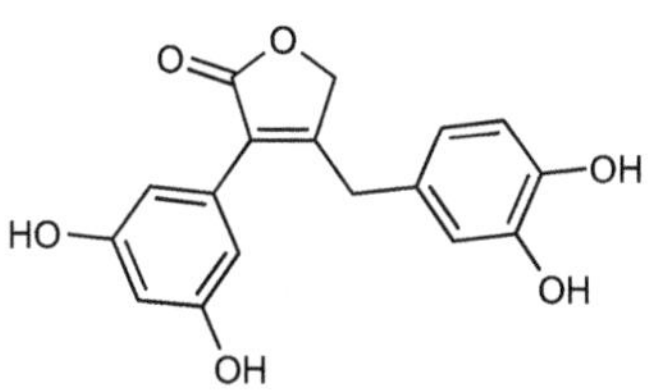

43 Flavalactone 1

Type: Butanolides. $C_{21}H_{38}O_3$ Needles (MeOH), mp 67 °C. Source: Gorgonian *Euplexaura flava*. Pharm: Anti-inflammatory. Ref: H. Kikuchi, et al, Chem. Lett., 1982, 233; H. Kikuchi, et al, CPB, 1983, 31, 1172

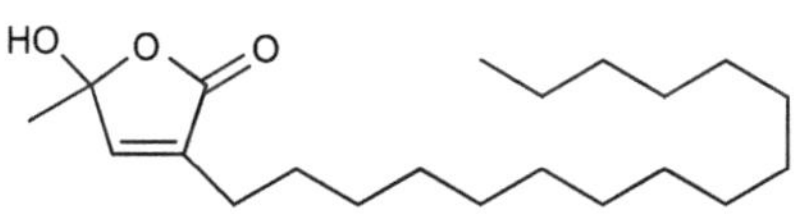

44 Flavalactone 2

Type: Butanolides. $C_{25}H_{42}O_3$ Oil. Source: Gorgonian *Euplexaura flava*. Pharm: Anti-inflammatory. Ref: H. Kikuchi, et al, Chem. Lett., 1982, 233; H. Kikuchi, et al, CPB, 1983, 31, 1172

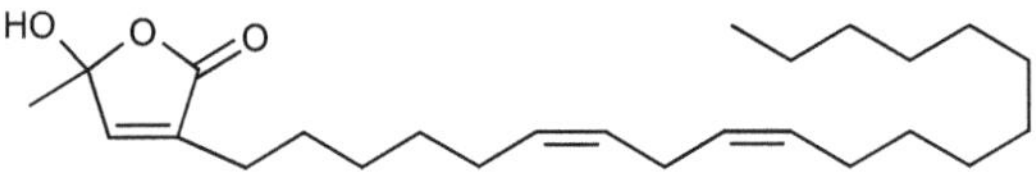

45 Flavalactone 3

Type: Butanolides. $C_{27}H_{44}O_3$ Oil. Source: Gorgonian *Euplexaura flava*. Pharm: Anti-inflammatory. Ref: H. Kikuchi, et al, Chem. Lett., 1982, 233; H. Kikuchi, et al, CPB, 1983, 31, 1172

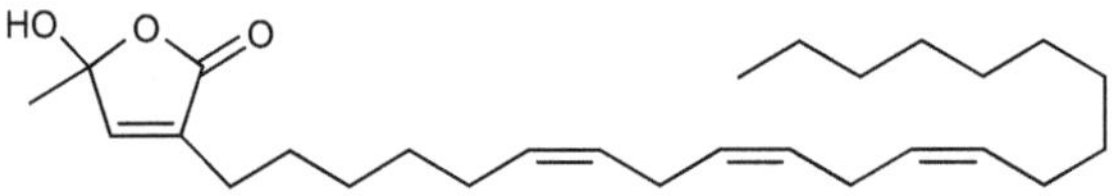

46 Flavipesin A

Type: Butanolides. $C_{22}H_{22}O_5$ Source: mangrove-derived fungus *Aspergillus flavipes* AIL8 (endophytic) from mangrove *Acanthus ilicifolius*. Pharm: Antibacterial (*Staphylococcus aureus*, MIC = 8.0 μg/mL, *Bacillus subtilis*, MIC = 0.25 μg/mL). Ref: Z. Q. Bai, et al, Fitoterapia, 2014, 95, 194

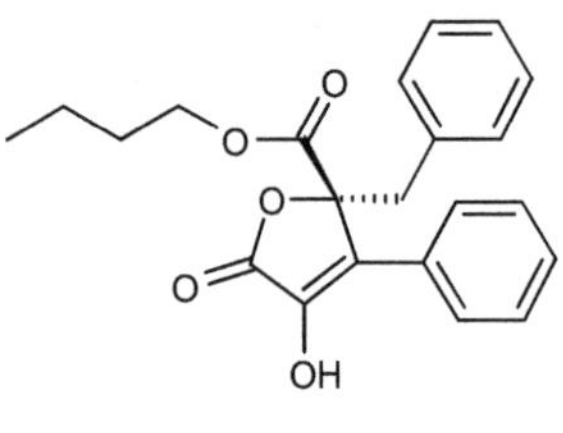

47 Harzialactone A

Type: Butanolides. $C_{11}H_{12}O_3$ Colorless oil, mp 82–84 °C, $[\alpha]_D^{32}$ = +33.3° (*c* = 0.3, $CHCl_3$). Source: Marine-derived fungus *Trichoderma harzianum* OUPS-N115 from sponge *Halichondria okadai*. Pharm: Cytotoxic (cultured P_{388}, ED_{50} > 100 μg/mL). Ref: T. Amagata, et al, J. Antibiot., 1998, 51, 33

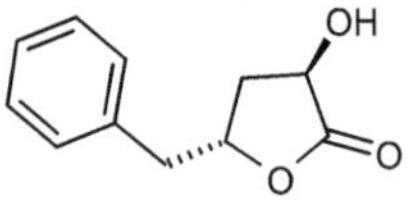

48 Honaucin A

Type: Butanolides. $C_8H_9NO_4$ Source: Cyanobacterium *Leptolyngbya crossbyana* (Hōnaunau reef, Hawaii). Pharm: Inhibits NO production and expression of several pro-infiammatory cytokines (RAW264.7 cells); inhibits bioluminescence (*Vibrio harveyi*). Ref: H. Choi, et al, Chem. Biol., 2012, 19,589

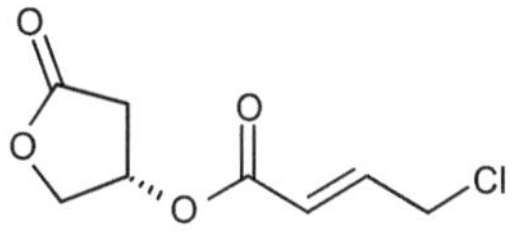

49 (*S*)-2-(2′-Hydroxyethyl)-4-methyl-γ-butyrolactone

Type: Butanolides. $C_7H_{10}O_3$ Source: Deep-sea fungus *Aspergillus* sp. 16-02-1 from sediment (Lau Basin hydrothermal vent, depth of 2255 m, temperature 114 °C). Pharm: Cytotoxic (MTT method, 100 μg/mL: HL60, InRt = 14.2%, control Docetaxol, InRt = 49.9%; HeLa, InRt = 15.7%, Docetaxol, InRt = 45.1%). Ref: X. Chen, et al, Mar. Drugs, 2014, 12, 3116

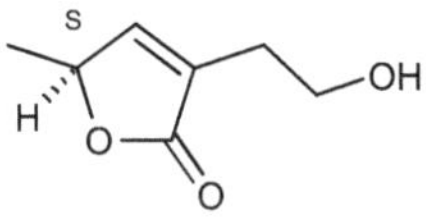

50 Isoaspulvinone E

Type: Butanolides. $C_{17}H_{12}O_5$ Source: Mangrove-derived fungus *Aspergillus terreus* GWQ-48. Pharm: Antiviral (influenza a viral H1N1, IC_{50} = 101.23 μmol/L). Ref: H. Gao, et al, Bioorg. Med. Chem. Lett. 2013, 23, 1776

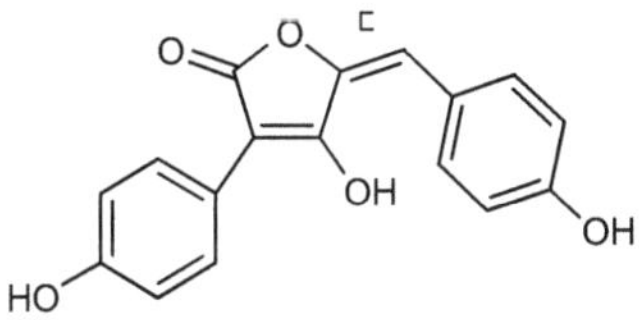

51 Isobutyrolactone II

Type: Butanolides. $C_{18}H_{16}O_6$ Source: Marine-derived fungus *Aspergillus terreus* SCSGAF0162. Pharm: Antiviral (HSV virus, IC_{50} = 62.08 μmol/L). Ref: X. -H. Nong, et al, Mar. Drugs, 2014, 12, 6113

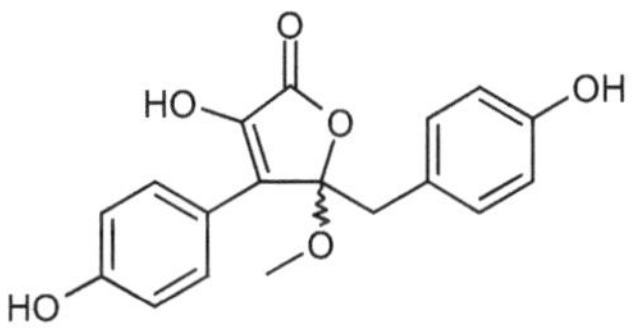

52 Isocladospolide B

NG 261 Type: Butanolides. $C_{12}H_{20}O_4$ Amorph. solid, $[\alpha]_D = -90°$ (c = 0.23, MeOH). Source: An unidentified marine-derived fungus I96S215 from both unidentified sponge and fungus *Cladosporium cladosporioides* sp. TF-0380. Pharm: Nerve growth factor agonist. Ref: C. J. Smith, et al, JNP, 2000, 63, 142 | S. Gesner, et al, JNP, 2005, 68, 135

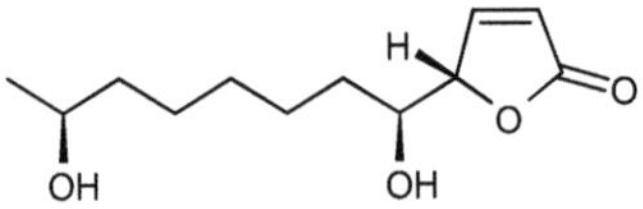

53 Lissoclinolide

Tetrenolin Type: Butanolides. $C_{11}H_{12}O_4$ Yellow cryst. ($CHCl_3$), mp 126–128 °C. Source: Ascidian *Lissoclinum patella*. Pharm: Antibacterial (gram-negative bacterium *Escherichia coli*). Ref: B. S. Davidson, et al, JNP, 1990, 53, 1036 | R. Rossi, et al, Tet. Lett., 1998, 39, 7799 | C. Xu, et al, Tet. Lett., 1999, 40, 431

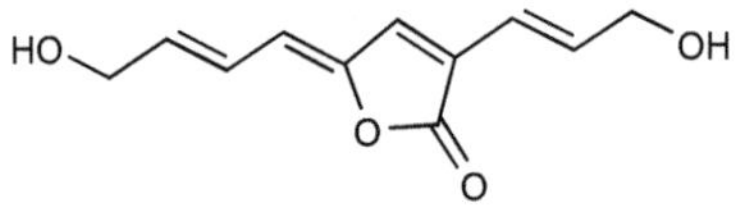

54 Nodulisporacid A

Type: Butanolides. $C_{16}H_{20}O_6$ Equilib. mixt. of *E*- and *Z*-isomers (1:1), amorph. yellow solid, $[\alpha]_D^{27} = -20.6°$ (c = 1.1, MeOH). Source: Marine-derived fungus *Nodulisporium* sp. CRIF 1. Pharm: Antiplasmodial (*Plasmodium falciparum* 94 strain, IC_{50} = 1–10 μmol/L). Ref: C. Kasettrathat, et al, Phytochemistry, 2008, 69, 2621 | T. Sumiya, et al, Tet. Lett., 2010, 51, 2765

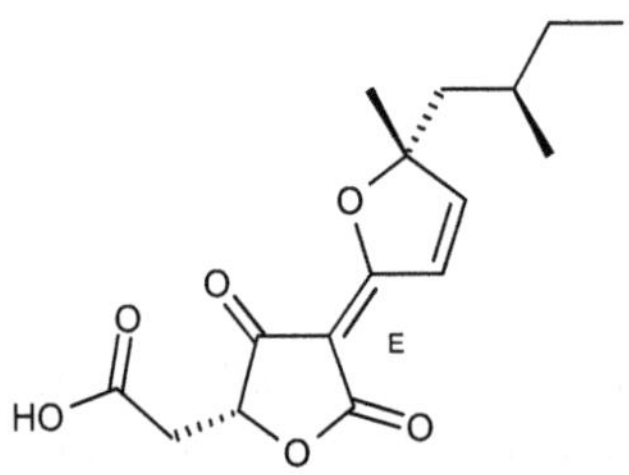

55 Nodulisporacid A benzyl ester

Type: Butanolides. $C_{22}H_{24}O_6$ Source: Marine-derived fungus *Nodulisporium* sp. CRIF 1. Pharm: Cytotoxic (HuCCA-1, IC_{50} = 2.10 μg/mL; KB, IC_{50} = 3.20 μg/mL; HeLa, IC_{50} = 2.60 μg/mL; MDA-MB-231, IC_{50} = 0.38 μg/mL; T47D, IC_{50} = 0.14 μg/mL; HepG2, IC_{50} = 2.00 μg/mL; A549, IC_{50} = 2.20 μg/mL; HCC-S102, IC_{50} = 4.80 μg/mL; HL60, IC_{50} = 1.18 μg/mL; P_{388}, IC_{50} = 0.70 μg/mL; control Etoposide: HuCCA-1,

IC_{50} = 5.30 μg/mL; KB, IC_{50} = 0.46 μg/mL; HeLa, IC_{50} = 0.40 μg/mL; MDA-MB-231, IC_{50} = 0.40 μg/mL; T47D, IC_{50} = 0.04 μg/mL; HL60, IC_{50} = 0.77 μg/mL; P_{388}, IC_{50} = 0.10 μg/mL; control Doxorubicind: HepG2, IC_{50} = 0.19 μg/mL; A549, IC_{50} = 0.48 μg/mL; HCC-S102, IC_{50} = 1.20 μg/mL). Ref: C. Kasettrathat, et al, Phytochemistry, 2008, 69, 2621

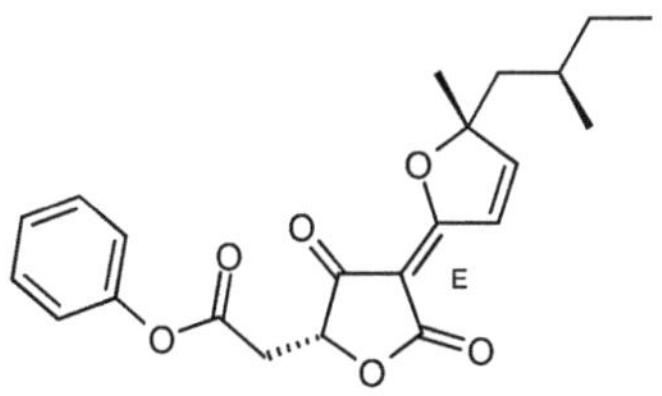

56 Nodulisporacid A methyl ester

Type: Butanolides. $C_{17}H_{22}O_6$ Source: Marine-derived fungus *Nodulisporium* sp. CRIF 1. Pharm: Cytotoxic (HuCCA-1, IC_{50} = 2.30 μg/mL; KB, IC_{50} = 2.20 μg/mL; HeLa, IC_{50} = 2.70 μg/mL; MDA-MB-231, IC_{50} = 2.50 μg/mL; T47D, IC_{50} = 1.70 μg/mL; HepG2, IC_{50} = 2.30 μg/mL; A549, IC_{50} = 7.50 μg/mL; HCC-S102, IC_{50} = 6.00 μg/mL; HL60, IC_{50} = 1.01 μg/mL; P_{388}, IC_{50} = 0.77 μg/mL; control Etoposide: HuCCA-1, IC_{50} = 5.30 μg/mL; KB, IC_{50} = 0.46 μg/mL; HeLa, IC_{50} = 0.40 μg/mL; MDA-MB-231, IC_{50} = 0.40 μg/mL; T47D, IC_{50} = 0.04 μg/mL; HL60, IC_{50} = 0.77 μg/mL; P_{388}, IC_{50} = 0.10 μg/mL; control Doxorubicind: HepG2, IC_{50} = 0.19 μg/mL; A549, IC_{50} = 0.48 μg/mL; HCC-S102, IC_{50} = 1.20 μg/mL). Ref: C. Kasettrathat, et al, Phytochemistry, 2008, 69, 2621

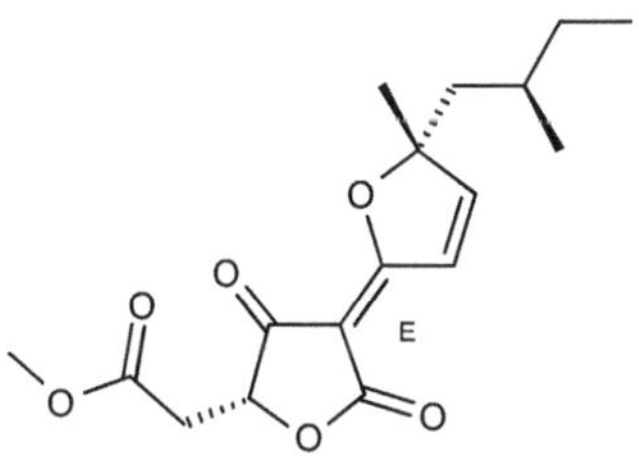

57 Penicilactone

Type: Butanolides. $C_7H_{10}O_4$ Gum, $[\alpha]_D^{25}$ = +23° (c = 0.12, MeOH). Source: Marine-derived fungus *Penicillium* sp. PSU-F44 from gorgonian sea fan *Annella* sp. (Thailand). Pharm: Antibacterial (MRSA SK1). Ref: K. Trisuwan, et al, CPB, 2009, 57, 1100

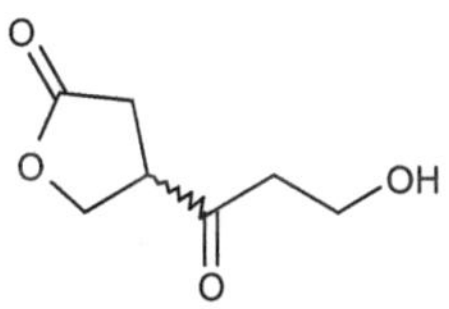

58 Pestalolide

Type: Butanolides. $C_{12}H_{20}O_2$ Colorless gum. Source: Mangrove-derived fungus *Pestalotiopsis* sp. from mangrove *Rhizophora apiculata* (branch, Satun, Thailand). Pharm: Antifungal (*Candida albicans* NCPF3153 and *Cryptococcus neoformans* ATCC 90112, MIC = 128 μg/mL). Ref: S. Klaiklay, et al, Tetrahedron, 2012, 68, 2299

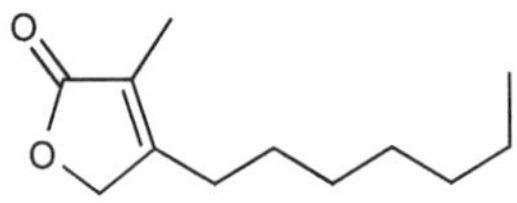

59 Plakilactone C

Type: Butanolides. $C_{16}H_{24}O_3$ Source: Sponge *Plakinastrella mamillaris* (Fiji). Pharm: Antidiabetic and anti-arthrosclerosis (due to itscovalent binding to peroxisome proliferator-activated receptor *γ* (PPAR*γ*), inhibiting adipocyte gene transcription). Ref: C. Festa, et al, JMC, 2012, 55, 8303

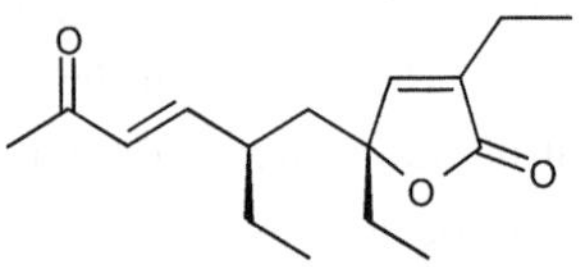

60 Plakorsin D

Type: Butanolides. $C_{15}H_{24}O_3$ Source: Sponge *Plakortis simplex* (Yongxing I., South China Sea, China). Pharm: Cytotoxic (HeLa, IC_{50} > 100 μmol/L; K562, IC_{50} > 100 μmol/L; A549, IC_{50} > 100 μmol/L; Bel7402, IC_{50} > 100 μmol/L, control Adriamycin: HeLa, IC_{50} = (0.6 ± 0.0)μmol/L; K562, IC_{50} = (0.3 ± 0.0)μmol/L; A549, IC_{50} = (0.2 ± 0.0)μmol/L). Ref: J. Zhang, et al, JNP, 2013, 76, 600

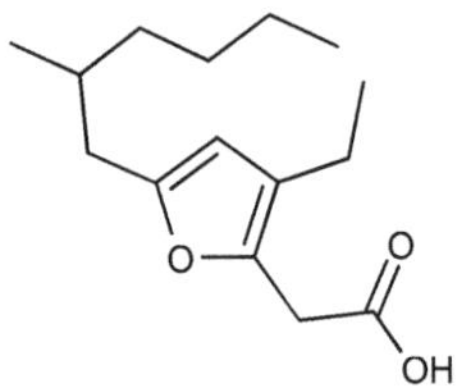

61 Plakortone G

Type: Butanolides. $C_{18}H_{30}O_2$ Oil, $[\alpha]_D$ = −25.9° (*c* = 0.008, $CHCl_3$). Source: Sponge *Plakortis* sp. (subclass Homoscleromorpha, order Homosclerophorida, famiily Plakinidae, Jamaica). Pharm: Antiplasmodial (*Plasmodium falciparum* D6 clone, *in vitro*, IC_{50} = 4200 ng/mL, control Artemlsinin, IC_{50} = 12 ng/mL; CRPF W2 clone, *in vitro*, IC_{50} > 4760 ng/mL, control Artemlsinin, IC_{50} = 7 ng/mL); cytotoxic (HT29, IC_{50} > 1 μg/mL, control Tamoxifen, IC_{50} = 1.86 μg/mL; A549, IC_{50} > 1 μg/mL, control

Tamoxifen, IC_{50} = 1.86 µg/mL; MEL28, IC_{50} > 10 µg/mL, control Tamoxifen, IC_{50} = 1.86 µg/mL; hmn primary tumor cells, IC_{50} = 4.7 µg/mL, control Doxorubicin, IC_{50} = 25 nmol/L); anti-hepatitis-B (EC_{50} > 100 µg/mL, control 3TC, EC_{50} = 0.062–0.065 µg/mL); antituberculosis (*Mycobacterium tuberculosis*, InRt = 4% at 6.25 µg/mL, control Rifampin, MIC = 0.25 µg/mL). Ref: D. J. Gochfeld, et al, JNP, 2001, 64, 1477 | S. Kowashi, et al, Tet. Lett., 2004, 45, 4393

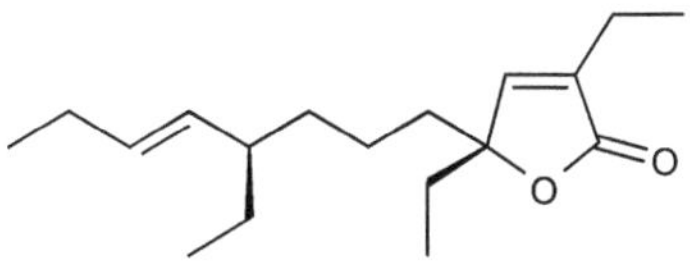

62 Plakortoxide A

Type: Butanolides. $C_{17}H_{28}O_3$ Colorless oil. Source: Sponge *Plakortis simplex* (Yongxing I., South China Sea, China). Pharm: Cytotoxic (HeLa, IC_{50} = (31.5 ± 2.9)µmol/L; K562, IC_{50} > 50 µmol/L; A549, IC_{50} > 100 µmol/L; Bel7402, IC_{50} > 100 µmol/L, control Adriamycin: HeLa, IC_{50} = (0.6 ± 0.0)µmol/L; K562, IC_{50} = (0.3 ± 0.0)µmol/L; A549, IC_{50} = (0.2 ± 0.0) µmol/L). Ref: J. Zhang, et al, JNP, 2013, 76, 600

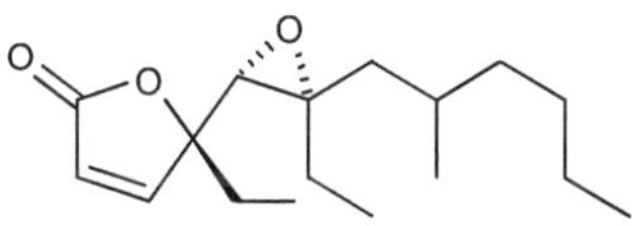

63 Plakortoxide B

Type: Butanolides. $C_{18}H_{30}O_3$ Source: Sponge *Plakortis simplex* (Yongxing I., South China Sea, China). Pharm: Cytotoxic (HeLa, IC_{50} > 50 µmol/L; K562, IC_{50} > 100 µmol/L; A549, IC_{50} > 100 µmol/L; Bel7402, IC_{50} > 100 µmol/L, control Adriamycin: HeLa, IC_{50} = (0.6 ± 0.0)µmol/L; K562, IC_{50} = (0.3 ± 0.0)µmol/L; A549, IC_{50} = (0.2 ± 0.0)µmol/L). Ref: J. Zhang, et al, JNP, 2013, 76, 600

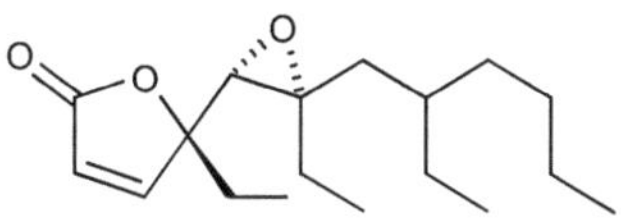

64 Rubrolide R

Type: Butanolides. $C_{22}H_{20}O_4$ Colorless triclinic crystal. Source: Marine-derived fungus *Aspergillus terreus* OUCMDZ-1925. Pharm: Antioxidant ($ABTS^{\bullet+}$ radicals, IC_{50} = 1.33 mmol/L); cytotoxic (K562, IC_{50} = 12.8 mmol/L). Ref: T. Zhu, et al, J. Antibiot., 2014, 67, 315

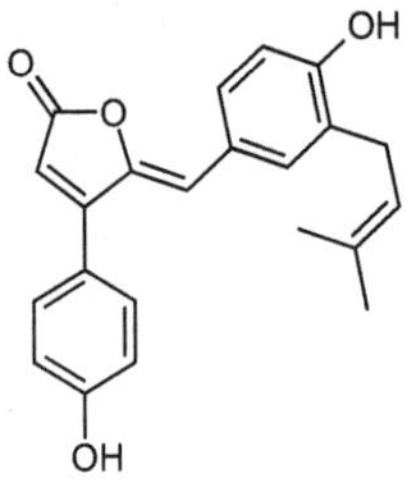

65 Rubrolide S

Type: Butanolides. $C_{22}H_{20}O_4$ Source: Marine-derived fungus *Aspergillus terreus* OUCMDZ-1925. Pharm: Antiviral (influenza virus (IFV) virus, IC_{50} = 87.1 μmol/L); cytotoxic (K562, IC_{50} = 12.8 mmol/L). Ref: T. Zhu, et al, J. Antibiot., 2014, 67, 315

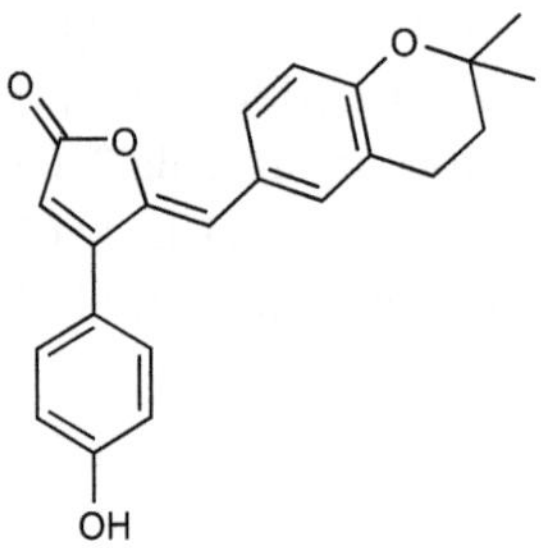

66 (5*E*)-Simplexolide

Type: Butanolides. $C_{18}H_{32}O_3$ Source: Sponge *Plakortis simplex* (Yongxing I., South China Sea, China). Pharm: Antifungal (weak). Ref: X. -F. Liu, et al, Tetrahedron, 2012, 68, 4635

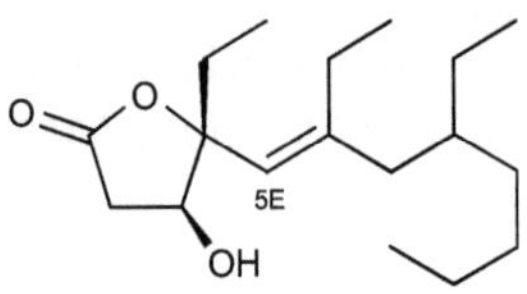

67 (5*E*)-Simplexolide A

Type: Butanolides. $C_{17}H_{30}O_3$ Source: Sponge *Plakortis simplex* (Yongxing I., South China Sea, China). Pharm: Antifungal (weak). Ref: X. -F. Liu, et al, Tetrahedron, 2012, 68, 4635

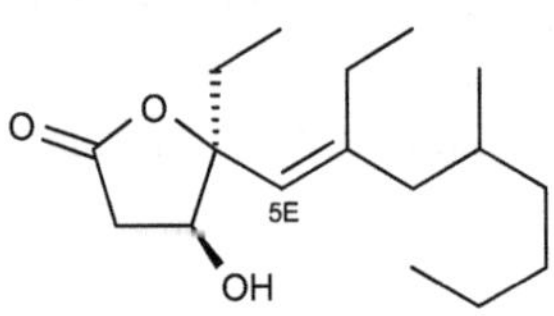

68 (5*Z*)-Simplexolide A

Type: Butanolides. $C_{17}H_{30}O_3$ Source: Sponge *Plakortis simplex* (Yongxing I., South China Sea, China). Pharm: Antifungal (weak). Ref: X. -F. Liu, et al, Tetrahedron, 2012, 68, 4635

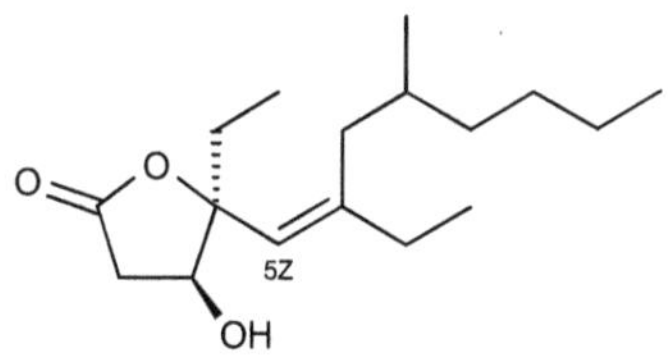

69 (5*Z*)-Simplexolide B

Type: Butanolides. $C_{17}H_{30}O_3$ Source: Sponge *Plakortis simplex* (Yongxing I., South China Sea, China). Pharm: Cytotoxic (HeLa, IC_{50} = (4.7 ± 0.5)μmol/L; K562, IC_{50} = (2.2 ± 0.2)μmol/L; A549, IC_{50} > 100 μmol/L; Bel7402, IC_{50} > 100 μmol/L, control Adriamycin: HeLa, IC_{50} = (0.6 ± 0.0)μmol/L; K562, IC_{50} = (0.3 ± 0.0)μmol/L; A549, IC_{50} = (0.2 ± 0.0)μmol/L). Ref: J. Zhang, et al, JNP, 2013, 76, 600

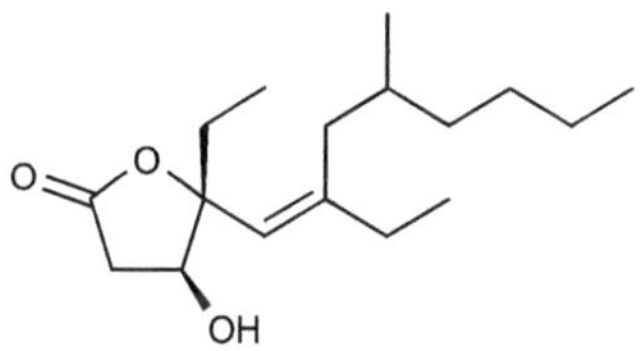

70 (5*E*)-Simplexolide E

Type: Butanolides. $C_{17}H_{30}O_3$ Source: Sponge *Plakortis simplex* (Yongxing I., South China Sea, China). Pharm: Cytotoxic (HeLa, IC_{50} = (31.2 ± 2.4)μmol/L; K562, IC_{50} = (49.2 ± 4.2)μmol/L; A549, IC_{50} > 100 μmol/L; Bel7402, IC_{50} > 100 μmol/L, control Adriamycin: HeLa, IC_{50} = (0.6 ± 0.0) μmol/L; K562, IC_{50} = (0.3 ± 0.0)μmol/L; A549, IC_{50} = (0.2 ± 0.0)μmol/L). Ref: J. Zhang, et al, JNP, 2013, 76, 600

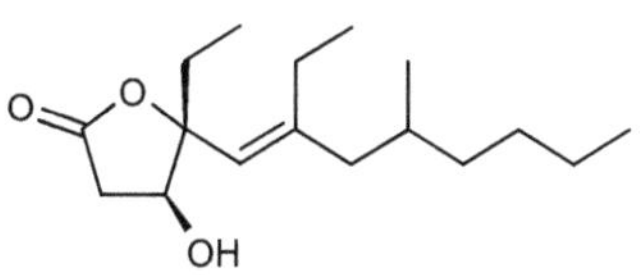

71 Simplextone A

Type: Butanolides. $C_{18}H_{32}O_4$ Colorless cryst., $[\alpha]_D^{23}$ = +28° (*c* = 0.085, CH_3OH). Source: Sponge *Plakortis simplex* (Yongxing I., South China Sea, China). Pharm: Cytotoxic (MTT assay, HCT116, IC_{50} = 26.3 μmol/L; SGC7901, IC_{50} = 57.4 μmol/L; HeLa, IC_{50} = 64.7 μmol/L; SW480, IC_{50} = 60.6 μmol/L). Ref: X. -F. Liu, et al, Org. Lett., 2011, 13, 3154

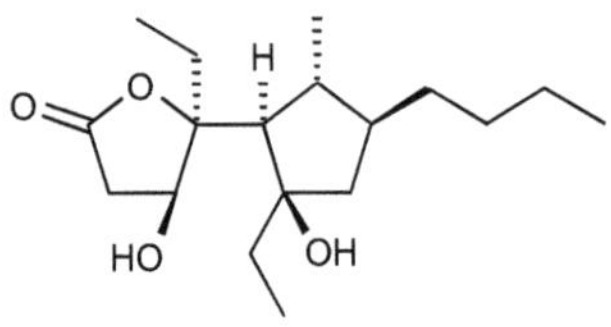

72 Simplextone B

Type: Butanolides. $C_{17}H_{30}O_4$ Colorless oil, $[\alpha]_D^{23}$ = +21° (c = 0.085, CH_3OH). Source: Sponge *Plakortis simplex* (Yongxing I., South China Sea, China). Pharm: Cytotoxic (MTT assay, HCT116, IC_{50} = 23.7 µmol/L; SGC7901, IC_{50} = 45.8 µmol/L; HeLa, IC_{50} = 66.2 µmol/L; SW480, IC_{50} = 61.1 µmol/L). Ref: X. -F. Liu, et al, Org. Lett., 2011, 13, 3154

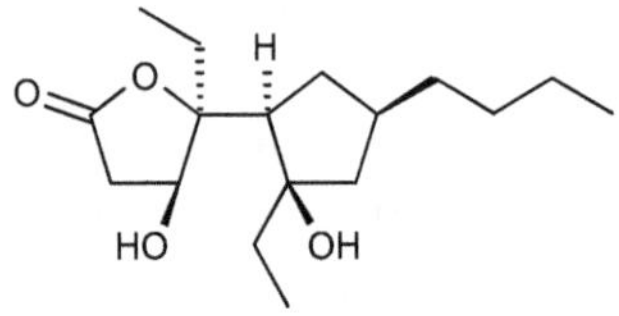

73 Simplextone C

Type: Butanolides. $C_{18}H_{30}O_4$ Source: Sponge *Plakortis simplex* (Yongxing I., South China Sea, China). Pharm: Cytotoxic (HeLa, IC_{50} = (29.0 ± 2.0)µmol/L; K562, IC_{50} = (19.4 ± 1.0)µmol/L; A549, IC_{50} = (40.4 ± 3.5)µmol/L; Bel7402, IC_{50} > 100 µmol/L, control Adriamycin: HeLa, IC_{50} = (0.6 ± 0.0)µmol/L; K562, IC_{50} = (0.3 ± 0.0)µmol/L; A549, IC_{50} = (0.2 ± 0.0)µmol/L). Ref: J. Zhang, et al, JNP, 2013, 76, 600

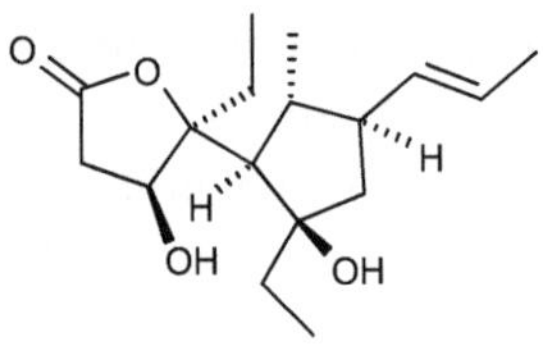

74 Simplextone D

4 Type: Butanolides. $C_{17}H_{30}O_4$ Source: Sponge *Plakortis simplex* (Yongxing I., South China Sea, China). Pharm: Cytotoxic (HeLa, IC_{50} > 50 µmol/L; K562, IC_{50} > 100 µmol/L; A549, IC_{50} > 100 µmol/L; Bel7402, IC_{50} > 100 µmol/L, control Adriamycin: HeLa, IC_{50} = (0.6 ± 0.0)µmol/L; K562, IC_{50} = (0.3 ± 0.0)µmol/L; A549, IC_{50} = (0.2 ± 0.0)µmol/L). Ref: J. Zhang, et al, JNP, 2013, 76, 600

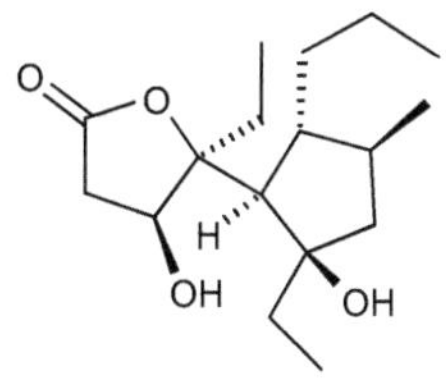

75 Spiculisporic acid B

Type: Butanolides. $C_{17}H_{26}O_6$ Source: Marine-derived fungus *Aspergillus* sp. HDf2 from urchin *Anthocidaris crassispina* (Qionghai, Hainan, China). Pharm: Antibacterial (*Staphylococcus aureus*, weak). Ref: R. Wang, et al, Molecules, 2012, 17, 13175

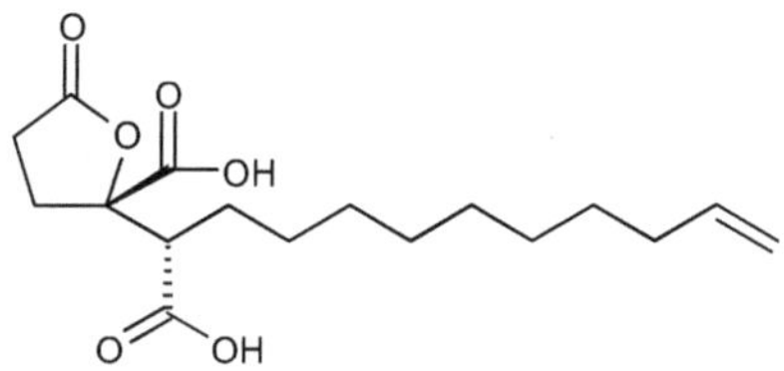

76 Spiculisporic acid D

Type: Butanolides. $C_{18}H_{30}O_6$ Source: Marine-derived fungus *Aspergillus* sp. HDf2 from urchin *Anthocidaris crassispina* (Qionghai, Hainan, China). Pharm: Antibacterial (*Staphylococcus aureus*, weak). Ref: R. Wang, et al, Molecules, 2012, 17, 13175

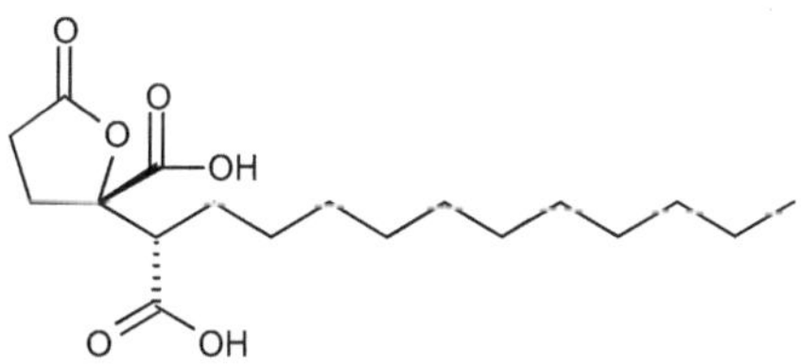

77 Tetracyclic salimabromide

Type: Butanolides. $C_{20}H_{20}Br_2O_3$ Source: Marine myxobacterium *Enhygromxya salina* (sediment, Prerow I., Germany). Pharm: Antibacterial (bacterium *Arthrobacter cristallopoietes*, moderate). Ref: S. Felder, et al, Chem. Eur. J., 2013, 19, 9319

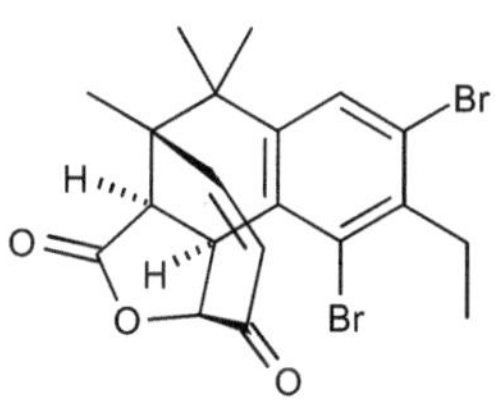

78 Tuberatolide A

Type: Butanolides. $C_{18}H_{26}O_3$ Source: Ascidian *Botryllus tuberatus* (Tong-Yeong City, R. O. Korea waters). Pharm: Farnesoid X receptor antagonist (potent). Ref: H. Choi, et al, JNP, 2011, 74, 90

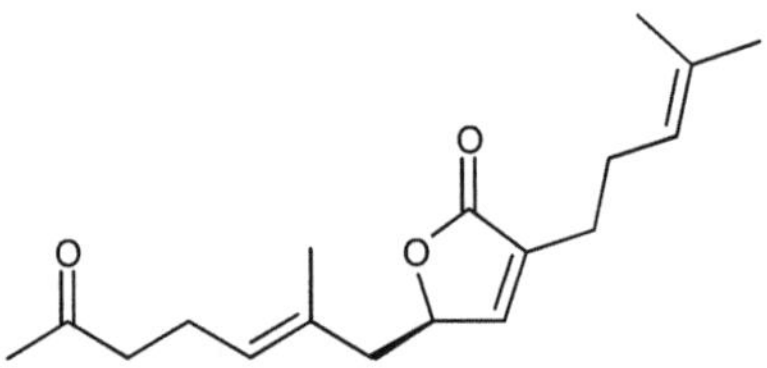

79 Tuberatolide B

Type: Butanolides. $C_{27}H_{34}O_4$ Source: Ascidian *Botryllus tuberatus* (Tong-Yeong City, R. O. Korea waters). Pharm: Farnesoid X receptor antagonist (potent). Ref: H. Choi, et al, JNP, 2011, 74, 90

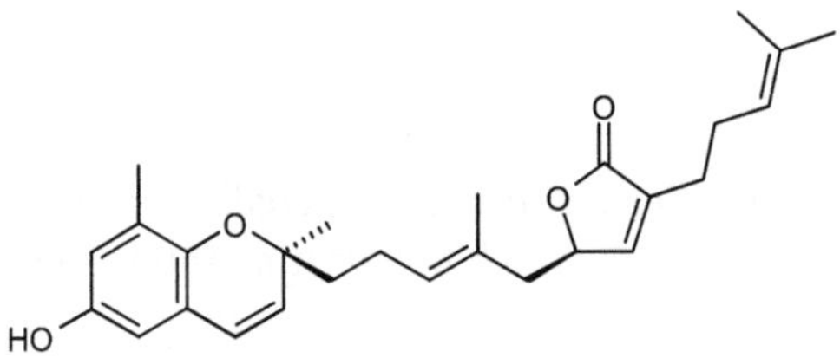

80 2′-*epi*-Tuberatolide B

Type: Butanolides. $C_{27}H_{34}O_4$ Source: Ascidian *Botryllus tuberatus* (Tong-Yeong City, R. O. Korea waters). Pharm: Farnesoid X receptor antagonist (potent). Ref: H. Choi, et al, JNP, 2011, 74, 90

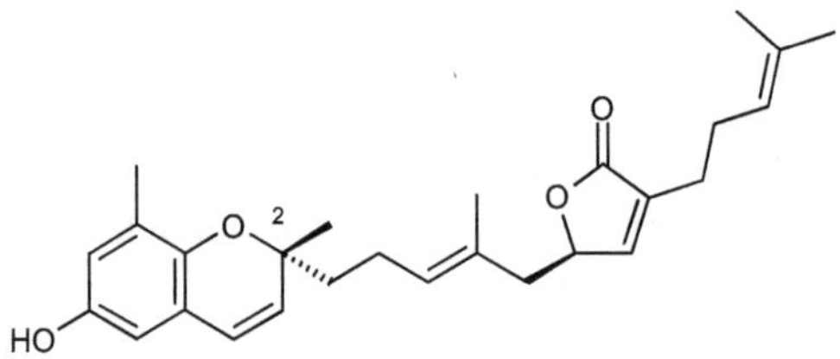

81 Tubingenoic anhydride A

Type: Butanolides. $C_{11}H_{16}O_3$ Source: Marine-derived fungus *Aspergillus tubingensis* OY907. Pharm: Antifungal (inhibits *Neurospora crassa* growth, MIC = 330 μmol/L, affected hyphal morphology). Ref: L. Koch, et al, Mar. Drugs, 2014, 12, 4713

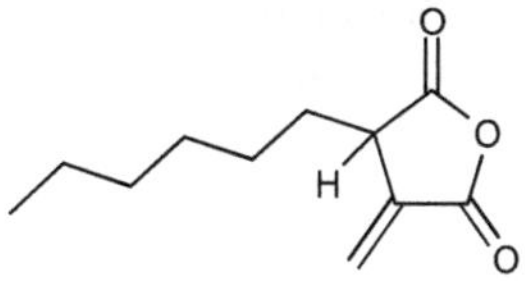

82 Zooxanthellactone

Type: Butanolides. $C_{22}H_{30}O_2$ Oil, $[\alpha]_D^{27}$ = +64.6° (c = 0.24, $CHCl_3$). Source: Dinoflagellates *Symbiodinium* spp. Pharm: Cytotoxic. Ref: K. Onodera, et al, Biosci., Biotechnol., Biochem., 2004, 68, 848-

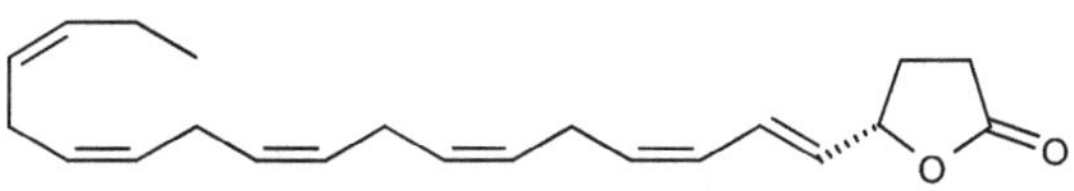

1.4 Pyrans

83 Mycalamide C

Type: Pyrans. $C_{20}H_{35}NO_8$ Source: Sponge *Stylinos* sp. Pharm: Cytotoxic (P_{388}, IC_{50} = 95 ng/mL). Ref: J. S. Simpson, et al, JNP, 2000, 63, 704

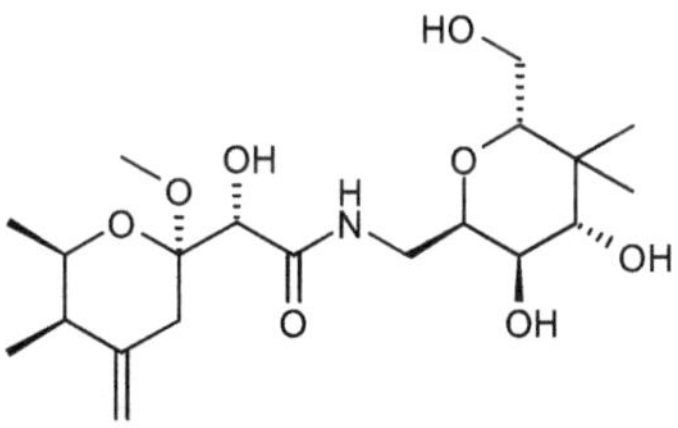

84 NSC 646282

Type: Pyrans. $C_{30}H_{50}N_2O_8$ Oil, $[\alpha]_D$ = −3.1° (c = 0.005, MeOH). Source: Marine-derived bacterium *Alteromonas* sp. from sponge *Darwinella rosacea*. Pharm: Antimicrobial. Ref: D. B. Stierle, et al, Experientia, 1992, 48, 1165

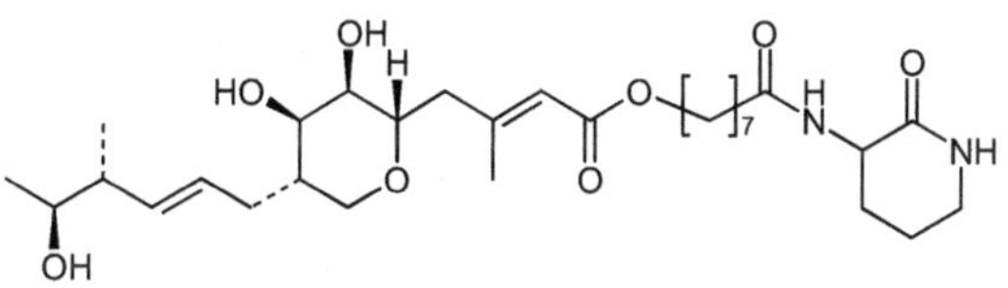

85 8-[(2-Oxo-3-piperidinyl)amino]-8-oxooctyl-5,9-anhydro-2,3,8-trideoxy-8-(5-hydroxy-4-methyl-2-hexenyl)-3-methyl-*DL*-gly- cero-*LD*-*allo*-non-2-enoate

Type: Pyrans. $C_{30}H_{50}N_2O_9$ Oil, $[\alpha]_D = -1.8°$ ($c = 0.003$, MeOH). Source: Marine-derived bacterium *Alteromonas* sp. from sponge *Darwinella rosacea*. Pharm: Antimicrobial. Ref: D. B. Stierle, et al, Experientia, 1992, 48, 1165

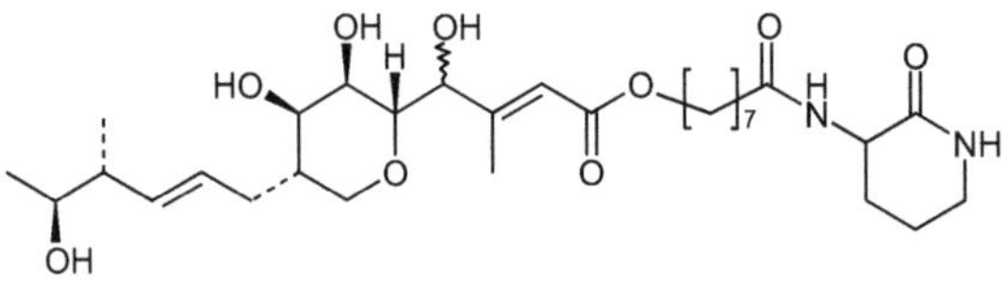

86 (+)-Pederin

Type: Pyrans. $C_{25}H_{45}NO_9$ mp 113 °C. Source: Insect *Paederus fuscipes* (toxic principle). Pharm: Dermatoxic; vesicant (hmn and animal skin). Ref: J. Piel, et al, JNP, 2005, 68, 472 (rev) | S. J. Robinson, et al, JNP, 2007, 70, 1002

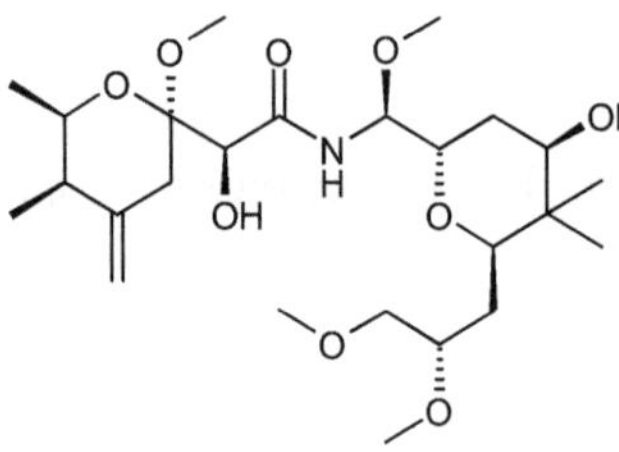

87 Vermelhotin

3-[6- (1-Propenyl)-2*H*-pyran-2-ylidene]-2,4-pyrrolidinedione Type: Pyrans. $C_{12}H_{11}NO_3$ Equilib. 1:1 mixture of *E*– and *Z*–isomers, red needles (CH_2Cl_2/MeOH), mp 212–214 °C, mixt. of *E*– and *Z*–isomers (1:2) from CRI247–01. Source: Marine-derived fungus *Nodulisporium* sp. CRI247-01, an unidentified fungus IFM 52672. Pharm: Antiplasmodial (hmn erythrocytes type O infected by CRPF 94, IC_{50} = 1–10 μmol/L, control Chloroquine hydrochloride, IC_{50} = 0.29 μmol/L); cytotoxic (HuCCA-1, IC_{50} = 2.90 μg/mL; control Etoposide, IC_{50} = 5.30 μg/mL; KB, IC_{50} = 0.50 μg/mL, Etoposide, IC_{50} = 0.46 μg/mL; HeLa, IC_{50} = 0.33 μg/mL, Etoposide, IC_{50} = 0.40 μg/mL; MDA-MB-231, IC_{50} = 0.31 μg/mL, Etoposide, IC_{50} = 0.40 μg/mL; T47D, IC_{50} = 1.25 μg/mL, Etoposide, IC_{50} = 0.04 μg/mL; H69AR, IC_{50} = 2.50 μg/mL, Etoposide, IC_{50} = 36.0 μg/mL; HepG2, IC_{50} = 2.50 μg/mL, control Doxorubicin, IC_{50} = 0.19 μg/mL; A549, IC_{50} = 8.20 μg/mL, Doxorubicin, IC_{50} = 0.48 μg/mL; HCC-H102, IC_{50} = 13.5 μg/mL, Doxorubicin, IC_{50} = 1.20 μg/mL; HL60, IC_{50} = 1.60 μg/mL, Etoposide, IC_{50} = 0.77 μg/mL; P_{388}, IC_{50} = 1.23 μg/mL Etoposide, IC_{50} = 0.10 μg/mL). Ref: T. Hosoe, et al, Heterocycles, 2006, 68, 1949 | C. Kasettrathat, et al, Phytochemistry, 2008, 69, 2621

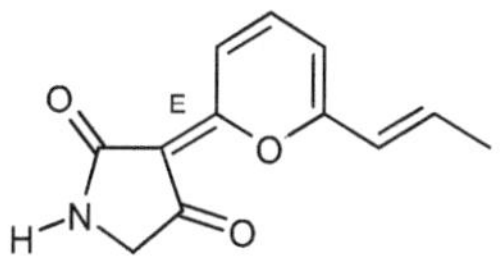

88 Xestin A

Type: Pyrans. $C_{27}H_{46}O_4$ Cryst. (MeOH), mp 55–56 °C, $[\alpha]_D^{20}$ = +26.5° (*c* = 0.37, CH_2Cl_2). Source: Sponge *Xestospongia* sp. Pharm: Cytotoxic (P_{388}, IC_{50} = 0.3 μg/mL); cytotoxic (5 μg/mL, strong activity for A549, HCT8 and MDA-MB). Ref: E. Quinoa, et al, JOC, 1986, 51, 4260

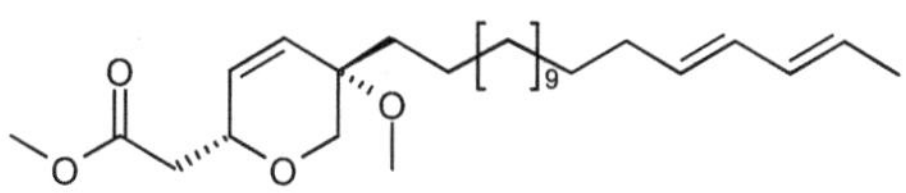

89 Xestin B

Type: Pyrans. $C_{27}H_{46}O_4$ $[\alpha]_D^{20}$ = +19.6° (*c* = 0.11, CH_2Cl_2). Source: Sponge *Xestospongia* sp. Pharm: Cytotoxic (P_{388}, IC_{50} = 3 μg/mL). Ref: E. Quinoa, et al, JOC, 1986, 51, 4260

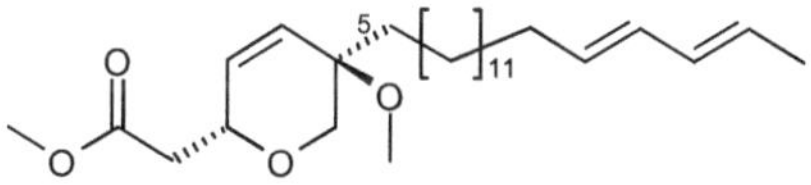

1.5 2-Pyrones (Pyranones; Pentanolides)

90 Albidopyrone

Type: 2-Pyrones. $C_{14}H_{13}NO_4$ Powder. Source: Marine-derived streptomycete *Streptomyces* sp. NTK 227. Pharm: Protein tyrosine phosphatase B inhibitor (moderate activity). Ref: C. Hohmann, et al, J. Antibiot., 2009, 62, 75

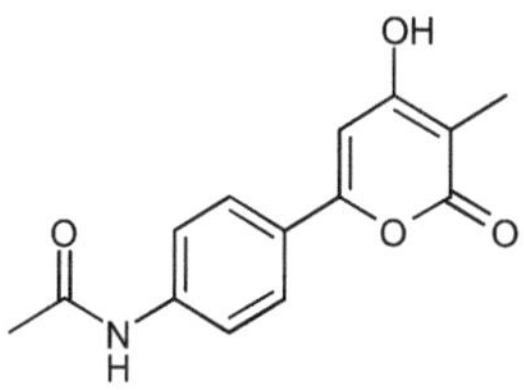

91 Asnipyrone A

Type: 2-Pyrones. $C_{21}H_{22}O_3$ Yellow cryst. (EtOH), mp 161–163 °C. Source: Mangrove-derived fungus *Alternaria niger* from mangrove *Avicennia marina* (Dongzhai Harbour, Hainan, China). Pharm: Cytotoxic (A549, IC_{50} = 62 μmol/L, control Fluorouracil,

IC_{50} = 52 μmol/L). Ref: Guang-yi, L., et al, Heterocycles, 1989, 28, 899 | D. Liu, et al, JNP, 2011, 74, 1787

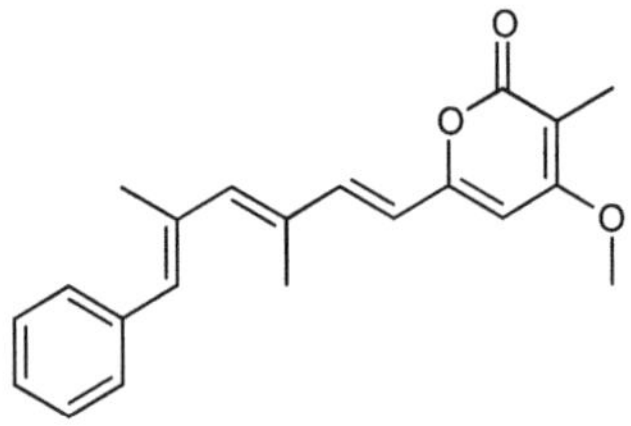

92 Asnipyrone B

Type: 2-Pyrones. $C_{20}H_{20}O_3$ Yellow cryst. (MeOH/CH_2Cl_2), mp 158–160 °C. Source: Mangrove-derived fungus *Alternaria niger* from mangrove *Avicennia marina* (Dongzhai Harbour, Hainan, China). Pharm: Mycotoxin. Ref: Guang-yi, L., et al, Heterocycles, 1989, 28, 899 | D. Liu, et al, JNP, 2011, 74, 1787

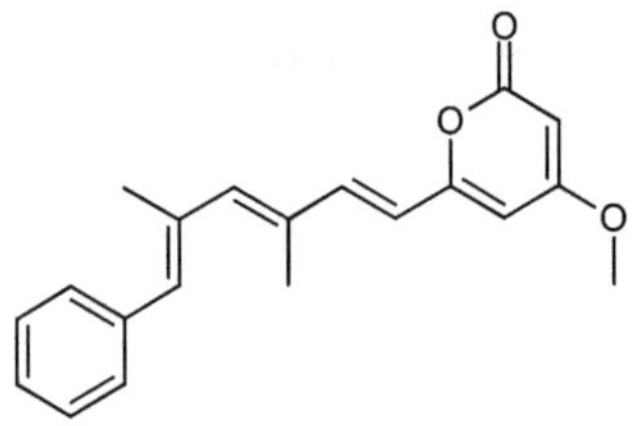

93 Aspyrone

Type: 2-Pyrones. $C_9H_{12}O_4$ Cryst. (C_6H_6 or Me_2CO/petrol), mp 112–112.5 °C, $[\alpha]_D$ = −14° (*c* = 1, EtOH). Source: Marine-derived fungus *Exophiala* sp., terrestrial fungi (*Aspergillus* spp.). Pharm: Antibiotic (broad-spectrum, weak). Ref: CRC Press, DNP on DVD, 2012, version 20.2

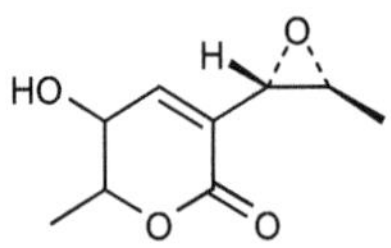

94 Aspyronol

Type: 2-Pyrones. $C_{10}H_{16}O_5$ Colorless oil (MeOH), $[\alpha]_D^{25}$ = −41.6° (*c* = 0.22, MeOH). Source: Deep-sea fungus *Aspergillus* sp. 16-02-1 from sediment (Lau Basin hydrothermal vent, depth of 2255 m, temperature 114 °C). Pharm: Cytotoxic (MTT method, HL60, IC_{50} = 52.1 μg/mL (241.2 μmol/L); 100 μg/mL: HL60, InRt = 67.2%, control Docetaxol, InRt = 49.9%; HeLa, InRt = 14.0%, Docetaxol, InRt = 45.1%; K562, InRt = 27.9%, Docetaxol, InRt = 55.6%). Ref: X. Chen, et al, Mar. Drugs, 2014, 12, 3116

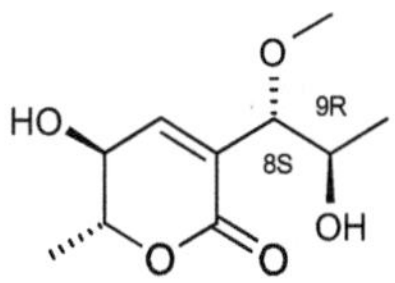

95 Bitungolide A

Type: 2-Pyrones. $C_{25}H_{33}ClO_5$ Needles (MeOH aq), mp 179–182 °C, $[\alpha]_D^{27}$ = +30° (*c* = 0.38, MeOH). Source: Lithistid sponge *Theonella* cf. *swinhoei* (along Lembeh Strait off Bitung, Sulawesi Island, Indonesia). Pharm: Dual-specificity protein phosphatase inhibitor (vaccinia H1-related (VHR), weak). Ref: S. Sirirath, et al, JNP, 2002, 65, 1820 | P. L. Winder, et al, Mar. Drugs, 2011, 9, 2644 (rev)

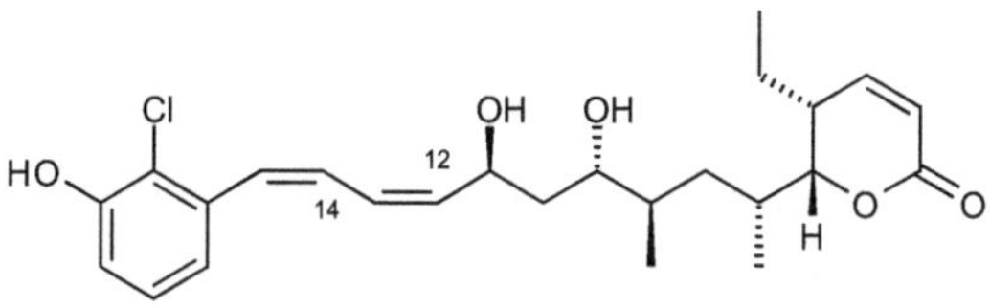

96 Bitungolide B

Type: 2-Pyrones. $C_{25}H_{33}ClO_5$ Amorph. solid, $[\alpha]_D^{27}$ = +42° (*c* = 4.2, $CHCl_3$). Source: Lithistid sponge *Theonella* cf. *swinhoei* (along Lembeh Strait off Bitung, Sulawesi Island, Indonesia). Pharm: Dual-specificity protein phosphatase inhibitor (vaccinia H1-related (VHR), weak). Ref: S. Sirirath, et al, JNP, 2002, 65, 1820 | P. L. Winder, et al, Mar. Drugs, 2011, 9, 2644 (rev)

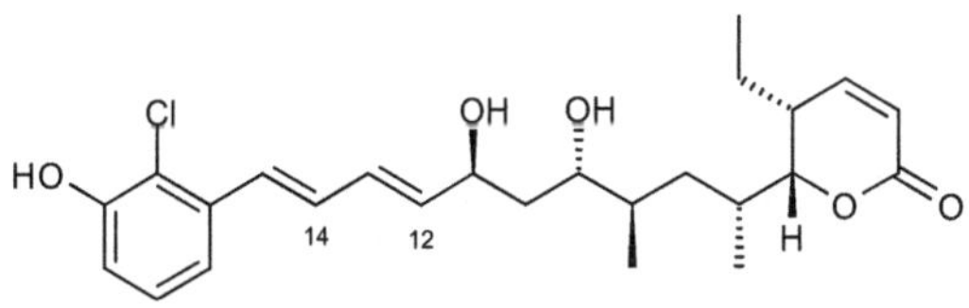

97 Bitungolide C

Type: 2-Pyrones. $C_{25}H_{33}ClO_5$ Amorph. solid, $[\alpha]_D^{27}$ = +89° (*c* = 0.11, $CHCl_3$). Source: Lithistid sponge *Theonella* cf. *swinhoei* (along Lembeh Strait off Bitung, Sulawesi Island, Indonesia). Pharm: Dual-specificity protein phosphatase inhibitor (vaccinia H1-related (VHR), weak). Ref: S. Sirirath, et al, JNP, 2002, 65, 1820 | P. L. Winder, et al, Mar. Drugs, 2011, 9, 2644 (rev)

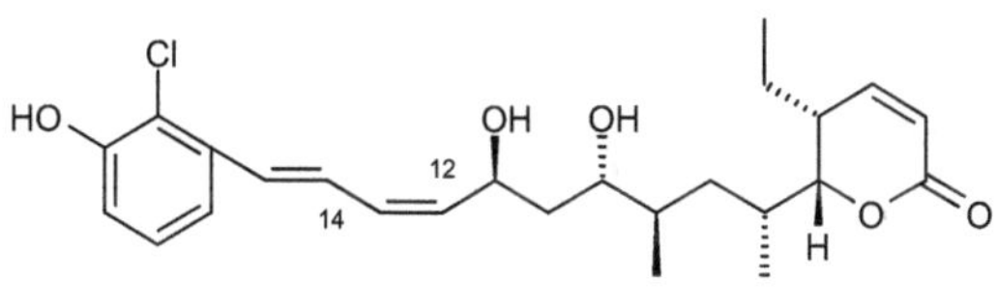

98 Bitungolide D

Type: 2-Pyrones. $C_{25}H_{33}ClO_5$ Amorph. solid, $[\alpha]_D^{27}$ = +66° (*c* = 0.58, $CHCl_3$). Source: Lithistid sponge *Theonella* cf. *swinhoei* (along Lembeh Strait off Bitung, Sulawesi Island, Indonesia). Pharm: Dual-specificity protein phosphatase inhibitor (vaccinia H1-related (VHR), weak). Ref: S. Sirirath, et al, JNP, 2002, 65, 1820 | P. L. Winder, et al, Mar. Drugs, 2011, 9, 2644 (rev)

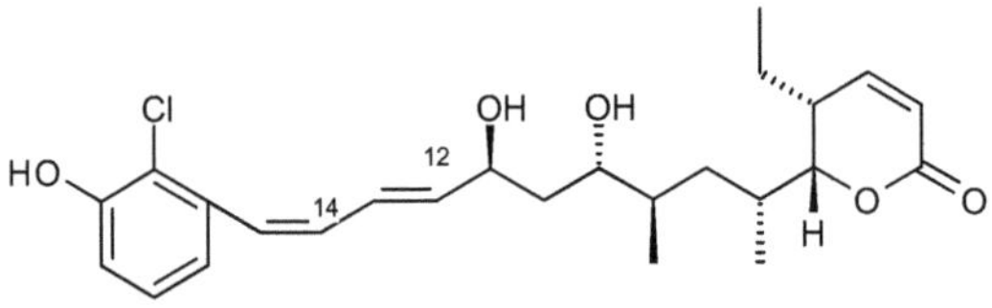

99 Bitungolide E

Type: 2-Pyrones. $C_{25}H_{34}O_4$ Pale yellow glass, $[\alpha]_D^{27}$ = +107° (*c* = 1.3, $CHCl_3$). Source: Lithistid sponge *Theonella* cf. *swinhoei* (along Lembeh Strait off Bitung, Sulawesi Island, Indonesia). Pharm: Dual-specificity protein phosphatase inhibitor (vaccinia H1-related (VHR), weak). Ref: S. Sirirath, et al, JNP, 2002, 65, 1820 | P. L. Winder, et al, Mar. Drugs, 2011, 9, 2644 (rev)

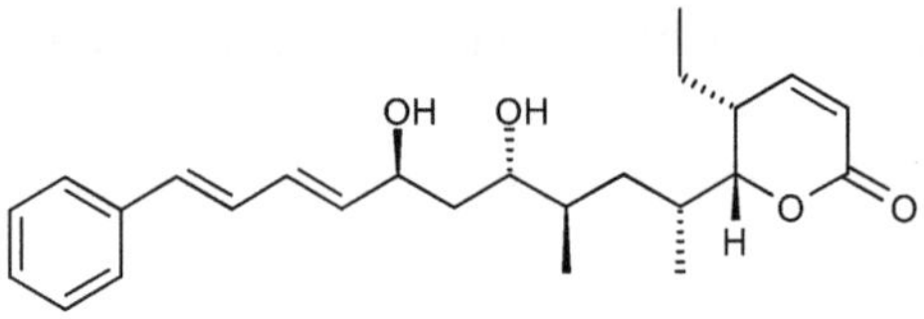

100 Bitungolide F

Type: 2-Pyrones. $C_{24}H_{32}O_4$ Pale yellow glass, $[\alpha]_D^{27}$ = +43° (*c* = 0.85, $CHCl_3$). Source: Lithistid sponge *Theonella* cf. *swinhoei* (along Lembeh Strait off Bitung, Sulawesi Island, Indonesia). Pharm: Dual-specificity protein phosphatase inhibitor (vaccinia H1-related (VHR), weak). Ref: S. Sirirath, et al, JNP, 2002, 65, 1820 | P. L. Winder, et al, Mar. Drugs, 2011, 9, 2644 (rev)

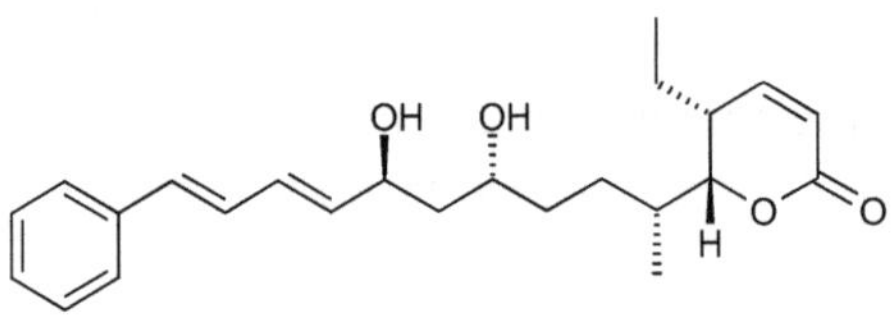

101 Bromomethylchlamydosporol A

Type: 2-Pyrones. $C_{12}H_{15}BrO_5$ Source: Marine-derived fungus *Fusarium tricinctum* (adding calcium bromide to cultures) from brown alga *Sargassum ringgoldianum* (GeoMun

I., Yeosu, R. O. Korea). Pharm: Antibacterial (three strains of *Staphylococcus aureus*, MIC = 15.6 μg/mL). Ref: V. Nenkep, et al, JNP, 2010, 73, 2061

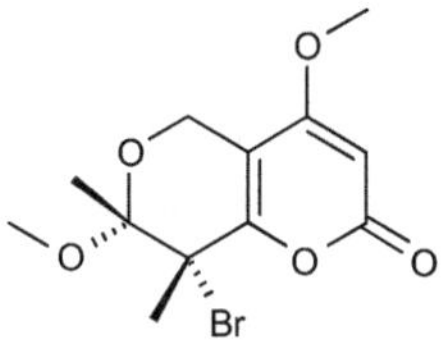

102 Bromomethylchlamydosporol B

Type: 2-Pyrones. $C_{12}H_{14}Br_2O_5$ Source: Marine-derived fungus *Fusarium tricinctum* (adding calcium bromide to cultures) from brown alga *Sargassum ringgoldianum* (GeoMun I., Yeosu, R. O. Korea). Pharm: Antibacterial (three strains of *Staphylococcus aureus*, MIC = 15.6 μg/mL). Ref: V. Nenkep, et al, JNP, 2010, 73, 2061

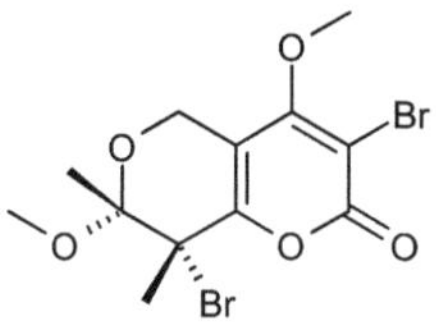

103 Callystatin A

Type: 2-Pyrones. $C_{29}H_{44}O_4$ Oil, $[\alpha]_D$ = −107° (c = 0.1, MeOH). Source: Sponge *Callyspongia truncata*. (Japan waters). Pharm: Cytotoxic (KB, IC_{50} = 0.01 ng/mL). Ref: M. Kobayashi, et al, Tet. Lett., 1997, 38, 2859 | N. Murakami, et al, Tet. Lett., 1997, 38, 5533

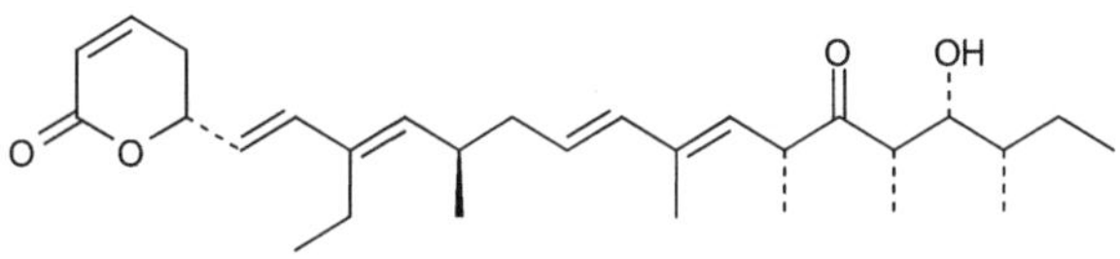

104 Chaetoquadrin F

Type: 2-Pyrones. $C_9H_{12}O_4$ Source: Deep-sea fungus *Aspergillus* sp. 16-02-1 from sediment (Lau Basin hydrothermal vent, depth of 2255 m, temperature 114 °C). Pharm: Cytotoxic (MTT method, 100 μg/mL: HeLa, InRt = 13.5%, control Docetaxol, InRt = 45.1%). Ref: X. Chen, et al, Mar. Drugs, 2014, 12, 3116

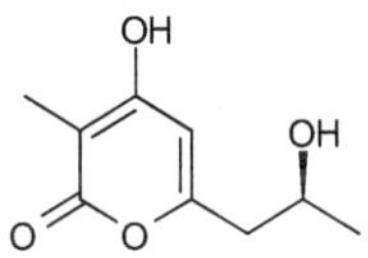

105 Chlamydosporol

Type: 2-Pyrones. $C_{11}H_{14}O_5$ Prisms (EtOAc), mp 172–174 °C. Source: Marine-derived fungus *Fusarium tricinctum* (adding calcium bromide to cultures) from brown alga *Sargassum ringgoldianum* (GeoMun I., Yeosu, R. O. Korea), terrestrial fungi (*Fusarium acuminatum* and *Fusarium chlamydosporum*). Pharm: Antibacterial (moderate). Ref: M. E. Savard, et al, Mycopathologia, 1990, 110, 177 | J. F. Grove, et al, JCS Perkin I, 1991, 997 | V. Nenkep, et al, JNP, 2010, 73, 2061

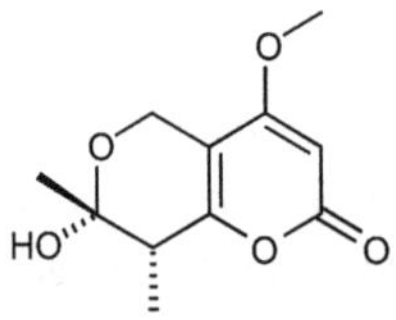

106 Chlorohydroaspyrone A

Type: 2-Pyrones. $C_9H_{13}ClO_4$ Oil, $[\alpha]_D^{20}$ = −110° (*c* = 0.1, MeOH). Source: Marine-derived fungus *Exophiala* sp. from sponge *Halichondria panicea* (surface, Korea waters). Pharm: Antibacterial (mild: *Staphylococcus aureus*, MRSA, MDRSA). Ref: D. Zhang, et al, JNP, 2008, 71, 1458

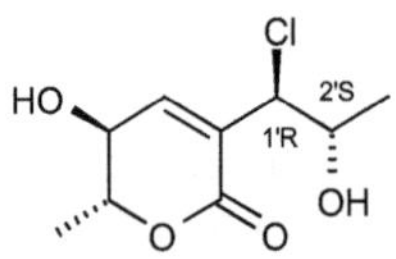

107 Chlorohydroaspyrone B

Type: 2-Pyrones. $C_9H_{13}ClO_4$ Oil, $[\alpha]_D^{20}$ = +70° (*c* = 0.1, MeOH). Source: Marine-derived fungus *Exophiala* sp. from sponge *Halichondria panicea* (surface, Korea waters). Pharm: Antibacterial (mild: *Staphylococcus aureus*, MRSA, MDRSA). Ref: D. Zhang, et al, JNP, 2008, 71, 1458

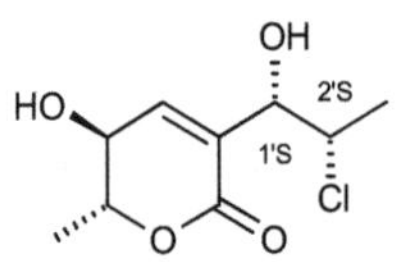

108 9-Chloro-8-hydroxy-8,9-deoxyaspyrone

Type: 2-Pyrones. $C_{10}H_{15}ClO_3$ $[\alpha]_D^{26}$ = +17.2° (*c* = 0.17, $CHCl_3$). Source: Marine-derived fungus *Aspergillus ostianus* TUF 01F313 from an unidentified sponge (Pohnpei I., Federated States of Micronesia). Pharm: Antibacterial (25 μg/disk, inhibits growth of *Ruegeria atlantica*, inhibition zone = 10.5 mm). Ref: M. Namikoshi, et al, J. Antibiot., 2003, 56, 755 | M. Saleem, et al, NPR, 2007, 24, 1142 (rev)

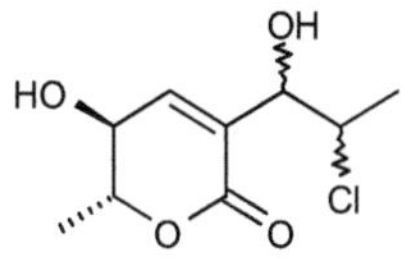

109 Coibacin A

Type: 2-Pyrones. $C_{19}H_{24}O_2$ Source: Cyanobacterium *Oscillatoria* sp. (Coiba National Park, Panama). Pharm: Antileishmanial (selective); anti-inflammatory (potent). Ref: M. J. Balunas, et al, Org. Lett., 2012, 14, 3878

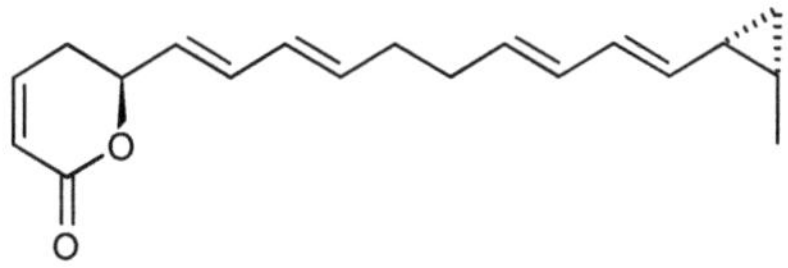

110 Coibacin B

Type: 2-Pyrones. $C_{17}H_{22}O_2$ Source: Cyanobacterium *Oscillatoria* sp. (Coiba National Park, Panama). Pharm: Antileishmanial (selective); anti-inflammatory (potent). Ref: M. J. Balunas, et al, Org. Lett., 2012, 14, 3878

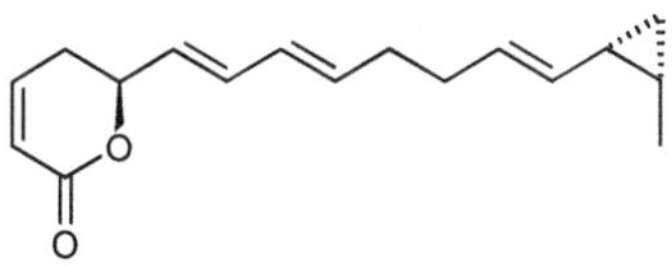

111 Coibacin C

Type: 2-Pyrones. $C_{15}H_{19}ClO_2$ Source: Cyanobacterium *Oscillatoria* sp. (Coiba National Park, Panama). Pharm: Antileishmanial (selective); anti-inflammatory (potent). Ref: M. J. Balunas, et al, Org. Lett., 2012, 14, 3878

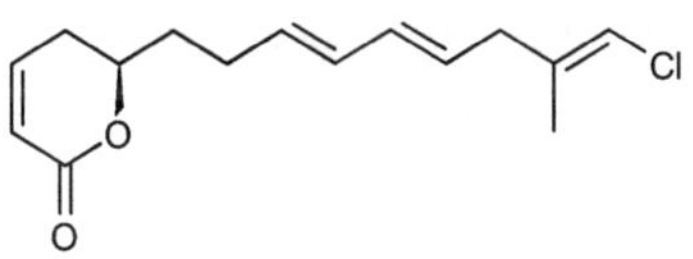

112 Coibacin D

Type: 2-Pyrones. $C_{15}H_{21}ClO_2$ Source: Cyanobacterium *Oscillatoria* sp. (Coiba National Park, Panama). Pharm: Antileishmanial (selective); anti-inflammatory (potent). Ref: M. J. Balunas, et al, Org. Lett., 2012, 14, 3878

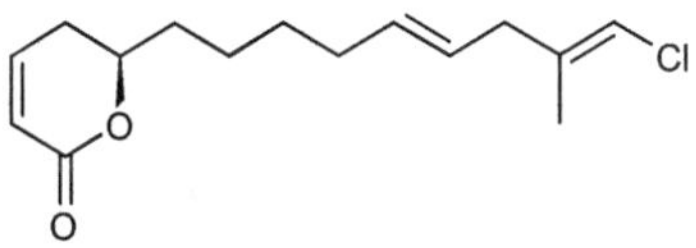

113 9(13)-Cyclodiscodermolide

Type: 2-Pyrones. $C_{33}H_{53}NO_7$ Solid, $[\alpha]_D^{21}$ = +24° (c = 0.01, MeOH). Source: Lithistid sponge *Discodermia* sp. (numerous places, Bahamas, using Johnson-Sea-Link submersible). Pharm: Cytotoxic (P_{388}, IC_{50} = 5043 nmol/L; A549, IC_{50} = 4487 nmol/L). Ref: S. P. Gunasekera, et al, JNP, 2002, 65, 1643 | P. L. Winder, et al, Mar. Drugs, 2011, 9, 2644 (rev)

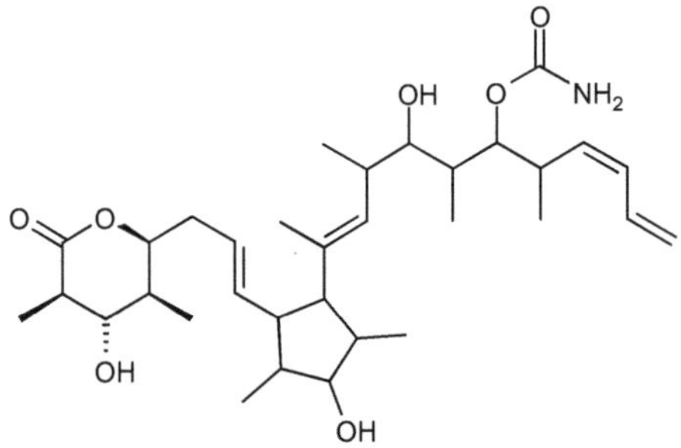

114 Cyercene 1

4-Methoxy-3-methyl-6- (3-methyl-1,3-pentadienyl)-2*H*-pyran-2-one Type: 2-Pyrones. $C_{13}H_{16}O_3$ Source: Sacoglossan *Cyerce cristallina*. Pharm: Ichthyotoxic. Ref: R. R. Vardaro, et al, Tetrahedron, 1991, 47, 5569

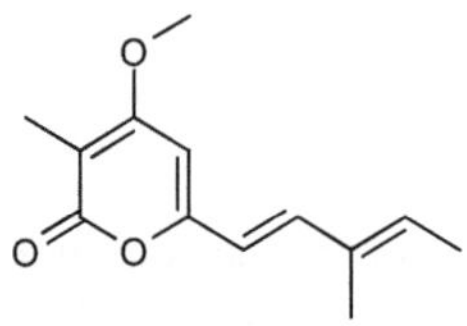

115 Cyercene 2

4-Methoxy-3-methyl-6-(3-methyl-1,3-hexadienyl)-2*H*-pyran-2-one Type: 2-Pyrones. $C_{14}H_{18}O_3$ Source: Sacoglossan *Cyerce cristallina*. Pharm: Ichthyotoxic. Ref: R. R. Vardaro, et al, Tetrahedron, 1991, 47, 5569

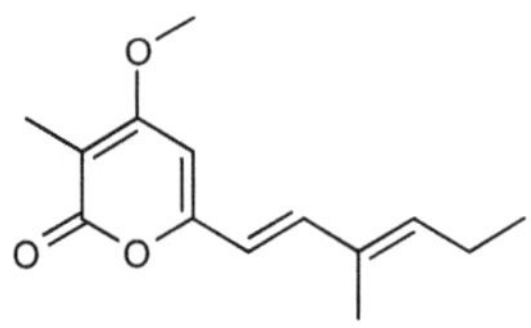

116 Cyercene 3

6- (3,5-Dimethyl-1,3-hexadienyl)-4-methoxy-3-methyl-2*H*-pyran-2-one Type: 2-Pyrones. $C_{15}H_{20}O_3$ Source: Sacoglossan *Cyerce cristallina*. Pharm: Ichthyotoxic. Ref: R. R. Vardaro, et al, Tetrahedron, 1991, 47, 5569

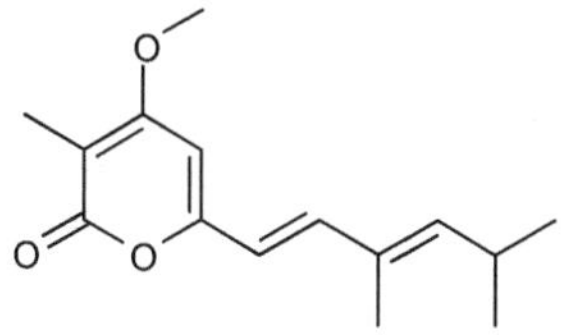

117 Cyercene 4

6-(1,3-Dimethyl-1,3-pentadienyl)-4-methoxy-5-methyl-2*H*-pyran-2-one Type: 2-Pyrones. $C_{14}H_{18}O_3$ Source: Sacoglossan *Cyerce cristallina*. Pharm: Ichthyotoxic. Ref: R. R. Vardaro, et al, Tetrahedron, 1991, 47, 5569

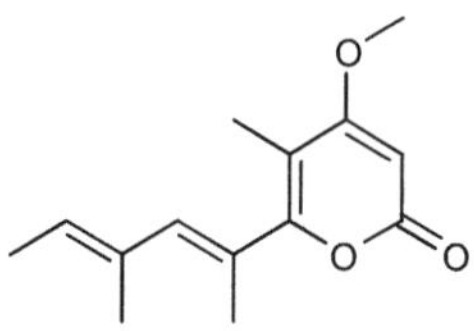

118 Cyercene 5

6-(1,3-Dimethyl-1,3-hexadienyl)-4-methoxy-5-methyl-2*H*-pyran-2-one Type: 2-Pyrones. $C_{15}H_{20}O_3$ Source: Sacoglossan *Cyerce cristallina*. Pharm: Ichthyotoxic. Ref: V. Roussis, et al, Experientia, 1990, 46, 327 | R. R. Vardaro, et al, Tetrahedron, 1991, 47, 5569

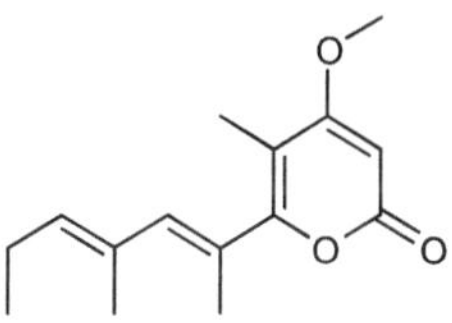

119 19-Deaminocarbonyldiscodermolide

Type: 2-Pyrones. $C_{32}H_{54}O_7$ Solid, $[\alpha]_D^{21}$ = +18° (*c* = 0.1, MeOH). Source: Lithistid sponge *Discodermia* sp. (numerous places, Bahamas, using Johnson-Sea-Link submersible). Pharm: Cytotoxic (P_{388}, IC_{50} = 128 nmol/L; A549, IC_{50} = 74 nmol/L). Ref: S. P. Gunasekera, et al, JNP, 2002, 65, 1643 | P. L. Winder, et al, Mar. Drugs, 2011, 9, 2644 (rev)

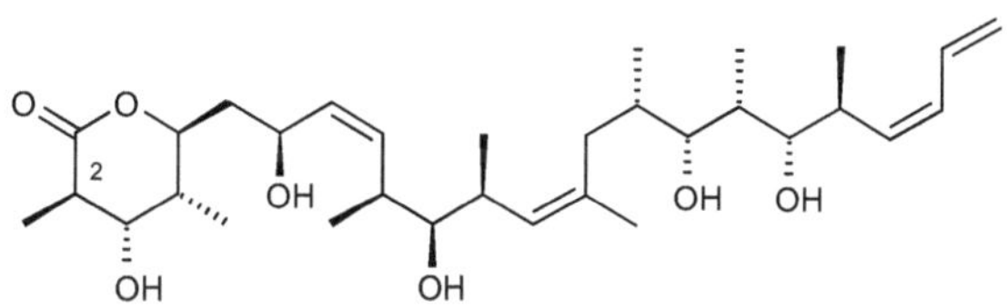

120 8,9-Dehydroxylarone

Type: 2-Pyrones. $C_{13}H_{16}O_3$ Source: Sacoglossan *Aplysiopsis formosa* (herbivorous, Azores Is.). Pharm: Cytotoxic (Colo320, IC_{50} = 25 μg/mL; L_{1210}, C_{50} = 25 μg/mL; HL60, IC_{50} = 50 μg/mL); antibacterial (*Micrococcus luteus,* 100 μg/mL). Ref: A. Schuffler, et al, Z. Naturforsch. C., 2007, 62c, 169

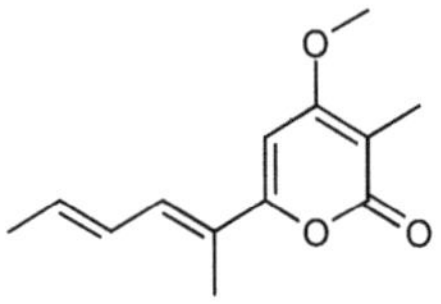

121 2-Demethyldiscodermolide

Type: 2-Pyrones. $C_{32}H_{53}NO_8$ Solid, $[\alpha]_D^{21}$ = +10.2° (*c* = 0.1, MeOH). Source: Lithistid sponge *Discodermia* sp. (numerous places, Bahamas, using Johnson-Sea-Link submersible). Pharm: Cytotoxic (P_{388}, IC_{50} = 172 nmol/L; A549, IC_{50} = 120 nmol/L). Ref: S. P. Gunasekera, et al, JNP, 2002, 65, 1643 | P. L. Winder, et al, Mar. Drugs, 2011, 9, 2644 (rev)

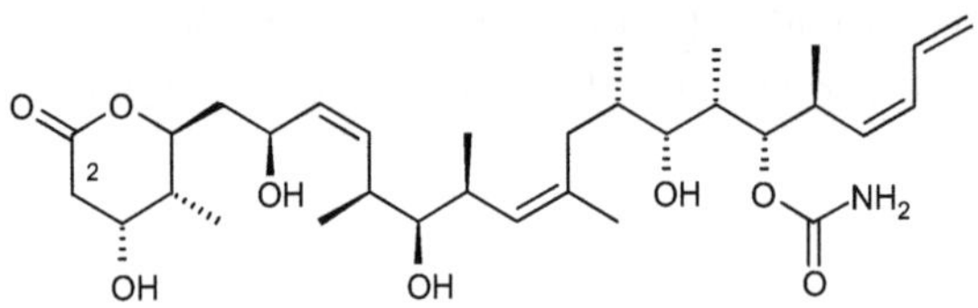

122 Diemenensin A

4-Hydroxy-3,5-dimethyl-6-(1,3,5,7-tetramethyl-1,3-decadienyl)-2*H*-pyran-2-one *Type*: 2-Pyrones. $C_{21}H_{32}O_3$ Oil, $[\alpha]_D$ = +77.3° (*c* = 4.7, MeOH). Source: Pulmonate limpet *Siphonaria diemenensis* (Australia). Pharm: Antibacterial (*Staphylococcus aureus* and *Bacillus subtilis*). Ref: J. E. Hochlowski, et al, Tet. Lett., 1983, 24, 1917

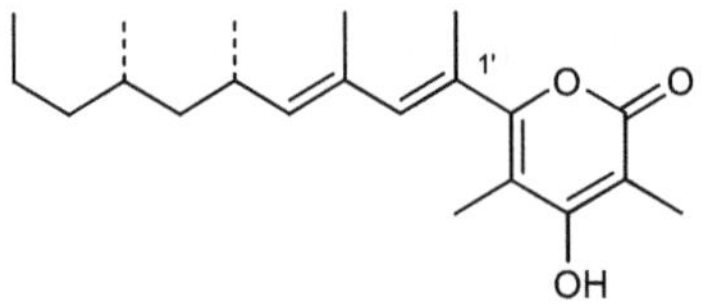

123 Diemenensin B

Type: 2-Pyrones. $C_{21}H_{32}O_3$ Oil, $[\alpha]_D$ = +32.4° (*c* = 1.15, MeOH). Source: Pulmonate limpet *Siphonaria diemenensis* (Australia). Pharm: Antibacterial. Ref: J. E. Hochlowski, et al, Tet. Lett., 1983, 24, 1917

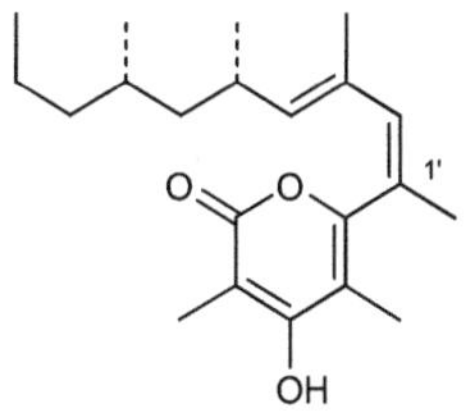

124 Dihydroaspyrone

Type: 2-Pyrones. $C_9H_{14}O_4$ $[\alpha]_D^{20}$ = +17.8° (*c* = 0.7, MeOH). Source: Marine-derived fungus *Aspergillus ostianus* 01F313 from an unidentified sponge (Pohnpei I., Federated States of Micronesia), deep-sea fungus *Aspergillus* sp. 16-02-1 from sediment (Lau Basin hydrothermal vent, depth of 2255 m, temperature 114 °C), terrestrial fungus (*Aspergillus ochraceus*). Pharm: Cytotoxic (MTT method, 100 μg/mL: HL60, InRt = 19.9%, control Docetaxol, InRt = 49.9%, HeLa, InRt = 10.1%, Docetaxol, InRt = 45.1%). Ref: J. Fuchser, et al, Liebigs Ann./Recl., 1997, 87 | K. Kito, et al, JNP, 2007, 70, 2022 | X. Chen, et al, Mar. Drugs, 2014, 12, 3116

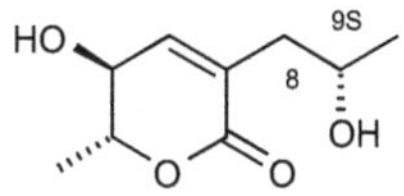

125 3′,4′-Dihydroinfectopyrone

Type: 2-Pyrones. $C_{14}H_{18}O_5$ Source: Marine-derived fungus *Petriella* sp. TUBS 7961 from sponge *Suberites domuncula* (Mediterranean Sea). Pharm: Cytotoxic (L5178Y, pronounced activity). Ref: P. Proksch, et al, Bot. Mar., 2008, 51, 209

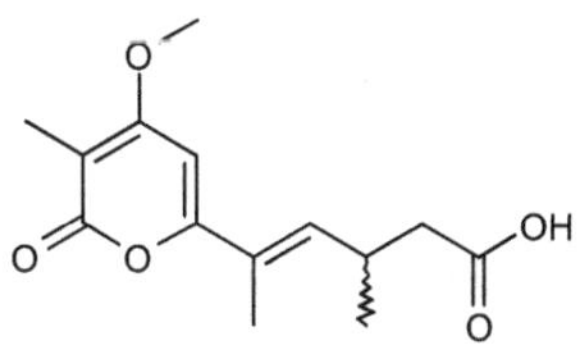

126 5,7-Dihydroxy-2-[[1-(4-methoxy-2-oxo-2*H*-pyran-6-yl)-2-phenylethyl]amino]-1,4-naphthoquinone

Type: 2-Pyrones. $C_{24}H_{19}NO_7$ Amorph. red powder, mp 118–120 °C, $[\alpha]_D^{25}$ = +54.9° (*c* = 0.19, $CHCl_3$). Source: Marine-derived fungus *Aspergillus niger* EN-13 from brown alga *Colpomenia sinuosa* (China waters). Pharm: Antifungal (*Candida albicans*, moderate). Ref: Y. Zhang, et al, Chin. Chem. Lett., 2007, 18, 951

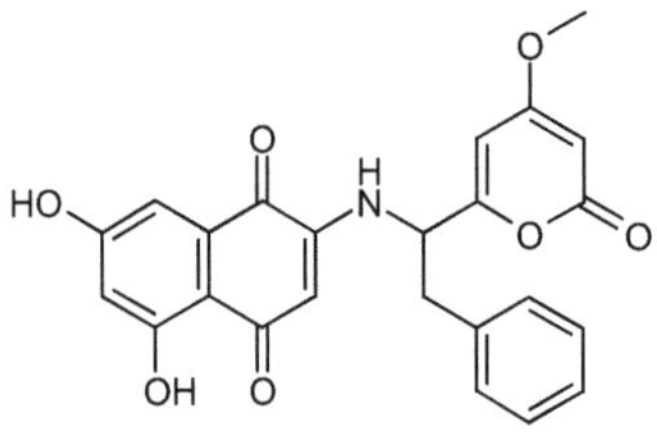

127 Discodermolide

Disermolide Type: 2-Pyrones. $C_{33}H_{55}NO_8$ Cryst., mp 115–116 °C, $[\alpha]_D^{22}$ = +7.2° (*c* = 0.72, MeOH). Source: Lithistid sponge *Discodermia dissoluta* (depth of 33 m, off Grand Bahama I., Caribbean Sea). Pharm: Arrests cells in G2/M phase of cell cycle; cytotoxic (breast cancer MDA-MB-231, GI_{50} = 0.029 μmol/L; non-small-cell lung cancer A549, GI_{50} = 0.020 μmol/L; colon cancer HT29, GI_{50} = 0.015 μmol/L); cytotoxic (P_{388} cell line, IC_{50} = 35 nmol/L; A549 cell line, IC_{50} = 13.5 nmol/L); cytotoxic (inhibits cell proliferation by polymerizing and hyperstabilizing tubulin, similar to paclitaxel but shows activity against paclitaxel resistant tumors); potent promotor of tubulin assembly (similar to paclitaxel); immunosuppressant. Ref: S. P. Gunasekera, et al, JOC, 1990, 55, 4912; 1991, 56, 1346 | R. E. Longley, et al, Ann. N.Y. Acad. Sci., 1993, 696, 94 | L. E. Broker, et al, Cancer Res., 2002, 62, 4081 | G. J. Florence, et al, Nat. Prod. Rep., 2008, 25, 342 | A. B. Smith, et al, Tetrahedron, 2008, 64, 261 | P. L. Winder, et al, Mar. Drugs, 2011, 9, 2644 (rev)

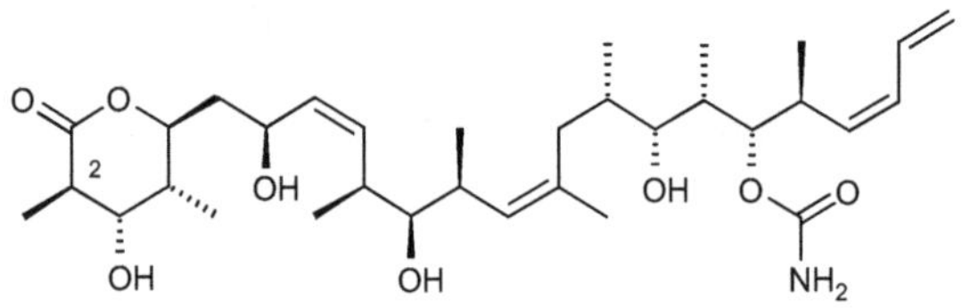

128 2-*epi*-Discodermolide

Type: 2-Pyrones. $C_{33}H_{55}NO_8$ Solid, $[\alpha]_D^{21}$ = +10.7° (*c* = 0.1, MeOH). Source: Lithistid sponge *Discodermia* sp. (numerous places, Bahamas, using Johnson-Sea-Link submersible). Pharm: Cytotoxic (P_{388}, IC_{50} = 134 nmol/L; A549, IC_{50} = 67 nmol/L). Ref: S. P. Gunasekera, et al, JNP, 2002, 65, 1643 | P. L. Winder, et al, Mar. Drugs, 2011, 9, 2644 (rev)

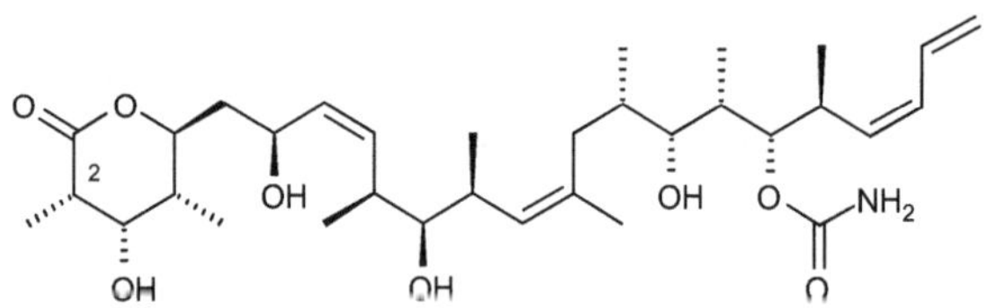

129 Harzialactone B

Type: 2-Pyrones. $C_7H_{10}O_3$ Oil, $[\alpha]_D^{32}$ = −23.5° (*c* = 1.76, $CHCl_3$). Source: Marine-derived fungus *Trichoderma harzianum* OUPS-N115 from sponge *Halichondria okadai*. Pharm: Cytotoxic (cultured P_{388}, ED_{50} = 60.0 μg/mL). Ref: T. Amagata, et al, J. Antibiot., 1998, 51, 33

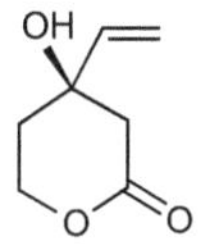

130 Helicascolide C

Type: 2-Pyrones. $C_{12}H_{18}O_3$ Source: Marine-derived fungi *Curvularia eschscholzii* from red alga *Gracilaria* sp. (South Sulawesi coast, Indonesia) and *Daldinia eschscholzii* KT32 (algicolous, Indonesia). Pharm: Antifungal (phytopathogenic fungus *Cladosporium cucumerinum*, 200 μg/disk, IZ = 5 mm). Ref: K. Tarman, et al, Phytochem. Lett., 2012, 5, 83

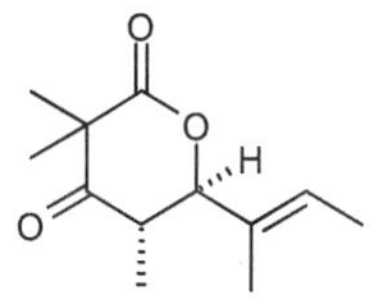

131 (*E*)-6-(1-Heptenyl)-2*H*-pyran-2-one

Type: 2-Pyrones. $C_{12}H_{16}O_2$ Oil. Source: Marine-derived fungus *Botrytis* sp. from red alga *Hyalosiphonia caespitosa* (Korea waters). Pharm: Tyrosinase inhibitor (IC_{50} = 4.5 μmol/L, more active than control kojic acid). Ref: D. Zhang, et al, Bull. Korea waters watersn Chem. Soc., 2007, 28, 887

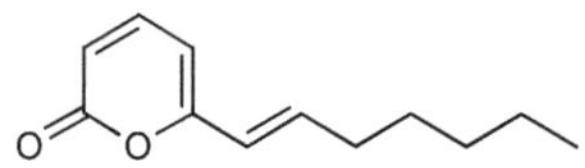

132 5-Hydroxy-4-(hydroxymethyl)-2*H*-pyran-2-one

Type: 2-Pyrones. $C_6H_6O_4$ Cryst. (MeOH), mp 138–140 °C. Source: Marine-derived fungus *Aspergillus flavus* C-F-3 from green alga *Enteromorpha tubulosa* (China waters). Pharm: Induces production of cAMP (GPR12 transfected CHO and HEK-293, in a dose-dependent manner, it might be possible ligand for GPR12, which can be significant molecular target for treating a variety of neurological disorders). Ref: A. Lin, et al, J. Antibiot., 2008, 61, 245

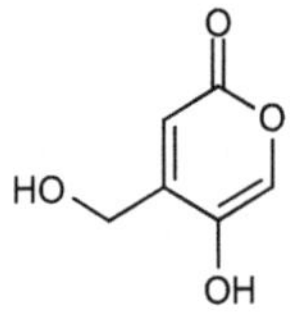

133 Hydroxypestalopyrone

Type: 2-Pyrones. $C_{10}H_{12}O_4$ Powder. Source: Marine-derived fungus *Nigrospora* sp. PSU-F5. Pharm: Antifungal. Ref: K. Trisuwan, et al, JNP, 2008, 71, 1323

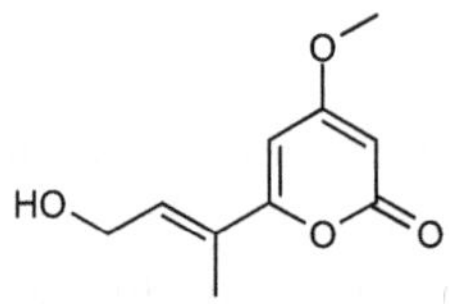

134 Isopectinatone

Type: 2-Pyrones. $C_{21}H_{34}O_3$ Oil, $[\alpha]_D^{25}$ = +35° (*c* = 0.1, $CHCl_3$). Source: Pulmonate limpet *Siphonaria pectinata* (Cádiz, Spain, W6°18´ N36°32´). Pharm: Cytotoxic (P_{388}, A549, HT29 and MEL28, ED_{50} = 2.5 µg/mL). Ref: M. C. Paul, et al, Tetrahedron, 1997, 53, 2303

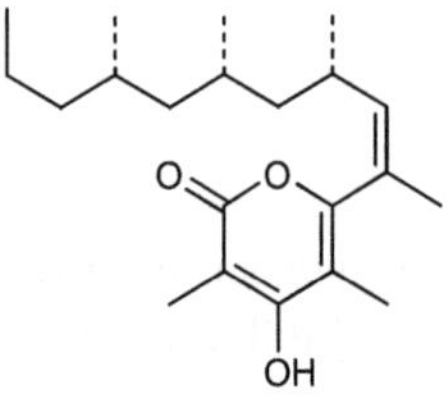

135 Lagunapyrone B

Type: 2-Pyrones. $C_{34}H_{52}O_5$ $[\alpha]_D^{20}$ = +10.9° (*c* = 4, MeOH). Source: Marine-derived actinomycete Actinomyces sp. CNB-984 (shallow water marine sediment, Southern California). Pharm: Cytotoxic (HCT116, ED_{50} = 3.5 µg/mL). Ref: T. Lindel, et al, Tet. Lett., 1996, 37, 1327

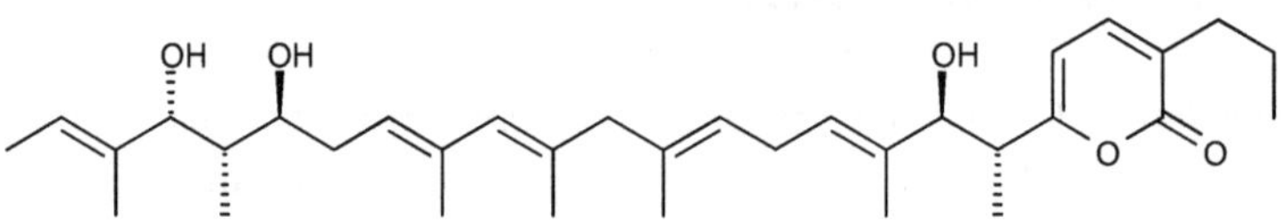

136 Lehualide E

Type: 2-Pyrones. $C_{24}H_{34}O_4$ Source: Sponge *Plakortis* sp. ('Eua I., Tonga). Pharm: The lehualides were moderately toxic to sensitised yeast cells but not to wild type and are therefore susceptible to efflux pump action. Ref: J. M. Barber,et al, JNP, 2011, 74, 809

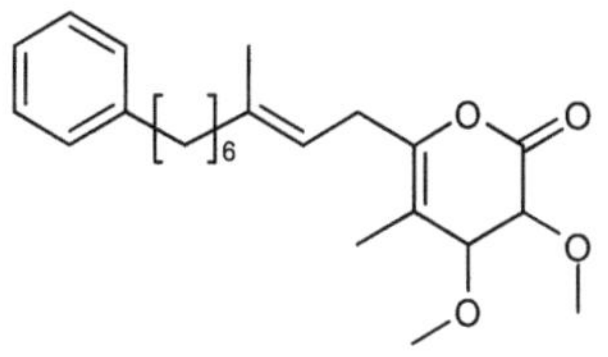

137 (−)-Malyngolide

Type: 2-Pyrones. $C_{16}H_{30}O_3$ Crystal, mp 36–38 °C, $[\alpha]_D^{20}$ = −13.0° (*c* = 0.9, $CHCl_3$); mp 36–37 °C, $[\alpha]_D$ = −13° (*c* = 2, $CHCl_3$). Source: Cyanobacterium *Lyngbya majuscula*. Pharm: Antibacterial (*Mycobacterium smegmatis, Streptococcus pyogenes*). Ref: J. H. Cardellina II, et al, JOC, 1979, 44, 4039 | S. Ohira, et al, J. Chem. Soc., Chem. Commun., 1993, 1299 | D. Enders, et al, Tetrahedron, 1996, 52, 5805 | H. Flörke, et al, Liebigs Ann. Chem., 1996, 147 | S. Sankaranarayanan, et al, Tetrahedron: Asymmetry, 1996, 7, 2639 | S. Ohira, et al, JCS Perkin I, 1998, 293 | E. Winter, et al, Tetrahedron, 1998, 54, 10329 | N. Maezaki, et al, Tetrahedron, 1998, 54, 13087 | K. Matsuo,, Heterocycles, 1998, 48, 1213

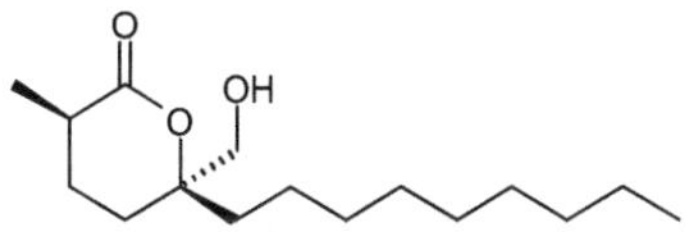

138 Malyngolide dimer

Type: 2-Pyrones. $C_{32}H_{60}O_6$ Source: Cyanobacterium *Lyngbya majuscula* (Coiba National Park, Panama). Pharm: Antiplasmodial (*in vitro*, CRPF W2, moderate); cytotoxic (H460). Ref: M. Guti-errez, et al, JNP, 2010, 73, 709

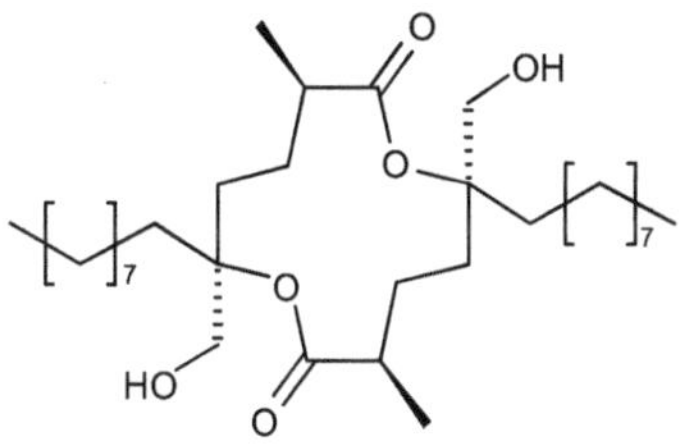

139 7-Methylcyercene 1

Type: 2-Pyrones. $C_{14}H_{18}O_3$ Solid. Source: Sacoglossan *Ercolania funerea* (Mediterranean Sea). Pharm: Phytotoxin. Ref: R. R. Vardaro, et al, Tetrahedron, 1991, 47, 5569; 1992, 48, 9561

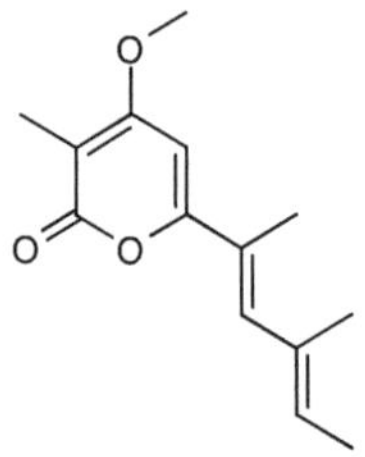

140 Nectriapyrone A

Type: 2-Pyrones. $C_{11}H_{14}O_3$ Cryst., mp 111–112 °C. Source: Marine-derived fungus from sponge *Stylotella* sp. Pharm: Melanin biosynthesis inhibitor. Ref: L. M. Abrell, et al, Tet. Lett., 1994, 35, 9159

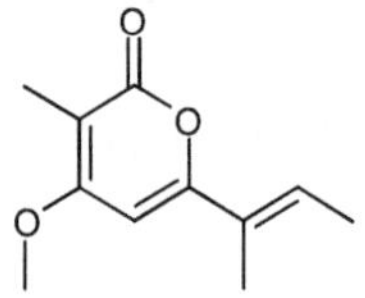

141 Neofusapyrone

Type: 2-Pyrones. $C_{34}H_{54}O_9$ Oil, $[\alpha]_D^{20} = -23.6°$ (c = 1, MeOH). Source: Marine-derived fungus *Fusarium* sp. FH-146 (driftwood, Japan waters). Pharm: Antifungal (*Aspergillus clavatus* F318a, MIC = 6.26 μg/mL; control Amphotericin B, MIC = 0.15 μg/mL *Candida albicans* ATCC 2019, MIC > 100 μg/mL; Amphotericin B, MIC = 0.62 μg/mL); antibacterial (*Pseudomonas aeruginosa* ATCC 15442, MIC > 100 μg/mL; control Chloramphenicol, MIC = 1.25 μg/mL; *Staphylococcus aureus* NBRC 13276, MIC > 100 μg/mL; Chloramphenicol, MIC = 5 μg/mL). Ref: F. Hiramatsu, et al, J. Antibiot., 2006, 59, 704

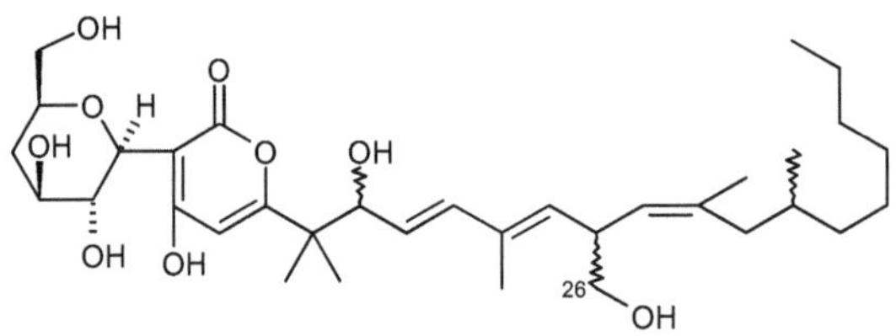

142 Neurymenolide A

Type: 2-Pyrones. $C_{24}H_{32}O_3$ Oil, $[\alpha]_D^{24} = -150°$ (c = 0.02, MeOH). Source: Red alga *Neurymenia fraxinifolia* (Taveuni I., Fiji). Pharm: Antibacterial (MRSA, IC_{50} = 2.1 μmol/L; VREF, IC_{50} = 4.5 μmol/L; tuberculosis, IC_{50} > 100 μmol/L); cytotoxic (mean of 12 cell lines, IC_{50} = 8.8 μmol/L; DU4475, IC_{50} = 3.9 μmol/L); antimalarial (IC_{50} = 68 μmol/L); antifungal (ABRCA, IC_{50} > 600 μmol/L). Ref: E. P. Stout, et al, Org. Lett., 2009, 11, 225; 1865

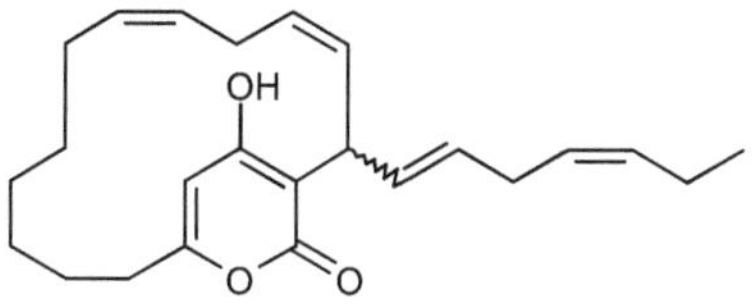

143 Neurymenolide B

Type: 2-Pyrones. $C_{26}H_{36}O_3$ Oil, $[\alpha]_D^{24}$ = −240° (*c* = 0.02, MeOH). Source: Red alga *Neurymenia fraxinifolia* (Taveuni I., Fiji). Pharm: Antibacterial (MRSA, IC_{50} = 7.8 μmol/L; VREF, IC_{50} = 31 μmol/L; tuberculosis, IC_{50} > 100 μmol/L); cytotoxic (mean of 12 cell lines, IC_{50} = 39 μmol/L; DU4475, IC_{50} = 19 μmol/L); antimalarial (IC_{50} > 100 μmol/L); antifungal (ABRCA, IC_{50} > 600 μmol/L). Ref: E. P. Stout, et al, Org. Lett., 2009, 11, 225; 1865

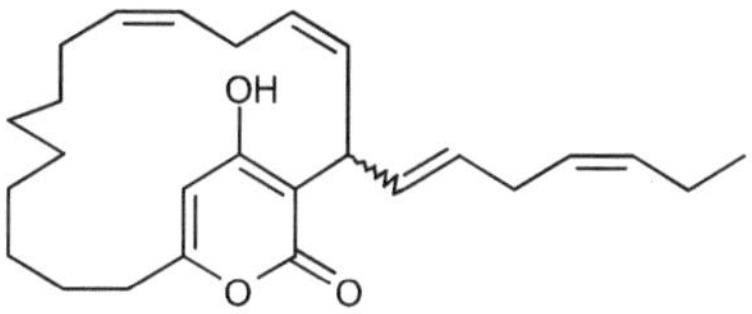

144 Nigerapyrone B

Type: 2-Pyrones. $C_{21}H_{20}O_3$ Yellowish amorph. powder. Source: Mangrove-derived fungus *Alternaria niger* from mangrove *Avicennia marina*, (Dongzhai Harbour, Hainan, China). Pharm: Cytotoxic (HepG2, IC_{50} = 62 μmol/L, control Fluorouracil, IC_{50} = 109 μmol/L). Ref: D. Liu, et al, JNP, 2011, 74, 1787

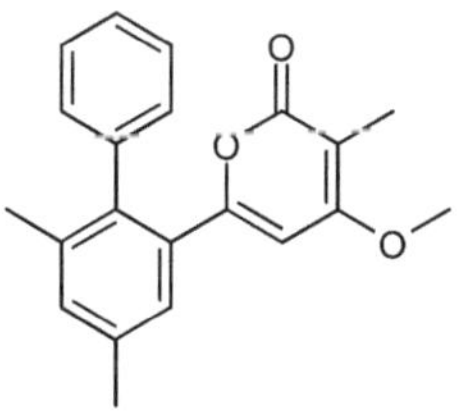

145 Nigerapyrone D

Type: 2-Pyrones. $C_{14}H_{16}O_4$ Yellowish amorph. powder. Source: Mangrove-derived fungus *Alternaria niger* from mangrove *Avicennia marina*, (Dongzhai Harbour, Hainan, China). Pharm: Cytotoxic (MCF7, IC_{50} = 121 μmol/L, control Fluorouracil, IC_{50} = 31 μmol/L; HepG2, IC_{50} = 81 μmol/L, Fluorouracil, IC_{50} = 109 μmol/L; A549, IC_{50} = 81 μmol/L, Fluorouracil, IC_{50} = 52 μmol/L). Ref: D. Liu, et al, JNP, 2011, 74, 1787

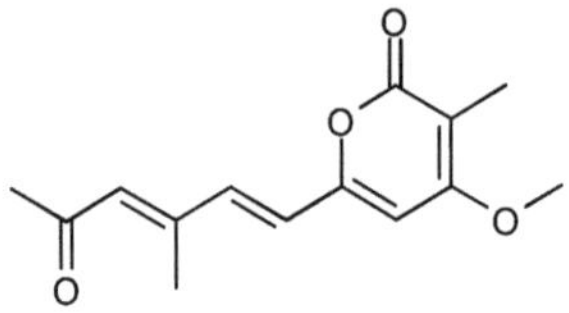

146 Nigerapyrone E

Type: 2-Pyrones. $C_{11}H_{12}O_4$ Yellowish amorph. powder. Source: Mangrove-derived fungus *Alternaria niger* from mangrove *Avicennia marina*, (Dongzhai Harbour, Hainan, China). Pharm: Cytotoxic (A549, IC_{50} = 43 μmol/L control Fluorouracil, IC_{50} = 52 μmol/L; MDA-MB-231, IC_{50} = 48 μmol/L, Fluorouracil, IC_{50} = 59 μmol/L; SW1990, IC_{50} = 38 μmol/L, Fluorouracil, IC_{50} = 121 μmol/L; MCF7, IC_{50} = 105 μmol/L, Fluorouracil, IC_{50} = 31 μmol/L; HepG2, IC_{50} = 86 μmol/L, Fluorouracil, IC_{50} = 109 μmol/L; DU145, IC_{50} = 86 μmol/L, Fluorouracil, IC_{50} = 3.3 μmol/L; NCI-H460, IC_{50} = 43 μmol/L, Fluorouracil, IC_{50} = 8.5 μmol/L; MDA-MB-231, IC_{50} = 48 μmol/L, Fluorouracil, IC_{50} = 59 μmol/L). Ref: D. Liu, et al, JNP, 2011, 74, 1787

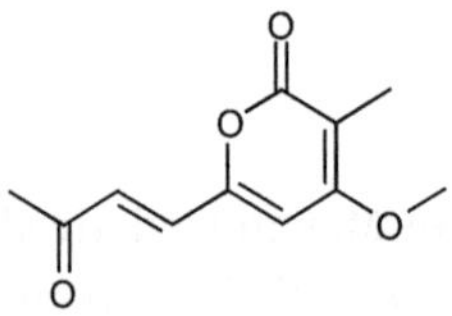

147 Nigrosporapyrone A

Type: 2-Pyrones. $C_{18}H_{22}O_5$ Gum, $[\alpha]_D^{29}$ = −254° (c = 0.1, $CHCl_3$). Source: Marine-derived fungus *Nigrospora* sp. PSU-F18 from gorgonian sea fan *Annella* sp. (Similan Is., Thailand). Pharm: Antibacterial (*Staphylococcus aureus* ATCC 25923, MIC = 128 μg/mL, MRSA, MIC = 128 μg/mL). Ref: K. Trisuwan, et al, Phytochemistry, 2009, 70, 554

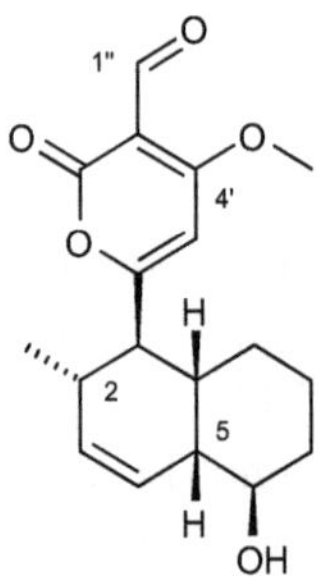

148 Pectinatone

4-Hydroxy-3,5-dimethyl-6-(1,3,5,7-tetramethyl-1-decenyl)-2*H* pyran-2-one Type: 2-Pyrones. $C_{21}H_{34}O_3$ Cryst. (CH_2Cl_2), mp 127–129 °C, $[\alpha]_D^{20}$ = +62° (c = 0.184, $CHCl_3$).

Source: Pulmonate limpets *Siphonaria pectinata* (Cádiz, Spain, W6°18´ N36°32´) and *Siphonaria grisea*. Pharm: Cytotoxic (P_{388}, A549, HT29 and MEL28, all ED_{50} = 5 µg/mL); antibacterial (gram-positive bacteria, *Bacillus subtilis*); antifungal (*Candida albicans*, yeast *Saccharomyces cerevisiae*). Ref: J. E. Biskupiak, et al, Tet. Lett., 1983, 24, 3055 | M. J. Garson, et al, JCS Perkin I, 1990, 805 | M. Norte, et al, Tetrahedron, 1990, 46, 1669 | M. C. Paul, et al, Tetrahedron, 1997, 53, 2303 | A. A. Birkbeck, et al, Tet. Lett., 1998, 39, 7823

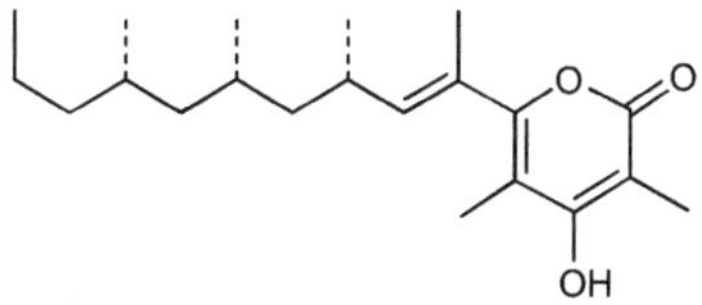

149 Penicitide A

Type: 2-Pyrones. $C_{18}H_{34}O_4$ Colorless oil, $[\alpha]_D^{20}$ = +42.9° (c = 0.14, MeOH). Source: Marine-derived fungus *Penicillium chrysogenum* QEN-24S [Syn. *Penicillium chrysogenum*] from red alga *Laurencia* sp. (Weizhou I., Guangxi, China). Pharm: Antifungal (20 µg/disk, *Aspergillus brassicae,* IZD = 6 mm, control Amphotericin B, IZD = 18 mm; *Aspergillus niger*, slight inhibition, Amphotericin B, IZD = 24 mm); Cytotoxic (HepG2, moderate). Ref: S. -S. Gao, et al, Mar. Drugs, 2011, 9, 59

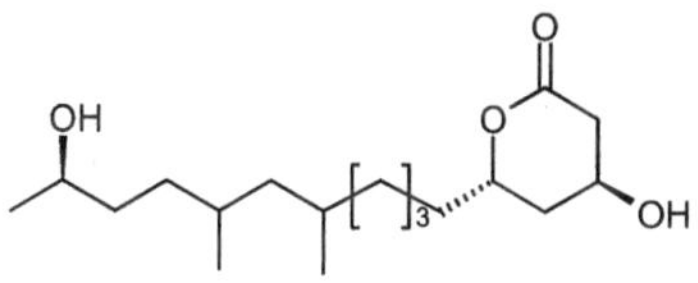

150 Peniphenone B

Type: 2-Pyrones. $C_{21}H_{18}O_8$ Source: Marine fungus *Penicillium dipodomyicola*. Pharm: Antituberculosis (*Mycobacterium tuberculosis*, strongly inhibits mycobacterial Protein tyrosine phosphatase B, MptpB). Ref: H. Li, et al, JNP 2014, 77, 800

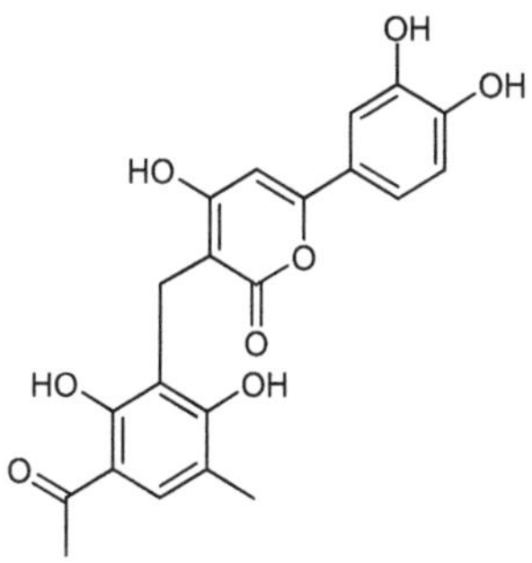

151 Phomalactone

Antibiotic LL-Z 1276 Type: 2-Pyrones. $C_8H_{10}O_3$ mp 50–53 °C, $[\alpha]_D^{22}$ = +179° ($CHCl_3$). Source: Marine-derived fungus *Nigrospora* sp. PSU-F18. Pharm: Antibacterial (*Pseudomonas* spp., *Trichophyton* spp.); phytotoxin; herbicide. Ref: K. Trisuwan, et al, Phytochemistry, 2009, 70, 554

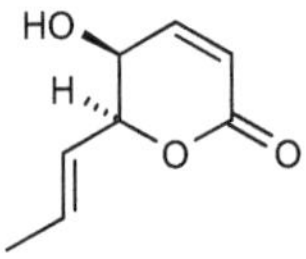

152 Phomapyrone C

Type: 2-Pyrones. $C_{10}H_{14}O_3$ Film, $[\alpha]_D$ = −16.1° (*c* = 0.28, $CHCl_3$). Source: Marine-derived fungus *Paecilomyces lilacinus*. Pharm: Phytotoxin. Ref: M. S. C. Pedras, et al, Phytochemistry, 1994, 36, 1315 | M. Elbandy, et al, Bull. Korea waters watersn Chem. Soc., 2009, 30, 188

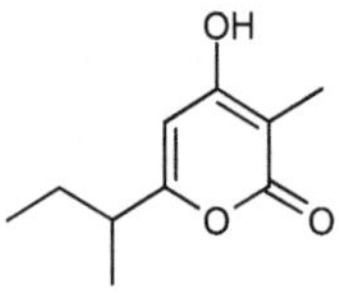

153 Phomopsis H76C

Type: 2-Pyrones. $C_{30}H_{16}N_2O_{12}$ Yellow powder, mp 317–320 °C. Source: Mangrove-derived fungus *Phomopsis* sp. ZSU-H76 from mangrove *Excoecaria agallocha* (China waters). Pharm: Blood vessel growth inhibitor. Ref: J. -X. Yang, et al, EurJOC, 2010, 3692

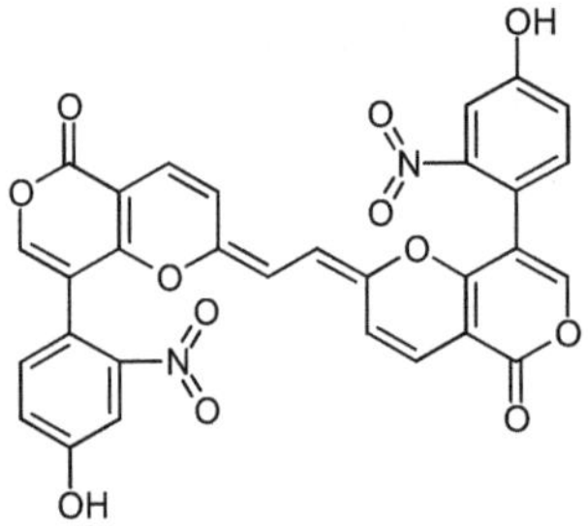

154 Pseudopyronine A

Type: 2-Pyrones. $C_{16}H_{26}O_3$ Powder. Source: Marine bacterium *Pseudomonas* sp. F92S91. Pharm: Antibacterial (including MRSA). Ref: M. P. Singh, et al, J. Antibiot., 2003, 56, 1033

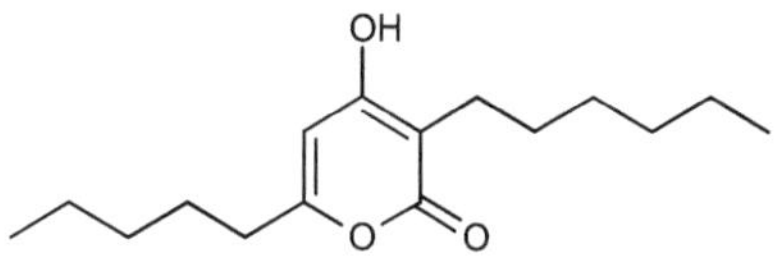

155 Pterocidin

Type: 2-Pyrones. $C_{23}H_{34}O_6$ Pale yellow oil, $[\alpha]_D^{20}$ = −27.7° (*c* = 0.46, $CHCl_3$). Source: Marine-derived streptomycetes *Streptomyces hygroscopicus* TP-A0451 and *Streptomyces* sp. (sediment, Otsuchi Bay, Japan waters). Pharm: Cytotoxic. Ref: Y. Igarashi, et al, J. Antibiot., 2006, 59, 193 | Y. Igarashi, et al, Tet. Lett., 2012, 53, 654

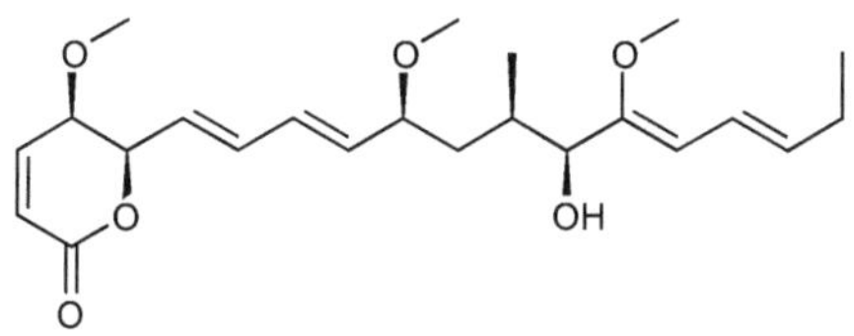

156 Pyrenocine A

Citreopyrone Type: 2-Pyrones. $C_{11}H_{12}O_4$ Cryst. (MeOH), mp 110.9–111.6 °C. Source: Marine-derived fungus *Penicillium waksmanii*. Pharm: Cytotoxic; plant growth inhibitor. Ref: T. Amagata, et al, J. Antibiot., 1998, 51, 432

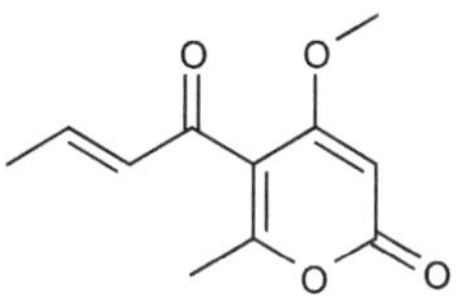

157 Pyrenocine B

Type: 2-Pyrones. $C_{11}H_{14}O_5$ Cryst. (Et_2O/hexane), mp 103–103.5 °C, $[\alpha]_D^{20}$ = +0.03° (*c* = 0.56, $CHCl_3$). Source: Marine-derived fungus *Penicillium waksmanii*. Pharm: Cytotoxic; phytotoxin. Ref: T. Amagata, et al, J. Antibiot., 1998, 51, 432

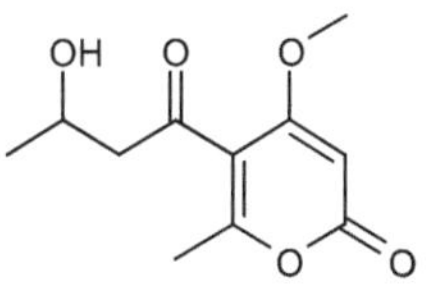

158 Pyrenocine E

Type: 2-Pyrones. $C_{12}H_{16}O_5$ Powder, mp 82–84 °C. Source: Marine-derived fungus *Penicillium waksmanii* OUPS-N133 from brown alga *Sargassum ringgoldianum*.

Pharm: Cytotoxic (P_{388}, ED_{50} = 1.30 μg/mL). Ref: T. Amagata, et al, J. Antibiot., 1998, 51, 432

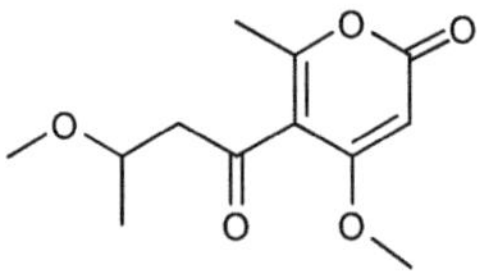

159 Salinipyrone A

Type: 2-Pyrones. $C_{17}H_{24}O_4$ Yellow oil, $[\alpha]_D$ = −87° (*c* = 0.33, MeOH). Source: Marine-derived actinomycete *Salinispora pacifica* CNS-237. Pharm: Anti-inflammatory (mouse splenocyte IL-5 inhibitor, IC_{50} = 10 μg/mL). Ref: D. -C. Oh, et al, JNP, 2008, 71, 570

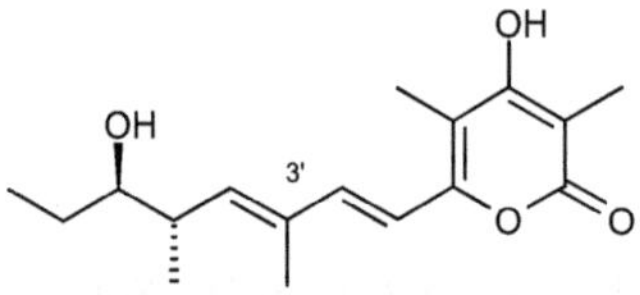

160 Simplactone A

Type: 2-Pyrones. $C_7H_{12}O_3$ Amorph. solid, $[\alpha]_D^{25}$ = −3° (*c* = 0.002, $CHCl_3$). Source: Sponge *Plakortis simplex* (Caribbean Sea). Pharm: Cytotoxic (WEHI-164, IC_{50} = 20 μg/mL). Ref: F. Cafieri, et al, Tetrahedron, 1999, 55, 13831

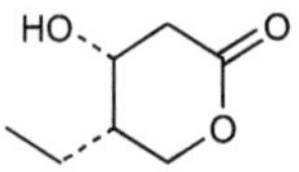

161 Simplactone B

Type: 2-Pyrones. $C_7H_{12}O_3$ Amorph. solid, $[\alpha]_D^{25}$ = −6° (*c* = 0.001, $CHCl_3$). Source: Sponge *Plakortis simplex* (Caribbean Sea). Pharm: Cytotoxic (WEHI-164, IC_{50} = 20 μg/mL). Ref: F. Cafieri, et al, Tetrahedron, 1999, 55, 13831

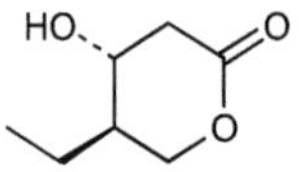

162 Solanapyrone A

Type: 2-Pyrones. $C_{18}H_{22}O_4$ Oil, $[\alpha]_D$ = −67.3° (*c* = 2.26, $CHCl_3$). Source: Marine-derived fungus *Nigrospora* sp. PSU-F18. Pharm: Phytotoxin. Ref: A. Ichihara, et al, Tet. Lett., 1983, 24, 5373; 1985, 26, 2453; 1987, 28, 1175

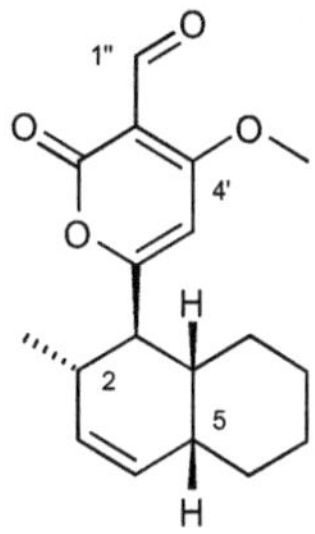

163 Solanapyrone C

Type: 2-Pyrones. $C_{19}H_{25}NO_4$ Oil, $[\alpha]_D = -5°$ ($c = 0.88$, $CHCl_3$). Source: An unidentified marine-derived fungus from green alga *Halimeda monile* (surface). Pharm: Antialgal (green alga *Dunaliella* sp., EC = 100 μg/mL); phytotoxin. Ref: A. Ichihara, et al, Tet. Lett., 1983, 24, 5373; 1985, 26, 2453; 1987, 28, 1175 | K. M. Jenkins, et al, Phytochemistry, 1998, 49, 2299

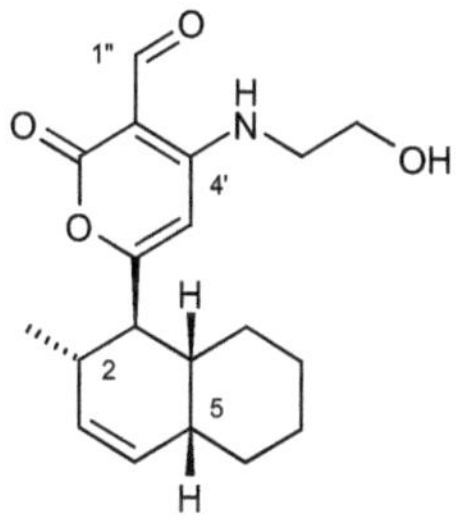

164 Solanapyrone E

Type: 2-Pyrones. $C_{18}H_{24}O_4$ Oil, $[\alpha]_D^{22} = -76.4°$ ($c = 1$, $CHCl_3$). Source: An unidentified marine fungus. Pharm: Antialgal (green alga *Dunaliella* sp., EC = 100 μg/mL). Ref: K. M. Jenkins, et al, Phytochemistry, 1998, 49, 2299

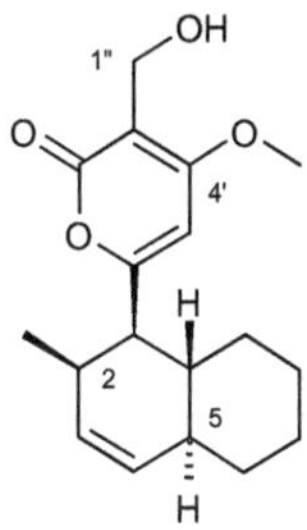

165 7*β*-Solanapyrone F

Type: 2-Pyrones. $C_{17}H_{21}NO_4$ Solid, $[\alpha]_D^{25} = -16.4°$ ($c = 0.28$, MeOH). Source: An unidentified marine-derived fungus CNC-159 cultured from green alga *Halimeda monile* (surface). Pharm: Algicide (green alga *Dunaliella* sp., EC = 100 μg/mL). Ref: K. M. Jenkins, et al, Phytochemistry, 1998, 49, 2299

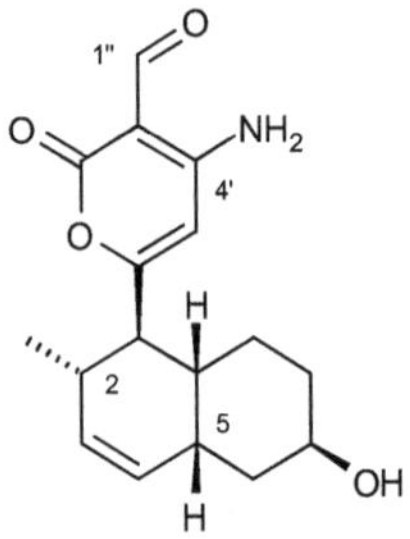

166 Solanapyrone F

Type: 2-Pyrones. $C_{17}H_{21}NO_4$ Solid, $[\alpha]_D^{25} = -53.2°$ (c = 0.3, MeOH). Source: An unidentified marine-derived fungus CNC-159 cultured from green alga *Halimeda monile* (surface). Pharm: Antialgal (green alga *Dunaliella* sp., EC = 100 μg/mL). Ref: K. M. Jenkins, et al, Phytochemistry, 1998, 49, 2299

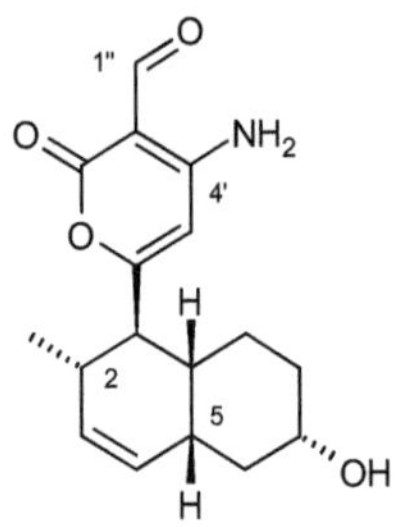

167 Solanapyrone G

Type: 2-Pyrones. $C_{17}H_{21}NO_3$ Solid. Source: An unidentified marine-derived fungus from green alga *Halimeda monile* (surface). Pharm: Antialgal (green alga *Dunaliella* sp., EC = 100 μg/mL). Ref: K. M. Jenkins, et al, Phytochemistry, 1998, 49, 2299

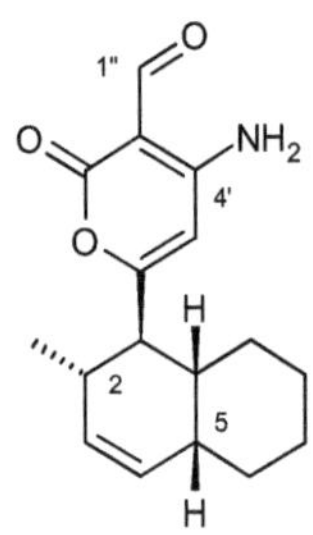

168 (+)-Tanikolide

Type: 2-Pyrones. $C_{17}H_{32}O_3$ Oil, $[\alpha]_D^{25}$ = +2.3° (*c* = 0.65, $CHCl_3$). Source: Cyanobacterium *Lyngbya majuscula* (Madagascar). Pharm: Antifungal (paper disk-agar plate assay, *Candida albicans*, 100 µg/disk, IZD = 13 mm); molluscacidal (snail vector *Biomphalaria glabrata*, LD_{50} = 31.6 µmol/L); LD_{50} (brine shrimp) = 3.6 µg/mL, (snail) = 9.0 µg/mL. Ref: I. P. Singh, et al, JNP, 1999, 62, 1333 | R. M. Kanada, et al, Synlett, 2000, 1019 | A. R. Pereira, et al, JNP, 2010, 73, 217

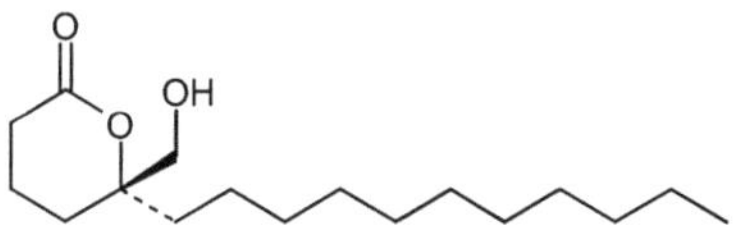

169 Verrucosidin

Type: 2-Pyrones. $C_{24}H_{32}O_6$ Plates (Et_2O), mp 90–91 °C, $[\alpha]_D^{26}$ = +92.4° (*c* = 0.25, MeOH). Source: Marine-derived fungus *Penicillium aurantiogriseum* (mud, Bohai Sea, China), terrestrial fungi (*Penicillium verrucosum* var. *cyclopium* and *Penicillium variabile*). Pharm: Neurotoxin (potent). Ref: M. Ganguli, et al, JOC, 1984, 49, 3762 | S. Nishiyama, et al, Tet. Lett., 1986, 27, 723 | A. A. El-Banna, et al, Microbiol., Aliments, Nutr., 1987, 5, 191; 1989, 7, 161 | J. K. Cha, et al, Tet. Lett., 1987, 28, 5473 | K. Yu, et al, Mar. Drugs, 2010, 8, 2744

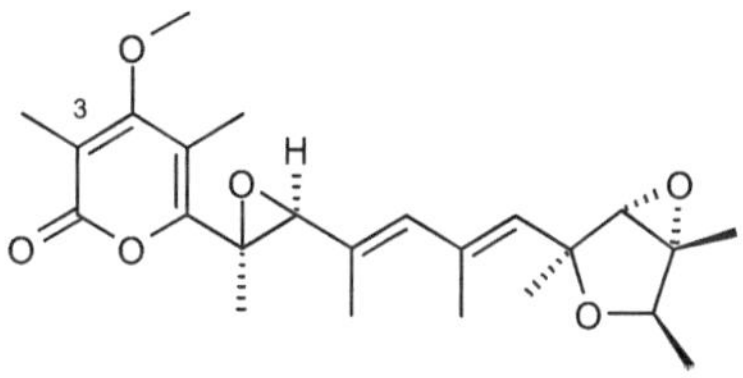

170 Xylarone

Aplysiopsene C Type: 2-Pyrones. $C_{13}H_{18}O_3$ Oil. Source: Sacoglossan *Aplysiopsis formosa* (herbivorous, Azores Is.). Pharm: Cytotoxic (Colo320, IC_{50} = 40–50 µg/mL; L_{1210}, C_{50} = 50 µg/mL). Ref: A. Schüffler, et al, Z. Naturforsch., C, 2007, 62, 169 | M. L. Ciavatta, et al, Tet. Lett., 2009, 50, 527

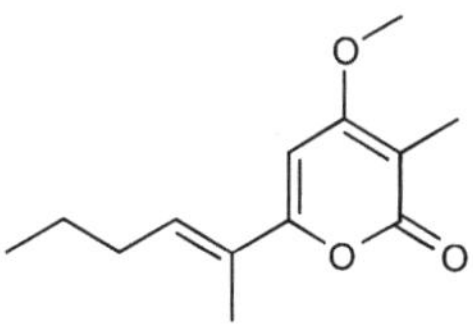

1.6 Nonadrides

171 (–)-Byssochlamic acid

Type: Nonadrides. $C_{18}H_{20}O_6$ Cryst. (EtOAc/petrol or $CHCl_3$), mp 164–165 °C, $[\alpha]_D^{23}$ = –104° (*c* = 0.1, $CHCl_3$). Source: An unidentified mangrove-derived fungus k38 (South China Sea). Pharm: Cytotoxic (Hep2, IC_{50} = 37 μg/mL, HepG2, IC_{50} = 35 μg/mL). Ref: C. Y. Li, et al, Chem. Nat. Compd., 2006, 42, 290 | C. -Y. Li, et al, J. Asian Nat. Prod. Res., 2007, 9, 285

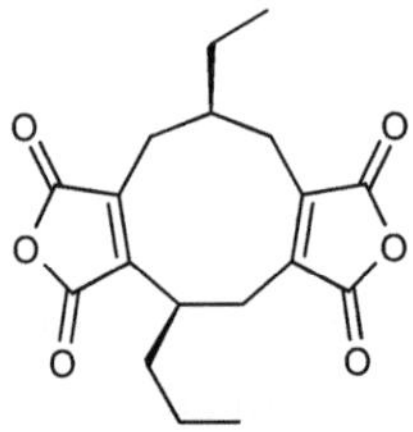

1.7 4-Pyrones (γ-Pyrones)

172 Auripyrone A

Type: 4-Pyrones. $C_{33}H_{50}O_7$ Cryst. (pentane), mp 172–176 °C, $[\alpha]_D^{26}$ = +28° (*c* = 0.08, $CHCl_3$). Source: Sea hare *Dolabella auricularia*. Pharm: Cytotoxic. Ref: K. Suenaga, et al, Tet. Lett., 1996, 37, 5151 | I. Hayakawa, et al, Angew. Chem., Int. Ed., 2010, 49, 2401

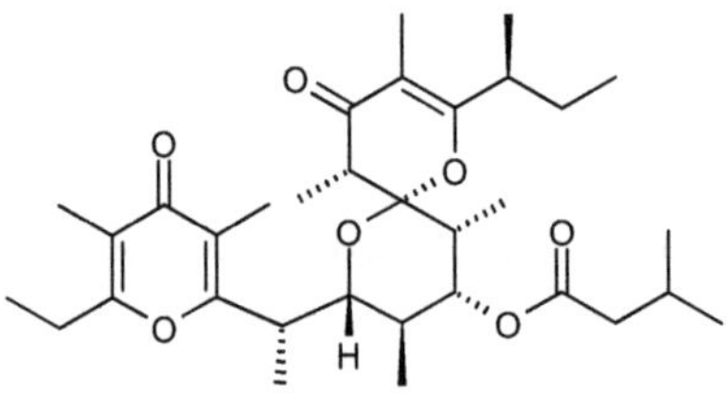

173 Auripyrone B

Type: 4-Pyrones. $C_{33}H_{50}O_7$ Cryst. (pentane), mp 126–128 °C, $[\alpha]_D^{26}$ = +39° (*c* = 0.14, $CHCl_3$). Source: Sea hare *Dolabella auricularia*. Pharm: Cytotoxic. Ref: K. Suenaga, et al, Tet. Lett., 1996, 37, 5151 | I. Hayakawa, et al, Angew. Chem., Int. Ed., 2010, 49, 2401

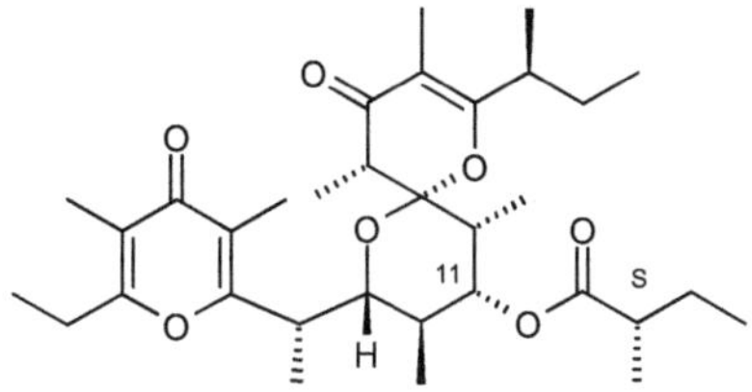

174 Bissetone

Type: 4-Pyrones. $C_9H_{14}O_5$ Oil, $[\alpha]_D = -43.6°$ (c = 4.25, EtOH). Source: Gorgonian *Briareum polyanthes*. Pharm: Antimicrobial; insecticide; plant growth regulator. Ref: J. H. Cardellina, et al, Tet. Lett., 1987, 28, 727

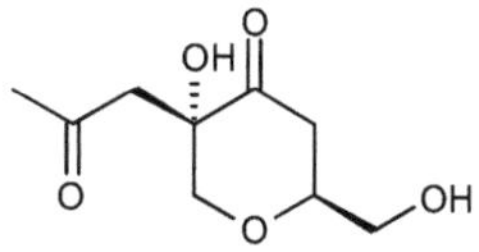

175 Carbonarone A

6-Benzyl-4-oxo-4*H*-pyran-3-carboxamide Type: 4-Pyrones. $C_{13}H_{11}NO_3$ Amorph. yellow powder. Source: Marine-derived fungus *Aspergillus carbonarius* WZ-4-11. Pharm: Cytotoxic (K562, IC_{50} = 56.0 μg/mL). Ref: Y. P. Zhang, et al, J. Antibiot., 2007, 60, 153

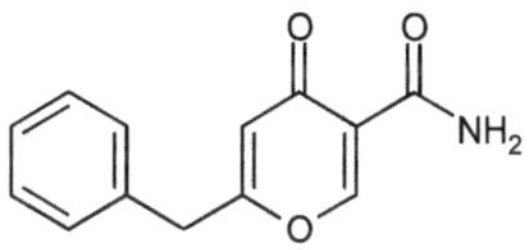

176 Carbonarone B

Type: 4-Pyrones. $C_{13}H_{11}NO_3$ Needles, mp 209–209.5 °C. Source: Marine-derived fungus *Aspergillus carbonarius* WZ-4-11. Pharm: Cytotoxic (K562, IC_{50} = 50.6 μg/mL, with antiproliferative activity). Ref: Y. Zhang, et al, J. Antibiot., 2007, 60, 153

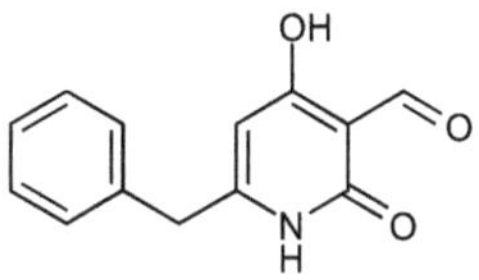

177 Chromanone

Type: 4-Pyrones. $C_{18}H_{18}O_6$ Source: Marine-derived streptomycete *Streptomyces sundarbansensis* (endophytic) from brown alga *Fucus* sp. (Bejaia, Algeria). Pharm: Antibacterial (MRSA, modest but selective). Ref: I. Djinni, et al, Mar. Drugs, 2013, 11, 124

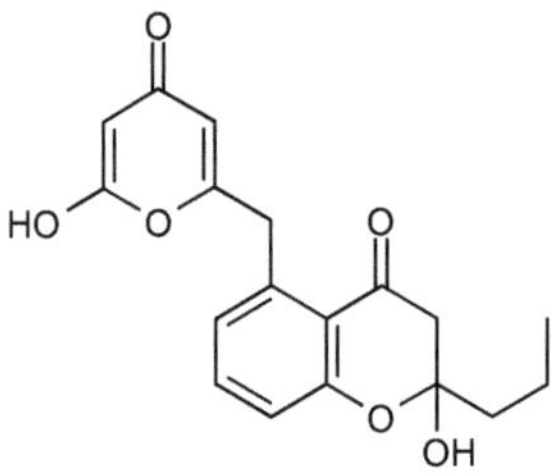

178 Crispatene

Type: 4-Pyrones. $C_{25}H_{34}O_4$ Oil, $[\alpha]_D = -92.8°$ (c = 0.12, $CHCl_3$). Source: Sacoglossan *Elysia crispata* [Syn. *Tridachia crispata*]. Pharm: Cytotoxic (PS, ED_{50} = 7.2 μg/mL). Ref: M. B. Ksebati, et al, JOC, 1985, 50, 5637

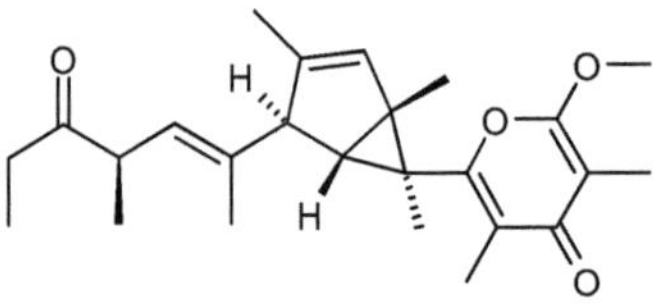

179 Crispatone

Type: 4-Pyrones. $C_{25}H_{34}O_5$ Cryst. (hexane), mp 164.5–166.5 °C, $[\alpha]_D = -84.7°$ (c = 0.03, $CHCl_3$). Source: Sacoglossan *Elysia crispata* [Syn. *Tridachia crispata*]. Pharm: Cytotoxic (PS, ED_{50} = 3.7 μg/mL). Ref: M. B. Ksebati, et al, JOC, 1985, 50, 5637

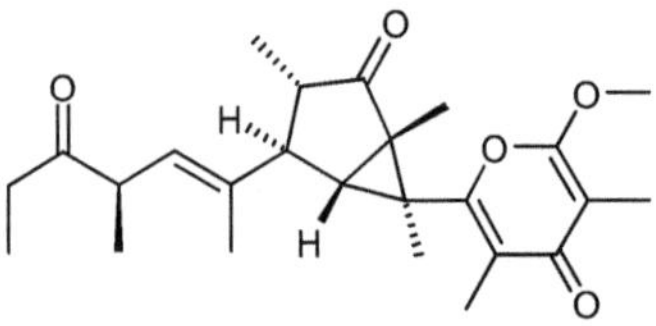

180 Cyercene A

Type: 4-Pyrones. $C_{16}H_{22}O_3$ Source: Sacoglossan *Cyerce cristallina*. Pharm: Ichthyotoxic, tissue regeneration stimulant. Ref: R. R. Vardaro, et al, Tetrahedron, 1991, 47, 5569

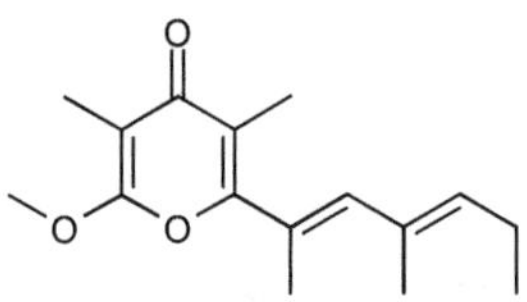

181 Cyercene B

Type: 4-Pyrones. $C_{14}H_{18}O_3$ Source: Sacoglossans *Cyerce cristallina* and *Ercolania funerea*. Pharm: Ichthyotoxic. Ref: R. R. Vardaro, et al, Tetrahedron, 1991, 47, 5569; 1992, 48, 9561

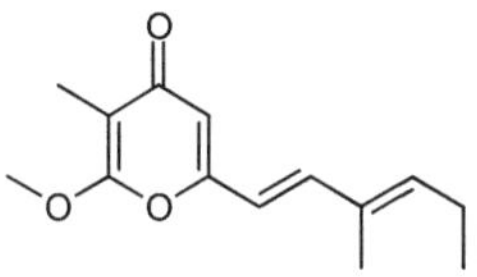

182 Deoxyfusapyrone

Type: 4-Pyrones. $C_{34}H_{54}O_8$ Oil, $[\alpha]_D^{20}$ = −23.3° (*c* = 0.23, MeOH). Source: Marine-derived fungus *Fusarium* sp. FH-146 (driftwood, Japan waters). Pharm: Antifungal (*Aspergillus clavatus* F318a, MIC = 3.12 μg/mL; control Amphotericin B, MIC = 0.15 μg/mL *Candida albicans* ATCC 2019, MIC > 100 μg/mL; Amphotericin B, MIC = 0.62 μg/mL); antibacterial (*Pseudomonas aeruginosa* ATCC 15442, MIC = 12.5 μg/mL; control Chloramphenicol, MIC = 1.25 μg/mL; *Staphylococcus aureus* NBRC 13276, MIC > 100 μg/mL; Chloramphenicol, MIC = 5 μg/mL). Ref: F. Hiramatsu, et al, J. Antibiot., 2006, 59, 704

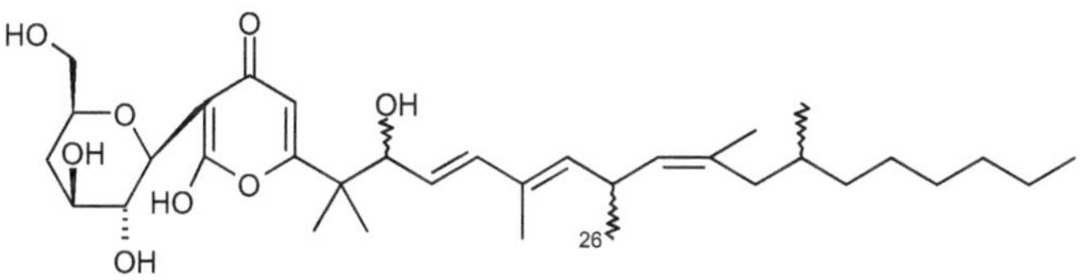

183 Fusapyrone

Type: 4-Pyrones. $C_{34}H_{54}O_9$ Oil, $[\alpha]_D^{20}$ = −22.9° (*c* = 0.31, MeOH). Source: Marine-derived fungus *Fusarium* sp. FH-146 (driftwood, Japan waters). Pharm: Antifungal (*Aspergillus clavatus* F318a, MIC = 25 μg/mL; control Amphotericin B, MIC = 0.15 μg/mL *Candida albicans* ATCC 2019, MIC > 100 μg/mL; Amphotericin B, MIC = 0.62 μg/mL); antibacterial (*Pseudomonas aeruginosa* ATCC 15442, MIC = 50 μg/mL; control Chloramphenicol, MIC = 1.25 μg/mL; *Staphylococcus aureus* NBRC 13276, MIC > 100 μg/mL; Chloramphenicol, MIC = 5 μg/mL). Ref: F. Hiramatsu, et al, J. Antibiot., 2006, 59, 704

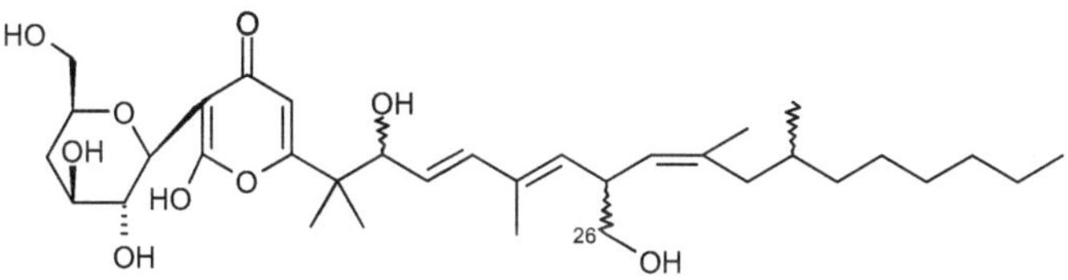

184 Himeic acid A

Type: 4-Pyrones. $C_{22}H_{29}NO_8$ $[\alpha]_D^{26}$ = −15° (*c* = 0.14, MeOH). Source: Marine-derived fungus *Aspergillus* sp. Pharm: Ubiquitin-activating enzyme inhibitor. Ref: S. Tsukamoto, et al, BoMCL, 2005, 15, 191

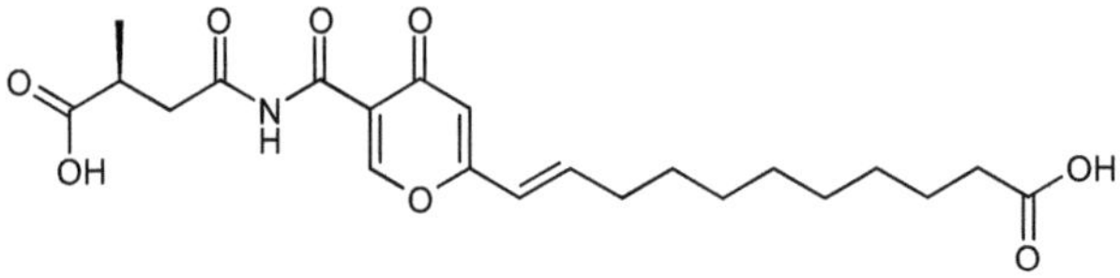

185 Hyalopyrone

6- (2-Hydroxypropyl)-3-methyl-2- (1-methylpropyl)-4*H*-pyran-4-one Type: 4-Pyrones. $C_{13}H_{20}O_3$ Pale yellow syrup, $[\alpha]_D^{27}$ = −16.6° (*c* = 1, MeOH). Source: Marine-derived fungus *Ascochyta salicorniae* from green alga *Ulva* sp. (North Sea coast, Germany). Pharm: Protein phosphatase inhibitor (MPtpB, IC_{50} = 87.8 µmol/L; Cdc25a, PTP1B, VHR, MPtpA and VE-PTP, all five inactive); phytotoxin. Ref: P. Venkatasubbaiah, et al, JNP, 1992, 55, 461 | S. F. Seibert, et al, Org. Biomol. Chem., 2006, 4, 2233

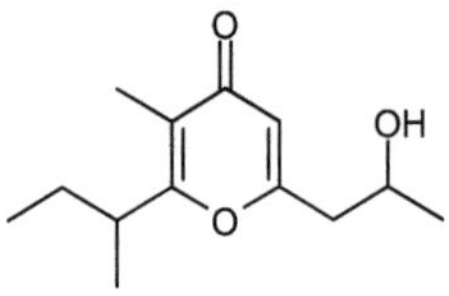

186 Ilikonapyrone

Type: 4-Pyrones. $C_{32}H_{48}O_7$ Cryst., mp 96–98 °C, $[\alpha]_D$ = −16° (*c* = 1.5, CH_2Cl_2). Source: Pulmonate *Onchidium verraculatum*. Pharm: Defence allomone of pulmonate *Onchidium verruculatum*. Ref: C. M. Ireland, et al, JOC, 1984, 49, 559 | H. Arimoto, et al, Tet. Lett., 1993, 34, 5781

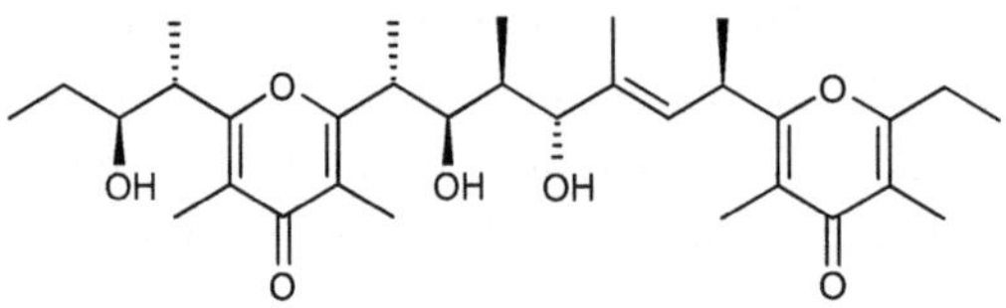

187 Isoplacidene A

Type: 4-Pyrones. $C_{16}H_{22}O_3$ Source: Sacoglossan *Placida dendritica*. Pharm: Ichthyotoxin. Ref: R. R, Vardaro, et al, Tet. Lett., 1992, 33, 2875

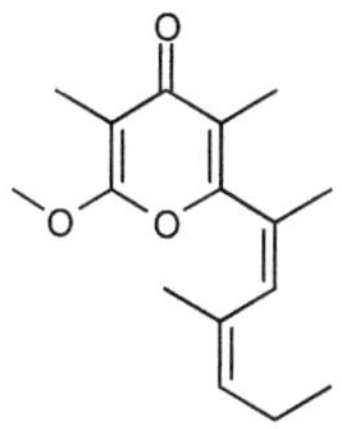

188 Isoplacidene B

Type: 4-Pyrones. $C_{15}H_{20}O_3$ Source: Sacoglossan *Placida dendritica*. Pharm: Ichthyotoxin. Ref: R. R, Vardaro, et al, Tet. Lett., 1992, 33, 2875

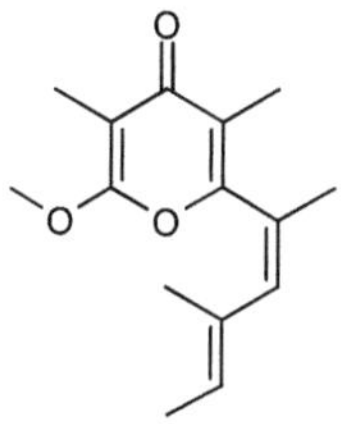

189 Kalkipyrone

Type: 4-Pyrones. $C_{20}H_{28}O_4$ Oil, $[\alpha]_D^{20}$ = +8.2° (*c* = 0.22, $CHCl_3$). Source: Cyanobacteria *Lyngbya majuscula* and *Tolypothrix* sp. (Curacao). Pharm: LD_{50} (brine shrimp) = 1 μg/mL, (goldfish) = 2 μg/mL. Ref: M. A. Graber, et al, JNP, 1998, 61, 677

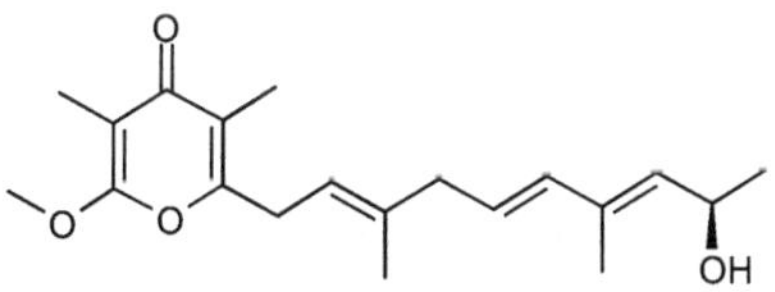

190 Lehualide B

Type: 4-Pyrones. $C_{28}H_{36}O_4$ Oil. Source: Sponge *Plakortis* sp. Pharm: Cytotoxic. Ref: N. Sata, et al, JNP, 2005, 68, 1400

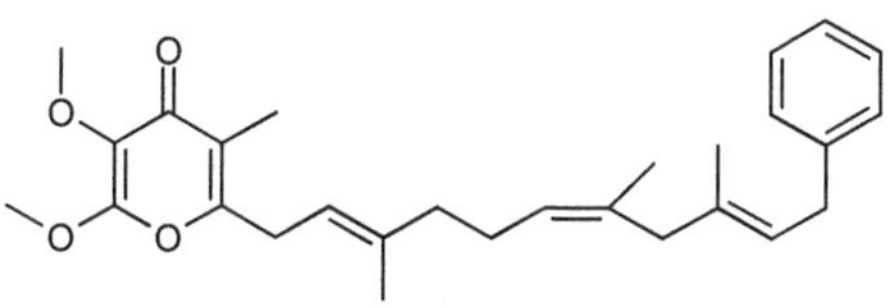

191 Lehualide D

Type: 4-Pyrones. $C_{21}H_{34}O_4S$ Oil. Source: Sponge *Plakortis* sp. Pharm: Cytotoxic. Ref: N. Sata, et al, JNP, 2005, 68, 1400

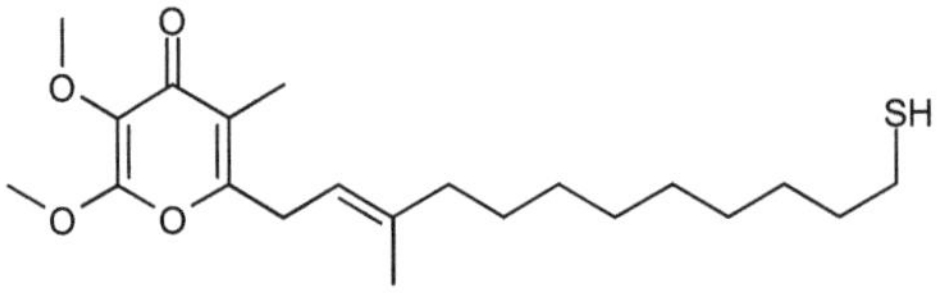

192 Marinactinone B

Type: 4-Pyrones. $C_{16}H_{26}O_3$ Source: Marine-derived bacterium *Marinactinospora thermotolerans* (sediment, South China Sea). Pharm: Cytotoxic (HTCLs, moderate); topoisomerase II inhibitor (weak). Ref: F. Wang, et al, J. Antibiot., 2011, 64, 189

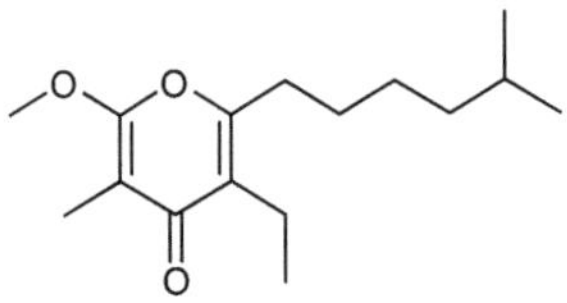

193 Marinactinone C

Type: 4-Pyrones. $C_{16}H_{26}O_3$ Source: Marine-derived bacterium *Marinactinospora thermotolerans* (sediment, South China Sea). Pharm: Cytotoxic (HTCLs, moderate). Ref: F. Wang, et al, J. Antibiot., 2011, 64, 189

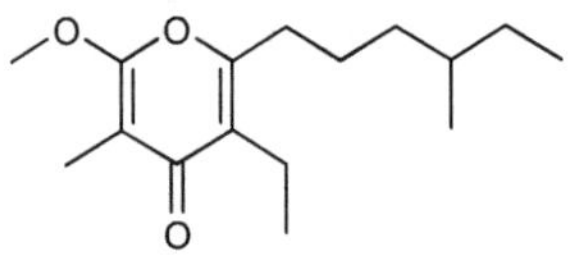

194 Maurapyrone A

Type: 4-Pyrones. $C_{26}H_{36}O_5$ Cryst., mp 110–112 °C. Source: Pulmonate limpet *Siphonaria maura*. Pharm: Antibacterial (*Vibrio anguillarum*). Ref: D. C. Manker, et al, JOC, 1986, 51, 814

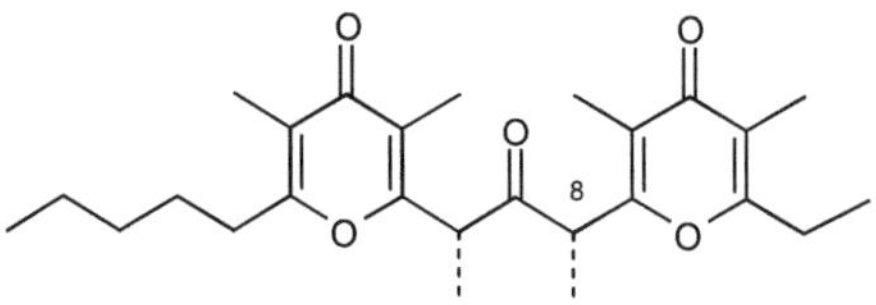

195 Maurapyrone C

Type: 4-Pyrones. $C_{25}H_{34}O_5$ Wax. Source: Pulmonate limpet *Siphonaria maura*. Pharm: Antibacterial (*Vibrio anguillarum*). Ref: D. C. Manker, et al, JOC, 1986, 51, 814

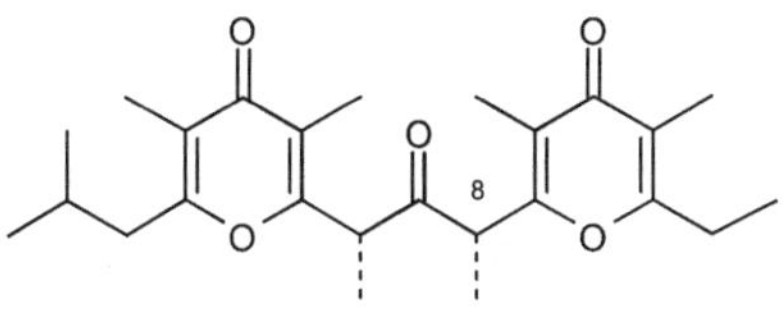

196 Maurapyrone D

Type: 4-Pyrones. $C_{25}H_{34}O_5$ Oil. Source: Pulmonate limpet *Siphonaria maura*. Pharm: Antibacterial (*Vibrio anguillarum*). Ref: D. C. Manker, et al, JOC, 1986, 51, 814

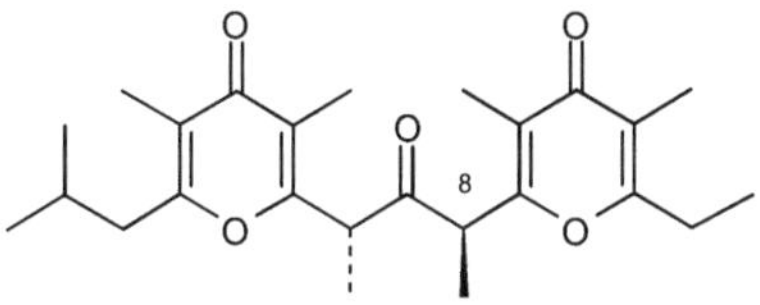

197 5′-Methoxyvermistatin

Type: 4-Pyrones. $C_{19}H_{18}O_7$ Cryst., mp 198–199 °C, $[\alpha]_D^{20} = -30°$ (c = 0.1, Me_2CO). Source: Mangrove-derived fungus *Guignardia* sp. 4382 from Mangrove *Kandelia candel* (South China Sea). Pharm: Cytotoxic (KB, IC_{50} = 20.0 μg/mL, control Cisplatin, IC_{50} = 0.56 μg/mL; KBV200, IC_{50} = 15.1 μg/mL, Cisplatin, IC_{50} = 0.78 μg/mL). Ref: X. K. Xia, et al, Helv. Chim. Acta, 2007, 90, 1925

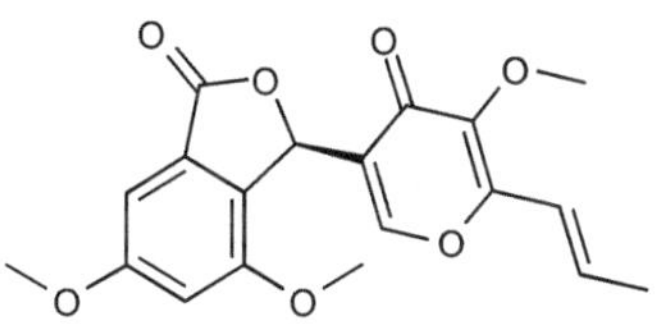

198 Onchitriol I

Type: 4-Pyrones. $C_{32}H_{48}O_7$ $[\alpha]_D = -20.0°$ (c = 0.01, CH_2Cl_2). Source: Pulmonate *Ohchidium* sp. Pharm: Cytotoxic (P_{388}, $IC_{50} \approx 10$ μg/mL; A549, $IC_{50} \approx 10$ μg/mL; HT29, $IC_{50} \approx 10$ μg/mL); antiviral (*Herpes simplex* virus HSV-1, IC_{50} = 10 μg/mL; *Vesicular stomatitis* virus VSV, IC_{50} = 20 μg/mL); cytotoxic (KB, 1–10 μg/mL, InRt = 90%–98%). Ref: J. Rodriguez, et al, Tet. Lett., 1992, 33, 1089 | J. Rodríguez, et al, JOC, 1992, 57, 4624 | H. Arimoto, et al, Tet. Lett., 1995, 36, 5357

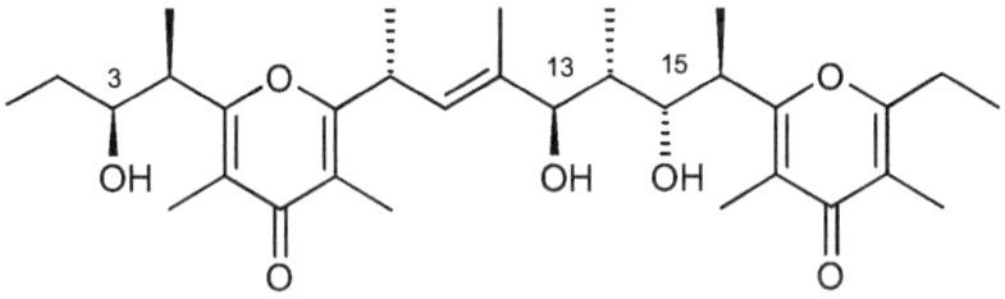

199 Onchitriol IA

Type: 4-Pyrones. $C_{36}H_{52}O_9$ $[\alpha]_D$ = −11.5° (*c* = 0.02, CH_2Cl_2). Source: Pulmonate *Onchidium* sp. (New Caledonia). Pharm: Cytotoxic (KB, 1–10 µg/mL, InRt = 90%–98%). Ref: J. Rodriguez, et al, Tet. Lett., 1992, 33, 1089 | J. Rodriguez, et al, JOC, 1992, 57, 4624

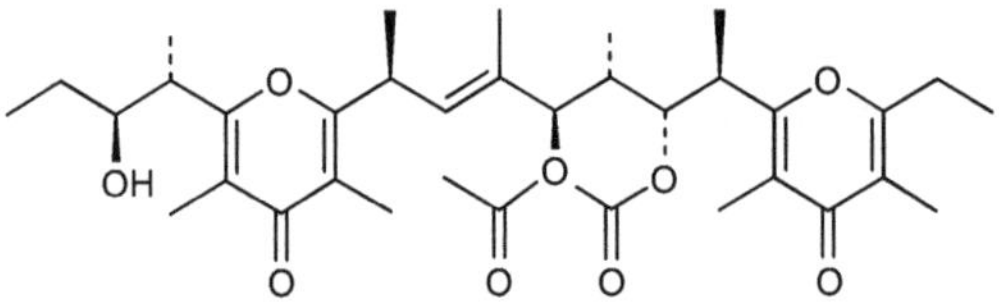

200 Onchitriol IB

Type: 4-Pyrones. $C_{37}H_{54}O_9$ $[\alpha]_D$ = −19.5° (*c* = 0.01, CH_2Cl_2). Source: Pulmonate *Onchidium* sp. (New Caledonia). Pharm: Cytotoxic (KB, 1–10 µg/mL, InRt = 90%–98%). Ref: J. Rodriguez, et al, Tet. Lett., 1992, 33, 1089 | J. Rodriguez, et al, JOC, 1992, 57, 4624

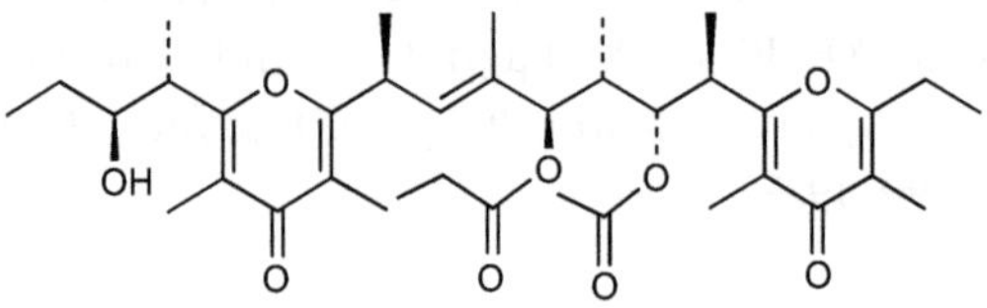

201 Onchitriol IC

Type: 4-Pyrones. $C_{37}H_{54}O_9$ $[\alpha]_D$ = −18.0° (*c* = 0.1, CH_2Cl_2). Source: Pulmonate *Onchidium* sp. (New Caledonia). Pharm: Cytotoxic (KB, 1–10 µg/mL, InRt = 90%–98%). Ref: J. Rodriguez, et al, Tet. Lett., 1992, 33, 1089 | J. Rodriguez, et al, JOC, 1992, 57, 4624

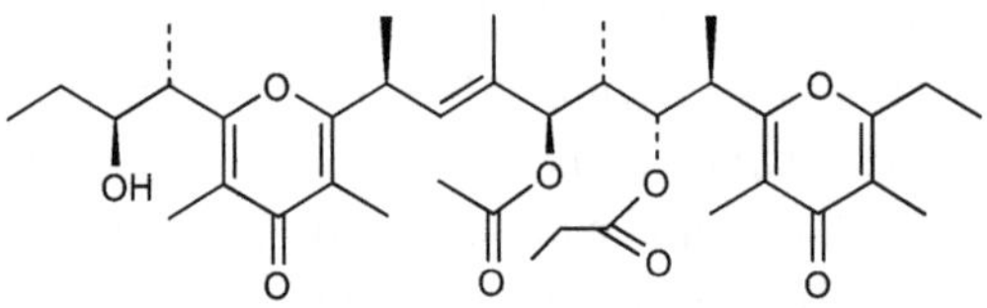

202 Onchitriol ID

Type: 4-Pyrones. $C_{40}H_{58}O_{10}$ $[\alpha]_D$ = −25.2° (*c* = 0.01, CH_2Cl_2). Source: Pulmonate *Onchidium* sp. (New Caledonia). Pharm: Cytotoxic (KB, 1–10 µg/mL, InRt = 90%–98%). Ref: J. Rodriguez, et al, Tet. Lett., 1992, 33, 1089 | J. Rodriguez, et al, JOC, 1992, 57, 4624

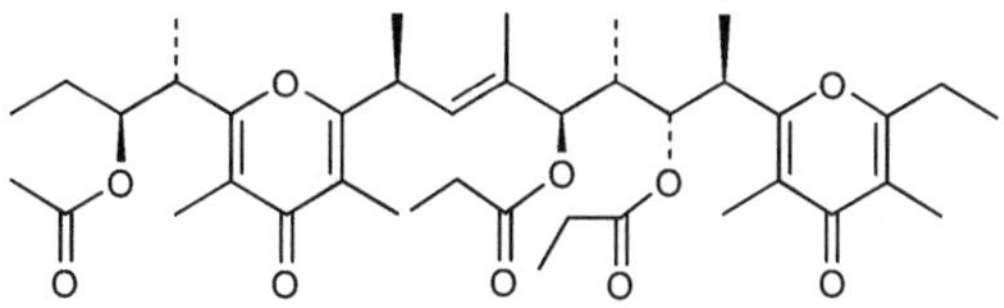

203 Onchitriol II

Type: 4-Pyrones. $C_{32}H_{48}O_7$ $[\alpha]_D$ = −33° (*c* = 0.01, $CHCl_3$). Source: Pulmonate *Onchidium* sp. Pharm: Cytotoxic (P_{388}, IC_{50} ≈ 20 μg/mL; A549, IC_{50} ≈ 20 μg/mL; HT29, IC_{50} ≈ 20 μg/mL); antiviral (*Herpes simplex* virus HSV-1, IC_{50} = 10 μg/mL; *Vesicular stomatitis* virus VSV, IC_{50} = 20 μg/mL); cytotoxic (KB, 1–10 μg/mL, InRt = 90%–98%). Ref: J. Rodríguez, et al, Tet. Lett., 1992, 33, 1089 | J. Rodríguez, et al, JOC, 1992, 57, 4624 | H. Arimoto, et al, Tet. Lett., 1994, 35, 9581

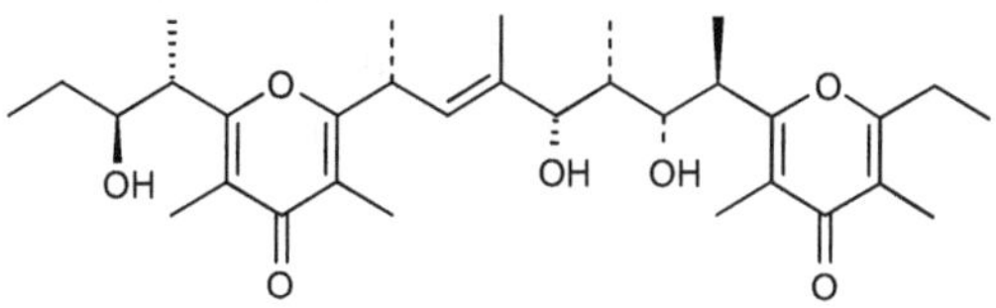

204 Onchitriol IIA

Type: 4-Pyrones. $C_{34}H_{50}O_8$ $[\alpha]_D$ = −26.0° (*c* = 0.1, CH_2Cl_2). Source: Pulmonate *Onchidium* sp. (New Caledonia). Pharm: Cytotoxic (KB, 1–10 μg/mL, InRt = 90%–98%). Ref: J. Rodriguez, et al, JOC, 1992, 57, 4624 | J. Rodriguez, et al, Tet. Lett., 1992, 33, 1089

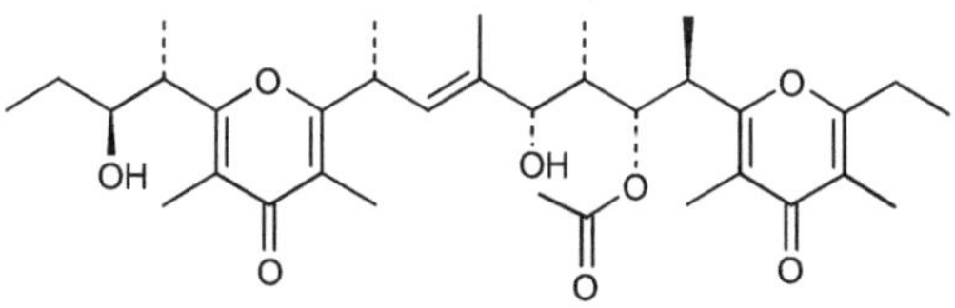

205 Onchitriol IIB

Type: 4-Pyrones. $C_{35}H_{52}O_8$ $[\alpha]_D$ = −25.2° (*c* = 0.01, CH_2Cl_2). Source: Pulmonate *Onchidium* sp. (New Caledonia). Pharm: Cytotoxic (KB, 1–10 μg/mL, InRt = 90%–98%). Ref: J. Rodriguez, et al, Tet. Lett., 1992, 33, 1089 | J. Rodriguez, et al, JOC, 1992, 57, 4624

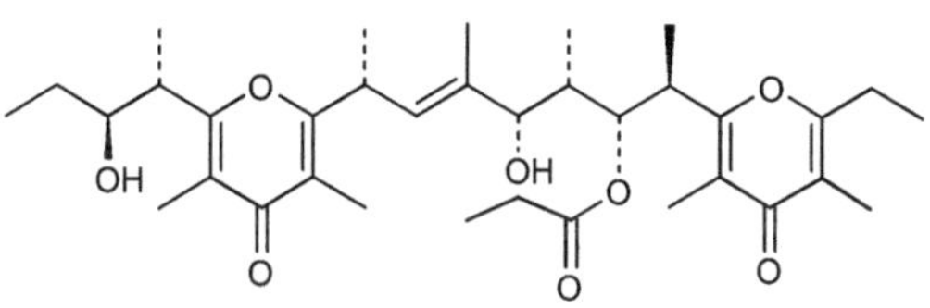

206 Onchitriol IIC

Type: 4-Pyrones. $C_{38}H_{54}O_{10}$ $[\alpha]_D$ = −26.0° (*c* = 0.01, CH_2Cl_2). Source: Pulmonate *Onchidium* sp. (New Caledonia). Pharm: Cytotoxic (KB, 1–10 μg/mL, InRt = 90%–98%). Ref: J. Rodriguez, et al, Tet. Lett., 1992, 33, 1089 | J. Rodriguez, et al, JOC, 1992, 57, 4624

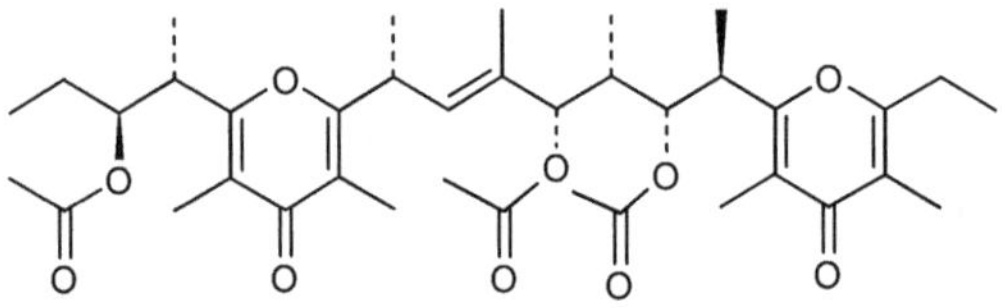

207 Onchitriol IID

Type: 4-Pyrones. $C_{36}H_{52}O_9$ $[\alpha]_D$ = −20.0° (*c* = 0.1, CH_2Cl_2). Source: Pulmonate *Onchidium* sp. (New Caledonia). Pharm: Cytotoxic (KB, 1–10 μg/mL, InRt = 90%–98%). Ref: J. Rodriguez, et al, Tet. Lett., 1992, 33, 1089 | J. Rodriguez, et al, JOC, 1992, 57, 4624

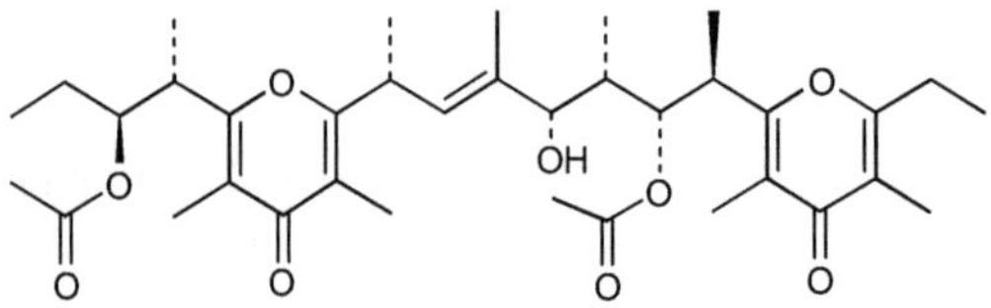

208 Peroniatriol I

Type: 4-Pyrones. $C_{32}H_{48}O_7$ $[\alpha]_D$ = −12.4° (*c* = 1.2, CH_2Cl_2). Source: Pulmonate *Peronia peronia*. Pharm: Cytotoxic. Ref: J. E. Biskupiak, et al, Tet. Lett., 1985, 26, 4307 | H. Arimoto, et al, Tet. Lett., 1993, 34, 5781

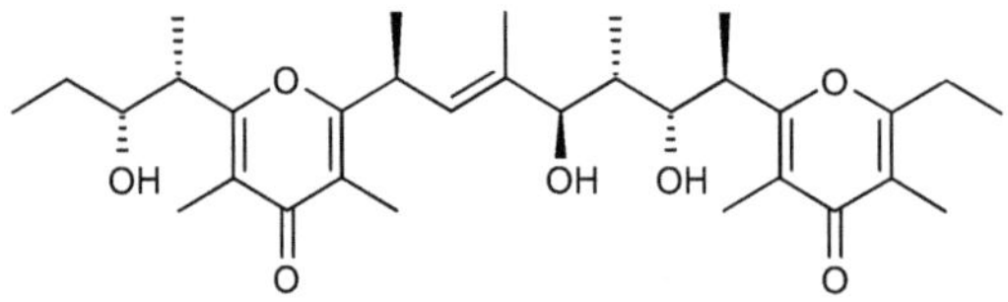

209 Peroniatriol II

Type: 4-Pyrones. $C_{32}H_{48}O_7$ $[\alpha]_D$ = +224.8° (*c* = 1, CH_2Cl_2). Source: Pulmonate *Peronia peronia*. Pharm: Cytotoxic. Ref: J. E. Biskupiak, et al, Tet. Lett., 1985, 26, 4307 | H. Arimoto, et al, Tet. Lett., 1993, 34, 5781

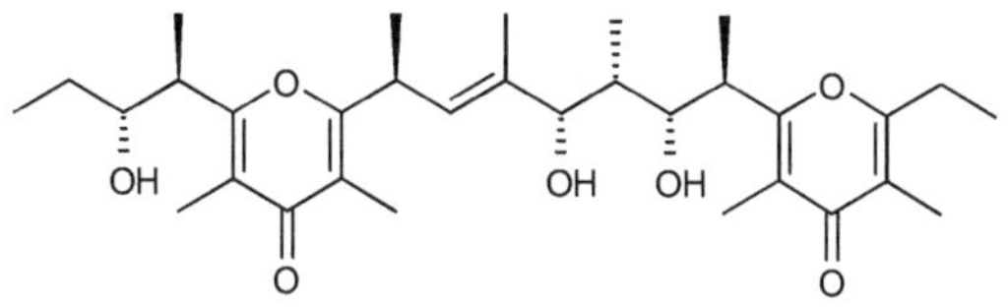

210 Photodeoxytridachione

Type: 4-Pyrones. $C_{22}H_{30}O_3$ $[\alpha]_D^{25}$ = +14.4° (c = 0.6, $CHCl_3$). Source: Sacoglossans *Elysia timida* (Mediterranean Sea) and *Placobranchus ocellatus*. Pharm: Ichthyotoxin. Ref: C. M. Ireland, et al, Science, 1979, 205, 922 | C. M. Ireland, et al, Tetrahedron, 1981, 37, 233

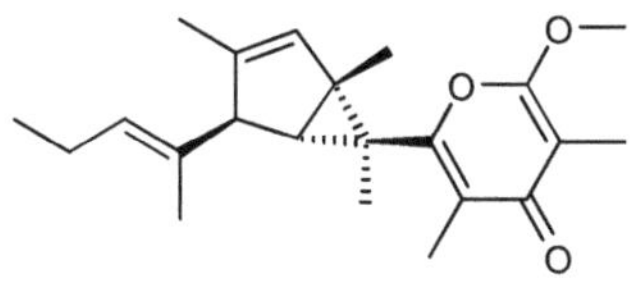

211 Pinnamine

Type: 4-Pyrones. $C_{13}H_{19}NO_2$ Oil. Source: Bivalve *Pinna muricata* (Okinawa). Pharm: Toxin; LD_{50} (mus, acute toxic) = 0.5 μg/mL; highly toxic; LD_{99} = 0.5 mg/kg (mus). Ref: N. Takada, et al, Tet. Lett., 2000, 41, 6425 | H. Kigoshi, et al, Tet. Lett., 2001, 42, 7469 | M. Kuramoto, et al, Mar. Drugs, 2004, 2, 39

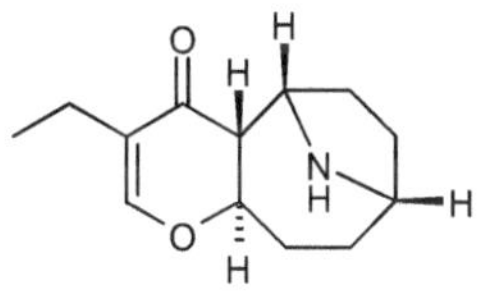

212 Placidene A

Type: 4-Pyrones. $C_{16}H_{22}O_3$ Source: Sacoglossan *Placida dendritica*. Pharm: Ichthyotoxin. Ref: R. R, Vardaro, et al, Tet. Lett., 1992, 33, 2875

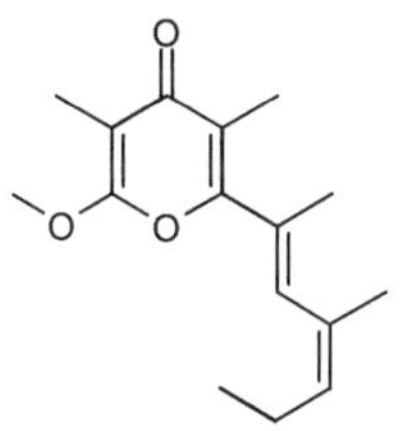

213 Placidene B

Type: 4-Pyrones. $C_{15}H_{20}O_3$ Source: Sacoglossan *Placida dendritica*. Pharm: Ichthyotoxin. Ref: R. R, Vardaro, et al, Tet. Lett., 1992, 33, 2875

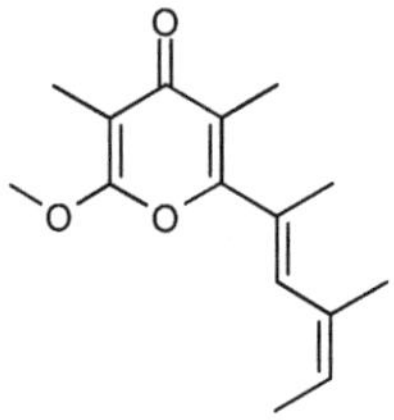

214 Polyporapyranone A

Type: 4-Pyrones. $C_{13}H_{12}O_4$ Source: Two marine-derived fungi *Polyporales* sp. from green alga *Thalassia hemprichii* (location unspecified, presumably Thailand). Pharm: Cytotoxic (Vero, moderate and weak). Ref: V. Rukachaisirikul, et al, Tetrahedron, 2013, 69, 6981

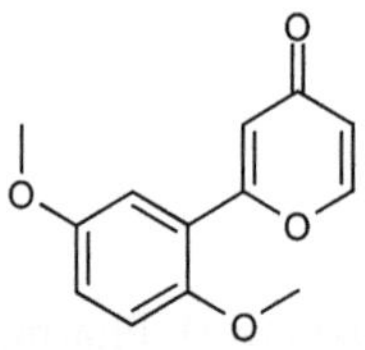

215 Polyporapyranone D

Type: 4-Pyrones. $C_{11}H_7ClO_3$ Source: Two marine-derived fungi *Polyporales* sp. from green alga *Thalassia hemprichii* (location unspecified, presumably Thailand). Pharm: Cytotoxic (Vero, moderate and weak). Ref: V. Rukachaisirikul, et al, Tetrahedron, 2013, 69, 6981

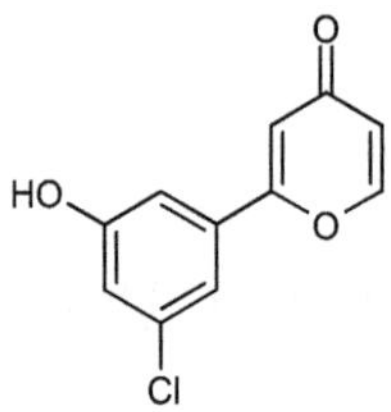

216 Tridachiapyrone A

Type: 4-Pyrones. $C_{25}H_{34}O_4$ Oil. Source: Sacoglossan *Elysia crispata* [Syn. *Tridachia crispata*]. Pharm: Cytotoxic (PS, ED_{50} = 7.2 μg/mL). Ref: M. B. Ksebati, et al, JOC, 1985, 50, 5637

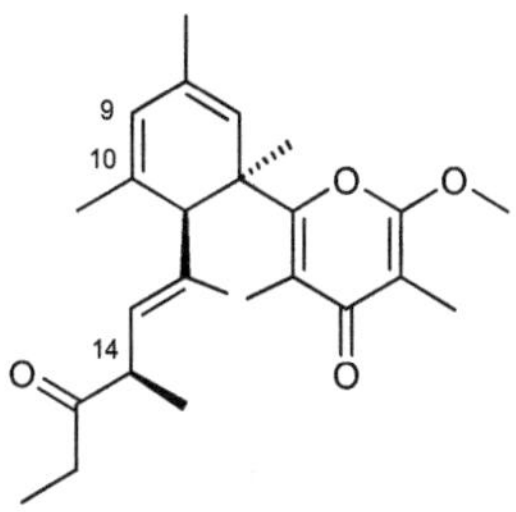

217 Tridachiapyrone B

Type: 4-Pyrones. $C_{25}H_{32}O_5$ Oil. Source: Sacoglossan *Elysia crispata* [Syn. *Tridachia crispata*]. Pharm: Cytotoxic (PS, ED_{50} = 6 μg/mL). Ref: M. B. Ksebati, et al, JOC, 1985, 50, 5637

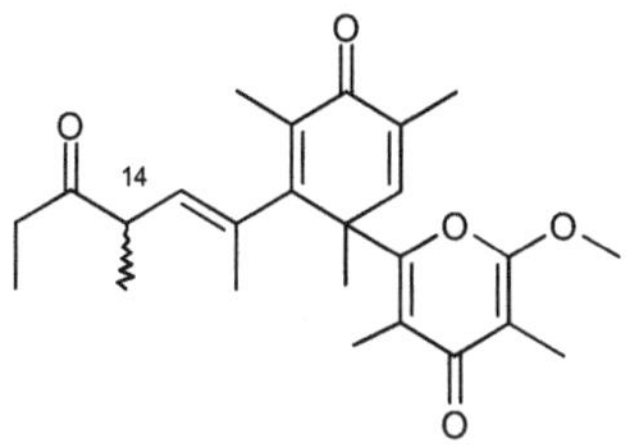

218 Tridachiapyrone D

Type: 4-Pyrones. $C_{25}H_{34}O_6$ Oil. Source: Sacoglossan *Elysia crispata* [Syn. *Tridachia crispata*]. Pharm: Cytotoxic (PS, ED_{50} = 3.1 μg/mL). Ref: M. B. Ksebati, et al, JOC, 1985, 50, 5637

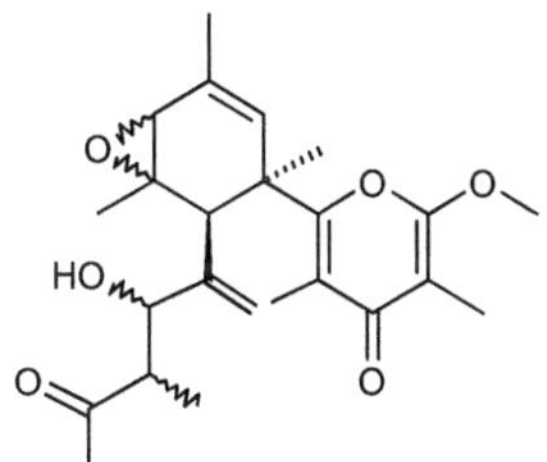

219 Vermistatin

Fijiensin Type: 4-Pyrones. $C_{18}H_{16}O_6$ Cryst., mp 213–214 °C, $[\alpha]_D^{20}$ = −8.5° (*c* = 0.2, $CHCl_3$). Source: Mangrove-derived fungus *Guignardia* sp. 4382 from Mangrove *Kandelia candel* (South China Sea). Pharm: Cytotoxic (EAC and P_{388}, RNA synthesis inhibitors as an elastase inhibitor); Cytotoxic (KB, IC_{50} = 90.2 μg/mL, control Cisplatin, IC_{50} = 0.56 μg/mL); phytotoxin (various banana cultivars). Ref: J. Fuska, et al, J. Antibiot., 1986, 39, 1605 | X. K. Xia, et al, Helv. Chim. Acta, 2007, 90, 1925

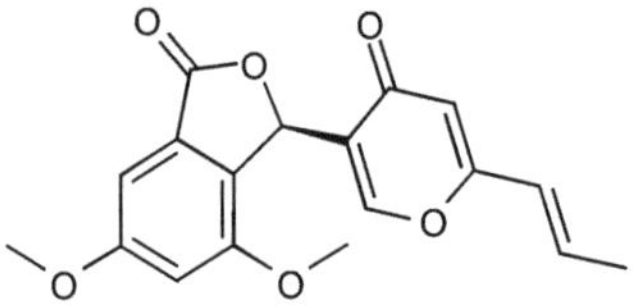

220 Yoshinone A

Type: 4-Pyrones. $C_{21}H_{32}O_5$ Source: Cyanobacterium *Leptolyngbya* sp. Pharm: Anti-obesity (inhibits differentiation of 3T3-L1 cells into adipocytes without accompanying cytotoxicity, suggesting potential as an anti-obesity lead). Ref: T. Inuzuka, et al, Tet. Lett., 2014, 55, 6711

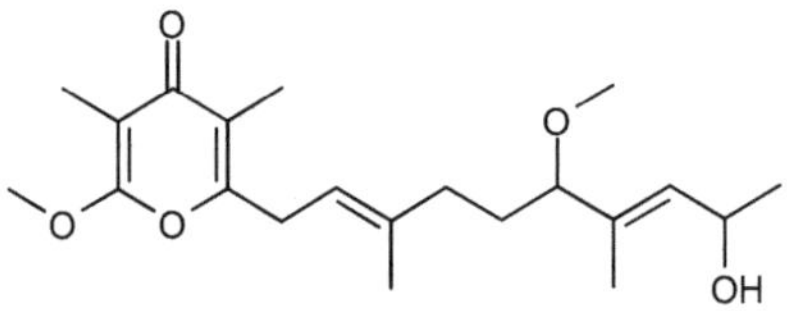

1.8 Spiroketals

221 Antibiotic H 668

Type: Spiroketals. $C_{36}H_{64}O_{12}$ Amorph. solid, $[\alpha]_D^{24}$ = +48.2° (c = 0.17, MeOH). Source: Marine-derived streptomycete *Streptomyces* sp. H668. Pharm: Antimalarial (*Plasmodium falciparum* D6 and W2, IC_{50} = 0.1–0.2 μg/mL). Ref: M. K. Na, et al, Tet. Lett., 2008, 49, 6282

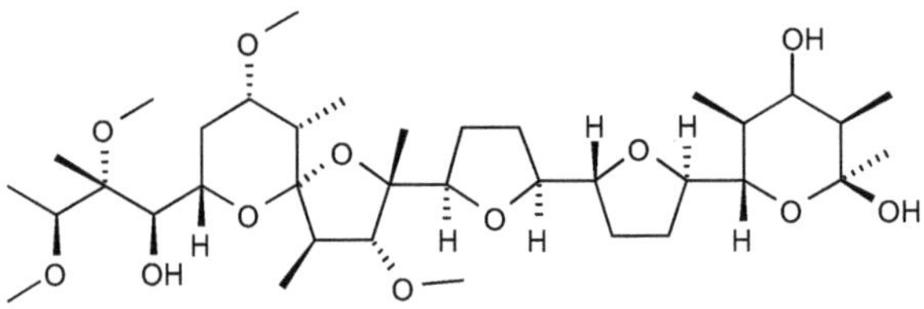

222 Ascochytatin

Type: Spiroketals. $C_{20}H_{14}O_7$ Needles (MeOH), mp 227–229 °C (dec), $[\alpha]_D^{25}$ = −153° (c = 0.2, MeOH). Source: Marine-derived fungus *Ascochyta* sp. NGB4. Pharm: Antibacterial (temperature-sensitive mutant *Bacillus subtilis* CNM2000, MID = 0.3 μg, stronger than wild-type strain 168, suggesting ascochytatin inhibited function of YycG/YycF-TCS in bacterium); antibacterial (gram-positive bacteria, strongly); antifungal (*Candida albicans*, strongly); cytotoxic (two mammalian cancer cell lines, IC_{50} values in lower μmol/L range). Ref: K. Kanoh, et al, J. Antibiot., 2008,

61, 142 | A. M. S. Mayer et al, Comparative Biochemistry and Physiology, Part C 153, 2011, 191 (rev)

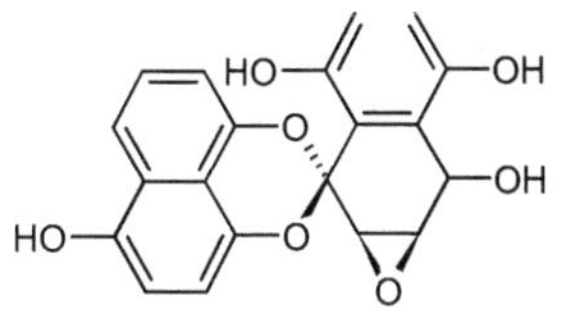

223 Attenol A

Type: Spiroketals. $C_{22}H_{38}O_5$ Source: Bivalve *Pinna attenuata* (China waters). Pharm: Cytotoxic (P_{388}, IC_{50} = 24 µg/mL). Ref: N. Takada, et al, Chem. Lett., 1999, 1025 | K. Suenaga, et al, Org. Lett., 2001, 3, 527

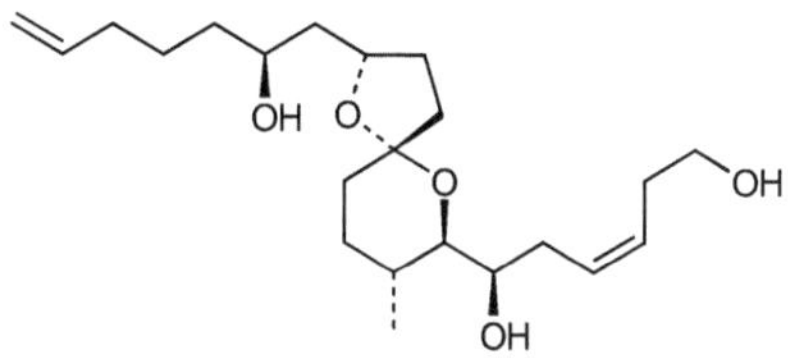

224 Azaspirofuran A

Type: Spiroketals. $C_{22}H_{21}NO_7$ Pale yellow oil, $[\alpha]_D^{20}$ = −19.7° (*c* = 0.295, $CHCl_3$). Source: Marine-derived fungi *Aspergillus sydowi* (sediment, Jiaozhou Bay, China) and *Aspergillus fumigatus* from fish *Mugil cephalus* (Katsuura Bay, Japan waters). Pharm: Cytotoxic (MTT assay, A549, IC_{50} = 10 µmol/L). Ref: K. Mizoue, et al, Eur. Pat. Appl., 1987, EP 216607 | T. Yamada, et al, JOC, 2010, 75, 4146 | H. Ren, et al, Arch. Pharmacal Res., 2010, 33, 499

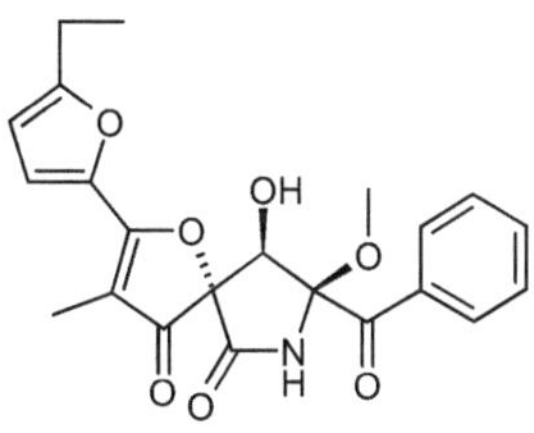

225 Bistramide A

Bistratene A; BST-A Type: Spiroketals. $C_{40}H_{68}N_2O_8$ Amorph. solid, $[\alpha]_D^{20}$ = +10° (*c* = 0.05, CH_2Cl_2). Source: Ascidians *Lissoclinum bistratum* and *Trididemnum cyclops*. Pharm: Cytotoxic (fibroblast cell MRC5CVl and bladder carcinoma cell T-24, IC_{50} = 0.07 µg/mL); cytotoxic (KB, IC_{50} = 0.53 µg/mL; P_{388}, IC_{50} = 0.20 µg/mL; P_{388}/dox, IC_{50} = 0.05 µg/mL; B16, IC_{50} = 0.10 µg/mL; HT29, IC_{50} = 0.32 µg/mL; NSCLC-N6, IC_{50} = 0.03 µg/mL; control 6-Mercaptopurine: KB, IC_{50} = 0.55 µg/mL; P_{388},

IC_{50} = 0.70 μg/mL; P_{388}/dox, IC_{50} = 0.26 μg/mL; B16, IC_{50} = 0.80 μg/mL; HT29, IC_{50} = 0.87 μg/mL; NSCLC-N6, IC_{50} = 0.79 μg/mL); inhibits cell cycle (significant decreases *S* phase with partial blocks in the G phase). Ref: D. Gouiffes, et al, Tetrahedron 1988, 44, 451 | B. M. Degnan, et al, JMC, 1989, 32, 1354 | M. P. Foster, et al, JACS, 1992, 114, 1110 (structure revised) | R. Dunkel, et al, Anal. Chem., 1992, 64, 3150 | J. -F. Biard, et al, JNP, 1994, 57, 1336

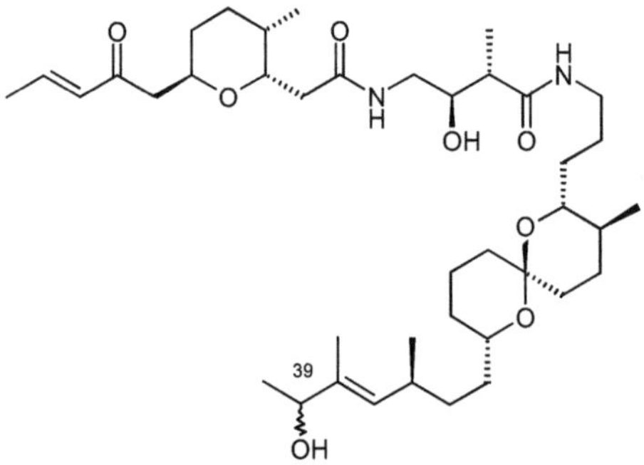

226 Bistramide B

BST-B Type: Spiroketals. $C_{40}H_{70}N_2O_8$ $[\alpha]_D^{20}$ = +10° (*c* = 0.01, CH_2Cl_2). Source: Ascidian *Lissoclinum bistratum* (New Caledonia and Fiji). Pharm: Cytotoxic (KB, IC_{50} = 2.10 μg/mL; P_{388}, IC_{50} = 0.20 μg/mL; P_{388}/Dox, IC_{50} = 1.16 μg/mL; B16, IC_{50} = 1.20 μg/mL; HT29, IC_{50} = 0.71 μg/mL; NSCLC-N6, IC_{50} = 0.32 μg/mL; control 6-Mercaptopurine: KB, IC_{50} = 0.55 μg/mL; P_{388}, IC_{50} = 0.70 μg/mL; P_{388}/Dox, IC_{50} = 0.26 μg/mL; B16, IC_{50} = 0.80 μg/mL; HT29, IC_{50} = 0.87 μg/mL; NSCLC-N6, IC_{50} = 0.79 μg/mL); inhibits cell cycle (significant decreases S phase with partial blocks in the G phase); antiparasitic. Ref: J. -F. Biard, et al, JNP, 1994, 57, 1336

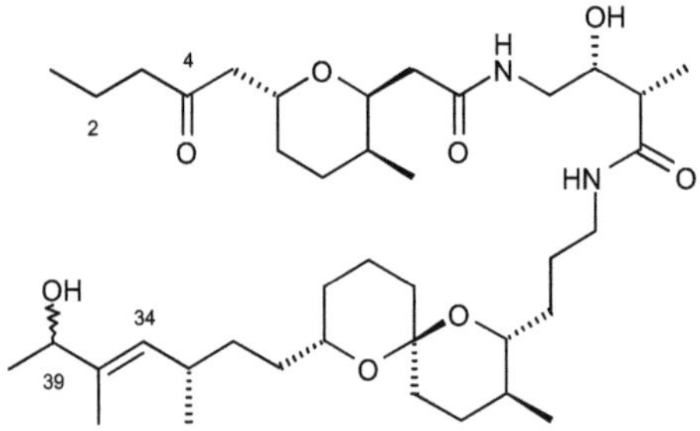

227 Bistramide C

BST-C Type: Spiroketals. $C_{40}H_{66}N_2O_8$ Amorph. solid, $[\alpha]_D^{20}$ = +10° (*c* = 0.05, CH_2Cl_2). Source: Ascidian *Lissoclinum bistratum* (New Caledonia and Fiji). Pharm: Cytotoxic (KB, IC_{50} – 0.65 μg/mL; P_{388}, IC_{50} = 0.02 μg/mL; P_{388}/Dox, IC_{50} = 0.05 μg/mL; B16, IC_{50} = 0.06 μg/mL; HT29, IC_{50} = 0.50 μg/mL; NSCLC-N6, IC_{50} = 0.05 μg/mL; control

6-Mercaptopurine: KB, IC_{50} = 0.55 μg/mL; P_{388}, IC_{50} = 0.70 μg/mL; P_{388}/Dox, IC_{50} = 0.26 μg/mL; B16, IC_{50} = 0.80 μg/mL; HT29, IC_{50} = 0.87 μg/mL; NSCLC-N6, IC_{50} = 0.79 μg/mL); inhibits cell cycle (significant decreases S phase with partial blocks in the G phase); antiparasitic; antiproliferative. Ref: J. -F. Biard, et al, JNP, 1994, 57, 1336 | G. Zuber, et al, Org. Lett., 2005, 7, 5269

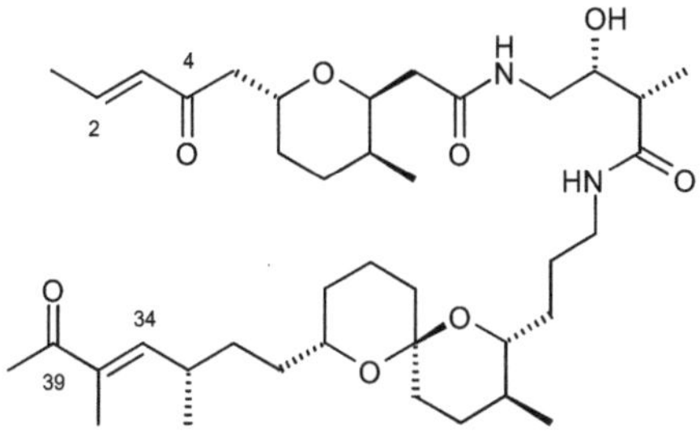

228 Bistramide D

BST-D Type: Spiroketals. $C_{40}H_{77}N_2O_8$ Amorph. solid, $[\alpha]_D^{20}$ = +8° (*c* = 0.04, CH_2Cl_2). Source: Ascidians *Lissoclinum bistratum* (New Caledonia and Fiji) and *Tridemnum cyclops*. Pharm: Cytotoxic (KB, IC_{50} = 10.0 μg/mL; P_{388}, IC_{50} = 0.36 μg/mL; P_{388}/Dox, IC_{50} = 5.82 μg/mL; B16, IC_{50} = 0.10 μg/mL; HT29, IC_{50} = 2.76 μg/mL; NSCLC-N6, IC_{50} = 3.43 μg/mL; control 6-Mercaptopurine: KB, IC_{50} = 0.55 μg/mL; P_{388}, IC_{50} = 0.70 μg/mL; P_{388}/dox, IC_{50} = 0.26 μg/mL; B16, IC_{50} = 0.80 μg/mL; HT29, IC_{50} = 0.87 μg/mL; NSCLC-N6, IC_{50} = 0.79 μg/mL); inhibits cell cycle (significant decreases S phase with partial blocks in the G phase); antiparasitic. Ref: J. -F. Biard, et al, JNP, 1994, 57, 1336 | C. Bauder, et al, Org. Biomol. Chem., 2006, 4, 1860

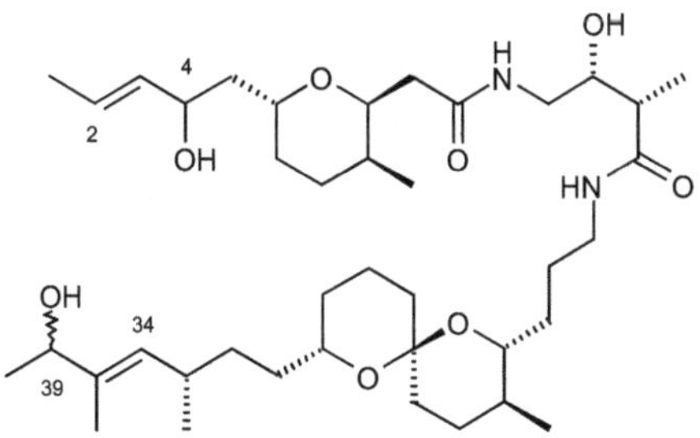

229 Bistramide K

Type: Spiroketals. $C_{40}H_{70}N_2O_8$ Amorph. solid, $[\alpha]_D^{20}$ = +20° (*c* = 0.02, CH_2Cl_2). Source: Ascidian *Lissoclinum bistratum* (New Caledonia and Fiji). Pharm: Cytotoxic (KB, IC_{50} > 10.0 μg/mL; P_{388}, IC_{50} = 0.57 μg/mL; P_{388}/Dox, IC_{50} > 10.0 μg/mL; B16, IC_{50} = 1.90 μg/mL; HT29, IC_{50} = 5.60 μg/mL; NSCLC-N6, IC_{50} = 3.23 μg/mL; control 6-Mercaptopurine: KB, IC_{50} = 0.55 μg/mL; P_{388}, IC_{50} = 0.70 μg/mL; P_{388}/Dox, IC_{50} = 0.26 μg/mL; B16, IC_{50} = 0.80 μg/mL; HT29, IC_{50} = 0.87 μg/mL; NSCLC-N6,

IC_{50} = 0.79 μg/mL); inhibits cell cycle (completely blocks NSCLC-N6 cells in G phase after 48 h of growth); antiparasitic. Ref: J. -F. Biard, et al, JNP, 1994, 57, 1336

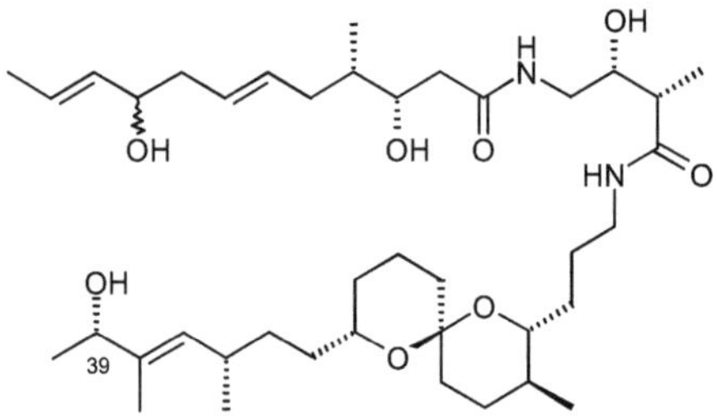

230 Bistratene B

Type: Spiroketals. $C_{42}H_{70}N_2O_9$ Source: Ascidian *Lissoclinum bistratum* Pharm: Cytotoxic (MRC5CVl and T-24, IC_{50} = 0.09 μg/mL). Ref: B. M. Degnan, et al, JMC, 1989, 32, 1354 | M. P. Foster, et al, JACS, 1992, 114, 1110 (struc. revised) | R. Dunkel, et al, Anal. Chem., 1992, 64, 3150

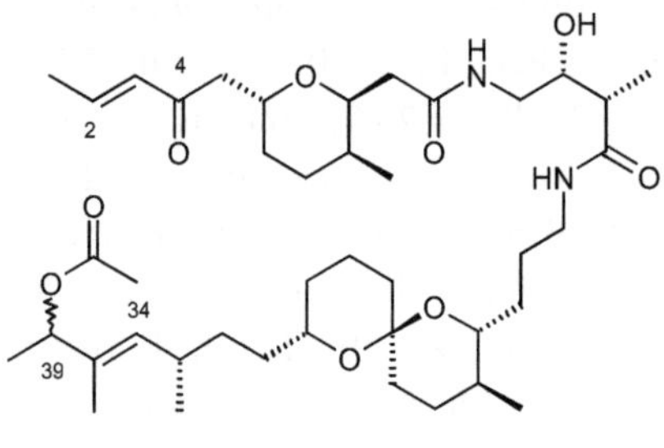

231 Didemnaketal A

Type: Spiroketals. $C_{44}H_{72}O_{14}$ Oil, $[\alpha]_D$ = −11.0° (c = 0.8, $CHCl_3$). Source: Ascidian *Didemnum* sp. (Palau, Oceania). Pharm: HIV-1 protease inhibitor (IC_{50} = 2 μg/mL). Ref: B. C. M. Potts, et al, JACS, 1991, 113, 6321 | C. E. Salomon,et al, Org. Lett., 2002, 4, 1699

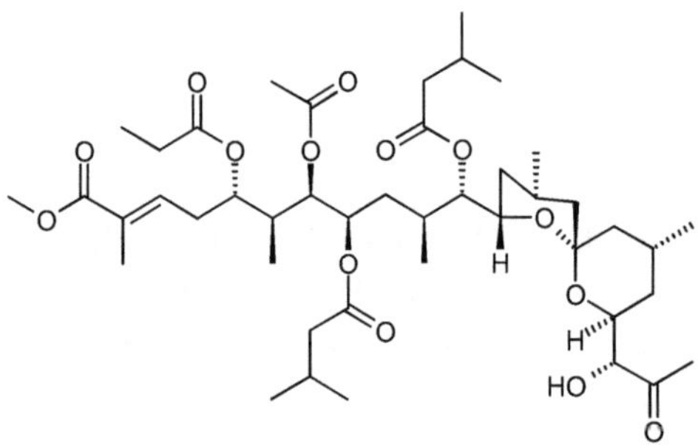

232 Didemnaketal B

Type: Spiroketals. $C_{52}H_{86}O_{15}$ Oil. Source: Ascidian *Didemnum* sp. (Palau, Oceania). Pharm: HIV-1 protease inhibitor (IC_{50} = 10 µg/mL). Ref: B. C. M. Potts, et al, JACS, 1991, 113, 6321 | C. E. Salomon,et al, Org. Lett., 2002, 4, 1699

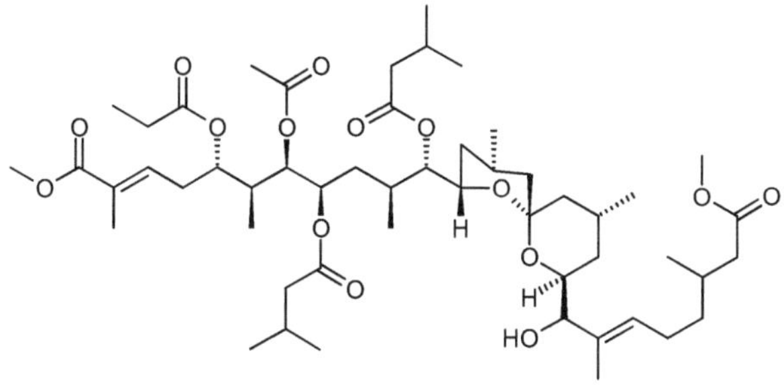

233 Gymnastatin A

Type: Spiroketals. $C_{23}H_{31}Cl_2NO_4$ Powder, mp 74.2–76.0 °C, $[\alpha]_D$ = −3.8° (*c* = 0.73, $CHCl_3$). Source: Marine-derived fungus *Gymnascella dankaliensis* from sponge *Halichondria japonica*. Pharm: Cytotoxic (P_{388}, ED_{50} = 18 ng/mL). Ref: Numata, et al, Tet. Lett., 1997, 38, 5675 | T. Amagata, et al, JCS Perkin I, 1998, 3585 | M. K. Gurjar, et al, Heterocycles, 2000, 53, 143

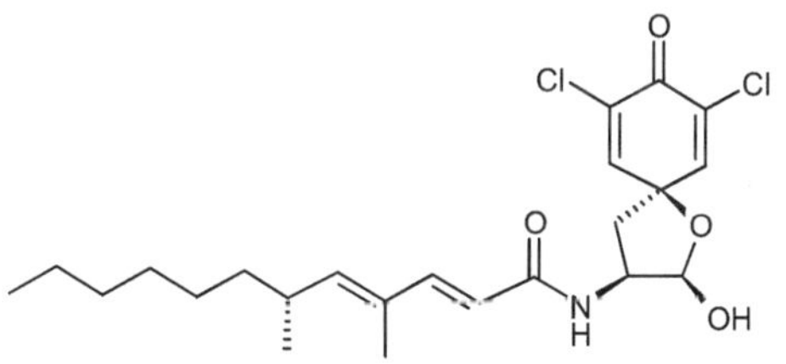

234 Gymnastatin B

Type: Spiroketals. $C_{24}H_{35}Cl_2NO_5$ Powder, mp 73.5–77.5 °C, $[\alpha]_D$ = −122.1° (*c* = 0.3, $CHCl_3$). Source: Marine-derived fungus *Gymnascella dankaliensis* from sponge *Halichondria japonica*. Pharm: Cytotoxic (P_{388}, ED_{50} = 108 ng/mL). Ref: T. Amagata, et al, JCS Perkin I, 1998, 3585

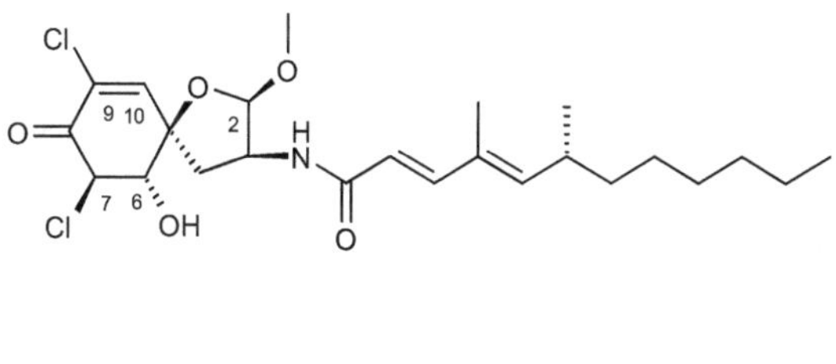

235 Gymnastatin C

Type: Spiroketals. $C_{24}H_{37}Cl_2NO_6$ Powder, mp 104.7–107.5 °C, $[\alpha]_D = -101.2°$ ($c = 0.1$, $CHCl_3$). Source: Marine-derived fungus *Gymnascella dankaliensis* from sponge *Halichondria japonica*. Pharm: Cytotoxic (P_{388}, ED_{50} = 106 ng/mL). Ref: T. Amagata, et al, JCS Perkin I, 1998, 3585

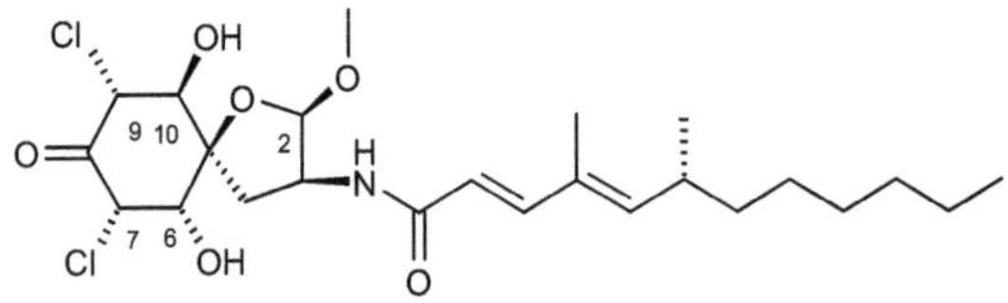

236 Gymnastatin D

Type: Spiroketals. $C_{23}H_{34}ClNO_5$ Amorph. powder, mp 86.4–88.2 °C, $[\alpha]_D = -8.9°$ ($c = 0.45$, $CHCl_3$). Source: Marine-derived fungus *Gymnascella dankaliensis* from sponge *Halichondria japonica*. Pharm: Cytotoxic (P_{388}, ED_{50} = 10.5 μg/mL). Ref: T. Amagata, et al, JCS Perkin I, 1998, 3585

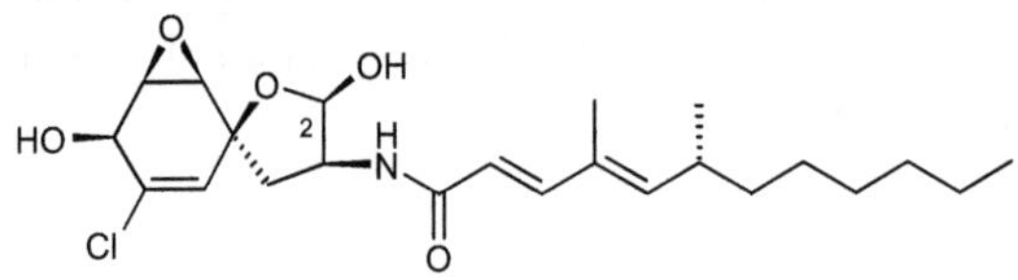

237 Gymnastatin E

Type: Spiroketals. $C_{23}H_{34}ClNO_5$ Amorph. powder, mp 87.3–88 °C, $[\alpha]_D = -8.5°$ ($c = 0.52$, $CHCl_3$). Source: Marine-derived fungus *Gymnascella dankaliensis* from sponge *Halichondria japonica*. Pharm: Cytotoxic (P_{388}, ED_{50} = 10.8 μg/mL). Ref: T. Amagata, et al, JCS Perkin I, 1998, 3585

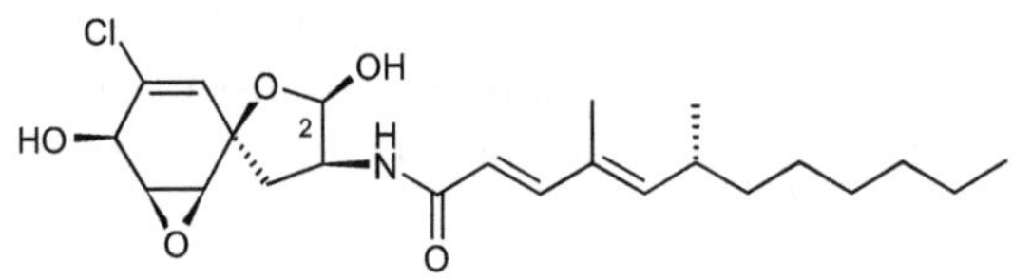

238 Gymnastatin I

Type: Spiroketals. $C_{23}H_{31}Br_2NO_4$ Source: Marine-derived fungus *Gymnascella dankaliensis*. Pharm: Cytotoxic (panel of 39 hmn cancer cell lines); cytotoxic (P_{388}, potent). Ref: T. Amagata, et al, Heterocycles, 2010, 81, 897

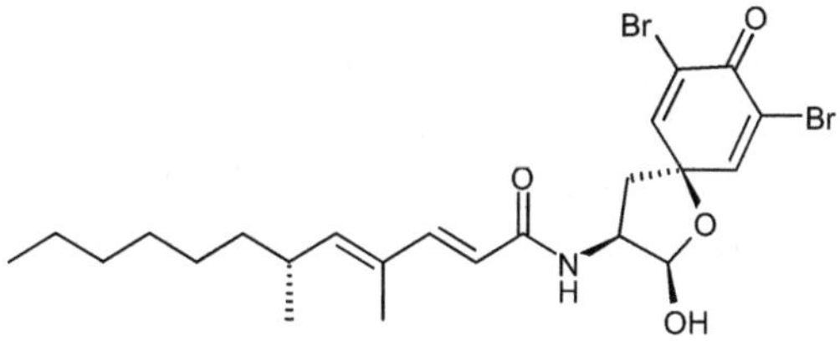

239 Gymnastatin J

Type: Spiroketals. $C_{24}H_{35}Br_2NO_6$ Source: Marine-derived fungus *Gymnascella dankaliensis*. Pharm: Cytotoxic (panel of 39 hmn cancer cell lines); cytotoxic (P_{388}, potent). Ref: T. Amagata, et al, Heterocycles, 2010, 81, 897

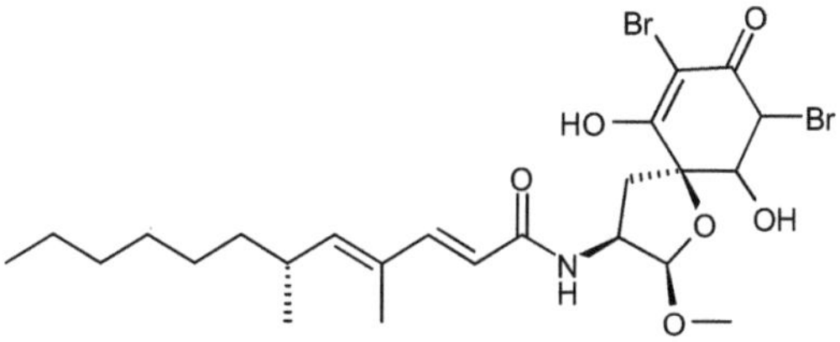

240 Gymnastatin K

Type: Spiroketals. $C_{24}H_{37}Br_2NO_6$ Source: Marine-derived fungus *Gymnascella dankaliensis*. Pharm: Cytotoxic (P_{388}, potent). Ref: T. Amagata, et al, Heterocycles, 2010, 81, 897

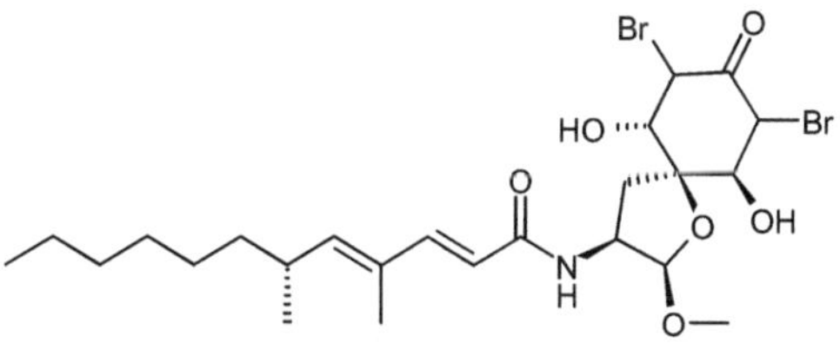

241 Marinisporolide A

Type: Spiroketals. $C_{38}H_{58}O_{10}$ Amorph. yellow powder. Source: Marine-derived actinomycete *Marinispora* sp. CNQ-140 (sediment, culture, La Jolla, California). Pharm: Antifungal (yeast *Candida albicans,* modest). Ref: H. C. Kwon, et al, JOC, 2009, 74, 675

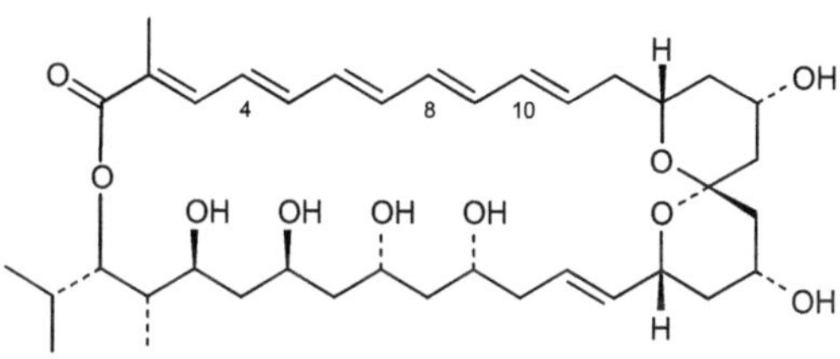

242 30-Methyloscillatoxin D

Type: Spiroketals. $C_{32}H_{44}O_8$ Cryst. (MeOH), mp 166 °C. Source: Cyanobacterium *Lyngbya majuscula* (deepwater); a mixed collection of *Schizothrix calcicola* and *Oscillatoria nigroviridis*. Pharm: Antileukemia; toxin. Ref: M. Entzeroth, et al, JOC, 1985, 50, 1255 | H, -J, Knölker, et al, Tet. Lett., 1995, 36, 5339 | H. Toshima, et al, Tet. Lett., 1995, 36, 3373

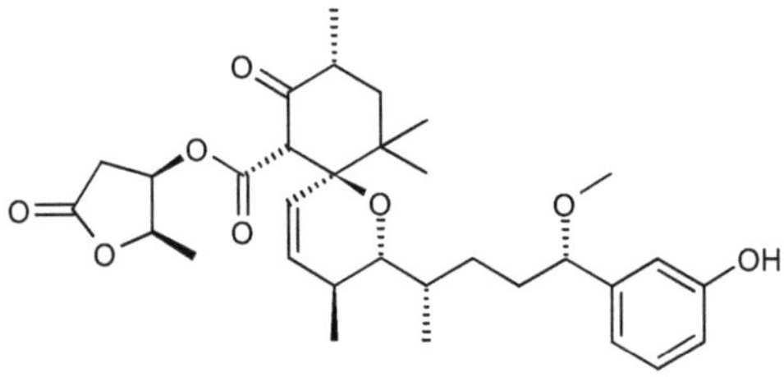

243 Oscillatoxin D

Type: Spiroketals. $C_{31}H_{42}O_8$ Gum. Source: Cyanobacteria *Schizothrix calcicola* and *Oscillatoria nigroviridis* (mixture). Pharm: Antileukemia; toxin. Ref: M. Entzeroth, et al, JOC, 1985, 50, 1255 | H, -J, Knölker, et al, Tet. Lett., 1995, 36, 5339 | H. Toshima, et al, Tet. Lett., 1995, 36, 3373

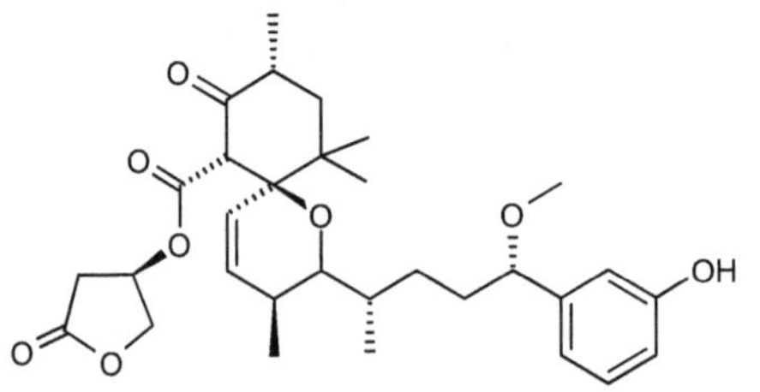

244 39-Oxobistramide K

Type: Spiroketals. $C_{40}H_{68}N_2O_8$ Amorph. powder, $[\alpha]_D^{25} = -72°$ (c = 0.05, MeOH). Source: Ascidian *Trididemnum cyclops* (Madagascar). Pharm: Cytotoxic (A2780, IC_{50} = 0.34 μmol/L). Ref: B. T. Murphy, et al, JNP, 2009, 72, 1338

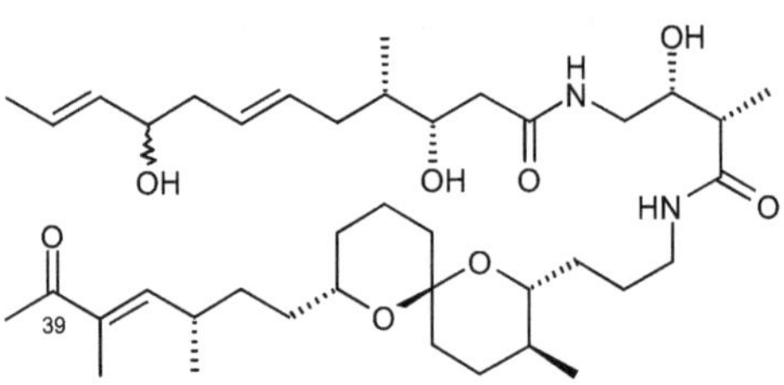

245 Paecilospirone

Type: Spiroketals. $C_{32}H_{44}O_5$ $[\alpha]_D^{25}$ = +202.5° (*c* = 0.37, MeOH). Source: Marine-derived fungus *Paecilomyces* sp. (from a coral reef, Yap, Federated States of Micronesia). Pharm: Microtubule assemblage inhibitor. Ref: M. Namikoshi, et al, Chem. Lett., 2000, 308 | M. Namikoshi, et al, CPB, 2000, 48, 1452

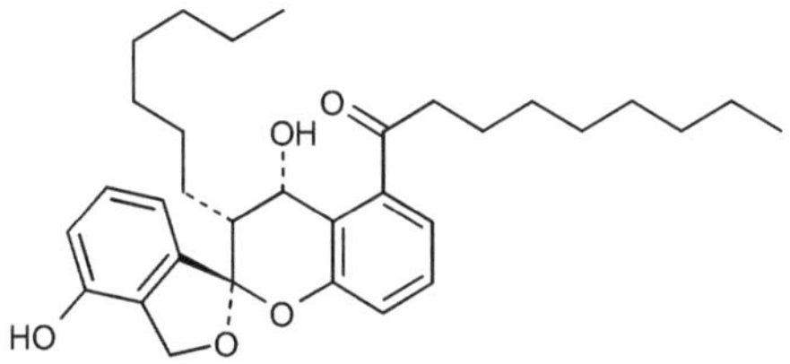

246 Penisporolide A

Type: Spiroketals. $C_{18}H_{30}O_5$ Pale yellow oil, $[\alpha]_D^{20}$ = +37° (*c* = 0.5, MeOH). Source: Mangrove-derived fungus *Penicillium* sp. HKI GT20022605 from mangrove *Kandelia candel* (China waters). Pharm: Xanthine oxidase inhibitor (mild); 3*α*-Hydroxysteroid dehydrogenase inhibitor (mild). Ref: X. Li, et al, J. Antibiot., 2007, 60, 191

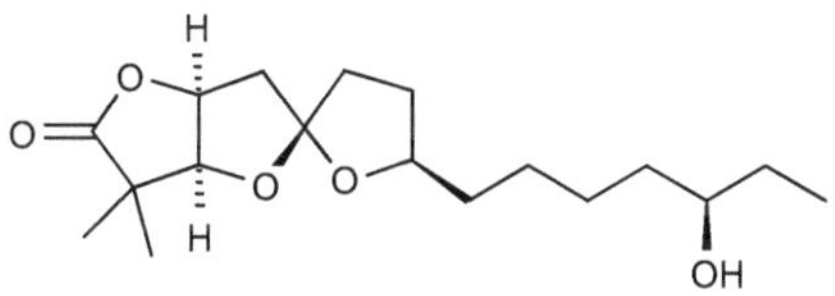

247 Penisporolide B

Type: Spiroketals. $C_{17}H_{26}O_5$ Pale yellow oil, $[\alpha]_D^{20}$ = +81° (*c* = 1, MeOH). Source: Mangrove-derived fungus *Penicillium* sp. HKI GT20022605 from mangrove *Kandelia candel* (China waters). Pharm: Xanthine oxidase inhibitor (mild); 3*α*-Hydroxysteroid dehydrogenase inhibitor (mild). Ref: X. Li, et al, J. Antibiot., 2007, 60, 191

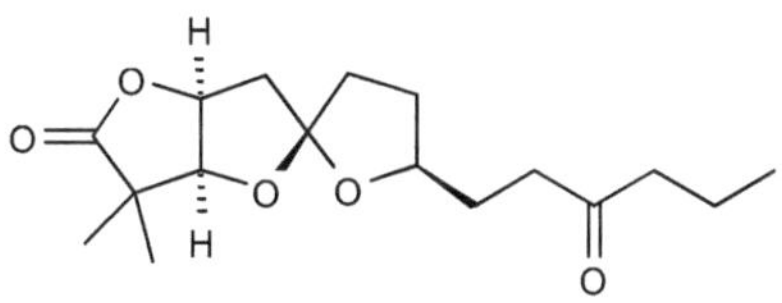

248 Prunolide A

Type: Spiroketals. $C_{34}H_{14}Br_8O_9$ Yellow plates (DMSO), mp 212–214 °C, $[\alpha]_D^{25}$ = 0° (*c* = 0.5, MeOH). Source: Ascidian *Synoicum prunum* (Australia). Pharm: Cytotoxic (inhibits growth of HeLa at 25 µmol/L, weak). Ref: A. R. Carroll, et al, JOC, 1999, 64, 2680

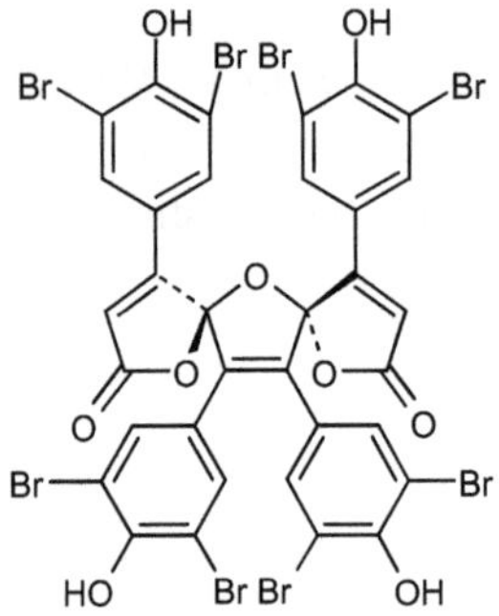

249 Prunolide B

Type: Spiroketals. $C_{34}H_{16}Br_6O_9$ Yellow gum, $[\alpha]_D^{25}$ = 0° (*c* = 0.1, MeOH). Source: Ascidian *Synoicum prunum* (Australia). Pharm: Cytotoxic (weak). Ref: A. R. Carroll, et al, JOC, 1999, 64, 2680

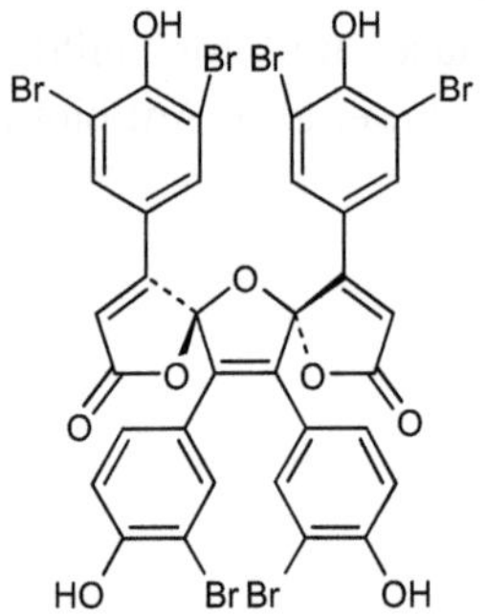

250 Prunolide C

Type: Spiroketals. $C_{34}H_{22}O_9$ Yellow gum, $[\alpha]_D^{25}$ = 0° (*c* = 0.4, MeOH). Source: Ascidian *Synoicum prunum* (Australia). Pharm: Cytotoxic (inhibits growth of HeLa at 15 μmol/L, weak). Ref: A. R. Carroll, et al, JOC, 1999, 64, 2680

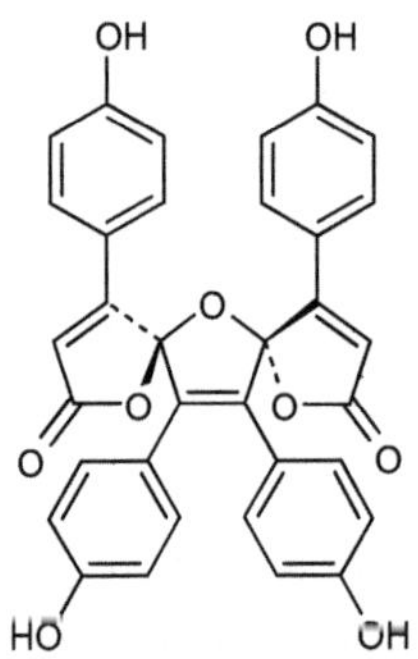

251 Spirastrellolide A

Type: Spiroketals. $C_{52}H_{81}ClO_{17}$ Oil (Me ester), $[\alpha]_D^{25}$ = +27° (c = 0.16, CH_2Cl_2) (Me ester). Source: Sponge *Spirastrella coccinea*, sponge *Epipolasis* sp. (Nagannu I., Japan, East China Sea). Pharm: Antimitotic; protein phosphatase inhibitor (selective); cytotoxic (free acids are more active than their methyl esters). Ref: D. E. Williams, et al, JACS, 2003, 125, 5296 | D. E. Williams, et al, Org. Lett., 2004, 6, 2607 | M. Suzuki, et al, JNP, 2012, 75, 1192

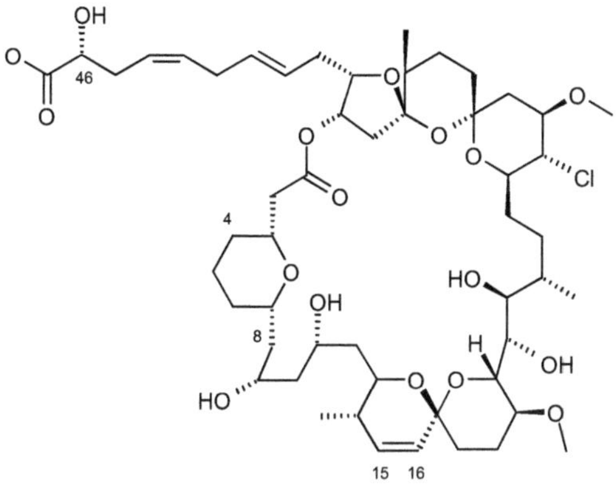

252 Spirastrellolide B

Spirostrellolide E Type: Spiroketals. $C_{52}H_{84}O_{17}$ Oil (Me ester), $[\alpha]_D^{25}$ = +44.7° (c = 0.5, MeOH) (Me ester). Source: Sponge *Spirastrella coccinea*, sponge *Epipolasis* sp. (Nagannu I., Japan, East China Sea). Pharm: Cytotoxic (free acids are more active than their methyl esters). Ref: K. Warabi, et al, JACS, 2007, 129, 508 | M. Suzuki, et al, JNP, 2012, 75, 1192

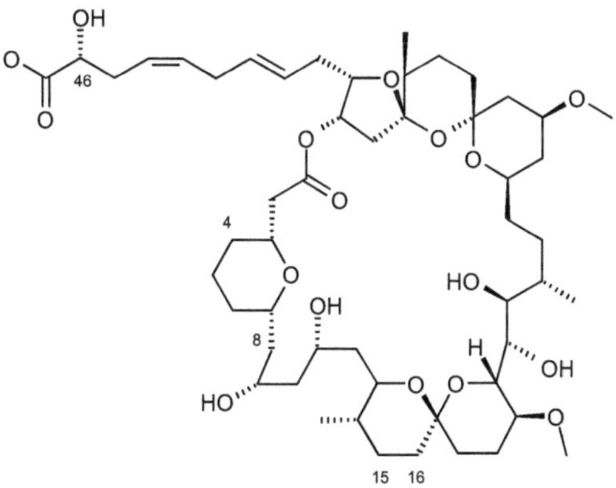

253 Spirastrellolide D

Type: Spiroketals. $C_{52}H_{80}Cl_2O_{17}$ Oil (Me ester), $[\alpha]_D^{25}$ = +45.9° (*c* = 0.5, MeOH) (Me ester). Source: Sponge *Spirastrella coccinea*. Pharm: Inhibits premature mitosis (IC_{50} = 0.4–0.7 μmol/L, MMOA: protein phosphatase 2A inhibition). Ref: D. E. Williams, et al, JOC, 2007, 72, 9842

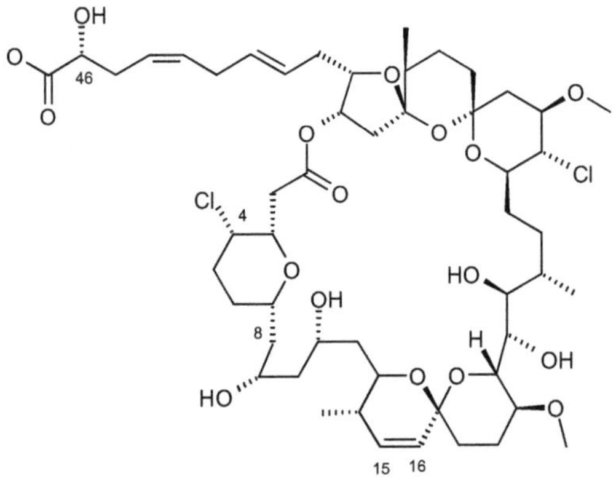

254 Spirastrellolide E

Type: Spiroketals. $C_{52}H_{82}O_{17}$ Oil (Me ester), $[\alpha]_D^{25}$ = +47.4° (*c* = 0.5, MeOH) (Me ester). Source: Sponge *Spirastrella coccinea*. Pharm: Inhibits premature mitosis (IC_{50} = 0.4–0.7 μmol/L, MMOA: protein phosphatase 2A inhibition). Ref: D. E. Williams, et al, JOC, 2007, 72, 9842

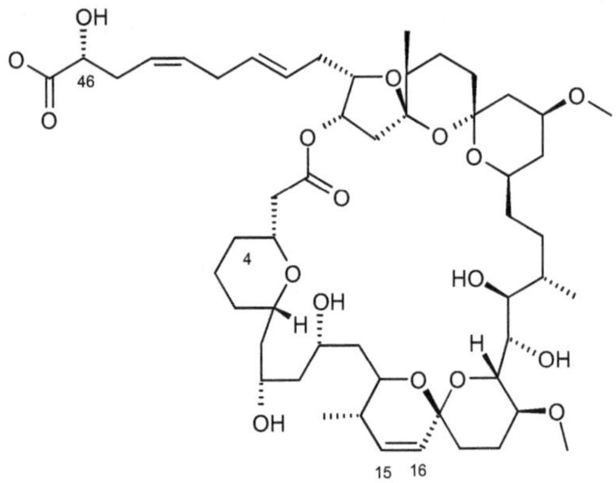

255 Spiroxin A

Type: Spiroketals. $C_{20}H_9ClO_8$ $[\alpha]_D^{25}$ = −644° (*c* = 0.18, MeOH). Source: An unidentified marine-derived fungus LL-37H248 from an unidentified soft coral. Pharm: Cytotoxic (25 diverse cell lines, IC_{50} = 0.09 μg/mL); antineoplastic (*in vivo* ovarian carcinoma in nude mice, after 21 days, 59% growth inhibition).

Ref: L. A. McDonald, et al, Tet. Lett., 1999, 40, 2489 | T. Wang, et al, Can. J. Chem., 2001, 79, 1786 | K. Krohn, Prog. Chem. Org. Nat. Prod., 2003, 85, 1-49 (rev)

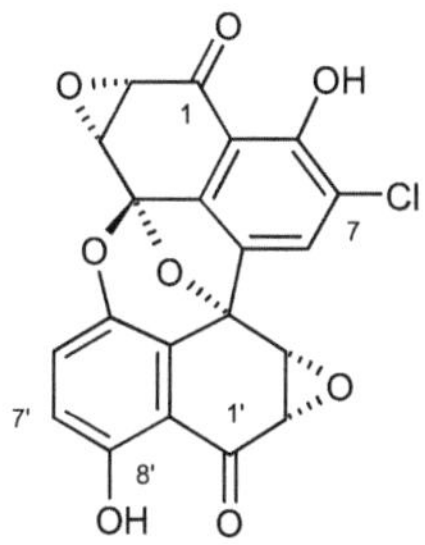

256 Spiroxin B

Antibiotic F 12517 Type: Spiroketals. $C_{20}H_8Cl_2O_8$ $[\alpha]_D^{25}$ = −475° (*c* = 0.21, MeOH). Source: An unidentified marine-derived fungus LL-37H248 from an unidentified soft coral (orange, Vancouver Island). Pharm: DNA cleaving agent (antineoplastic, antibiotics); cytotoxic (mus xenograft model, hmn ovarian carcinoma). Ref: L. A. McDonald, et al, Tet. Lett., 1999, 40. 2489 | K. Krohn, Prog. Chem. Org. Nat. Prod., 2003, 85, 1-49 (rev)

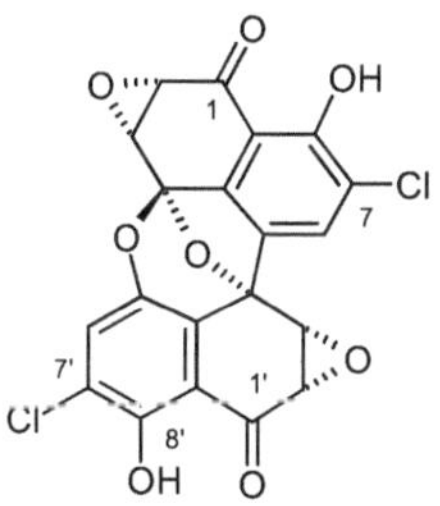

257 Spiroxin C

Type: Spiroketals. $C_{20}H_{10}O_7$ $[\alpha]_D^{25}$ = −706° (*c* = 0.26, MeOH). Source: An unidentified marine-derived fungus LL-37H248 from an unidentified soft coral (orange, Vancouver Island). Pharm: DNA cleaving agent (antineoplastic, antibiotics); cytotoxic (mus xenograft model, hmn ovarian carcinoma). Ref: L. A. McDonald, et al, Tet. Lett., 1999, 40, 2489 | K. Krohn, Prog. Chem. Org. Nat. Prod., 2003, 85, 1-49 (rev)

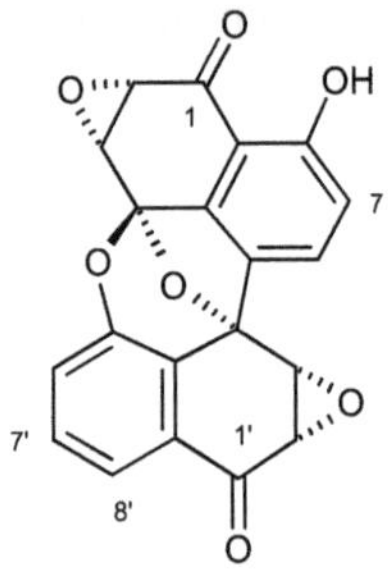

258 Spiroxin D

Type: Spiroketals. $C_{20}H_{12}O_7$ Source: An unidentified marine-derived fungus LL-37H248 from an unidentified soft coral (orange, Vancouver Island). Pharm: DNA cleaving agent (antineoplastic, antibiotics); cytotoxic (mus xenograft model, hmn ovarian carcinoma). Ref: L. A. McDonald, et al, Tet. Lett., 1999, 40, 2489 | K. Krohn, Prog. Chem. Org. Nat. Prod., 2003, 85, 1-49 (rev)

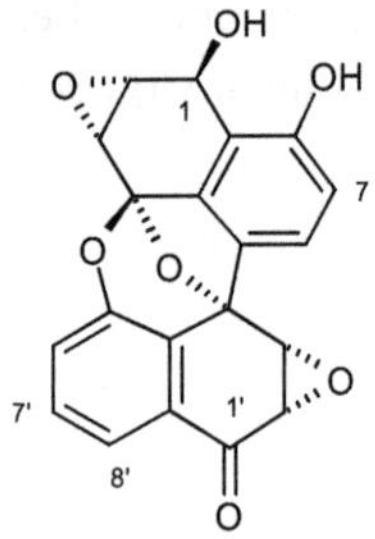

259 Spiroxin E

Type: Spiroketals. $C_{20}H_{10}Cl_2O_8$ Source: An unidentified marine-derived fungus LL-37H248 from an unidentified soft coral (orange, Vancouver Island). Pharm: DNA cleaving agent (antineoplastic, antibiotics); cytotoxic (mus xenograft model, hmn ovarian carcinoma). Ref: L. A. McDonald, et al, Tet. Lett., 1999, 40, 2489 | K. Krohn, Prog. Chem. Org. Nat. Prod., 2003, 85, 1-49 (rev)

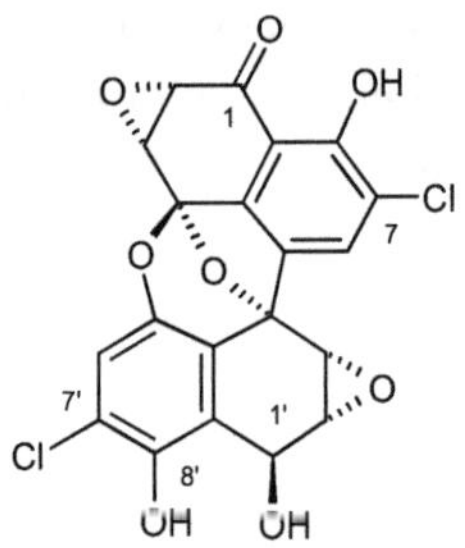

2 Aromatic Metabolites

2.1 Simple Benzene Derivatives

260 2,4-Dibromo-6-chlorophenol

Type: Phenol derivatives. $C_6H_3Br_2ClO$ mp 78–79 °C. Source: Marine-derived bacterium *Pseudoalteromonas luteoviolacea* from brown alga *Padina australis* (surface, Hawaii). Pharm: Antibacterial (MRSA and *Burkholderia cepacia*). Ref: Z. Jiang, et al, Nat. Prod. Lett., 2000, 14, 435

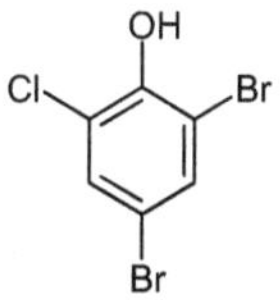

261 2,6-Dibromophenol

Type: Phenol derivatives. $C_6H_4Br_2O$ Needles (H_2O), mp 56–57 °C, bp_{21mmHg} 162 °C, p*K*a = 6.67 (25 °C), sublime. Source: Hemichordate acorn worm *Balanoglossus biminiensis*, phoronid tubeworm *Phoronopsis viridis*, occurs widespread in marine algae, fishes, molluscs and crustaceans, such as green alga *Ulva lactuca*. Pharm: Odour; flavour component (marine fish, molluscs and crustaceans). Ref: R. B. Ashworth, et al, Science, 1967, 155, 1588 | Y. M. Sheikh, et al, Experientia, 1975, 31, 265 | J. Buckingham (executive editor), Dictionary of Natural Products, Vol 2, pp1416, Chapman & Hall, 1994 | F. B. Whitfield, et al, J. Agric. Food Chem., 1997, 45, 4398; 1999, 47, 2367; 4756

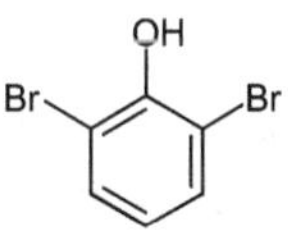

262 2,4-Dibromophenol

Type: Phenol derivatives. $C_6H_4Br_2O$ mp 40 °C, bp_{11mmHg} 154 °C. Source: Hemichordate acorn worms *Balanoglossus cornosus* and *Ptychodera* sp., marine algae, e.g. *Grateloupia elliptica*, and occurs in molluscs and crustaceans. Pharm: LD_{50} (mus, orl) = 282 mg/kg. Ref: T. Higa, et al, Comp. Biochem. Physiol., B, 1980, 65, 525 | F. B. Whitfield, et al, J. Agric. Food Chem., 1997, 45, 4398; 1999, 47, 2367

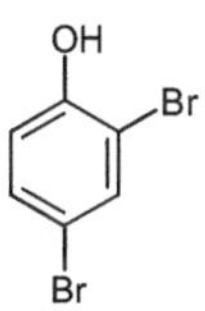

https://doi.org/10.1515/9783110655810-002

263 2,4-Dibromophenol

Type: Phenol derivatives. $C_6H_4Br_2O$ Source: Red alga *Grateloupia elliptica*. Pharm: Hypoglycemic (α-glucosidase inhibitor) (*in vitro* assay, *Bacillus stearothermophilus*, IC_{50} = 230.3 μmol/L; *Saccharomyces cerevisiae*, IC_{50} = 110.4 μmol/L; rat intestinal maltase, IC_{50} = 4.8 mmol/L; rat instestinal sucrase, IC_{50} = 3.6 mmol/L). Ref: K. Y. Kim, et al, Phytochemistry, 2008, 69, 2820

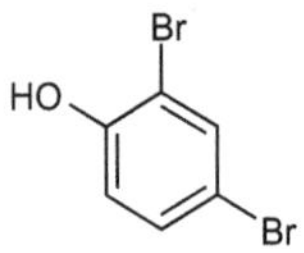

264 4-(1,1-Dimethyl-2-propenyl)-2-(3-methyl-2-butenyl)phenol

Type: Phenol derivatives. $C_{16}H_{22}O$ Oil. Source: Brown algae *Sporochnus comosus* (Shaw I., Queensland, Australia), *Perithalia caudata, Perithalia capillaris* and *Sporochnus pedunculatus*. Pharm: Cytotoxic (SF268, GI_{50} = 39 μmol/L; MCF7, GI_{50} = 27 μmol/L; H460, GI_{50} = 37 μmol/L; HT29, GI_{50} = 57 μmol/L; CHO-K1, GI_{50} = 17 μmol/L); antibacterial; antifungal. Ref: A. Blackman,et al, Aust. J. Chem., 1979, 32,2783 | L. S. Gunasekera, et al, J. Int. J. Pharmacogn., 1995, 33, 253 | C.E.Sansom, et al, JNP, 2007, 70, 2042 | S. P. B. Ovenden, et al, JNP, 2011, 74, 739

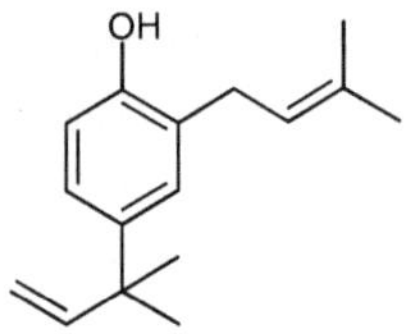

265 2,4-Diprenylphenol

Type: Phenol derivatives. $C_{16}H_{22}O$ Oil. Source: Brown alga *Encyothalia cliftonii* (Australia). Pharm: Antifeedant. Ref: V. Roussis, et al, Phytochemistry, 1993, 34, 107

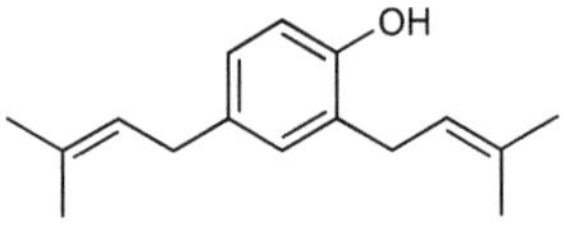

266 2,4,6-Tribromophenol

Bromol; Velcide Type: Phenol derivatives. $C_6H_3Br_3O$ Needles (EtOH), prisms (C_6H_6), mp 87–89 °C, subl. 95–96 °C Source: Phoronid tubeworm *Phoronopsis viridis*, hemichordate acorn worm *Ptychodera flava laysanica*, green alga *Ulva lactate*, red alga *Grateloupia elliptica*. Pharm: Hypoglycemic (α-glucosidase inhibitor) (*in vitro* assay, *Bacillus stearothermophilus*, IC_{50} = 103.3 μmol/L, *Saccharomyces cerevisiae*, IC_{50} = 60.3 μmol/L, rat intestinal maltase, IC_{50} = 5.0 mmol/L, rat instestinal sucrase,

IC_{50} = 4.2 mmol/L) (Kim, 2008); antiseptic (strong); irritant; LD_{50} (rat, orl) = 2000 mg/kg. Ref: Y. M. Sheikh, et al, Experientia, 1975, 31, 265 | C. Flodin, et al, Phytochemistry, 1999, 51, 249; 2000, 53, 77 | K. Y. Kim, et al, Phytochemistry, 2008, 69, 2820

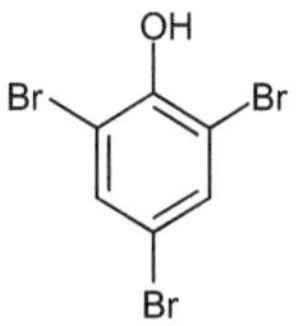

267 Citrinin H_2

Type: *m*-Benzenediol derivatives. $C_{12}H_{16}O_4$ Needles ($CHCl_3$), mp 139.5–140 °C, mp 131–133 °C. Source: Marine-derived fungus *Penicillium* sp. MFA446 from green alga *Ulva pertusa* (Korea waters). Pharm: Antioxidant (DPPH scavenger, moderate). Ref: M. Hirota, et al, Biosci. Biotechnol. Biochem., 2002, 66, 206 | D. Zhang, et al, J. Microbiol. Biotechnol., 2007, 17, 865

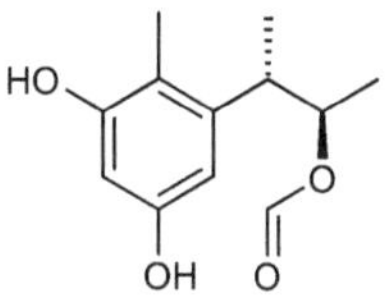

268 5,6-Dibromocatechol

4,5-Dibromo-1,3-benzenediol Type: *m*-Benzenediol derivatives. $C_6H_4Br_2O_2$ Source: Sponge *Dysidea* sp. (Indo-Pacific). Pharm: Inosine monophosphate dehydrogenase (IMPDH) inhibitor; guanosine monophosphate synthetase inhibitor; 15-lipoxygenase inhibitor. Ref: X. Fu, et al, JNP, 1995, 58, 1384

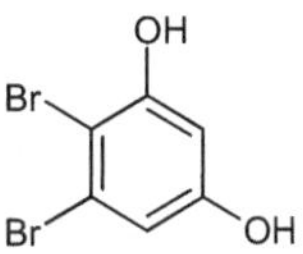

269 2,4-Dihydroxy-3,5,6-trimethylbenzaldehyde

Type: *m*-Benzenediol derivatives. $C_{10}H_{12}O_3$ Source: Deep-sea fungus *Aspergillus sydowi* YH11-2 (Guam, depth of 1000 m, E144°43′ N13°26′). Pharm: Cytotoxic (P_{388}, IC_{50} = 0.59 μmol/L, control CDDP, IC_{50} = 0.039 μmol/L). Ref: L. Tian, et al, Arch. Pharm. Res,. 2007, 30, 1051

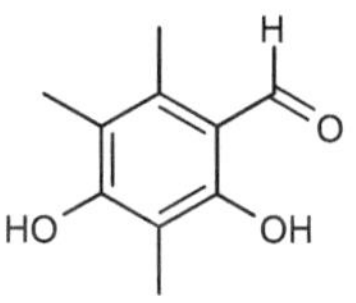

270 Phenol A acid

Type: *m*-Benzenediol derivatives. $C_{12}H_{16}O_5$ Source: Deep-sea fungus *Aspergillus* sp. SCSIOW3. Pharm: Anti-Aβ peptide aggregation inhibition (Aβ42 assembling activity, 100 μmol/L, 40.3%–72.3%). Ref: H. Liu, et al, Chin. J. Mar. Drugs, 2014, 33, 71 (in Chinese)

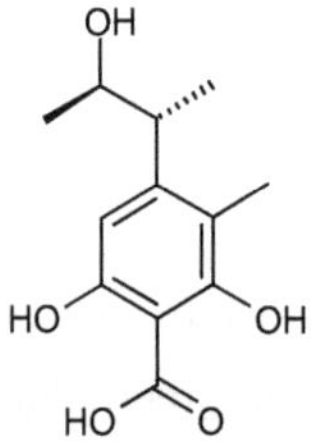

271 Acremonin A

7-Isopropenylbicyclo[4.2.0]octa-1,3,5-triene-2,5-diol Type: *p*-Benzenediol derivatives. $C_{11}H_{12}O_2$ Yellow-brown viscous oil, $[\alpha]_D^{23}$ = +93° (*c* = 1.4, Me_2CO). Source: Marine-derived fungus *Acremonium* sp. Pharm: Antioxidant (thiobarbituric acid reactive substance (TBARS) assay, 25.0 μg/mL, InRt = 85.5%). Ref: A. Abdel-Lateff, et al, JNP, 2002, 65, 1605 | M. Saleem, et al, NPR, 2007, 24, 1142 (rev)

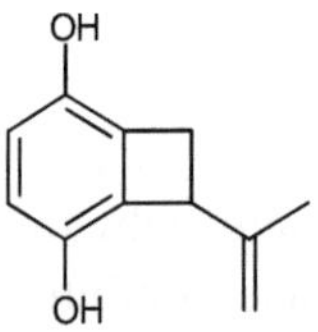

272 Acremonin A 5-β-glucopyranoside

7-Isopropenylbicyclo[4.2.0]octa-1,3,5-triene-2,5-diol-5-β-glucopyranoside Type: *p*-Benzenediol derivatives. $C_{17}H_{22}O_7$ Amorph. powder, $[\alpha]_D^{23}$ = +4.3° (*c* = 2, Me_2CO). Source: Marine-derived fungus *Acremonium* sp. Pharm: Antioxidant (thiobarbituric acid reactive substance (TBARS) assay, 25.0 μg/mL, InRt = 85.5%). Ref: A. Abdel-Lateff, et al, JNP, 2002, 65, 1605 | M. Saleem, et al, NPR, 2007, 24, 1142 (rev)

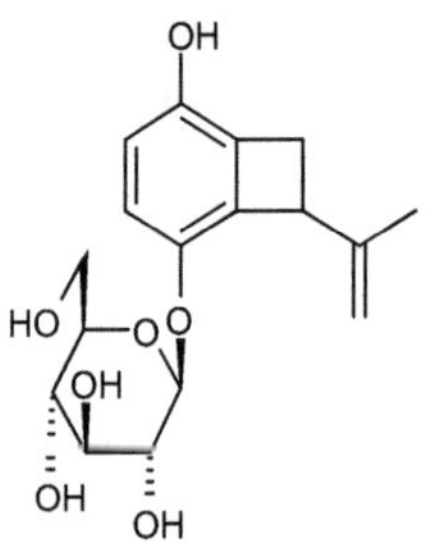

273 Comosone A

Type: *p*-Benzenediol derivatives. $C_{16}H_{24}O_2$ Source: Brown alga *Sporochnus comosus* (Shaw I., Queensland, Australia). Pharm: Cytotoxic (SF268, GI_{50} = 13 μmol/L; MCF7, GI_{50} = 14 μmol/L; H460, GI_{50} = 19 μmol/L; HT29, GI_{50} = 19 μmol/L; CHO-K1, GI_{50} = 17 μmol/L). Ref: S. P. B. Ovenden, et al, JNP, 2011, 74, 739

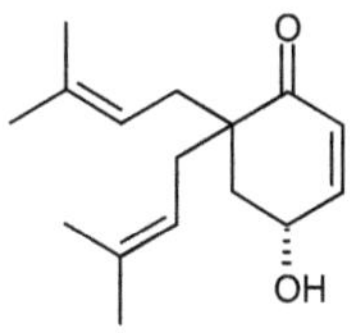

274 Comosusol A

Type: *p*-Benzenediol derivatives. $C_{16}H_{22}O_3$ Source: Brown alga *Sporochnus comosus* (Shaw I., Queensland, Australia). Pharm: Cytotoxic (SF268, GI_{50} = 55 μmol/L; MCF7, GI_{50} = 52 μmol/L; H460, GI_{50} = 55 μmol/L; HT29, GI_{50} = 53 μmol/L; CHO-K1, GI_{50} = 57 μmol/L). Ref: S. P. B. Ovenden, et al, JNP, 2011, 74, 739

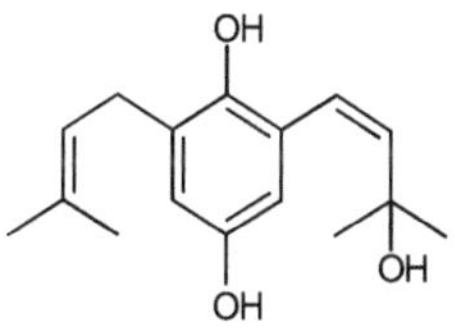

275 Comosusol B

Type: *p*-Benzenediol derivatives. $C_{16}H_{22}O_3$ Source: Brown alga *Sporochnus comosus* (Shaw I., Queensland, Australia). Pharm: Cytotoxic (SF268, GI_{50} = 5 μmol/L; MCF7, GI_{50} = 6 μmol/L; H460, GI_{50} = 6 μmol/L; HT29, GI_{50} = 6 μmol/L; CHO-K1, GI_{50} = 6 μmol/L). Ref: S. P. B. Ovenden, et al, JNP, 2011, 74, 739

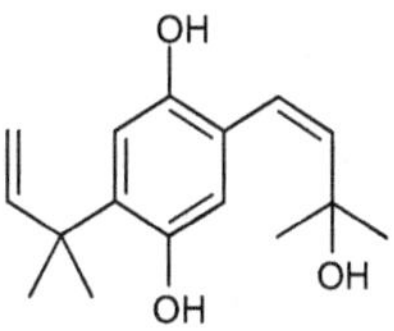

276 Comosusol C

Type: *p*-Benzenediol derivatives. $C_{16}H_{22}O_2$ Source: Brown alga *Sporochnus comosus* (Shaw I., Queensland, Australia). Pharm: Cytotoxic (SF268, GI_{50} = 35 μmol/L; MCF7, GI_{50} = 25 μmol/L; H460, GI_{50} = 29 μmol/L; HT29, GI_{50} = 43 μmol/L; CHO-K1, GI_{50} = 27 μmol/L). Ref: S. P. B. Ovenden, et al, JNP, 2011, 74, 739

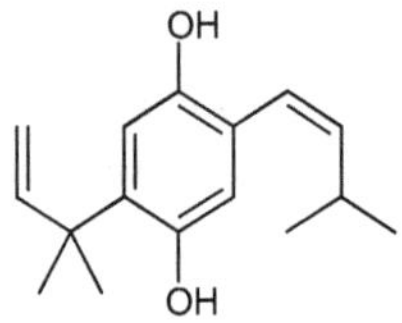

277 Comosusol D

Type: *p*-Benzenediol derivatives. $C_{16}H_{24}O_5$ Source: Brown alga *Sporochnus comosus* (Shaw I., Queensland, Australia). Pharm: Cytotoxic (SF268, GI_{50} = 59 μmol/L; MCF7, GI_{50} = 46 μmol/L; H460, GI_{50} = 54 μmol/L; HT29, GI_{50} = 51 μmol/L; CHO-K1, GI_{50} = 63 μmol/L). Ref: S. P. B. Ovenden, et al, JNP, 2011, 74, 739

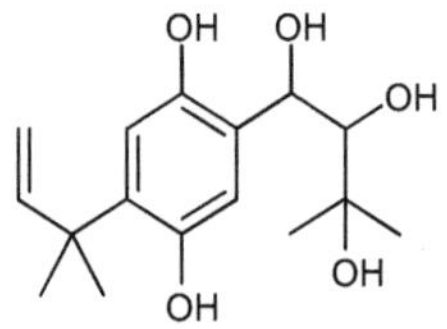

278 Geroquinol

Type: *p*-Benzenediol derivatives. $C_{16}H_{22}O_2$ Oil. Source: Ascidians *Aplidium* sp. (Spain) and *Aplidium savignyi*. Pharm: Cytotoxic (P_{388}, IC_{50} = 0.2 μg/mL; A549, IC_{50} = 1 μg/mL; HT29, IC_{50} = 1 μg/mL; MEL28, IC_{50} = 1 μg/mL); radioprotectant; antioxidant; contact allergen. Ref: A. Rueda, et al, Nat. Prod. Lett., 1998, 11, 127 | M. Aknin, et al, J. Agric. Food Chem., 1999, 47, 4175

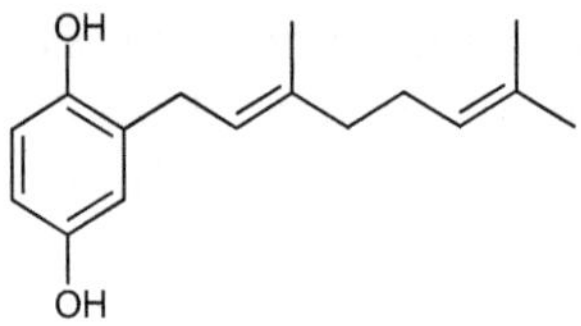

279 2-(7-Hydroxy-3,7-dimethyl-2-octenyl)-1,4-benzenediol

Type: *p*-Benzenediol derivatives. $C_{16}H_{24}O_3$ Oil. Source: Ascidian *Aplidium* sp. (Spain). Pharm: Cytotoxic (P_{388}, IC_{50} = 1.2 μg/mL; A549, IC_{50} = 5 μg/mL; HT29, IC_{50} = 2 μg/mL; MEL28, IC_{50} = 2 μg/mL). Ref: A. Rueda, et al, Nat. Prod. Lett., 1998, 11, 127

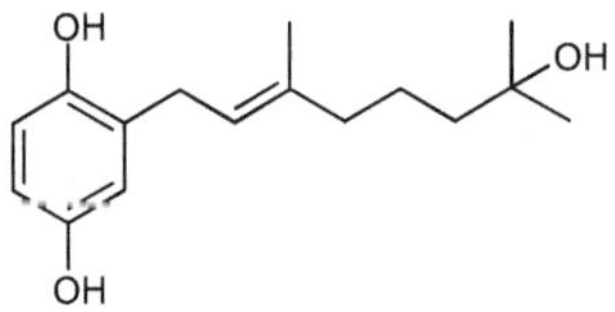

280 2-(3-Hydroxy-3,7-dimethyl-6-octenyl)-1,4-benzenediol

Type: *p*-Benzenediol derivatives. $C_{16}H_{24}O_3$ Oil. Source: Protochordate *Amaroucium multiplicatum* (marine-derived) from ascidian *Aplidium savignyi*. Pharm: Cytotoxic (significant); antioxidant. Ref: A. Sato, et al, JNP, 1989, 52, 975 | M. Aknin, et al, J. Agric. Food Chem., 1999, 47, 4175

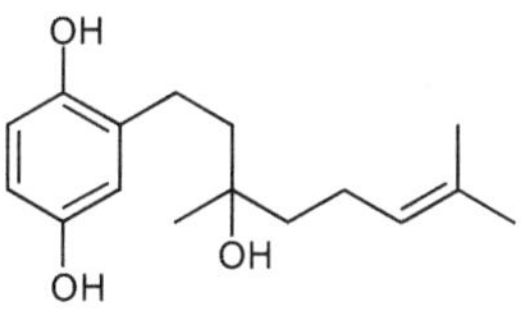

281 Mediterraneol A

Type: *p*-Benzenediol derivatives. $C_{27}H_{36}O_5$ Source: Brown alga *Cystoseira mediterranea*. Pharm: Cytotoxic (fertilized sea urchin eggs, mitotic cell division inhibitor). Ref: C. Francisco, et al, Tet. Lett., 1985, 26, 2629 | C. Francisco, et al, JOC, 1986, 51, 1115

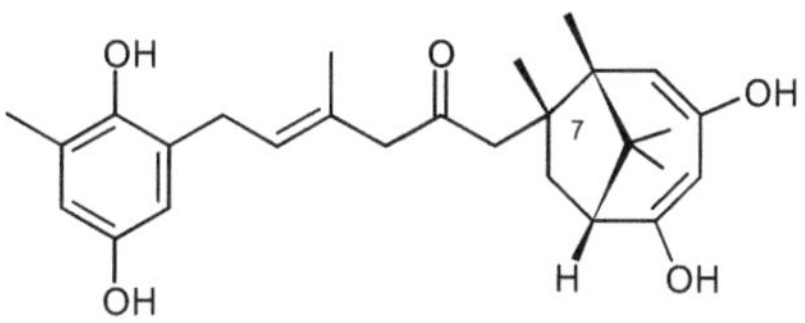

282 2-(Methoxymethyl)-1,4-benzenediol

Type: *p*-Benzenediol derivatives. $C_8H_{10}O_3$ Amorph. powder. Source: Marine-derived fungus *Penicillium terrestre* (from sediment sample, China). Pharm: Cytotoxic (HL60, IC_{50} = 58.9 μmol/L; Molt4, IC_{50} = 86.2 μmol/L; A549, IC_{50} > 100 μmol/L; Bel7402, IC_{50} > 100 μmol/L); antioxidant (DPPH scavenger, IC_{50} = 9.8 μmol/L; control Ascorbis acid, IC_{50} = 17.4 μmol/L). Ref: L. Chen, et al, JNP, 2008, 71, 66

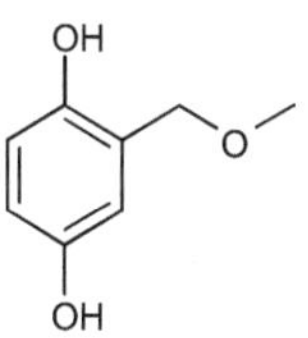

283 2-Pentaprenyl-1,4-benzenediol

Type: *p*-Benzenediol derivatives. $C_{31}H_{46}O_2$ Source: Sponges *Ircinia* sp. (New Caledonia) and *Spongia* sp. (Australia). Pharm: Tyrosine protein kinase (TPK) inhibitor; HIV-integrase inhibitor; binds to neuropeptide Y receptor Ref: D. Lumsdon, et al, Aust. J. Chem., 1992, 45, 1321 | G. Bifulco, et al, JNP, 1995, 58, 1444

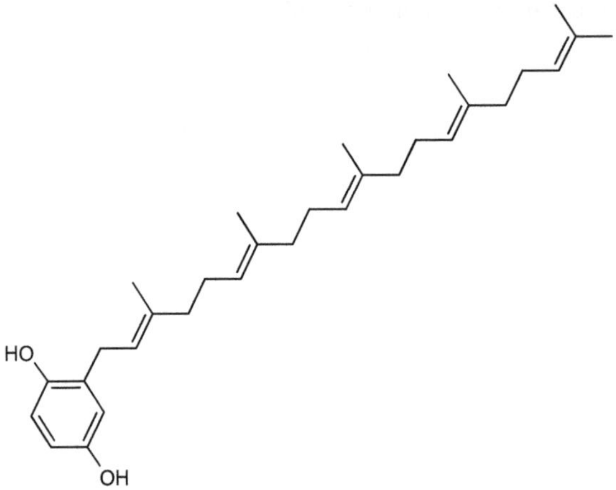

284 Phloroglucinol

1,3,5-Benzenetriol Type: *m*-Benzenetriol derivatives. $C_6H_6O_3$ Source: Brown algae *Ecklonia stolonifera, Eisenia bicyclis* and *Ecklonia maxima.* Pharm: Hypoglycemic (aldose reductase inhibitor) (rat lens aldose reductase *in vitro* assay, IC_{50} = 72.54 μmol/L) (Jung, 2008); hypoglycemic (α-glucosidase inhibitor) (*in vitro* assay, IC_{50} = 141.18 μmol/L) (Moon, 2011); hypoglycemic (PTP1B inhibitor) (*in vitro* assay, IC_{50} = 55.48 μmol/L) (Moon, 2011);hypoglycemic (α-glucosidase inhibitor) (*in vitro* assay, IC_{50} = 1991 μmol/L, control Acarbose, IC_{50} = 1013 μmol/L) (Rengasamy, 2013); antioxidant (DPPH scavenger, EC_{50} = 0.128 μmol/L, AAI (antioxidant activity index = final DPPH concentration/EC_{50}) = 390, control Ascorbic acid, EC_{50} = 0.011 μmol/L, AAI = 4356) (Rengasamy, 2013). Ref: H. A. Jung, et al, Fish. Sci., 2008, 74, 1363 | H. E. Moon, et al, Biosci. Biotechnol. Biochem., 2011, 75, 1472 | K. R. Rengasamy, et al, Food Chem., 2013, 141, 1412

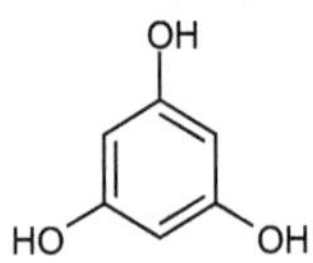

285 BDDE

Bis (2,3-dibromo-4,5-dihydroxybenzyl) ether Type: Benzyl alcohols. $C_{14}H_{10}Br_4O_5$ Pale yellow solid. Source: Red algae *Odonthalia corymbifera, Polyopes lancifolia, Rhodomela* spp. and *Symphyocladia latiuscula.* Pharm: Hypoglycemic (α-glucosidase inhibitor) (*in vitro* assay, *Saccharomyces cerevisiae,* IC_{50} = 0.098 μmol/L); hypoglycemic (α-glucosidase inhibitor) (*in vitro* assay, *Bacillus stearothermophilus,* IC_{50} = 0.12 μmol/L; *Saccharomyces cerevisiae,* IC_{50} = 0.098 μmol/L; rat intestinal maltase, IC_{50} = 1.20 mmol/L; rat intestinal sucrase, IC_{50} = 1.00 mmol/L); hypoglycemic (α-glucosidase inhibitor) (*in vitro* assay, IC_{50} = 0.03 μmol/L); hypoglycemic (sucrase inhibitor) (*in vitro* assay, IC_{50} = 2.4 mmol/L); hypoglycemic (maltase inhibitor) (*in vitro* assay, IC_{50} = 3.2 mmol/L); α-glucosidase inhibitor (IC_{50} = 25 μmol/L). Ref: H. Kurihara, et al, JNP, 1999, 62, 882 | H. Kurihara, et al, Fish. Sci. 1999, 65, 300 | K. Y. Kim, et al, J. Food Sci. 2010, 75, H145 | Y. Sharifuddin, et al, Mar. Drugs, 2015, 13, 5447 (rev)

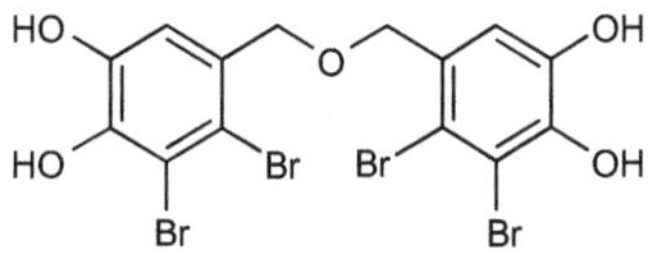

286 Bis(2,3-dibromo-4,5-dihydroxybenzyl)ether

Type: Benzyl alcohols. $C_{14}H_{10}Br_4O_5$ Source: Red alga *Rhodomela confervoides.* Pharm: Hypoglycemic (PTP1B inhibitor) (PTP1B and hypoglycaemic effect in streptozotocin-induced diabetic Wistar rats, IC_{50} = 1.5 μmol/L). Ref: D. Shi, et al, Chin. Sci. Bull., 2008, 53, 2476

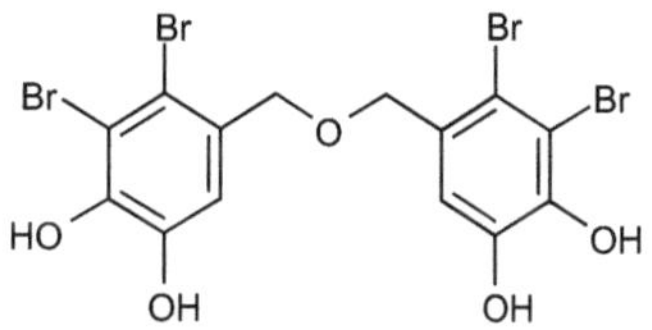

287 Bis(2,3,6-tribromo-4,5-dihydroxybenzyl)ether

4,4′-[Oxybis(methylene)]bis[3,5,6-tribromo-1,2-benzenediol] Type: Benzyl alcohols. $C_{14}H_8Br_6O_5$ Fine needles (C_6H_6/Me_2CO), mp 177–178 °C. Source: Red alga *Symphyocladia latiuscula.* Pharm: Hypoglycemic (PTP1B inhibitor) (*in vitro* assay, IC_{50} = 4.3 μmol/L); antifungal. Ref: K. Kurata, et al, Phytochemistry, 1980, 19, 141 | X. Liu, et al, Chin. J. Oceanol. Limnol.,2011, 29, 686

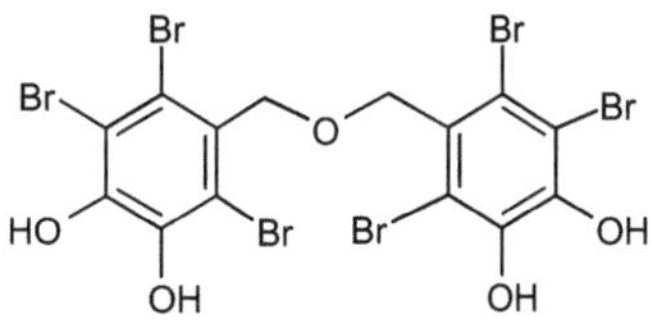

288 3-Chloro-4,5-dihydroxybenzyl alcohol

Type: Benzyl alcohols. $C_7H_7ClO_3$ Source: Marine-derived fungus *Aspergillus parasiticus* Pharm: Antioxidant (free radicals scavenger: DPPH radical, IC_{50} = 1.4 μmol/L, $ONOO^-$radical, IC_{50} = 2.2 μmol/L, $O_2^{\bullet-}$ radical, IC_{50} = 50 μmol/L inactive, $NO^\bullet$ radical, IC_{50} = 0.4 μmol/L). Ref: B. W. Son, et al, JNP, 2002, 65, 794

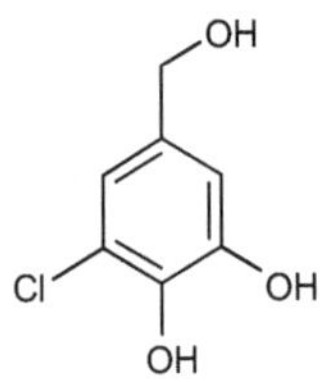

289 Chlorogentisyl alcohol

2-Chloro-6-(hydroxymethyl)-1,4-benzenediol Type: Benzyl alcohols. $C_7H_7ClO_3$ Needles (EtOAc/petrol), mp 147–148 °C, mp 140–141.5 °C. Source: Marine-derived fungus *Phoma herbarum* from red alga *Gloiopeitis tenax* (addition of halide salts to fermentation medium, Cheokpo, Tongnyeong, R. O. Korea), *Aspergillus varians* and *Ampelomyces* sp. Pharm: Antioxidant (DPPH scavenger, IC_{50} = 1.0 μmol/L, control *L*-Ascorbic acid, IC_{50} = 20 μmol/L); antibacterial. Ref: O. F. Smetanina, et al, Chem. Nat. Compd. (Engl. Transl.), 2005, 41, 243 | V. N. Nenkep, et al, J. Antibiot., 2010, 63, 199

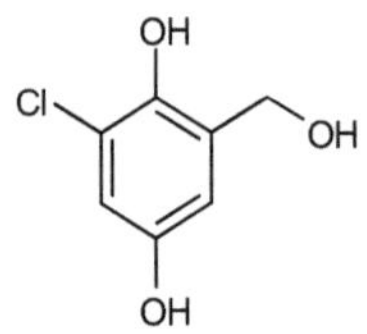

290 2-Chloro-6-(methoxymethyl)-1,4-benzenediol

Type: Benzyl alcohols. $C_8H_9ClO_3$ Amorph. powder. Source: Marine-derived fungus *Penicillium terrestre* (from sediment sample, China). Pharm: Cytotoxic (HL60, IC_{50} = 6.7 μmol/L; Molt4, IC_{50} = 64.7 μmol/L; A549, IC_{50} = 56.5 μmol/L; Bel7402, IC_{50} = 58.1 μmol/L); antioxidant (DPPH scavenger, IC_{50} = 8.5 μmol/L; control Ascorbis acid, IC_{50} = 17.4 μmol/L). Ref: L. Chen, et al, JNP, 2008, 71, 66

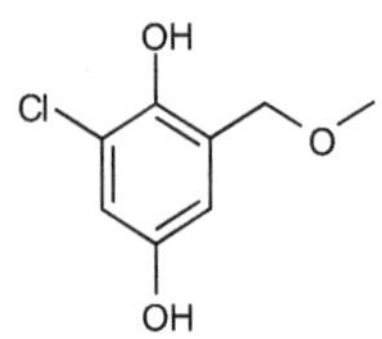

291 2,3-Dibromo-4,5-dihydroxybenzyl alcohol

Type: Benzyl alcohols. $C_7H_6Br_2O_3$ Cryst. (C_6H_6), mp 129–130 °C. Source: Red algae *Odonthalia corymbifera* and *Lenormandia prolifera*, red algae *Polysiphonia* spp., *Odonthalia* spp., and *Rhodomela* spp. Pharm: Hypoglycemic (α-glucosidase inhibitor) (*in vitro* assay, *Saccharomyces cerevisiae*, IC_{50} = 89.0 μmol/L). Ref: H. Kurihara, et al, Fish. Sci. 1999, 65, 300 | CRC Press, DNP on DVD, 2012, version 20.2 | Y. Sharifuddin, et al, Mar. Drugs, 2015, 13, 5447 (rev)

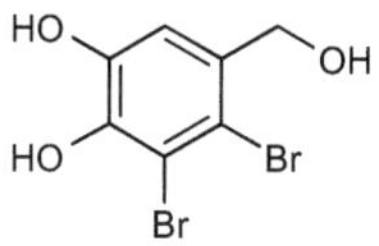

292 Gentisyl alcohol

2,5-Dihydroxybenzyl alcohol Type: Benzyl alcohols. $C_7H_8O_3$ Red-brown cryst. (Et_2O/petrol), mp 104–105 °C. Source: Marine-derived fungi *Phoma herbarum* from red alga *Gloiopeitis tenax* (addition of halide salts to fermentation medium, Cheokpo, Tongnyeong, R. O. Korea), *Aspergillus parasiticus* and *Aspergillus varians*. Pharm: Antioxidant (free radicals scavenger: DPPH radical, IC_{50} = 0.6 μmol/L, $ONOO^-$ radical, IC_{50} = 3.1 μmol/L, $O_2^{\bullet -}$radical, IC_{50} = 11.0 μmol/L, $NO^{\bullet}$ radical, IC_{50} = 0.5 μmol/L, control carboxy-PTIO, IC_{50} = 137.7 μmol/L); antioxidant (DPPH scavenger, IC_{50} = 7.0 μmol/L). Ref: B. W. Son, et al, JNP, 2002, 65, 794 | O. F. Smetanina, et al, Chem. Nat. Compd. (Engl. Transl.), 2005, 41, 243 | V. N. Nenkep, et al, J. Antibiot., 2010, 63, 199

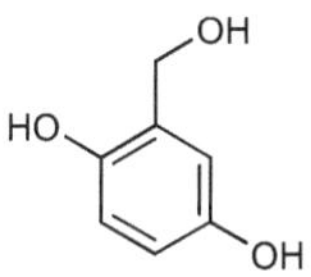

293 Terrestrol B

Type: Benzyl alcohols. $C_{14}H_{13}ClO_5$ Brown gum. Source: Marine-derived fungus *Penicillium terrestre* (from sediment sample, China). Pharm: Cytotoxic (HL60, IC_{50} = 6.1 μmol/L; Molt4, IC_{50} = 5.8 μmol/L; A549, IC_{50} = 18.3 μmol/L; Bel7402, IC_{50} = 62.3 μmol/L); antioxidant (DPPH scavenger, IC_{50} = 4.3 μmol/L; control Ascorbis acid, IC_{50} = 17.4 μmol/L). Ref: L. Chen, et al, JNP, 2008, 71, 66

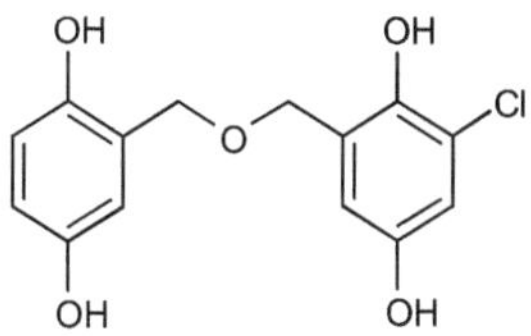

294 Terrestrol C

Type: Benzyl alcohols. $C_{14}H_{14}O_5$ Brown gum. Source: Marine-derived fungus *Penicillium terrestre* (from sediment sample, China). Pharm: Cytotoxic (HL60, IC_{50} = 5.5 μmol/L; Molt4, IC_{50} = 5.6 μmol/L; A549, IC_{50} = 18.2 μmol/L; Bel7402, IC_{50} = 57.3 μmol/L); antioxidant (DPPH scavenger, IC_{50} = 4.6 μmol/L; control Ascorbis acid, IC_{50} = 17.4 μmol/L). Ref: L. Chen, et al, JNP, 2008, 71, 66

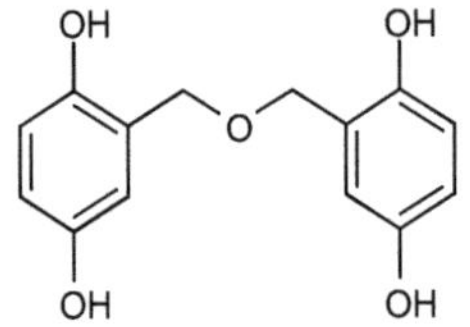

295 Terrestrol D

Type: Benzyl alcohols. $C_{14}H_{13}ClO_5$ Brown gum. Source: Marine-derived fungus *Penicillium terrestre* (from sediment sample, China). Pharm: Cytotoxic (HL60, IC_{50} = 5.3 μmol/L; Molt4, IC_{50} = 5.5 μmol/L; A549, IC_{50} = 14.3 μmol/L; Bel7402, IC_{50} = 38.5 μmol/L); antioxidant (DPPH scavenger, IC_{50} = 4.4 μmol/L; control Ascorbis acid, IC_{50} = 17.4 μmol/L). Ref: L. Chen, et al, JNP, 2008, 71, 66

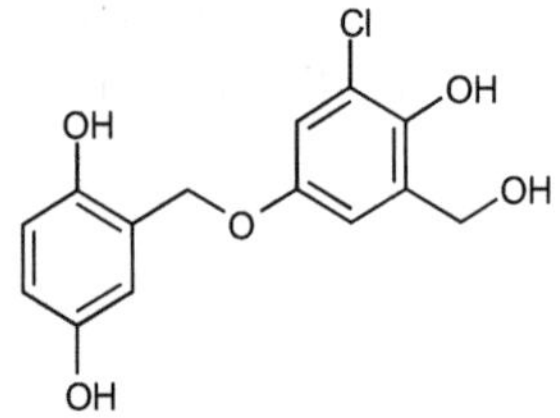

296 Terrestrol E

Type: Benzyl alcohols. $C_{14}H_{14}O_5$ Brown gum. Source: Marine-derived fungus *Penicillium terrestre* (from sediment sample, China). Pharm: Cytotoxic (HL60, IC_{50} = 54.7 μmol/L; Molt4, IC_{50} = 6.4 μmol/L; A549, IC_{50} = 9.6 μmol/L; Bel7402, IC_{50} = 59.0 μmol/L); antioxidant (DPPH scavenger, IC_{50} = 6.2 μmol/L; control Ascorbis acid, IC_{50} = 17.4 μmol/L). Ref: L. Chen, et al, JNP, 2008, 71, 66

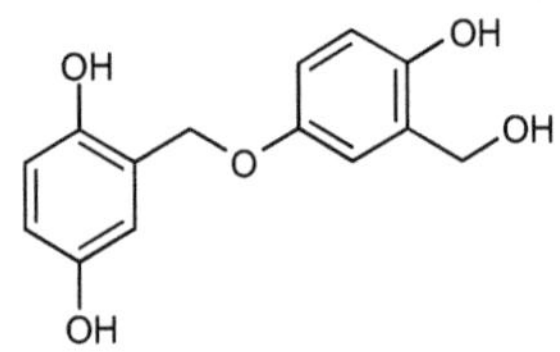

297 2,3,6-Tribromo-4,5-dihydroxybenzyl alcohol

Type: Benzyl alcohols. $C_7H_5Br_3O_3$ mp 128–130 °C. Source: Red algae *Symphyocladia latiuscula, Polysiphonia lanosa, Polysiphonia elongata* and *Rhodomela subfusca*. Pharm: Hypoglycemic (α-glucosidase inhibitor) (*in vitro* assay, IC_{50} = 11.0 μmol/L); hypoglycemic (sucrase inhibitor) (*in vitro* assay, IC_{50} = 4.2 mmol/L); hypoglycemic (maltase inhibitor) (*in vitro* assay, IC_{50} > 5.0 mmol/L); antibiotic; Antioxidant (DPPH radical scavenger). Ref. H. Kurihara, et al, Fish. Sci. 1999, 65, 300 | CRC Press, DNP on DVD, 2012, version 20.2

298 2,3,6-Tribromo-4,5-dihydroxybenzyl methyl ether

Type: Benzyl alcohols. $C_8H_7Br_3O_3$ Source: Red alga *Symphyocladia latiuscula*. Pharm: Hypoglycemic (PTP1B inhibitor) (*in vitro* assay, IC_{50} = 3.9 μmol/L). Ref: X. Liu, et al, Chin. J. Oceanol. Limnol., 2011, 29, 686

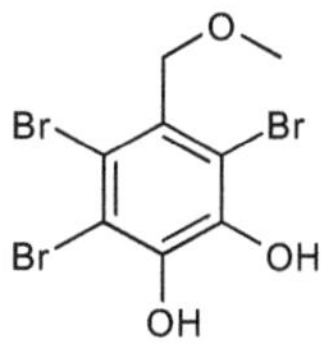

299 2,2′,3-Tribromo-3′,4,4′,5-tetrahydroxy-6′-ethyloxymethyldiphenylmethane

Type: Benzyl alcohols. $C_{16}H_{15}Br_3O_5$ Source: Red alga *Rhodomela confervoides*. Pharm: Hypoglycemic (PTP1B inhibitor) (PTP1B and hypoglycaemic effect in streptozotocin-induced diabetic Wistar rats, IC_{50} = 0.84 μmol/L). Ref: D. Shi, et al, Chin. Sci. Bull., 2008, 53, 2476

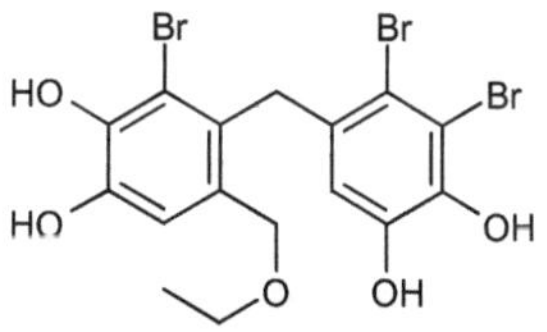

300 3-(2-Acetoxy-4,8-dimethyl-3,7-nonadienyl)benzaldehyde

Type: Benzaldehydes. $C_{20}H_{26}O_3$ Oil, $[\alpha]_D^{25}$ = +2.4° (c = 0.5, $CHCl_3$). Source: Green algae *Halimeda* spp. Pharm: Antimicrobial; cytotoxic. Ref: V. J. Paul, et al, Tetrahedron, 1984, 40, 3053

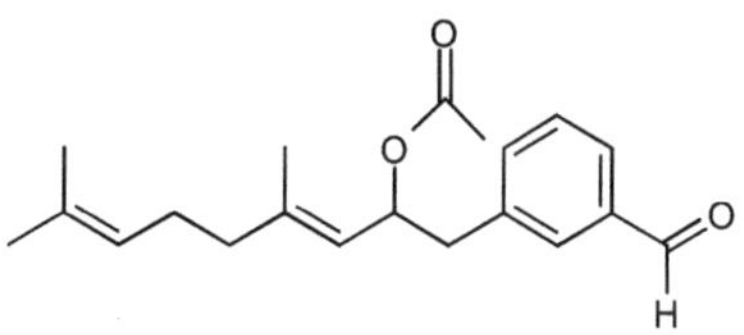

301 Chaetopyranin

5-Formyl-6-hydroxy-2-(3-hydroxy-1-butenyl)-7-prenylchroman Type: Benzaldehydes. $C_{19}H_{24}O_4$ Yellowish powder (Me_2CO), mp 136–138 °C, $[\alpha]_D^{22}$ = +8° (*c* = 0.5, MeOH). Source: Marine-derived fungus *Chaetomium globosum* from red alga *Polysiphonia urceolata* (China waters). Pharm: Antioxidant (DPPH scavenger, moderate); cytotoxic (SMMC-7721, A549, and HMEC, moderate to weak). Ref: S. Wang, et al, JNP, 2006, 69, 1622

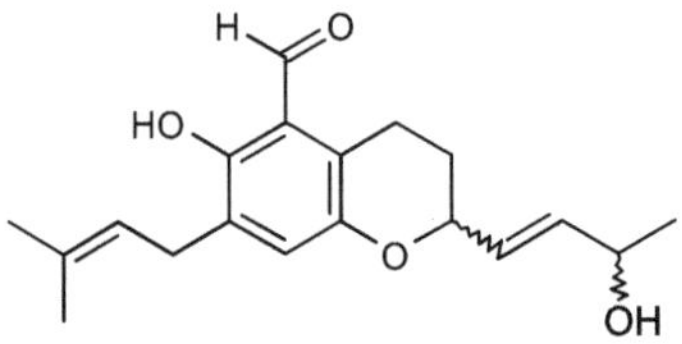

302 Cylindrocarpol

Type: Benzaldehydes. $C_{23}H_{34}O_5$ Gum, $[\alpha]_D^{23}$ = −11.7° (*c* = 0.3, MeOH). Source: Marine-derived fungus *Acremonium* sp. from sponge *Stelletta* sp. (Korea waters). Pharm: Farnesyl-protein transferase inhibitor. Ref: P. Zhang, et al, JNP, 2009, 72, 270 | S. B. Singh, et al, JOC, 1996, 61, 7727

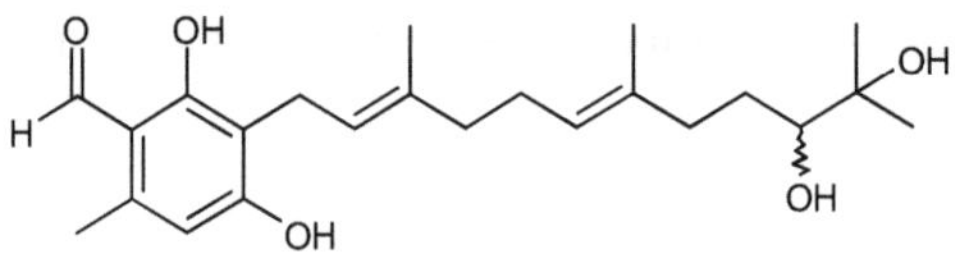

303 Flavoglaucin

Type: Benzaldehydes. $C_{19}H_{28}O_3$ Pale yellow cryst., mp 109–110 °C, mp 103 °C. Source: Marine-derived fungus *Microsporum* sp. from red alga *Lomentaria catenata* (Korea, surface). Pharm: Antioxidant (DPPH scavenger, IC_{50} = 11.3 µmol/L, control, Ascorbic acid, IC_{50} = 20 µmol/L); mycotoxin; antineoplastic. Ref: Y. Li, et al, CPB, 2006, 54, 882

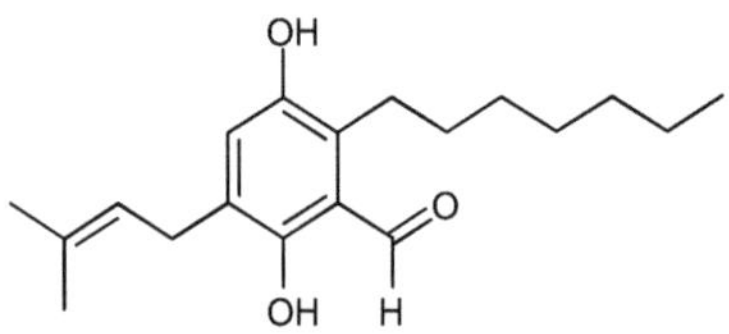

304 *p*-Hydroxybenzaldehyde

p-Formylphenol Type: Benzaldehydes. $C_7H_6O_2$ Needles (H_2O), mp 115–116 °C, p*K*a = 7.62 (25 °C, H_2O). Source: Mangrove-derived fungus *Pestalotiopsis* sp. from *Rhizophora* *mucronata* (China waters), green alga *Boodlea composita*. Pharm: Immunosuppressive; phytotoxic. Ref: S. F. Hinkley, et al, J. Antibiot., 1999, 52, 988

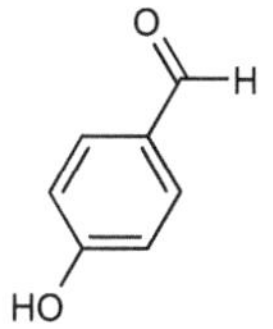

305 Isodihydroauroglaucin

Type: Benzaldehydes. $C_{19}H_{24}O_3$ Yellow needles (hexane), mp 114–115 °C. Source: Marine-derived fungus *Microsporum* sp. from red alga *Lomentaria catenata* (Korea, surface). Pharm: Antioxidant (DPPH scavenger, IC_{50} = 11.5 µmol/L, control, Ascorbic acid, IC_{50} = 20 µmol/L). Ref: Y. Li, et al, CPB, 2006, 54, 882

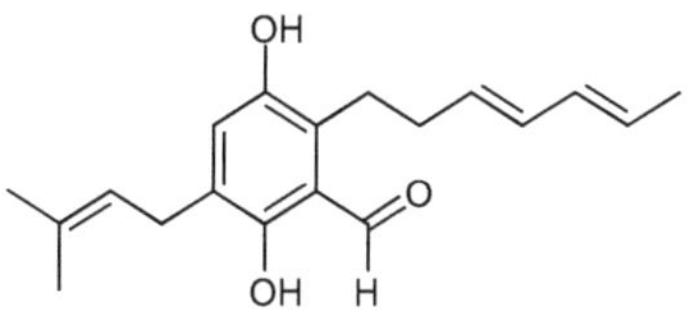

306 Redoxcitrinin

Type: Benzaldehydes. $C_{13}H_{16}O_4$ Source: Marine fungus *Penicillium* sp. MFA446 from green alga *Ulva pertusa* (Korea waters). Pharm: Antioxidant (DPPH scavenger, moderate). Ref: D. Zhang, et al, J. Microbiol. Biotechnol., 2007, 17, 865

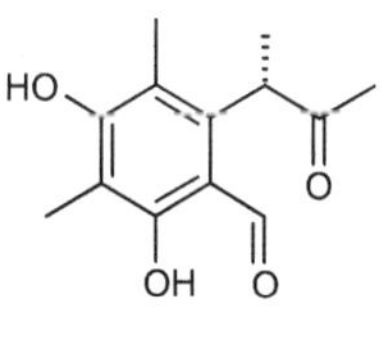

307 2,3,6-Tribromo-4,5-dihydroxybenzaldehyde

Type: Benzaldehydes. $C_7H_3Br_3O_3$ Amorph. yellowish solid, mp 134–135 °C. Source: Red alga *Symphyocladia latiuscula*. Pharm: Hypoglycemic (PTP1B inhibitor) (*in vitro* assay, IC_{50} = 19.40 µmol/L). Ref: W. Wang, et al, JNP, 2005, 68, 620 | X. Liu, et al, Chin. J. Oceanol. Limnol.,2011, 29, 686

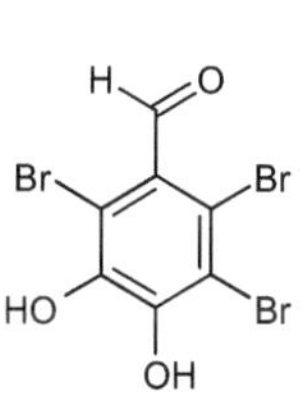

308 Acetylbenzoyl

Methyl phenyl diketone Type: Aryl ketones. $C_9H_8O_2$ Yellow oil with a pungent odour, bp 222 °C, bp_{12mmHg} 102 °C. Source: Marine-derived bacterium (family Rhodobacteraceae) *Oceanibulbus indolifex* HEL-4. Pharm: Flavouring ingredient. Ref: V. Thiel, et al, Org. Biomol. Chem., 2010, 234

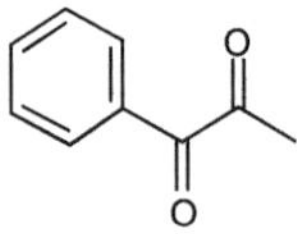

309 Communol A

Type: Aryl ketones. $C_{16}H_{16}O_6$ Source: Marine-derived fungus *Penicillium commune* 518 from gorgonian *Muricella abnormalis* (Danzhou, Hainan, China). Pharm: Antibacterial (*Escherichia coli*, MIC = 4.1 μmol/L; *Enterobacter aerogenes*, MIC = 16.4 μmol/L). Ref: J. Wang, et al, Chin. J. Chem., 2012, 30, 1236; 2880

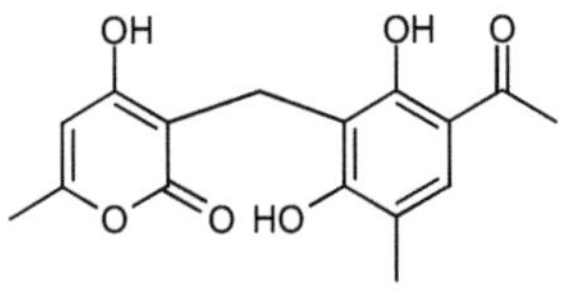

310 Communol F

Type: Aryl ketones. $C_{10}H_{10}O_4$ Source: Marine-derived fungus *Penicillium commune* 518 from a gorgonian, two marine fungi (Strains E33 and K38, co-culture broth). Pharm: Antibacterial (*Enterobacter coli*, MIC = 6.4 μmol/L, *Enterobacter aerogenes*, MIC = 25.8 μmol/L). Ref: J. Wang, et al, Chin. J. Chem., 2012, 30, 1236; 2880

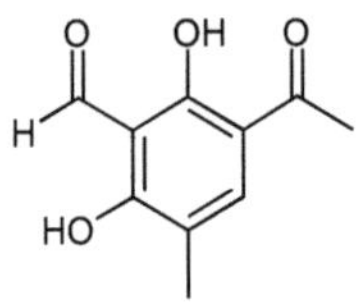

311 Communol G

Type: Aryl ketones. $C_{11}H_{14}O_4$ Source: Marine-derived fungus *Penicillium commune* 518 from a gorgonian, two marine fungi (Strains E33 and K38, co-culture broth). Pharm: Antibacterial (*Enterobacter coli*, MIC = 23.8 μmol/L, *Enterobacter aerogenes*, MIC = 23.8 μmol/L). Ref: J. Wang, et al, Chin. J. Chem., 2012, 30, 1236; 2880

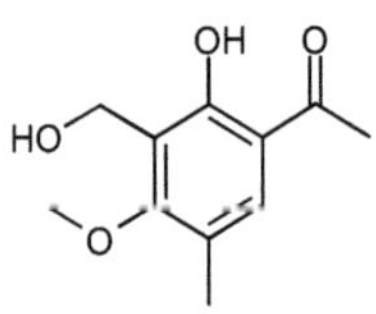

312 Cytosporone B

Type: Aryl ketones. $C_{18}H_{26}O_5$ Source: Mangrove-derived fungus *Leucostoma persoonii* from mangrove *Rhizophora mangle* (branch, Everglades, Florida), *Cytospora* sp. Pharm: Antibacterial (MRSA, MIC = 78 µmol/L, MBC_{90} = 93 µmol/L, $MBEC_{90}$, = 110 µmol/L; A549 IC_{50} = 170 µmol/L, IC_{90} = 190 µmol/L). Ref: M. P. Singh, et al, Mar. Drugs, 2007, 5, 71 | J. Beau, et al, Mar. Drugs, 2012, 10, 762

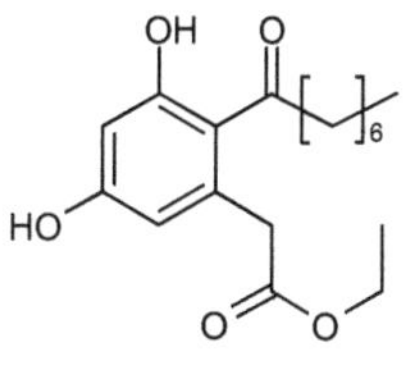

313 6-Demethylsorbicillin

1-(2,4-Dihydroxy-5-methylphenyl)-2,4-hexadien-1-one Type: Aryl ketones. $C_{13}H_{14}O_3$ Yellow solid (MeOH). Source: Marine-derived fungus *Trichoderma* sp. f-13 (sediment, Fujian province, China). Pharm: Cytotoxic (HL60 cell line, IC_{50} = 23.9 µmol/L, control VP-16, IC_{50} = 2.1 µmol/L; increases in cells percentage in sub-G_1 fraction, sub-G_1 = 66.4%, negative control MeOH, sub-G_1 = 2.3%). Ref: L. Du, et al, CPB, 2009, 57, 220

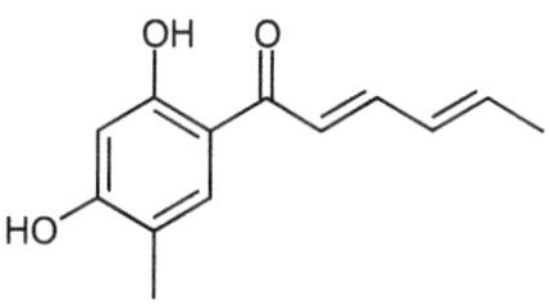

314 Ethyl 5-ethoxy-2-formyl-3-hydroxy-4-methylbenzoate

Type: Aryl ketones. $C_{13}H_{16}O_5$ Colorless needles. Source: Two marine fungi (Strains E33 and K38, co-culture broth). Pharm: Antifungal (0.25 mmol/L: *Blumeria graminearum*, IZD = 12.1 mm, *Gloeosporium musae*, IZD = 11.6 mm, *Rhizoctonia solani*, IZD = 10.2 mm, *Phytophthora sojae*, IZD = 8.5 mm). Ref: J. H. Wang, et al, Chem. Nat. Compd. 2013, 49, 799

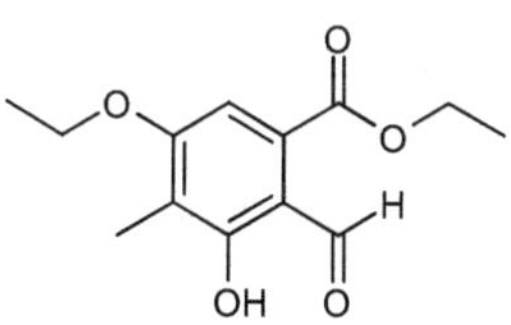

315 Isosorbicillin

(2*E*,4*E*)-1-(2,6-Dihydroxy-3,5-dimethyl-phenyl)hexa-2,4-dien-1-one Type: Aryl ketones. $C_{14}H_{16}O_3$ Yellow needles. Source: Marine-derived fungus *Penicillium* sp. M207142.

Pharm: Cytotoxic (HeLa and SW620, strong). Ref: S. Liu, et al, Chem. Nat. Compd., 2010, 46, 116

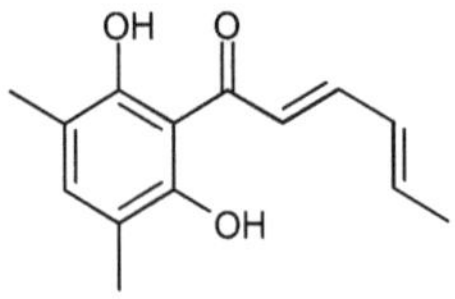

316 5-*O*-Methyldothiorelone A

Ethyl 2-(3-hydroxy-2-(7-hydroxyoctanoyl)-5-methoxyphenyl) acetate Type: Aryl ketones. $C_{19}H_{28}O_6$ Oil. Source: Mangrove-derived fungus *Phomopsis* sp. ZSU-H76 from mangrove *Excoecaria agallocha* (China waters). Pharm: Cytotoxic (MTT assay, Hep2, IC_{50} = 25 μg/mL; HepG2, IC_{50} = 30 μg/mL). Ref: Z. J. Huang, et al, Chem. Nat. Compd., 2009, 45, 625

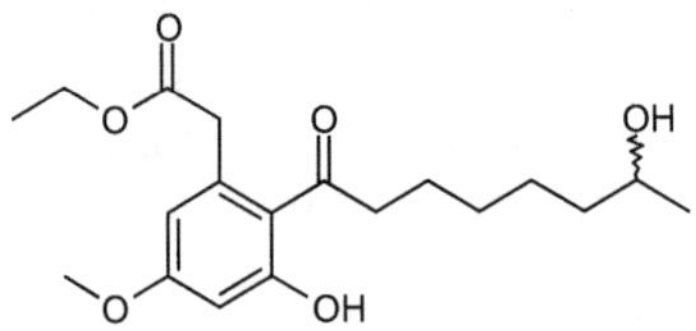

317 14,15-Secocurvularin

Type: Aryl ketones. $C_{16}H_{22}O_5$ Source: An unidentified marine-derived fungus 951014 from sponge *Spirastrella vagabunda* (Indo-Pacific). Pharm: Antibacterial (*Bacillus subtilis*, mild than tetracycline). Ref: L. M. Abrell, et al, Tet. Lett., 1996, 37, 8983

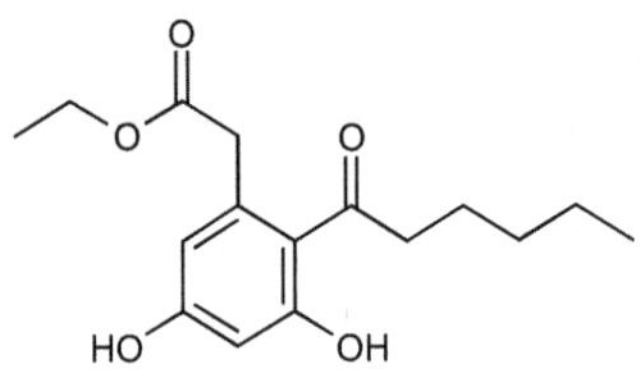

318 Sohirnone A

1-(2,4-Dihydroxy-5-methylphenyl)-4-hexen-1-one Type: Aryl ketones. $C_{13}H_{16}O_3$ Pale yellow solid. Source: Marine-derived fungus *Trichoderma* sp. f-13 (Chinese sediment sample, China waters). Pharm: Cytotoxic (HL60 cell line, IC_{50} > 50 μmol/L, control VP-16, IC_{50} = 2.1 μmol/L; increases in cells percentage in sub-G_1 fraction, sub-G_1 = 5.4%, negative control MeOH, sub-G_1 = 2.3%). Ref: L. Du, et al, CPB, 2009, 57, 220 | R. P. Maskey, et al, JNP, 2005, 68, 865

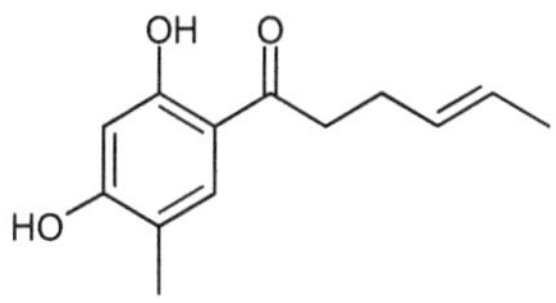

319 (4′Z)-Sorbicillin

Type: Aryl ketones. $C_{14}H_{16}O_7$ Source: Marine-derived fungus *Trichoderma* sp. from starfish *Acanthaster planci* (Hainan Sanya National Coral Reef Reserve, China). Pharm: Cytotoxic (several HTCLs cells, moderate). Ref: W. -J. Lan, et al, Nat. Prod. Commun., 2012, 7, 1337

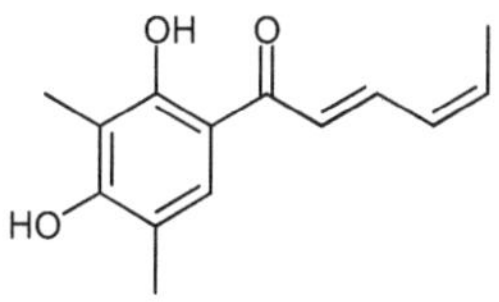

320 Ascochital

4-(1,3-Dimethyl-2-oxopentyl)-3-formyl-2,6-dihydroxybenzoic acid Type: Benzoic acids and esters. $C_{15}H_{18}O_6$ mp 122–124 °C. Source: Marine-derived fungus *Ascochyta salicorniae* from green alga *Ulva* sp. (North Sea coast, Germany), marine-derived fungus *Kirschsteiniothelia maritima*. Pharm: PPs inhibitors (MPtpB (mycobacterial protein tyrosine phosphatase B), IC_{50} = 61.2 μmol/L; Cdc25a, PTP1B, VHR, MPtpA and VE-PTP, all inactive); antibacterial (*Bacillus subtilis*, MIC = 500 ng/mL). Ref: C. Kusnick, et al, Pharmazie, 2002, 57, 510 | S. F. Siebert, et al, Org. Biomol. Chem., 2006, 4, 2233

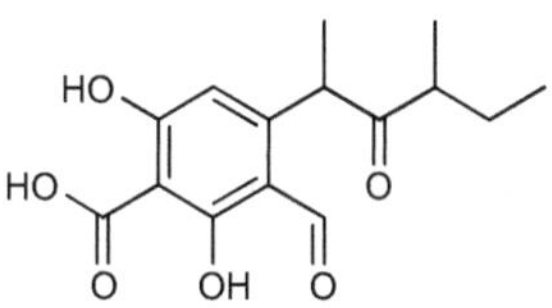

321 1,2-Benzenedicarboxylic acid2,12-diethyl-11-methylhexadecyl 2-ethyl-11-methylhexadecyl ester

Type: Benzoic acids and esters. $C_{48}H_{86}O_4$ Source: Fish (seahorse) *Hippocampus Kuda*. Pharm: Cathepsin B inhibitor (IC_{50} = 0.18–0.29 mmol/L). Ref: Y. Li, et al, BoMCL, 2008, 18, 6130

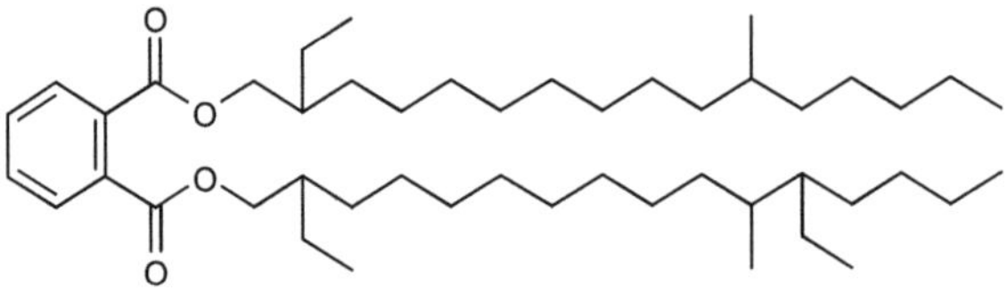

322 1,2-Benzenedicarboxylic acid2-Ethyldecyl 2-ethylundecyl ester

Type: Benzoic acids and esters. $C_{33}H_{56}O_4$ Source: Fish (seahorse) *Hippocampus Kuda*. Pharm: Cathepsin B inhibitor (IC_{50} = 0.18–0.29 mmol/L). Ref: Y. Li, et al, BoMCL, 2008, 18, 6130

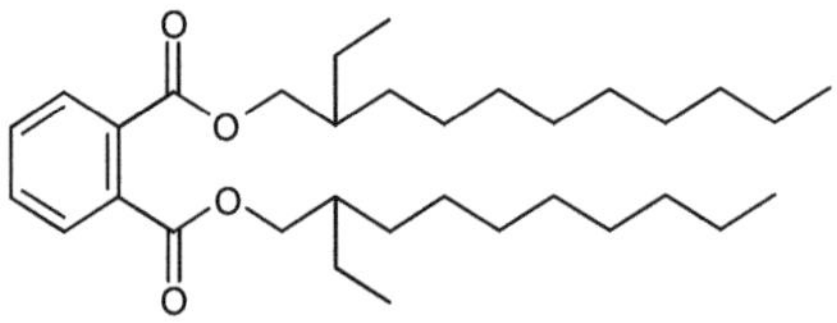

323 Bis(2-ethyldodecyl) phthalate

Type: Benzoic acids and esters. $C_{36}H_{62}O_4$ Source: Fish (seahorse) *Hippocampus Kuda*. Pharm: Cathepsin B inhibitor (IC_{50} = 0.18–0.29 mmol/L). Ref: Y. Li, et al, BoMCL, 2008, 18, 6130

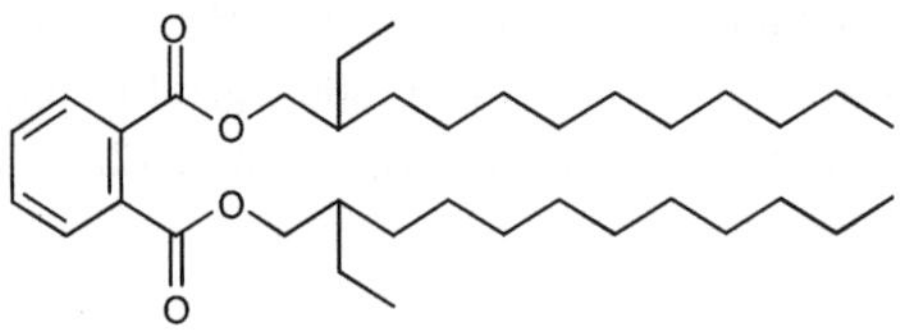

324 Butyl-isobutylphthalate

Type: Benzoic acids and esters. $C_{16}H_{22}O_4$ Source: Brown alga *Laminaria japonica* (rhizoids). Pharm: Hypoglycemic (*α*-glucosidase inhibitor) (*in vitro* assay, IC_{50} = 38.00 μmol/L); hypoglycaemic (*in vivo,* streptozotocin-induced diabetic mice). Ref: T. Bu, et al, Phytother. Res., 2010, 24, 1588

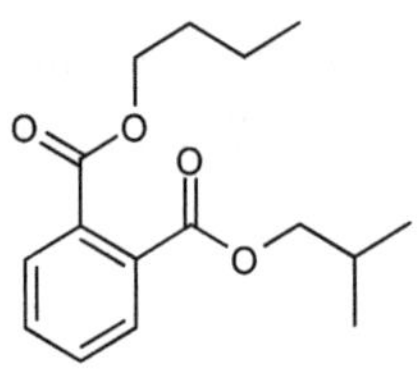

325 3,4-Dihydroxybenzoic acid

Hypogallicacid Type: Benzoic acids and esters. $C_7H_6O_4$ Source: Marine-derive fungus *Neosartorya fischeri* 1008F1, various higher plants. Pharm: Antiviral (TMV virus, IC_{50} = 630 μmol/L); antioxidant; free radical scavenger; dietary chemopreventive agent (inhibits development of neoplasms in animal models); inhibits LDL oxidation; platelet aggregation inhibitor. Ref: Q. -W. Tan, et al, Nat. Prod. Res., 2012, 26, 1402 | CRC press, DNP in DVD, 2012, version20.2

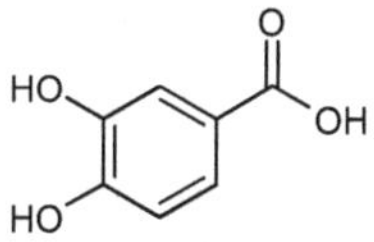

326 4-Hydroxybenzoic acid

Type: Benzoic acids and esters. $C_7H_6O_3$ Prisms (xylene/EtOH), cryst. + $1H_2O$ (EtOH aq or Me_2CO/EtOH), mp 213–214 °C. Source: Marine-derived fungus *Nigrospora* sp. PSU-F12 from gorgonian sea fan *Annella* sp. (Thailand). Pharm: Antifungal; eye irritant (eye, skin); LD_{50} (mus, orl) = 2200 mg/kg. Ref: V. Rukachaisirikul, et al, Arch. Pharmacal Res., 2010, 33, 375

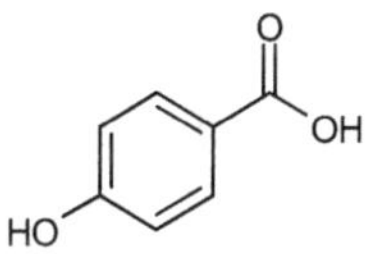

327 5-Hydroxy-2-methoxy benzoic acid

Type: Benzoic acids and esters. $C_8H_8O_4$ Source: Deep-sea fungus *Aspergillus* sp. CXCTD-06-6a. Pharm: Cytotoxic (1.68 μg/mL, HeLa, InRt = 7.29%). Ref: C. Ji, et al, Chin. J. Mar. Drugs, 2011, 30, 1

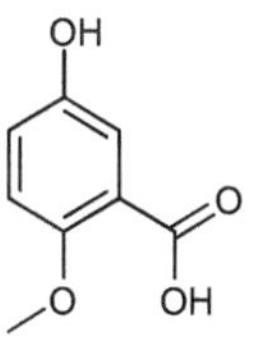

328 (2′*S*)-4-Methoxy-3-(2′-methyl-3′-hydroxy) propionyl-methyl benzoate

Type: Benzoic acids and esters. $C_{13}H_{16}O_6$ Source: Deep-sea fungus *Aspergillus* sp. 16-02-1 (sediment). Pharm: Cytotoxic (100 μg/mL: K562, InRt = 34.5%; HL60, InRt = 25.2%; HeLa, InRt = 3.2%; BGC823, InRt = 15.5%). Ref: X. Chen, et al, Chin. J. Mar. Drugs, 2013, 32, 1 (in Chinese)

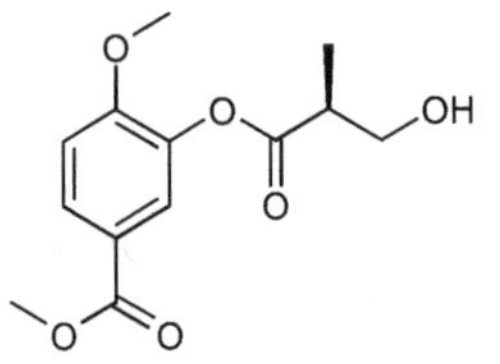

329 Methyl 4-hydroxy-3-(3-methylbut-2-enyloxy)benzoate

Type: Benzoic acids and esters. $C_{13}H_{16}O_4$ mp 49–50 °C. Source: Mangrove-derived unidentified fungus (from Chinese mangrove habitat). Pharm: Antibacterial (gram-positive bacterium *Staphylococcus aureus* ATCC27154, MIC = 6.25 μg/mL; gram-negative bacterium *Escherichia coli* ATCC 25922; MIC = 25 μg/mL, moderate); antifungal (*Fusarium oxysporum,* MIC = 12.5 μg/mL, *Candida albicans* ATCC 10231, inactive); cytotoxic (MTT assay, HepG2, IC_{50} = 10.0 μg/mL, moderate). Ref: C. L. Shao, et al, Chem. Nat. Compd., 2007, 43, 377

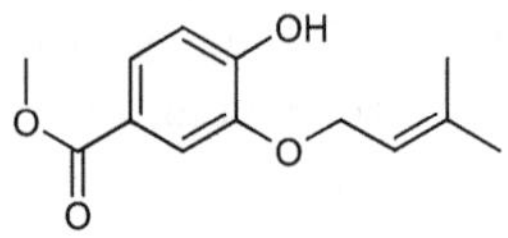

330 Tenellic acid C

Type: Benzoic acids and esters. $C_{23}H_{26}O_8$ Amorph. solid, mp 108–110 °C, $[\alpha]_D$ = −7.4° (*c* = 0.13, MeOH). Source: Mangrove-derived fungus *Talaromyces* sp. SBE-14 from mangrove *Kandelia candel* (Hong Kong). Pharm: Antibacterial (standard disk assays, 200 μg/disk: gram-positive bacterium *Bacillus subtilis* ATCC 6051, IZD = 12 mm; *Staphylococcus aureus* ATCC 29213, IZD = 14 mm; *Candida albicans* ATCC 14053, inactive). Ref: H. Oh, et al, JNP, 1999, 62, 580 | F. Liu, et al, Magn. Reson. Chem., 2009, 47, 453

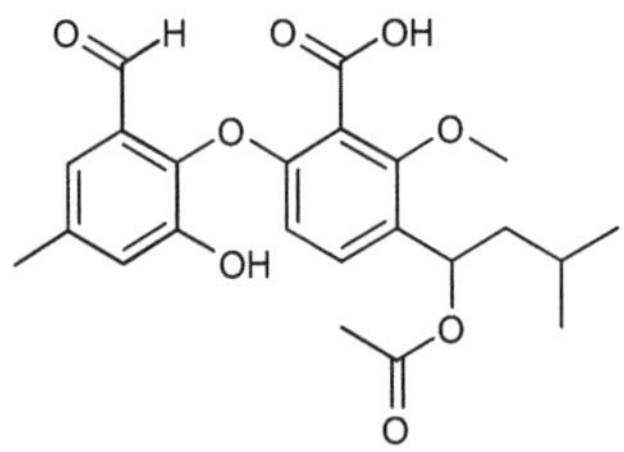

331 (+)-Aeroplysinin 1

3,5-Dibromo-1-cyanomethyl-4-methoxy-3,5-cyclohexadiene-1,2-diol Type: Phenylethanoids. $C_9H_9Br_2NO_3$ mp 120–121 °C, mp 112–113 °C, $[\alpha]_D$ = +193° (*c* = 0.63, Me_2CO), $[\alpha]_D^{20}$ = +185° (*c* = 0.17, MeOH). Source: Sponges *Aplysina aerophoba* [Syn. *Verongia aerophoba*], *Aplysina archeri, Aiolochroia crassa, Verongula rigida, Psammaplysilla purpurea* and *Ianthella* sp. Pharm: Antibacterial (*Photobacterium phosphoreum*, IC_{50} = 3.5 μmol/L) (García-Vilas, 2016); antimicroalgal (marine microalgae *Coscinodiscus wailesii*, IC_{50} = 5.6 μmol/L, *Prorocentrum minimum*, IC_{50} = 7.0 μmol/L) (García-Vilas, 2016); antiviral (HIV-1, IC_{50} = 14.6 μmol/L) (García-Vilas, 2016); anti-angiogenic (EVLC-2 Cell Line, 2.5 μmol/L, target MMP-2, effect decrease; HUVEC Cell Line, 2.5 μmol/L, target MMP-2, effect decrease; RF-24 Cell Line, 2.5 μmol/L, target MMP-2, effect decrease; BAEC Cell Line, 3 μmol/L, target MMP-2, effect decrease; BAEC Cell Line, 3 μmol/L, target PA, effect decrease; BAEC Cell Line, 3 μmol/L, target PAI, effect increase) (García-Vilas, 2016); anti-inflammatory (HUVEC Cell Line, 10 μmol/L, target MCP-1, effect decrease; HUVEC Cell Line, 10 μmol/L, target TSP-1, effect decrease; HUVEC Cell Line, 10 μmol/L, target COX-2, effect decrease; HUVEC Cell Line, 20 μmol/L, target Il-1*α*, effect decrease; HUVEC Cell Line, 20 μmol/L, target MMP-1, effect decrease) (García-Vilas, 2016); apoptogenic (BAEC Cell Line, 10 μmol/L, target Cleaved lamin-A, effect increase; BAEC Cell Line, 10 μmol/L, target Caspase-2, -3, -8, -9, effect increase; BAEC Cell Line, 10 μmol/L, target Cytochrome C, effect increase in cytoplasm; HUVEC Cell Line, 10 μmol/L, target p-Bad, effect increase) (García-Vilas, 2016); anti-angiogenic (HUVEC (hmn umbilical vein endothelial cells), IC_{50} = 4.7 μmol/L; EVLC-2 (SV40 large T-antigen immortalized hmn umbilical vein cells), IC_{50} = 3.0 μmol/L; RF-24 (papillomavirus 16 E6/E7 immortalized hmn umbilical vein cells), IC_{50} = 2.8 μmol/L; HMEC (hmn microvascular endothelial cells), IC_{50} = 2.6 μmol/L; BAEC (bovine aortic endothelial cells,), IC_{50} = 2.1 μmol/L) (García-Vilas, 2016); cytotoxic (HT1080, IC_{50} = 2.3 μmol/L, HTC116, IC_{50} = 4.7 μmol/L, HeLa, IC_{50} = 3.0 μmol/L, THP-1, IC_{50} = 10.0 μmol/L, NOMO-1, IC_{50} = 17.0 μmol/L, HL60, IC_{50} = 5.0 μmol/L) (García-Vilas, 2016); antibiotic (Gram-positive bacteria and several dinoflagellate microalgae causing toxic blooms) (García-Vilas, 2016); EGFR inhibitor (IC_{100} = 0.5 μmol/L); antitumor (strong effect on EGFR tumor cell lines, blocking proliferation of EGFR-dependent hmn breast cancer cell lines MCF7 and ZR-75-1); molluscacidal. Note: (+)-Aeroplysinin 1 has a wide spectrum of bio-activities. In preclinical studies, it has been shown to have promising anti-inflammatory, anti-angiogenic and anti-tumor effects. It might have a pharmaceutical interest for the treatment of different pathologies. Ref: T. N. Makarieva, et al, Comp. Biochem. Physiol., B, 1981, 68, 481 | M. -H. Kreuter, et al, Comp. Biochem. Physiol. B Biochem. Mol. Biol. 1990, 97, 151 | A. Koulman, et al, JNP, 1996, 59, 591 | M. Kita, et al, Tet. Lett., 2008, 49, 5383 | D. Skropeta, et al, Mar. Drugs, 2011, 9, 2131 (rev) | J. A. García-Vilas, et al, Mar. Drugs, 2016, 14, 1 (rev)

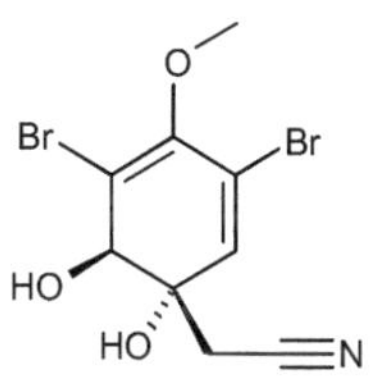

332 4-Hydroxy-3-methoxyphenylglyoxylic acid methyl ester

Vanilloylformic acid Type: Phenylethanoids. $C_{10}H_{10}O_5$ Yellow amorph. powder. Source: Ascidian *Polyandrocarpa zorritensis* (Bay of Taranto, Mediterranean Sea). Pharm: Cytotoxic (C6, IC_{50} > 1000 µmol/L; HeLa and H9c2, inactive). Ref: A. Aiello, et al, Mar. Drugs, 2011, 9, 1157

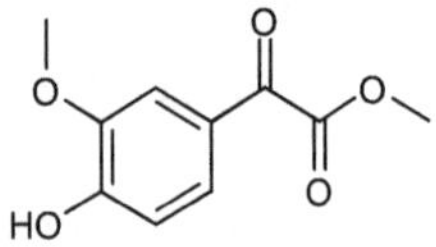

333 4-Hydroxyphenethyl 2-(4-hydroxyphenyl) acetate

Type: Phenylethanoids. $C_{16}H_{16}O_4$ Source: Mangrove-derived fungus *Penicillium griseofulvum* Y19-07 from mangrove *Lumnitzera racemosa*. Pharm: Antioxidant (radical scavenger, moderate); cytotoxic (weak). Ref: Y. N. Wang,et al, J. Asian Nat. Prod. Res., 2009, 11, 912

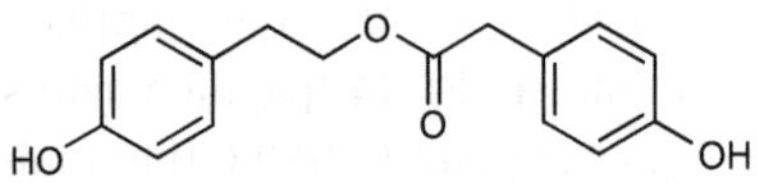

334 4-Hydroxyphenethyl methyl succinate

Type: Phenylethanoids. $C_{13}H_{16}O_5$ Oil. Source: Mangrove-derived fungus *Penicillium griseofulvum* Y19-07 from mangrove *Lumnitzera racemosa*. Pharm: Antioxidant (radical scavenger, moderate). Ref: Y. N. Wang,et al, J. Asian Nat. Prod. Res., 2009, 11, 912

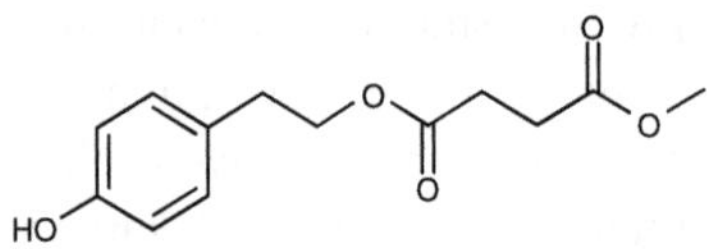

335 Isojaspisin

Type: Phenylethanoids. $C_8H_8O_6S$ Viscous oil (Na salt). Source: Sponge *Jaspis* sp. (Japan waters). Pharm: Inhibits hatching of sea urchin embryos. Ref: C. M. Cerda-Garcia-Rojas, et al, JNP, 1994, 57, 1758

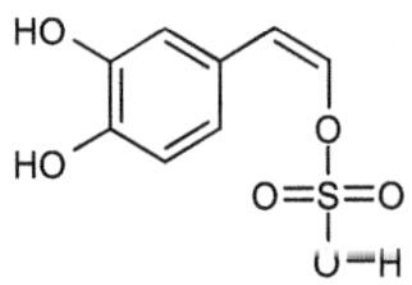

336 Jaspisin

3,4-Dihydroxystyryl sulfate Type: Phenylethanoids. $C_8H_8O_6S$ Viscous oil (Na salt). Source: Sponges *Poecillastra wondoensis* and *Jaspis* sp. (Japan waters). Pharm: Sea urchin gamete inhibitor; endoproteinase inhibitor; fertilisation inhibitor (*Asterina pectinifera*). Ref: C. M. Cerda-Garcia-Rojas, et al, JNP, 1994, 57, 1758 | S. Ohta, et al, Biosci. Biotechnol. Biochem., 1994, 58, 1752 | S. Ikegami, et al, J. Biol. Chem., 1994, 269, 23262 | Y. H. Chang, et al, JNP, 2008, 71, 779

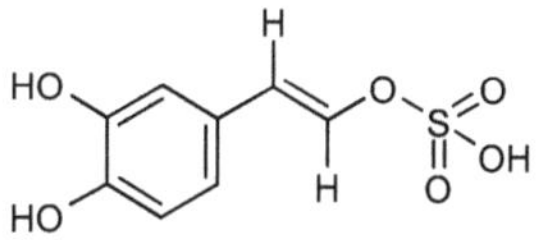

337 Methyl 4-hydroxyphenylacetate

Type: Phenylethanoids. $C_9H_{10}O_3$ Source: Marine fungus *Penicillium oxalicum* 0312f1. Pharm: Antiviral (TMV virus, EC_{50} = 829.15 μmol/L). Ref: S. Shen, et al, Acta Microbiol. Sinic., 2009, 49, 1240 | S. Shen, et al, Nat. Prod. Res., 2013, 27, 2286

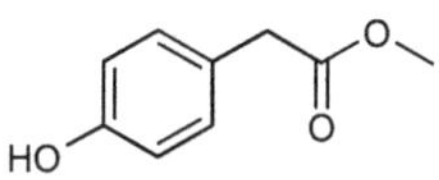

338 Subereaphenol B

4,6-Dibromo-homogentisic acid methyl ester Type: Phenylethanoids. $C_9H_8Br_2O_4$ Source: Sponges *Pseudoceratina* sp. (Vanuatu), *Suberea* sp. (Hurghada, Red Sea, Egypt) and *Suberea mollis*. Pharm: Pfnek-1 (a NIMA-related protein kinase of *Plasmodium falciparum*) inhibitor (IC_{50} = 1.8 μmol/L). Ref: M. I. Abou-Shoer, et al, JNP, 2008, 71, 1464 | N. Lcbouvier,et al, Mar. Drugs, 2009, 7, 640 | K. H. Shaker, et al, Chem. Biodiversity, 2010, 7, 2880 | D. Skropeta, et al, Mar. Drugs, 2011, 9, 2131 (rev)

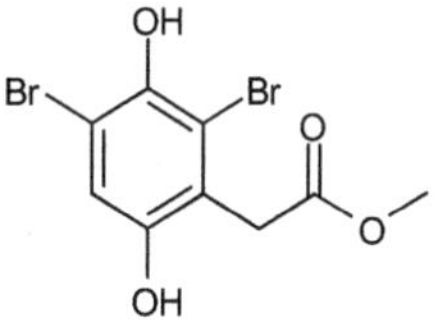

339 (3S)-(3,5-Dihydroxyphenyl)butan-2-one

Type: Phenylpropanoids. $C_{10}H_{12}O_3$ Oil, $[\alpha]_D^{20}$ = +124.0° (*c* = 0.12, $CHCl_3$). Source: Marine-derived fungus *Coniothyrium* sp. from sponge *Ectyoplasia ferox*. Pharm: Antifungal (*Ustilago violacea*). Ref: U. Höller, et al, JNP, 1999, 62, 114

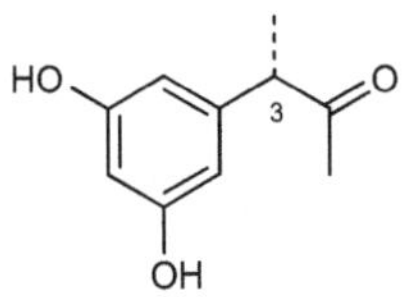

340 Ethyl 3,5-dibromo-4(3'*N,N*-dimethylaminopropyloxy)-cinnamate

Type: Phenylpropanoids. $C_{16}H_{21}Br_2NO_3$ Flakes, mp 67 °C. Source: Sponge *Pseudoceratina crassa*. Pharm: Antibacterial. Ref: K. E. Kassühlke, et al, Tetrahedron, 1991, 47, 1809

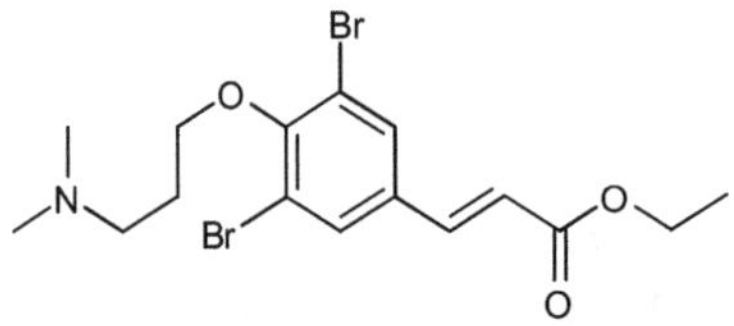

341 Fijiolide A

Type: Phenylpropanoids. $C_{34}H_{38}Cl_2N_2O_{10}$ Pale yellow oil, $[\alpha]_D^{21}$ = −440° (*c* = 0.5, MeOH). Source: Marine-derived actinomycete *Nocardiopsis* sp. CNS-653. Pharm: TNF-*α*-induced NF-κB activation inhibitor. Ref: S. -J. Nam, et al, JNP, 2010, 73, 1080

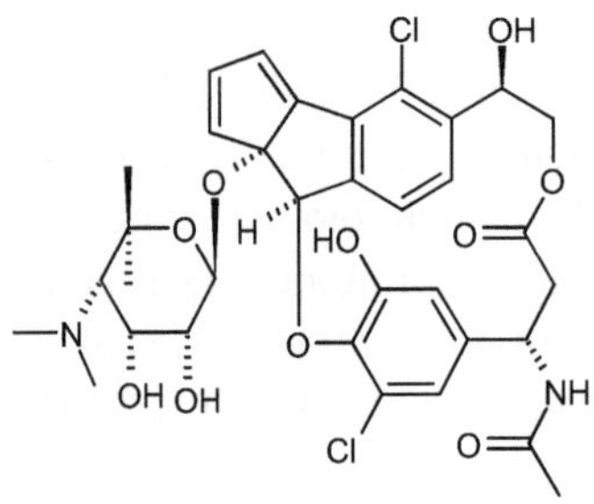

342 Hibiscusamide

N-trans-Feruloyl-3,5-dimethoxytyramine Type: Phenylpropanoids. $C_{20}H_{23}NO_6$ Needles (EtOH) or yellow solid, mp 126–128 °C, $[\alpha]_D^{25}$ = −18° (c = 0.68, MeOH). Source: Semimangrove *Hibiscus tiliaceus* (stem wood). Pharm: Cytotoxic. Ref: J. -J. Chen, et al, Planta Med., 2006, 72, 932

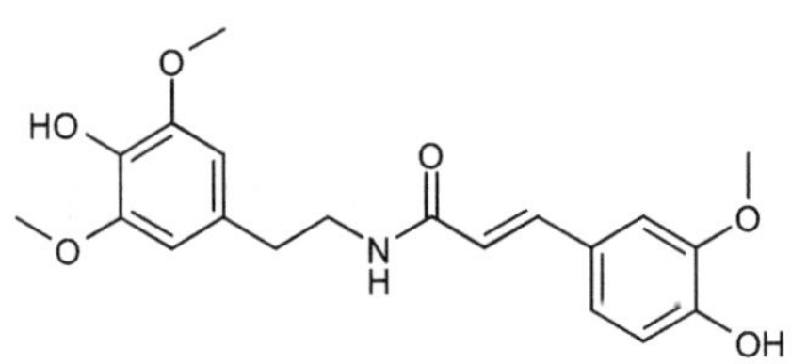

343 Phenol A

5-[(2*S*,3*R*)-3-Hydroxybutan-2-yl]-4-methylbenzene-1,3-diol Type: Phenylpropanoids. $C_{11}H_{16}O_3$ Source: Marine-derived fungus *Penicillium* sp. MFA446 from green alga *Ulva pertusa* (Korea waters), deep-sea fungus *Aspergillus sydowi* YH11-2 (Guam, depth of 1000 m, E144°43′ N13°26′). Pharm: Antioxidant (DPPH scavenger, moderate); cytotoxic (P_{388}, IC_{50} = 10.41 μmol/L, control CDDP, IC_{50} = 0.039 μmol/L). Ref: R. F. Curtis, et al, JCS Perkin I, 1968, 1, 85 | Zhang, D. et al, J. Microbiol. Biotechnol., 2007, 17, 865 | L. Tian, et al, Arch. Pharm. Res,. 2007, 30, 1051

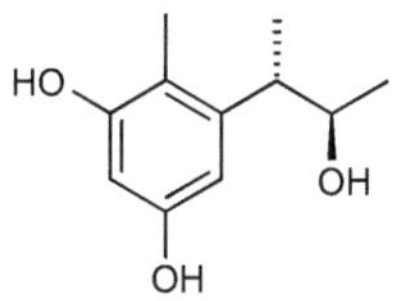

344 Sulfated cinnamic acid

Type: Phenylpropanoids. $C_9H_8O_6S$ Source: Marine monocotyledon *Zostera marina*. Pharm: Antifoulant. Ref: J. S. Todd, et al, Phytochemistry, 1993, 34, 401

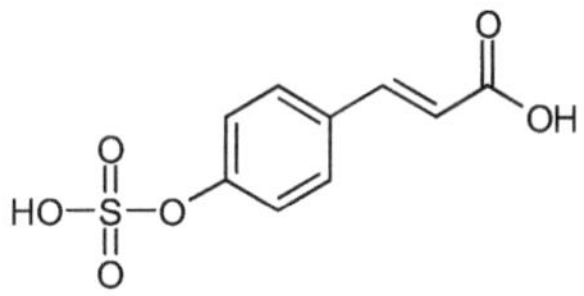

345 Aspergillusene A

Type: Other phenylalkanoids (C4–C8). $C_{15}H_{22}O_2$ Source: Marine-derived fungus *Aspergillus sydowii* ZSDS1-F6. Pharm: Antibacterial (*Klebsiella pneumonia*, MIC = 21.4 μmol/L). Ref: J. -F. Wang, et al, J. Antibiot., 2014, 67, 581

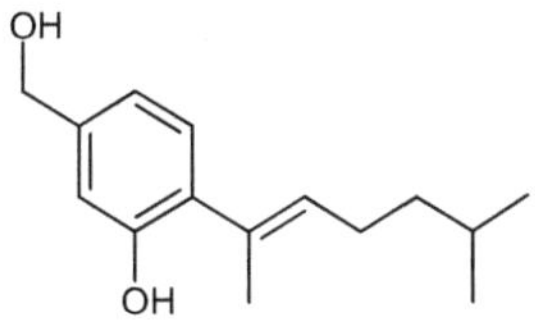

346 Caulerprenylol A

Type: Other phenylalkanoids (C4–C8). $C_{18}H_{26}O_2$ Source: Green alga *Caulerpa racemosa* (Zhanjiang coastline, China). Pharm: Antifungal (weak). Ref: A. -H. Liu, et al, Bioorg. Med. Chem. Lett., 2013, 23, 2491

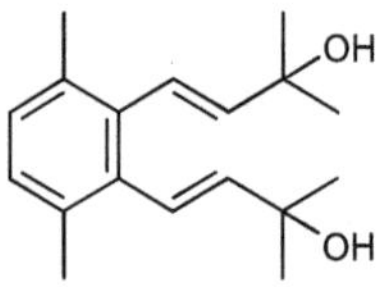

347 Caulerprenylol B

Type: Other phenylalkanoids (C4–C8). $C_{18}H_{26}O_2$ Source: Green alga *Caulerpa racemosa* (Zhanjiang coastline, China). Pharm: Antifungal (weak). Ref: A. -H. Liu, et al, Bioorg. Med. Chem. Lett., 2013, 23, 2491

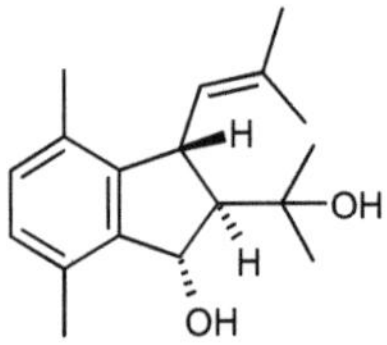

348 1,11-Dihydroxy-1,3,5,7*E*-bisabolatetren-15-oic acid

Engyodontiumone I Type: Other phenylalkanoids (C4–C8). $C_{15}H_{20}O_4$ Yellow gum. Source: Deep-sea fungus *Engyodontium album* DFFSCS021. Pharm: Antibacterial (*Bacillus subtilis*, 25 μg/disc, IZ = 10.0 mm, control PG, 10 μg/disc, IZ = 43.3 mm, MIC = 256.0 μg/mL, PG, MIC = 2.0 μg/mL). Ref: Q. Yao, et al, Mar. Drugs, 2014, 12, 5902

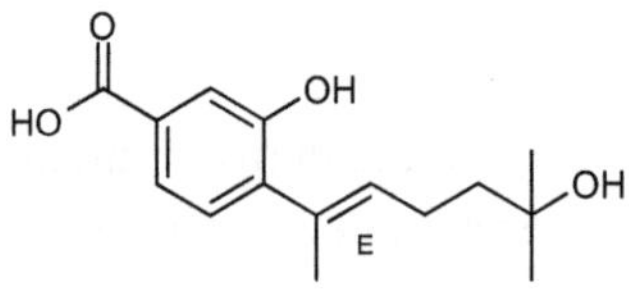

349 (Z)-5-(Hydroxymenthyl)-2-(6′-methylhept-2′-en-2′-yl)-phenol

Type: Other phenylalkanoids (C4–C8). $C_{15}H_{22}O_2$ Source: Marine-derived fungus *Aspergillus sydowii* ZSDS1-F6. Pharm: Antiviral (H3N2 virus, IC_{50} = 57.4 μmol/L); antibacterial (*Klebsiella pneumonia*, MIC = 10.7 μmol/L). Ref: J. -F. Wang, et al, J. Antibiot., 2014, 67, 581

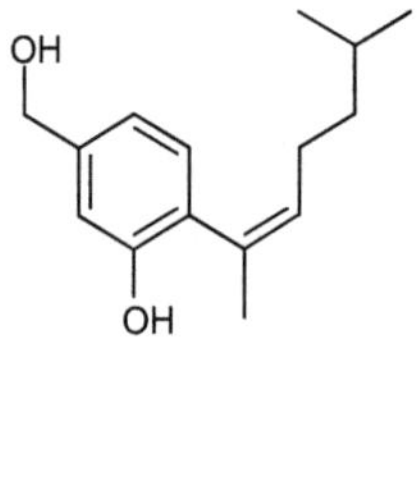

350 Lorneic acid A

4-[2-(1-Hexenyl)-4-methylphenyl]-3-butenoic acid Type: Other phenylalkanoids (C4–C8). $C_{17}H_{22}O_2$ Oil. Source: Marine-derived streptomycete *Streptomyces* sp. NPS-554 (sediment, Miyazaki Harbor, Japan waters). Pharm: Hmn platelet phosphodiesterase 5 inhibitor (PDE5) (strong). Ref: F. Iwata, et al, JNP, 2009, 72, 2046

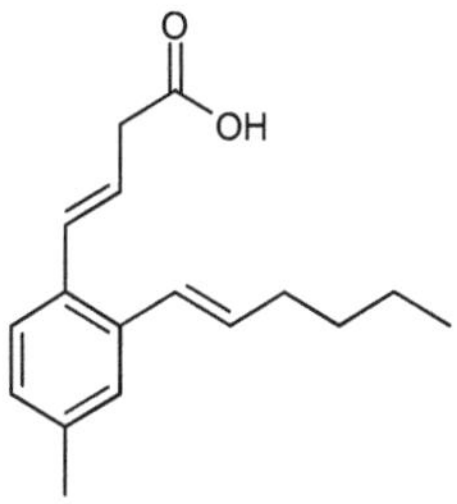

351 Lorneic acid B

Type: Other phenylalkanoids (C4–C8). $C_{17}H_{24}O_3$ Oil, [α]D = −22° (*c* = 0.2, MeOH). Source: Marine-derived streptomycete *Streptomyces* sp. NPS-554 (sediment, Miyazaki Harbor, Japan waters). Pharm: Hmn platelet phosphodiesterase 5 inhibitor (PDE5). Ref: F. Iwata, et al, JNP, 2009, 72, 2046

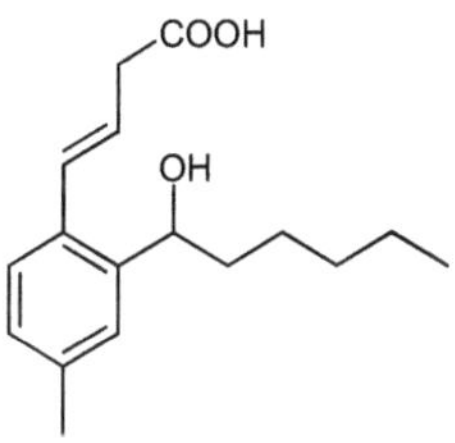

352 Nakitriol

Type: Other phenylalkanoids (C4–C8). $C_{11}H_{12}O_3$ Source: Cyanobacterium *Synechocystis* sp. growth on stony coral *Acropora* sp. (Okinawa). Pharm: Cytotoxic (DNA repair defective cell lines: EM9 (topoisomerase I sensitive Chinese hamster ovary line), XRS-6 (topoisomerase II sensitive Chinese hamster ovary line), UV20 (DNA cross-linking agent sensitive Chinese hamster ovary line), and BR1 (DNA repair competent Chinese hamster ovary line), all $LD_{50} \approx 20$ μg/mL). Ref: D. G. Nagle, et al, Tet. Lett., 1995, 36, 849

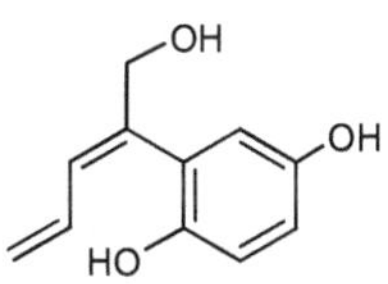

353 Rubrenoic acid A

Type: Other phenylalkanoids (C4–C8). $C_{16}H_{22}O_2$ Source: Marine bacterium *Alteromonas rubra*. Pharm: Bronchodilator. Ref: G. S. Holl, et al, Chem. Ind. (London), 1984, 850

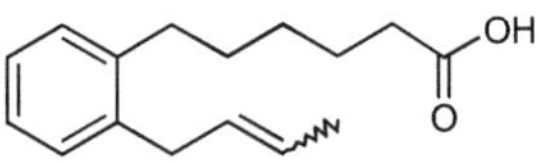

354 Rubrenoic acid B

Type: Other phenylalkanoids (C4–C8). $C_{16}H_{20}O_2$ Source: Marine bacterium *Alteromonas rubra*. Pharm: Bronchodilator. Ref: G. S. Holl, et al, Chem. Ind. (London), 1984, 850

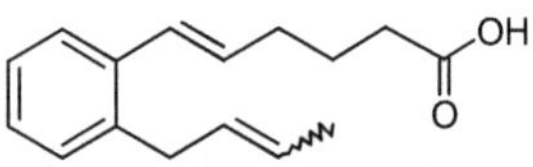

355 Rubrenoic acid C

Type: Other phenylalkanoids (C4–C8). $C_{16}H_{20}O_2$ Source: Marine bacterium *Alteromonas rubra*. Pharm: Bronchodilator. Ref: G. S. Holl, et al, Chem. Ind. (London), 1984, 850

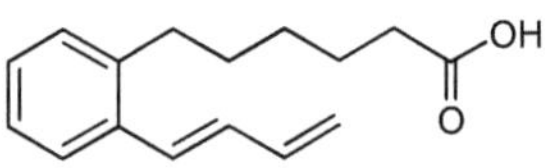

356 Strepchloritide A

Type: Other phenylalkanoids (C4–C8). $C_{10}H_8Cl_2O_4$ Source: Marine-derived streptomycete *Streptomyces* sp. from an unidentified soft coral (Weizhou I., Guangxi, China). Pharm: Cytotoxic (MCF7, modest). Ref: P. Fu, et al, Chin. J. Chem., 2013, 31, 100

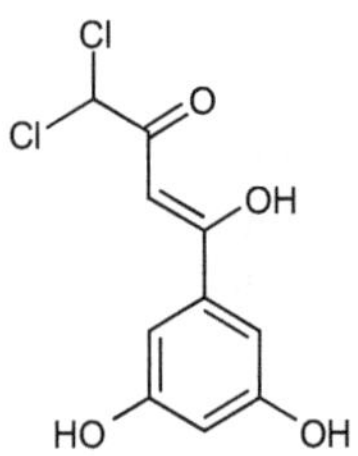

357 Strepchloritide B

Type: Other phenylalkanoids (C4–C8). $C_{10}H_7Cl_3O_4$ Source: Marine-derived streptomycete *Streptomyces* sp. from an unidentified soft coral (Weizhou I., Guangxi, China). Pharm: Cytotoxic (MCF7, modest). Ref: P. Fu, et al, Chin. J. Chem., 2013, 31, 100

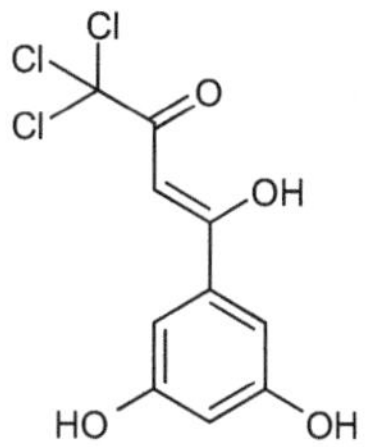

358 (ξ)-Sydonic acid

1,7-Dihydroxy-1,3,5-bisabolatrien-15-oic acid Type: Other phenylalkanoids (C4–C8). $C_{15}H_{22}O_4$ Needles (C_6H_6), mp 158–159 °C. Source: Deep-sea fungus *Engyodontium album* DFFSCS021, terrestrial and marine-derived *Aspergillus sydowi*, order Pleosporales sp. (fresh water). Pharm: Antibacterial (*Escherichia coli*, 25 μg/disc, IZ = 11.4 mm, control PG, 10 μg/disc, IZ = 31.8 mm, MIC = 64.0 μg/mL, PG, MIC = 2.0 μg/mL; *Bacillus subtilis*, 25 μg/disc, IZ = 13.6 mm, control PG, 10 μg/disc, IZ = 43.3 mm, MIC = 128.0 μg/mL, PG, MIC = 2.0 μg/mL). Ref: N. Ein-Gil, et al, ISME J., 2009, 3, 752 | T. Hosoe, et al, Heterocycles, 2010, 81, 2123 | Q. Yao, et al, Mar. Drugs, 2014, 12, 5902

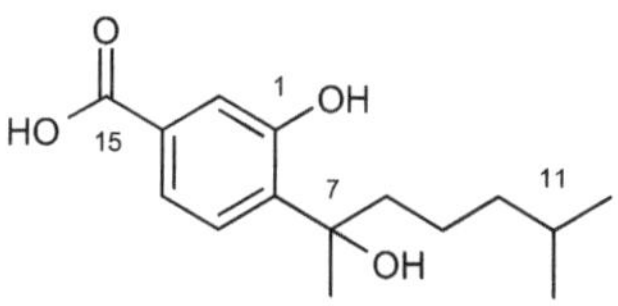

359 Varioxirane

Type: Other phenylalkanoids (C4–C8). $C_{15}H_{20}O_5$ Oil (tri-Ac), $[\alpha]_D^{25} = -28°$ (c = 0.31, $CHCl_3$) (tri-Ac). Source: Marine-derived fungus *Emericella variecolor*. Pharm: Antimicrobial. Ref: J. Malmstrøm, et al, JNP, 2002, 65, 364 | M. Saleem, et al, NPR, 2007, 24, 1142 (rev)

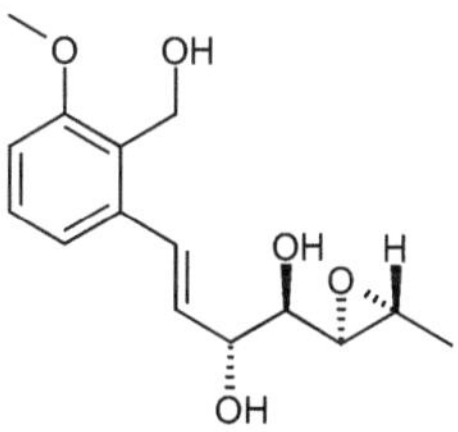

360 Ayamycin

Type: Nitrobenzene derivatives. $C_{14}H_{17}Cl_2NO_3$ Cryst., mp 115–117 °C. Source: Marine-derived bacterium *Nocardia* sp. ALAA 2000. Pharm: Antibacterial (gram-positive and -negative bacteria, MIC = 0.1 μg/mL). Ref: M. M. A. El-Gendy, et al,

J. Antibiot. (Tokyo) 2008, 61, 149 | M. M. A. El-Gendy, et al, J. Antibiot., 2008, 61, 379 | A. M. S. Mayer, et al, Comparative Biochemistry and Physiology, Part C 153, 2011, 191 (rev)

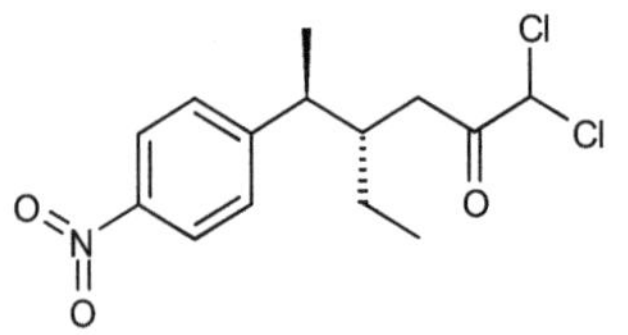

361 Bromochlorogentisylquinone A
Type: Benzoquinone derivatives. $C_7H_4BrClO_3$ Yellow solid. Source: Marine-derived fungus *Phoma herbarum* from red alga *Gloiopeitis tenax* (addition of halide salts to fermentation medium, Cheokpo, Tongnyeong, R. O. Korea). Pharm: Antioxidant (DPPH scavenger, IC_{50} = 3.8 μmol/L, control *L*-Ascorbic acid, IC_{50} = 20 μmol/L). Ref: V. N. Nenkep, et al, J. Antibiot., 2010, 63, 199

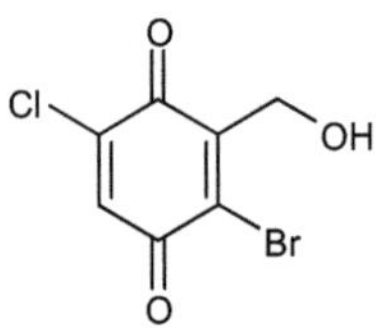

362 Bromochlorogentisylquinone B
Type: Benzoquinone derivatives. $C_7H_4BrClO_3$ Yellow solid. Source: Marine-derived fungus *Phoma herbarum* from red alga *Gloiopeitis tenax* (addition of halide salts to fermentation medium, Cheokpo, Tongnyeong, R. O. Korea). Pharm: Antioxidant (DPPH scavenger, IC_{50} = 3.9 μmol/L, control *L*-Ascorbic acid, IC_{50} = 20 μmol/L). Ref: V. N. Nenkep, et al, J. Antibiot., 2010, 63, 199

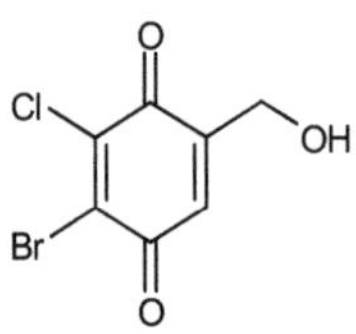

363 Chlorogentisylquinone
Type: Benzoquinone derivatives. $C_7H_5ClO_3$ Brown needles. Source: An unidentified marine-derived fungus. Pharm: Sphingomyelinase inhibitor (rat brain membrane). Ref: R. Uchida, et al, J. Antibiot., 2001, 54, 882

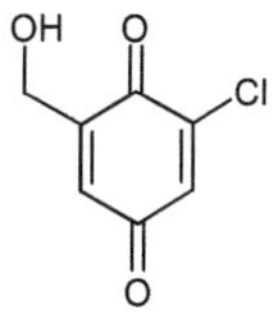

364 Coenzyme Q_9

Ubiquinone 45 Type: Benzoquinone derivatives. $C_{54}H_{82}O_4$ Orange cryst. (EtOH), mp 45.2 °C. Source: Marine-derived actinomycete *Nocardia* sp. KMM 3749 (psychrophilic, cold water, expedition in Kuril and Simushir Is., Sea of Okhotsk). Pharm: Cytotoxic (mouse erythrocytes, MIC = 30–50 μg/mL; fertilized eggs of sea urchin *Strongylocentrotus intermedius*, MIC = 40 μg/mL). Ref: T. A. Kuznetsova, et al, Russ. Chem. Bull., 2002, 51, 1951 | M.D. Lebar, et al, NPR, 2007, 24, 774 (rev) | CRC Press, DNP on DVD, 2012, version 20.2

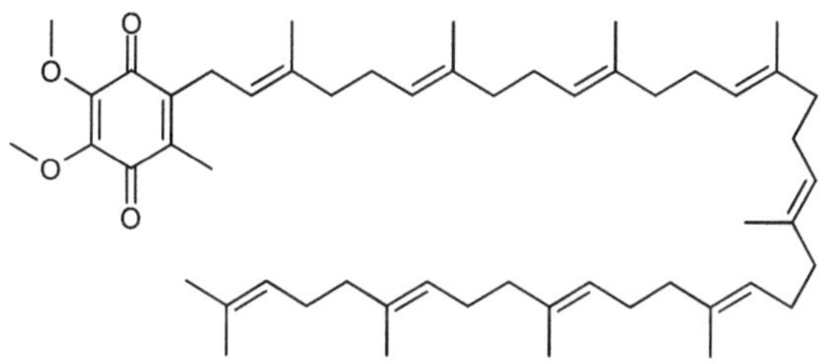

365 2,6-Dimethoxy-1,4-benzoquinone

Type: Benzoquinone derivatives. $C_8H_8O_4$ Yellow cryst. (AcOH), mp 256 °C, bp 311.1 °C Source: Marine-derived bacterium *Salegentibacter* sp. T436; marine-derived fungus *Dendryphiella salina* (grown on waste sulfite liquor), mangrove *Bruguiera sexangula* var. *rhynchopetala*. Pharm: Antibacterial; causes dermatitis; mutagenic. Ref: S. Bao, et al, Helv. Chim. Acta, 2005, 88, 2757 | I. Schuhmann, et al, J. Antibiot., 2009, 62, 453

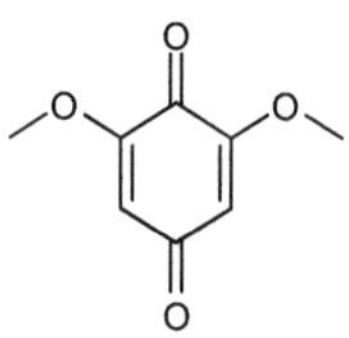

366 2-(1,1-Dimethyl-2-propenyl)-5-(3-methyl-2-butenyl)-1,4-benzoquinone

Type: Benzoquinone derivatives. $C_{16}H_{20}O_2$ Yellow cryst. (petrol), mp 27.5–29 °C. Source: Brown algae *Sporochnus comosus* (Shaw I., Queensland, Australia) and *Perithalia capillaris*. Pharm: Cytotoxic (SF268, GI_{50} = 17 μmol/L; MCF7, GI_{50} = 26 μmol/L; H460, GI_{50} = 41 μmol/L; HT29, GI_{50} = 37 μmol/L; CHO-K1, GI_{50} = 29 μmol/L); anti-inflammatory (hmn neutrophil free radical release inhibition *in vitro*,

IC_{50} = 2.1 μmol/L, MMOA: superoxide anion inhibition). Ref: S. P. B. Ovenden, et al, JNP, 2011, 74, 739 | C. E. Sansom, et al, JNP 2007, 70, 2042

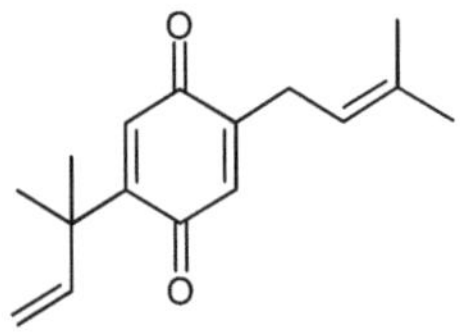

367 Glabruquinone A

Desmethylubiquinone Q_2 Type: Benzoquinone derivatives. $C_{18}H_{24}O_4$ Yellow oil. Source: Ascidian *Aplidium glabrum* (psychrophilic, cold water). Pharm: Cancer-preventive activity (anchorage-independent transformation assay, mouse JB6 P^+ CI41 cells that had been transformed with an epidermal growth factor); cytotoxic (HCT116, IC_{50} = 12.7 μmol/L; MEL28, IC_{50} = 17.5 μmol/L; HT460, IC_{50} = 50.5 μmol/L); contact allergen. Ref: L. K. Shubina, et al, Tet. Lett., 2005, 46, 559 | M. D. Lebar, et al, NPR, 2007, 24, 774 (rev)

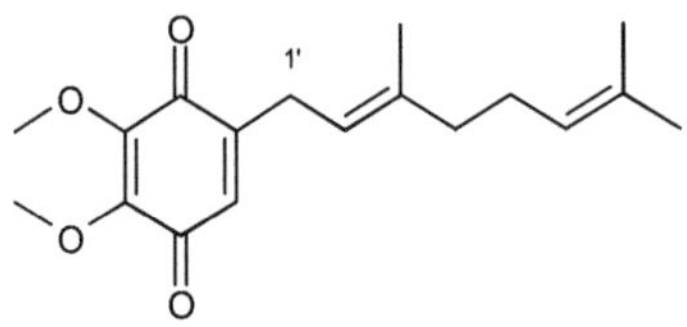

368 2-Hydroxy-3,6-dimethyl-5-(1-oxo-4-hexenyl)-1,4-benzoquinone

Type: Benzoquinone derivatives. $C_{14}H_{16}O_4$ Orange needles, mp 127–129 °C. Source: Marine-derived fungus *Penicillium terrestre*. Pharm: Cytotoxic (MTT method, P_{388} and A549). Ref: W. Liu, et al, J. Antibiot., 2005, 58, 441 | M. Saleem, et al, NPR, 2007, 24, 1142 (rev)

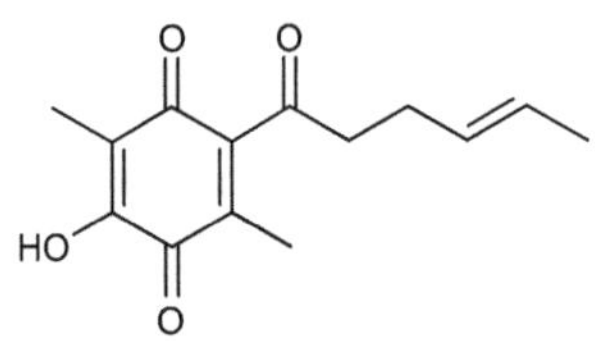

369 2-Hydroxy-3,5-dimethyl-6-(2-oxopropyl)-1,4-benzoquinone

Type: Benzoquinone derivatives. $C_{11}H_{12}O_4$ Orange needles, mp 118 °C (dec). Source: Marine-derived fungus *Penicillium terrestre*. Pharm: Cytotoxic (MTT method, P_{388} and A549). Ref: W. Liu, et al, J. Antibiot., 2005, 58, 441 | M. Saleem, et al, NPR, 2007, 24, 1142 (rev)

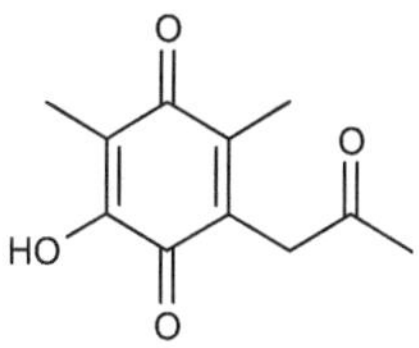

370 Ilimaquinone

Type: Benzoquinone derivatives. $C_{22}H_{30}O_4$ Cryst. (hexane), mp 113–114 °C, $[\alpha]_D^{23}$ = −23.2° (*c* = 1.12, $CHCl_3$). Source: Sponges *Hippospongia* sp. (Palau, Oceania), *Spongia* sp. (Australia), *Hippospongia metachromia, Smenospongia* spp., *Polyfibrospongia australis, Hyat* spp., *Dactylospongia elegans* and *Petrosaspongia metachromia.* Pharm: Cytotoxic (differentiation-inducing activity to K562 cells into erythroblast, lowest effective concentration = 15 µmol/L); cytotoxic (NCI-H460, $HepG_2$, SF268, MCF7, HeLa, and HL60 hmn tumor cell lines, inhibitory effects on maturation of starfish oocytes, and cell cycle arrest in $HepG_2$ cell line were evaluated); cytotoxic (SF268, GI_{50} = 2.7 µmol/L; MCF7, GI_{50} = 3.9 µmol/L; H460, GI_{50} = 1.8 µmol/L; HT29, GI_{50} = 5.4 µmol/L; CHO-K1, GI_{50} = 2.0 µmol/L); antistaphylococcal; HIV reverse transcriptase inhibitor; anti-HIV; induces vesiculation of Golgi membranes. Ref: R. T. Luibrand, et al, Tetrahedron, 1979, 35, 609 | R. J. Capon, et al, JOC, 1987, 52, 5059 | S. Loya, Antimicrob. Agents Chemother., 1990, 34, 2009 | P. A. Takizawa, et al, Cell (Cambridge, Mass.), 1993, 73, 1079 | U. Acharya, et al, J. Cell Biol., 1995, 129, 577 | N. K. Utkina, et al, Tetrahedron Lett. 2003, 44, 101 | S. Aoki, et al, CPB, 2004, 52, 935 | H. Liu, et al, Pharm. Biol. 2006, 44, 522 | M. Gordaliza, et al, Mar. Drugs, 2010, 8, 2849 (rev) | S. P. B. Ovenden, et al, JNP, 2011, 74, 65

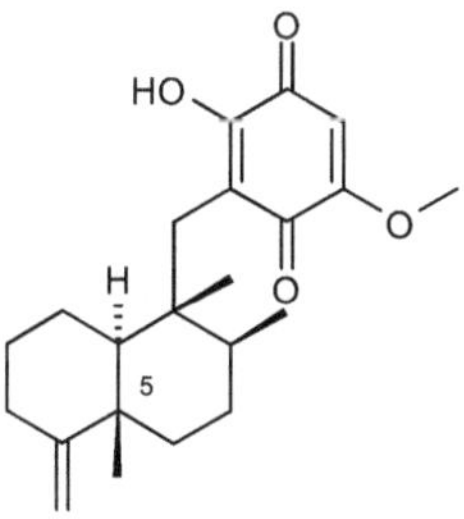

371 5-*epi*-Ilimaquinone

Type: Benzoquinone derivatives. $C_{22}H_{30}O_4$ $[\alpha]_D$ = +29.8° (*c* = 0.4, $CHCl_3$). Source: Sponges *Hippospongia* sp. (Palau, Oceania), *Fenestraspongia* sp., *Dactylospongia elegans* and *Hyrtios* sp. Pharm: Cytotoxic (P_{388}, IC_{50} = 2.2 µg/mL; A549, IC_{50} = 0.9 µg/mL; HT29, IC_{50} = 3.4 µg/mL; B16-F-10, IC_{50} = 1.1 µg/mL); cytotoxic (NCI-H460, HepG2, SF268, MCF7, HeLa, and HL60 hmn tumor cell lines, inhibitory effects on maturation of starfish oocytes, and cell cycle arrest in HepG2 cell line were evaluated); cytotoxic (differentiation-inducing activity to K562 cells into erythroblast, lowest effective concentration = 15 µmol/L); toxic (brine shrimp *Artemia larvae*). Ref: B. Carté, et al, JOC, 1985, 50, 2785 | J. Rodriguez, et al, Tetrahedron 1992, 48, 6667 | S. Aoki, et al, CPB, 2004, 52, 935 | H. Liu, et al, Pharm. Biol. 2006, 44, 522 | M. Gordaliza, et al, Mar. Drugs, 2010, 8, 2849 (rev)

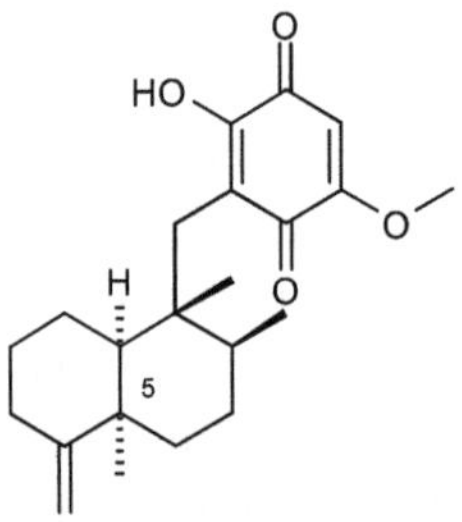

372 Longithorone A

Type: Benzoquinone derivatives. $C_{42}H_{46}O_5$ Cryst. (MeOH/CH_2Cl_2), mp 195–196 °C, $[\alpha]_D$ = −87.5° (*c* = 2.7, $CHCl_3$). Source: Ascidian *Aplidium longithorax*. Pharm: Cytotoxic (P_{388}, ED_{50} = 10 µg/mL, weak). Ref: X. Fu, et al, JACS, 1994, 116, 12125 (erratum: 1995, 117, 9381)

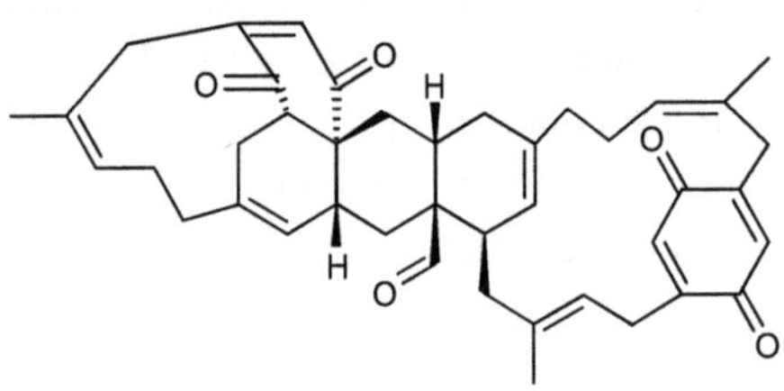

373 Sarcophytonone

Type: Benzoquinone derivatives. $C_{20}H_{30}O_5$ Yellow oil, $[\alpha]_D^{25}$ = +5.8° (*c* = 0.4, $CHCl_3$). Source: Soft coral *Sarcophyton crassocaule* (Lingshui Bay, Hainan), Pharm: Toxic (brine shrimp *Artemia salina*, mild). Ref: L. Li, et al, J. Asian Nat. Prod. Res., 2009, 11, 851

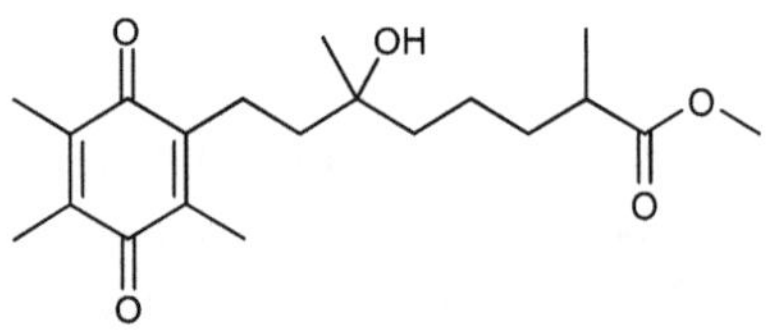

374 Sargaquinoic acid

Type: Benzoquinone derivatives. $C_{27}H_{36}O_4$ Yellow oil. Source: Brown alga *Sargassum serratifolium*. Pharm: Nervous system activity (butyrylcholinesterase inhibitor,

IC_{50} = 26 nmol/L); anticholinesterase activity of plastoquinones (lead compounds for Alzheimer's disease therapy). Ref: T. Kusumi, et al, Chem. Lett., 1979, 277 | B. W. Choi, et al, Phytother. Res. 2007, 21, 423

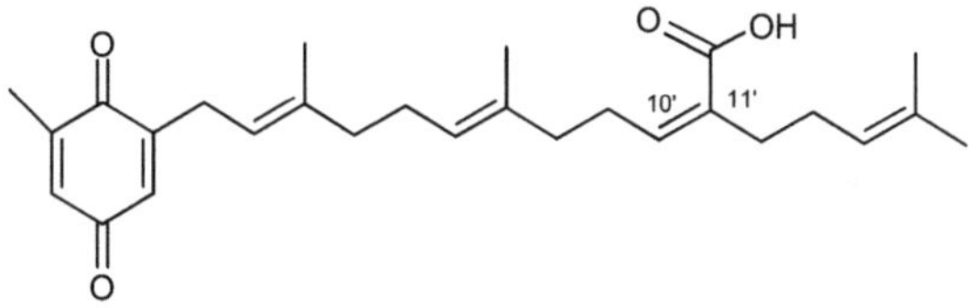

375 Smenoquinone

Type: Benzoquinone derivatives. $C_{21}H_{28}O_4$ Cryst., mp 350 °C. Source: Sponge *Smenospongia* sp. Pharm: Cytotoxic (L_{1210} cell line, IC_{50} = 2.5 μg/mL). Ref: M. L. Kondracki, et al, Tetrahedron, 1989, 45, 1995 | M. Gordaliza, et al, Mar. Drugs, 2010, 8, 2849 (rev)

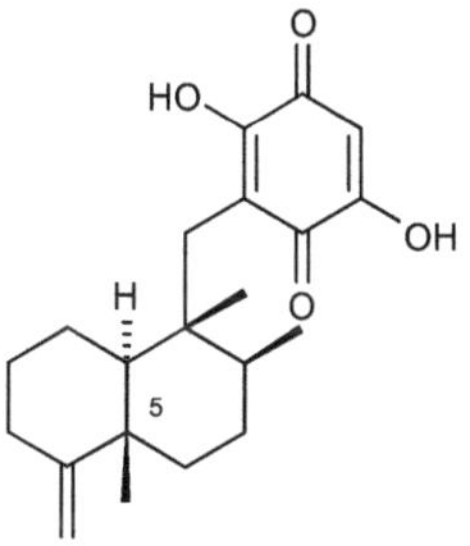

2.2 Diaryl Derivatives

376 Avrainvilleol

Type: Diphenylmethanoids. $C_{14}H_{14}Br_2O_4$ Viscous oil. Source: Green alga *Avrainvillea longicaulis*. Pharm: Ichthyotoxin; IMP dehydrogenase inhibitor; antifeedant (reef fish); toxin. Ref: H. H. Sun, et al, Phytochemistry, 1983, 22, 743

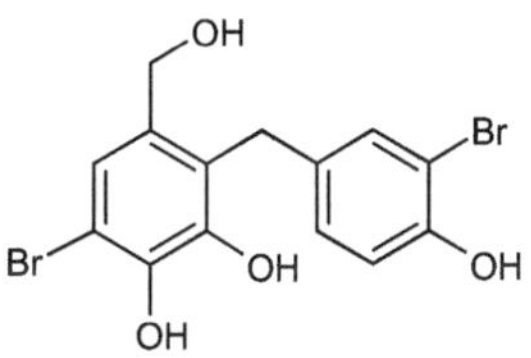

377 3-Bromo-4,5-bis(2,3-dibromo-4,5-dihydroxybenzyl)pyrocatechol

Type: Diphenylmethanoids. $C_{20}H_{13}Br_5O_6$ Source: Red alga *Rhodomela confervoides*. Pharm: Hypoglycemic (PTP1B inhibitor) (PTP1B and hypoglycaemic effect in streptozotocin-induced diabetic Wistar rats, IC_{50} = 1.7 μmol/L). Ref: D. Shi, et al, Chin. Sci. Bull., 2008, 53, 2476

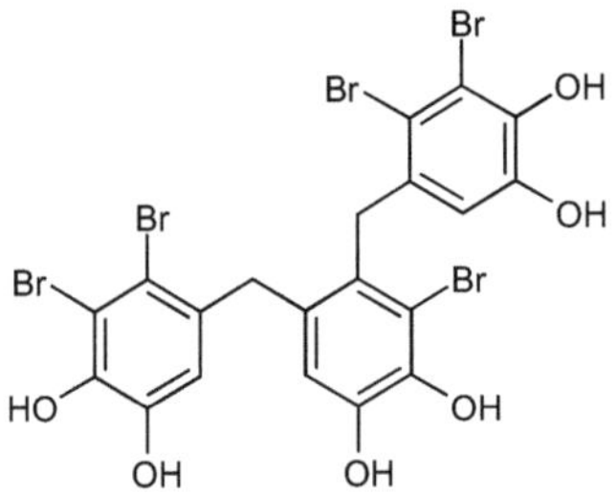

378 Isorawsonol

Type: Diphenylmethanoids. $C_{28}H_{22}Br_4O_7$ Source: Green alga *Avrainvillea rawsonii*. Pharm: Mosine 5′-monophosphate dehydrogenase inhibitor. Ref: J. L. Chen, et al, JNP, 1994, 57, 947

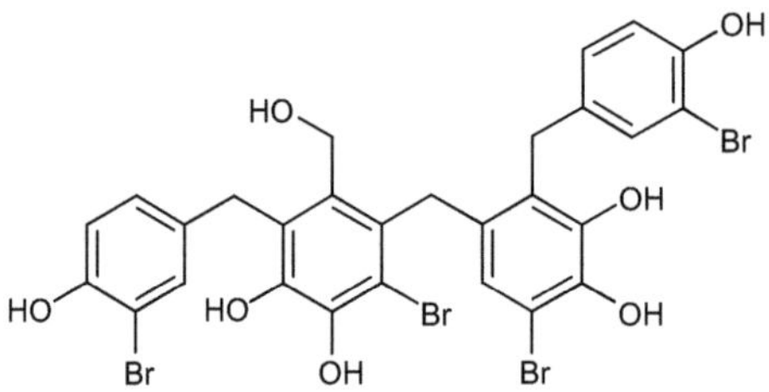

379 Peniphenone C

Type: Diphenylmethanoids. $C_{18}H_{18}O_6$ Source: Marine fungus *Penicillium dipodomyicola*. Pharm: Antituberculosis (*Mycobacterium tuberculosis*, strongly inhibits mycobacterial Protein tyrosine phosphatase B, MptpB). Ref: H. Li, et al, JNP 2014, 77, 800

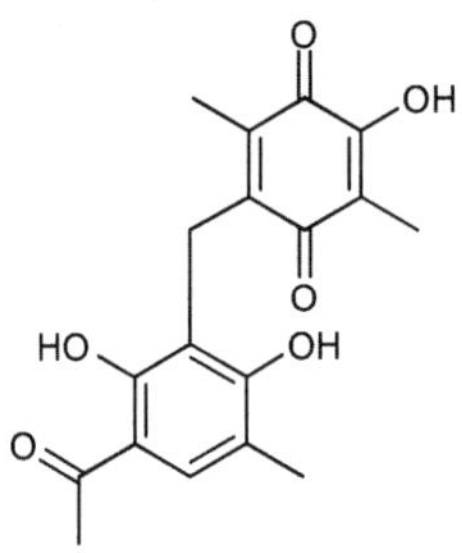

380 Rawsonol

Type: Diphenylmethanoids. $C_{29}H_{24}Br_4O_7$ Source: Green alga *Avrainvillea rawsonii*. Pharm: HMG-CoA reductase inhibitor (IC_{50} = 5 μmol/L, control fungal metabolite Mevinolin, (IC_{50} = 2 nmol/L)). Ref: B. K. Corte, et al, Phytochemistry, 1989, 28, 2917

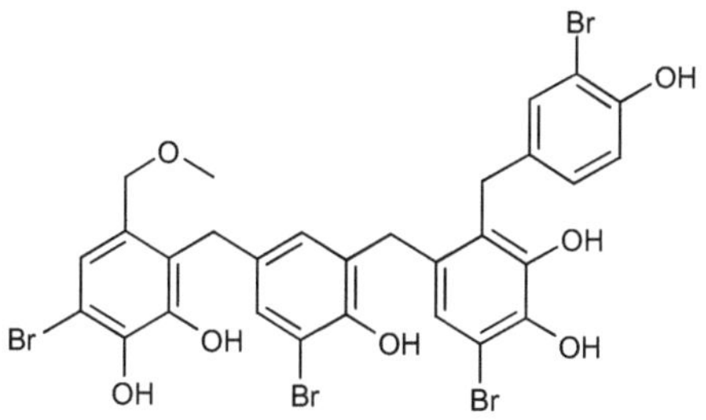

381 Terrestrol A

Type: Diphenylmethanoids. $C_{22}H_{22}O_7$ Brown gum. Source: Marine-derived fungus *Penicillium terrestre* (from sediment sample, China). Pharm: Cytotoxic (HL60, IC_{50} = 33.3 μmol/L; Molt4, IC_{50} = 5.5 μmol/L; A549, IC_{50} = 23.5 μmol/L; Bel7402, IC_{50} = 57.0 μmol/L); antioxidant (DPPH scavenger, IC_{50} = 2.6 μmol/L; control Ascorbis acid, IC_{50} = 17.4 μmol/L). Ref: L. Chen, et al, JNP, 2008, 71, 66

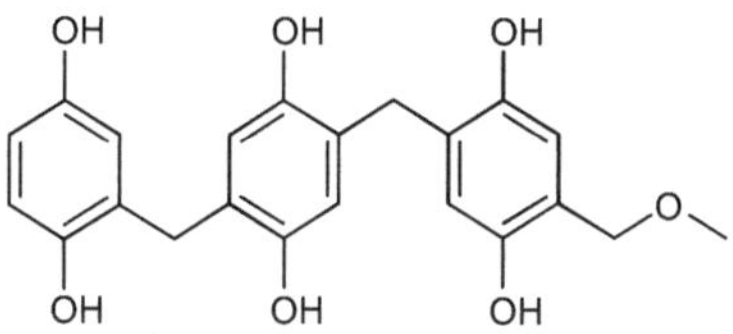

382 Terrestrol F

Type: Diphenylmethanoids. $C_{14}H_{13}ClO_5$ Brown gum. Source: Marine-derived fungus *Penicillium terrestre* (from sediment sample, China). Pharm: Cytotoxic (HL60, IC_{50} = 55.0 μmol/L; Molt4, IC_{50} = 58.1 μmol/L; A549, IC_{50} = 13.8 μmol/L; Bel7402, IC_{50} = 63.2 μmol/L); antioxidant (DPPH scavenger, IC_{50} = 5.2 μmol/L; control Ascorbis acid, IC_{50} = 17.4 μmol/L). Ref: L. Chen, et al, JNP, 2008, 71, 66

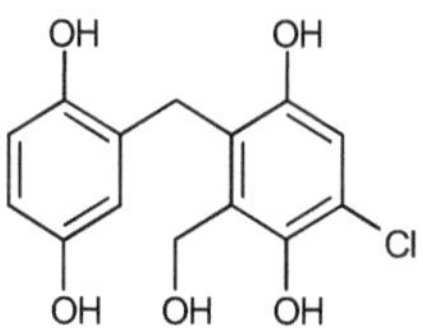

383 Terrestrol G

Type: Diphenylmethanoids. $C_{14}H_{13}ClO_5$ Brown gum. Source: Marine-derived fungus *Penicillium terrestre* (from sediment sample, China). Pharm: Cytotoxic (HL60, IC_{50} = 5.1 μmol/L; Molt4, IC_{50} = 6.5 μmol/L; A549, IC_{50} = 5.7 μmol/L; Bel7402, IC_{50} = 6.0 μmol/L); antioxidant (DPPH scavenger, IC_{50} = 4.1 μmol/L; control Ascorbis acid, IC_{50} = 17.4 μmol/L); PTK (protein tyrosine kinases) inhibitor (Src, 10 μmol/L, InRt = 35.9%, control PP2, 1 μmol/L, InRt = 89.4%; KDR, 10 μmol/L,

InRt = 31.8%, control SU11248, 10 μmol/L, InRt = 82.1%). Ref: L. Chen, et al, JNP, 2008, 71, 66

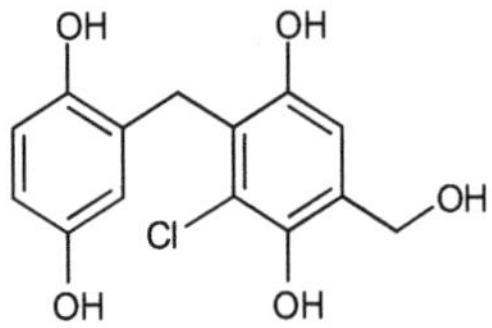

384 Terrestrol H

Type: Diphenylmethanoids. $C_{15}H_{16}O_5$ Brown gum. Source: Marine-derived fungus *Penicillium terrestre* (from sediment sample, China). Pharm: Cytotoxic (HL60, IC_{50} = 6.3 μmol/L; Molt4, IC_{50} = 5.8 μmol/L; A549, IC_{50} = 33.8 μmol/L; Bel7402, IC_{50} = 61.9 μmol/L); antioxidant (DPPH scavenger, IC_{50} = 6.3 μmol/L; control Ascorbis acid, IC_{50} = 17.4 μmol/L). Ref: L. Chen, et al, JNP, 2008, 71, 66

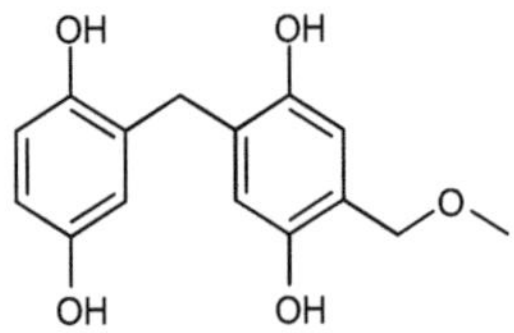

385 2,2′,3,3′-Tetrabromo-4,4′,5,5′-tetrahydroxydiphenylmethane

Type: Diphenylmethanoids. $C_{13}H_8Br_4O_4$ Cryst. (C_6H_6/MeOH), mp 200–201 °C. Source: Red algae *Rhodomela larix* and *Rhodomela confervoides*. Pharm: Hypoglycemic (PTP1B inhibitor) (PTP1B and hypoglycaemic effect in streptozotocin-induced diabetic Wistar rats, IC_{50} = 2.4 μmol/L). Ref: K. Kurata, et al, Chem. Lett., 1977, 1435 | D. Shi, et al, Chin. Sci. Bull., 2008, 53, 2476

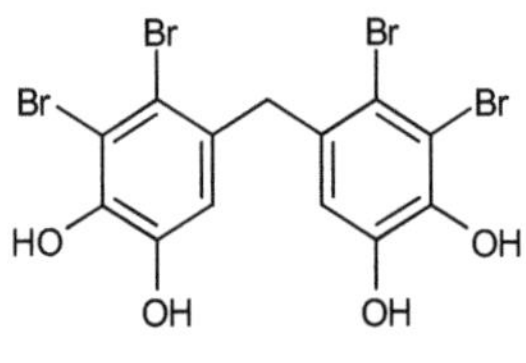

386 Arugosin H

Type: Benzophenonoids. $C_{22}H_{20}O_6$ Bright orange solid. Source: Marine-derived fungus *Emericella nidulans* var. *acristata* from an unidentified green alga (Mediterranean Sea). Pharm: Antifungal (*Mycotypha microspore*, 50 μg/disk, IZD = 3 mm); antialgal (*Chlorella fusca*, 50 μg/disk, IZD = 2 mm). Ref: A. Kralj, et al, JNP, 2006, 69, 995

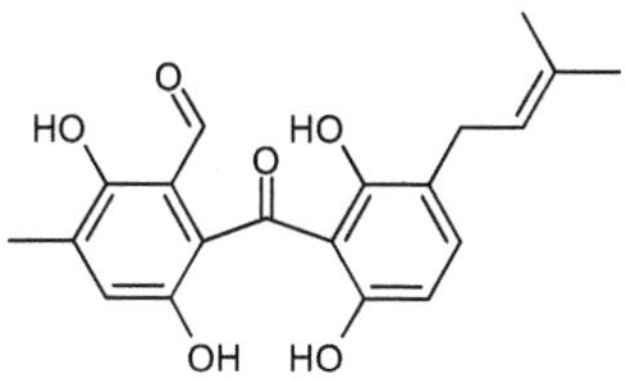

387 Isomonodictyphenone

Type: Benzophenonoids. $C_{15}H_{12}O_6$ Source: Mangrove-derived fungus *Penicillium* sp. MA-37 (rhizospheric soil). Pharm: Antibacterial (*Aeromonas hydrophilia*, MIC = 8 µg/mL, potent). Ref: H. Luo, et al, Phytochem. Lett., 2014, 9, 22 | Y. Zhang, et al, JNP, 2012, 75, 1888

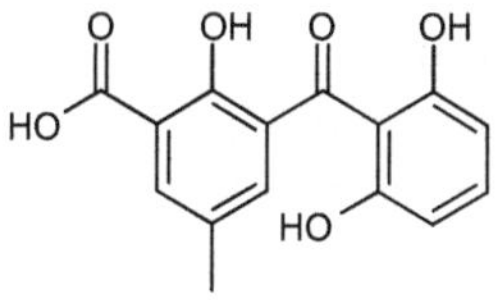

388 Monodictyphenone

Type: Benzophenonoids. $C_{15}H_{12}O_6$ Yellow solid. Source: Marine-derived fungus *Monodictys putredinis* 187/195 15 I. Pharm: Cytochrome P450 1A inhibitor (50 µmol/L, InRt = 28%); inducer of NAD(P)H quinone reductase (Hepa1c1c7, CD > 50 µmol/L, IC_{50} > 50 µmol/L); aromatase Cyp1A inhibitor (50 µmol/L, InRt = 25%). Ref: A. Krick, et al, JNP, 2007, 70, 353

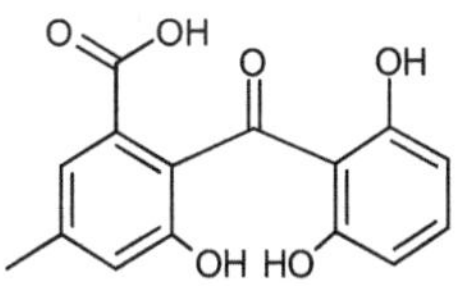

389 2′,5′,6′,5,6-Pentabromo-3′,4′,3,4-tetramethoxybenzo-phenone

Type: Benzophenonoids. $C_{17}H_{13}Br_5O_5$ Yellowish amorphous solid. Source: Red alga *Laurencia similis*. Pharm: Hypoglycemic (PTP1B inhibitor) (*in vitro* assay, IC_{50} = 2.66 µg/mL, control HD, IC_{50} = 0.80 µg/mL). Ref: J. Qin, et al, Bioorganic Med. Chem. Lett., 2010, 20, 7152

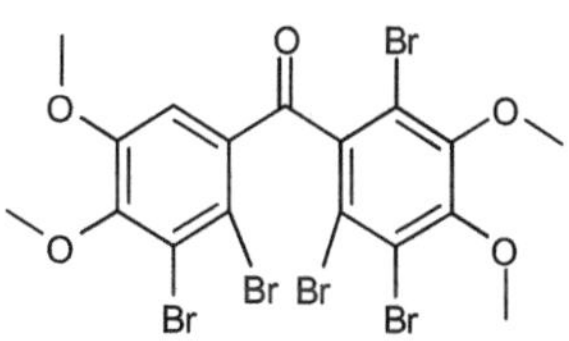

390 Pestalone

Type: Benzophenonoids. $C_{21}H_{20}Cl_2O_6$ Yellow cryst., mp 153–155 °C. Source: Marine-derived fungus *Pestalotia* sp. (cultured in presence of a bacterial antagonist) from brown alga *Rosenvingea* sp. (Bahamas). Pharm: Antibacterial (MRSA, MIC = 37 ng/mL; VREF, MIC = 78 ng/mL); cytotoxic. Ref: M. Cueto, et al, JNP, 2001, 64, 1444

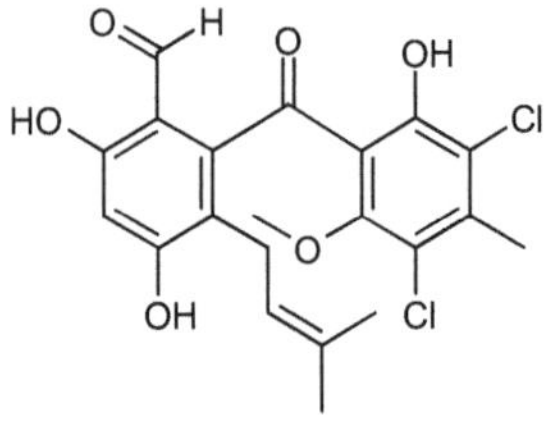

391 1,7-Dihydroxy-9-methyldibenzo[*b,d*] furan-3-carboxylic acid

Type: Dibenzofurans. $C_{14}H_{10}O_5$ Source: Sponge *Acanthella cavernosa* (Fiji). Pharm: Kinase EGFR inhibitor (100 µmol/L, InRt = 58.8%, control Genistein, InRt = 80%). Ref: M. E. Rateb, Bot. Mar. 2010, 53, 499 | D. Skropeta, et al, Mar. Drugs, 2011, 9, 2131 (rev)

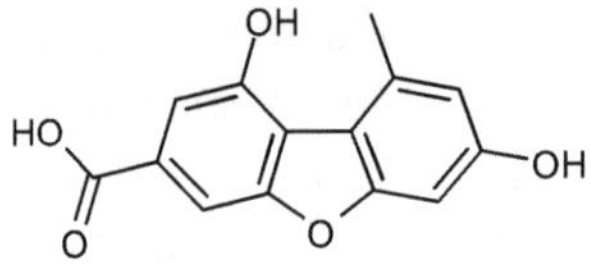

392 3,9-Dimethyldibenzo[*b,d*]furan-1,7-diol

Type: Dibenzofurans. $C_{14}H_{12}O_3$ Source: Sponge *Acanthella cavernosa* (Fiji). Pharm: Kinase EGFR inhibitor (100 µmol/L, InRt = 32.5%, control Genistein, InRt = 80%). Ref: M. E. Rateb, Bot. Mar. 2010, 53, 499 | D. Skropeta, et al, Mar. Drugs, 2011, 9, 2131 (rev)

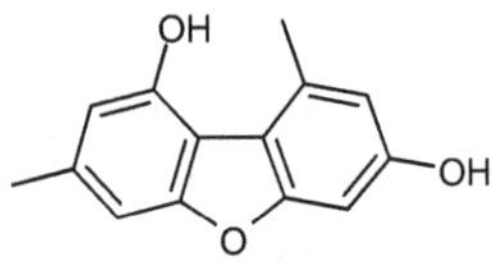

393 3-(Hydroxymethyl)-9-methyldibenzo[*b,d*]furan-1,7-diol

Type: Dibenzofurans. $C_{14}H_{12}O_4$ Source: Sponge *Acanthella cavernosa* (Fiji). Pharm: Kinase EGFR inhibitor (100 µmol/L, InRt = 42.5%, control Genistein, InRt = 80%). Ref: M. E. Rateb, Bot. Mar. 2010, 53, 499 | D. Skropeta, et al, Mar. Drugs, 2011, 9, 2131 (rev)

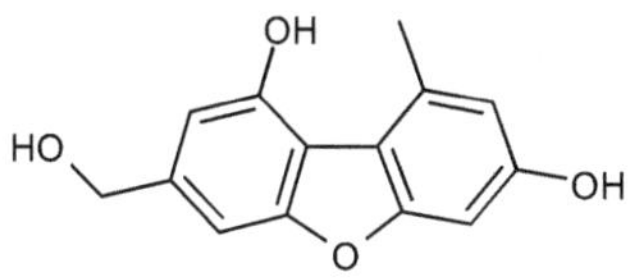

394 Isopopopolohuanone E

Type: Dibenzofurans. $C_{42}H_{54}O_6$ Deep purple solid. Source: Sponge *Dysidea* sp. (New Zealand). Pharm: Cytotoxic (P_{388}, IC_{50} = 25 µg/mL); antibacterial (*Bacillus subtilis*); antifungal (*Trichophyton mentagrophytes*). Ref: M. Stewart, et al, Aust. J. Chem., 1997, 50, 341

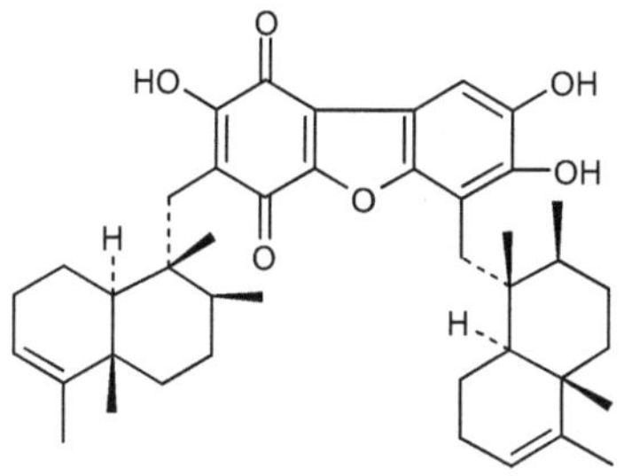

395 5,5'-*diepi-Δ3,Δ3'*-Popolohuanone E

Type: Dibenzofurans. $C_{42}H_{54}O_6$ Deep purple amorph. solid. Source: Sponge *Dysidea* sp. (New Zealand). Pharm: Cytotoxic (P_{388}, IC_{50} = 50 µg/mL); antibacterial (*Bacillus subtilis*); antifungal (*Trichophyton mentagrophytes*). Ref: M. Stewart, et al, Aust. J. Chem., 1997, 50, 341

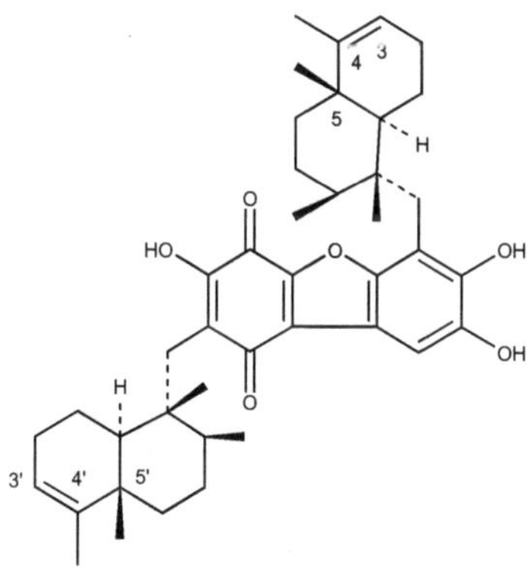

396 Popolohuanone E

Type: Dibenzofurans. $C_{42}H_{54}O_6$ Dark purple solid. Source: Sponge *Dysidea* sp. (Pohnpei I., Federated States of Micronesia). Pharm: Topoisomerase II inhibitor (with selective cytotoxicity against A549 non-small cell hmn lung cancer cell line).

Ref: J. R. Carney, et al, Tet. Lett., 1993, 34, 3727 | M. Gordaliza, et al, Mar. Drugs, 2010, 8, 2849 (rev)

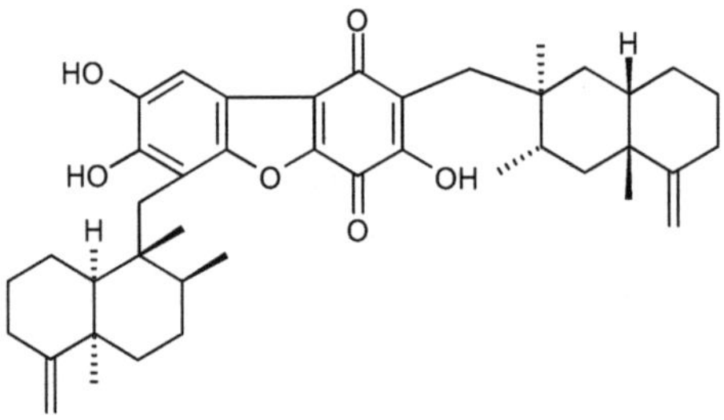

397 Porric acid D

Type: Dibenzofurans. $C_{15}H_{12}O_6$ Source: Marine-derived fungus *Alternaria* sp. (Bohai Sea, Tianjin, China). Pharm: Antibacterial (*Staphylococcus aureus*, weak). Ref: X. Xu, et al, Chem. Nat. Compd., 2012, 47, 893

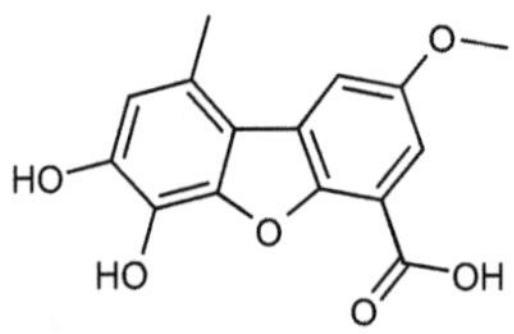

398 Cymobarbatol

Type: Dibenzo[*b*,*e*]pyrans. $C_{16}H_{20}Br_2O_2$ Cryst., mp 166 °C, $[\alpha]_D^{23}$ = −15.4°. Source: Green alga *Cymopolia barbata*. Pharm: Antimutagenic. Ref: M. E. Wall, et al, JNP, 1989, 52, 1092

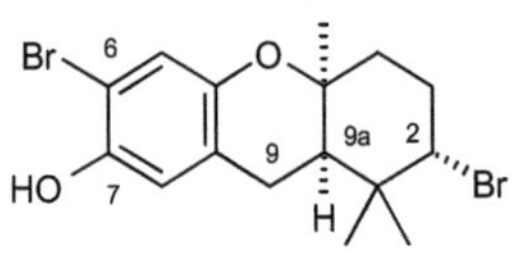

399 Debromoisocymobarbatol

Type: Dibenzo[*b*,*e*]pyrans. $C_{16}H_{21}BrO_2$ Viscous oil. Source: Green alga *Cymopolia barbata* (Florida Keys, USA). Pharm: Antifeedant (fish). Ref: M. Park, et al, Phytochemistry, 1992, 31, 4115

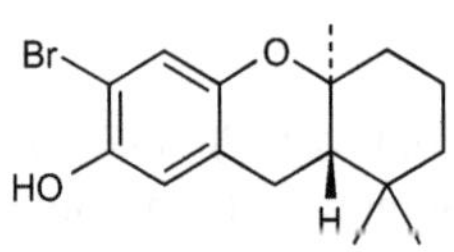

400 Isocymobarbatol

Type: Dibenzo[*b*,*e*]pyrans. $C_{16}H_{20}Br_2O_2$ Cryst., mp 147 °C, $[\alpha]_D^{23}$ = −51.4°. Source: Green alga *Cymopolia barbata*. Pharm: Antimutagenic. Ref: M. E. Wall, et al, JNP, 1989, 52, 1092

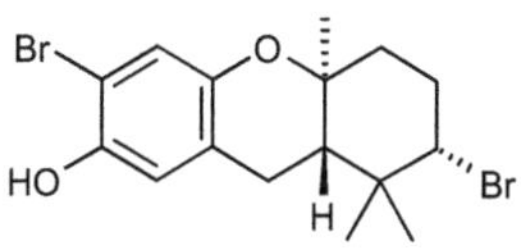

401 Anomalin A

2,3,6,8-Tetrahydroxy-1-methylxanthone Type: Xanthone derivatives. $C_{14}H_{10}O_6$ Yellow-brown powder. Source: Marine-derived fungus *Arthrinium* sp. from sponge *Geodia cydonium* (Adriatic Sea, Italy), marine fungi *Apiospora montagnei* and *Wardomyces anomalus*. Pharm: Protein kinases inhibitors (16 different protein kinases: AKT1, IC_{50} > 100 μmol/L; ALK, IC_{50} = 1.1 μmol/L; ARK5, IC_{50} = 15.6 μmol/L; Aurora-B, IC_{50} = 0.5 μmol/L; AXL, IC_{50} = 6.6 μmol/L; FAK, IC_{50} = 15.2 μmol/L; IGF1-R, IC_{50} = 8.6 μmol/L; MEK1 wt, IC_{50} > 100 μmol/L; MET wt, IC_{50} = 4.4 μmol/L; NEK2, IC_{50} = 83.9 μmol/L; NEK6, IC_{50} = 67.8 μmol/L; PIM1, IC_{50} = 0.3 μmol/L; PLK1, IC_{50} = 8.8 μmol/L; PRK1, IC_{50} = 45.3 μmol/L; SRC, IC_{50} = 12.9 μmol/L; VEGF-R2, IC_{50} = 0.8 μmol/L); cytotoxic (L5178Y, IC_{50} = 0.40 μmol/L, control Kahalalide F, IC_{50} = 4.30 μmol/L; K562, IC_{50} > 365 μmol/L, control Cisplatin (CDDP), IC_{50} = 7.80 μmol/L; A2780, IC_{50} = 4.34 μmol/L, Cisplatin (CDDP), IC_{50} = 0.80 μmol/L; A2780CisR, IC_{50} = 26.0 μmol/L, Cisplatin (CDDP), IC_{50} = 8.40 μmol/L); inhibits VEGF-A dependent endothelial cell sprouting (cellularangiogenesis assay, IC_{50} = 1.80 μmol/L, control Sunitinib, IC_{50} = 0.12 μmol/L); antioxidant (25.0 μg/mL, 94.7%). Ref: A. Abdel-Lateff, et al, JNP, 2003, 66, 706 | M. Saleem, et al, NPR, 2007, 24, 1142 (rev) | S. S. Ebada, et al, BoMC, 2011, 19, 4644

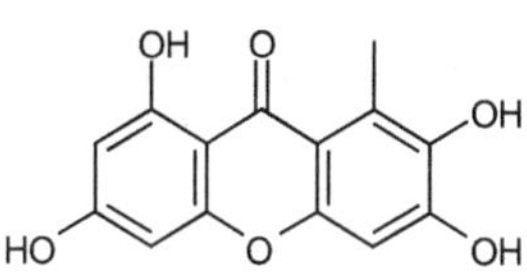

402 Antibiotic AGI-B4

Type: Xanthone derivatives. $C_{16}H_{14}O_7$ Yellow powder, mp 166–168 °C, $[\alpha]_D^{25}$ = −1.6° (*c* = 0.4, MeOH). Source: Marine-derive fungus *Neosartorya fischeri* 1008F1, terrestrial *Aspergillus* sp. Y80118. Pharm: Antiviral (TMV virus, IC_{50} = 260 μmol/L); inhibitor of VEGF-induced endothelial cell growth. Ref: H. S. Kim, et al, J. Antibiot., 2002, 55, 669 | Q. -W. Tan, et al, Nat. Prod. Res., 2012, 26, 1402

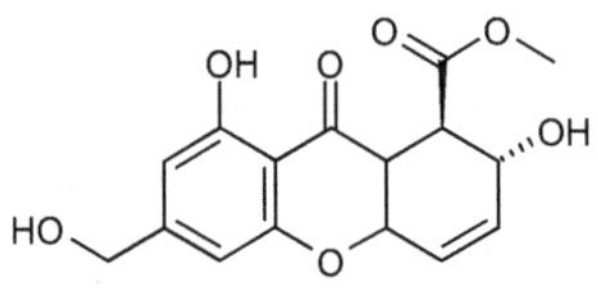

403 Aspergillusone B

1,2,8-Trihydroxy-6-(hydroxymethyl)-9-oxo-9*H*-xanthene-1-carboxylic acid methyl ester Type: Xanthone derivatives. $C_{16}H_{14}O_8$ Source: Marine-derived fungus *Aspergillus sydowii* from gorgonian sea fan *Annella* sp. (Surat Thani Province, Thailand), deep-sea fungus *Engyodontium album* DFFSCS021. Pharm: Antibacterial (*Escherichia coli,* 25 μg/disc, IZ = 11.0 mm, control PG, 10 μg/disc, IZ = 31.8 mm, MIC = 64.0 μg/mL, PG, MIC = 2.0 μg/mL; *Bacillus subtilis*, 25 μg/disc, IZ = 14.4 mm, control PG, 10 μg/disc, IZ = 43.3 mm, MIC = 64.0 μg/mL, PG, MIC = 2.0 μg/mL). Ref: K. Trisuwan, et al, JNP, 2011, 74, 1663 | Q. Yao, et al, Mar. Drugs, 2014, 12, 5902

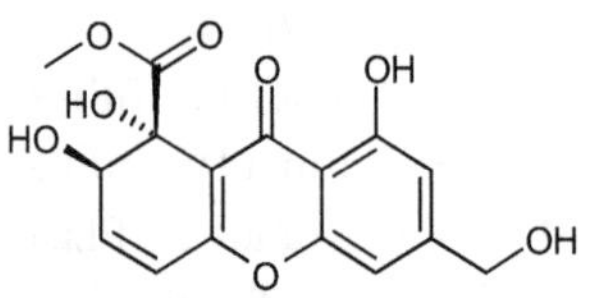

404 Chaetoxanthone A

Type: Xanthone derivatives. $C_{20}H_{18}O_7$ Yellow solid, $[\alpha]_D^{24}$ = −118° (*c* = 0.19, $CHCl_3$). Source: Marine-derived fungus *Chaetomium* sp. from an unidentified alga (Greece waters). Pharm: Antitrypanosomal (*Trypanosoma brucei rhodesiense* strain STIB 900, IC_{50} = 4.7 μg/mL, control Melarsoprol, IC_{50} = 0.004 μg/mL; *Trypanosoma cruzi* strain Tulahuen C4 IC_{50} > 10 μg/mL, control Benznidazole, IC_{50} = 0.317 μg/mL); anti-leishmanial (*Leishmania donovani* strain MHOM-ET-67/L82, IC_{50} = 5.3 μg/mL, control Miltefosine, IC_{50} = 0.125 μg/mL); antimalarial (*Plasmodium falciparum* strain K1, IC_{50} = 3.5 μg/mL, control Chloroquine, IC_{50} = 0.079 μg/mL); cytotoxic (L-6, IC_{50} = 59.1 μg/mL, control Podophyllotoxin, IC_{50} = 0.006 μg/mL). Ref: A. Pontius, et al, JNP, 2008, 71, 1579

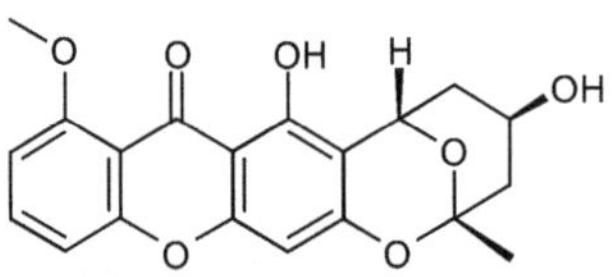

405 Chaetoxanthone B

Type: Xanthone derivatives. $C_{20}H_{18}O_6$ Yellow solid, $[\alpha]_D^{24}$ = +1.7° (*c* = 0.15, $CHCl_3$). Source: Marine-derived fungus *Chaetomium* sp. from an unidentified alga (Greece waters). Pharm: Antitrypanosomal (*Trypanosoma brucei rhodesiense* strain STIB 900, IC_{50} = 9.3 μg/mL, control Melarsoprol, IC_{50} = 0.004 μg/mL; *Trypanosoma cruzi* strain Tulahuen C4 IC_{50} = 7.1 μg/mL, control Benznidazole, IC_{50} = 0.317 μg/mL); antileishmanial (*Leishmania donovani* strain MHOM-ET-67/L82, IC_{50} = 3.4 μg/mL, control Miltefosine, IC_{50} = 0.125 μg/mL); antimalarial (*Plasmodium falciparum* strain K1, IC_{50} = 0.5 μg/mL, control Chloroquine, IC_{50} = 0.079 μg/mL); cytotoxic (L-6, IC_{50} > 90 μg/mL, control Podophyllotoxin, IC_{50} = 0.006 μg/mL). Ref: A. Pontius, et al, JNP, 2008, 71, 1579

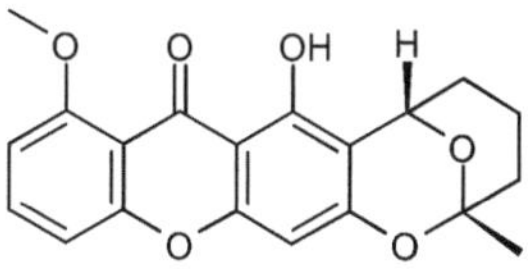

406 Chaetoxanthone C

Type: Xanthone derivatives. $C_{20}H_{19}ClO_6$ Yellow solid, $[\alpha]_D^{24}$ = +88° (*c* = 0.2, $CHCl_3$). Source: Marine-derived fungus *Chaetomium* sp. from an unidentified alga (Greece waters). Pharm: Antitrypanosomal (*Trypanosoma brucei rhodesiense* strain STIB 900, IC_{50} = 42.6 μg/mL, control Melarsoprol, IC_{50} = 0.004 μg/mL; *Trypanosoma cruzi* strain Tulahuen C4 IC_{50} = 1.5 μg/mL, control Benznidazole, IC_{50} = 0.317 μg/mL); antileishmanial (*Leishmania donovani* strain MHOM-ET-67/L82, IC_{50} = 3.1 μg/mL, control Miltefosine, IC_{50} = 0.125 μg/mL); antimalarial (*Plasmodium falciparum* strain K1, IC_{50} = 4.0 μg/mL, control Chloroquine, IC_{50} = 0.079 μg/mL); cytotoxic (L-6, IC_{50} = 46.7 μg/mL, control Podophyllotoxin, IC_{50} = 0.006 μg/mL). Ref: A. Pontius, et al, JNP, 2008, 71, 1579

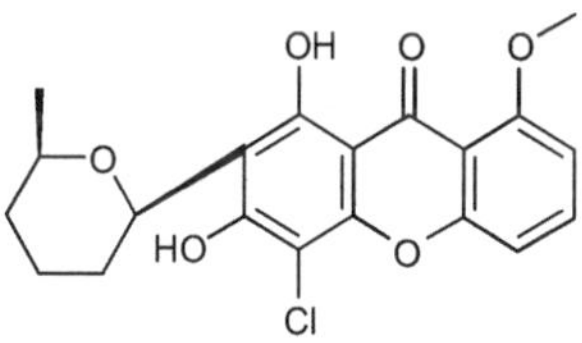

407 12-*O*-Deacetyl-phomoxanthone A

Type: Xanthone derivatives. $C_{36}H_{36}O_{15}$ Source: Mangrove-derived fungus *Phomopsis* sp. from mangrove *Rhizophora mucronata* (Muara Angke, Jakarta, Indonesia). Pharm: Antibacterial (several gram-positive bacteria, moderate). Ref: Y. Shiono, et al, Nat. Prod. Commun., 2013, 8, 1735

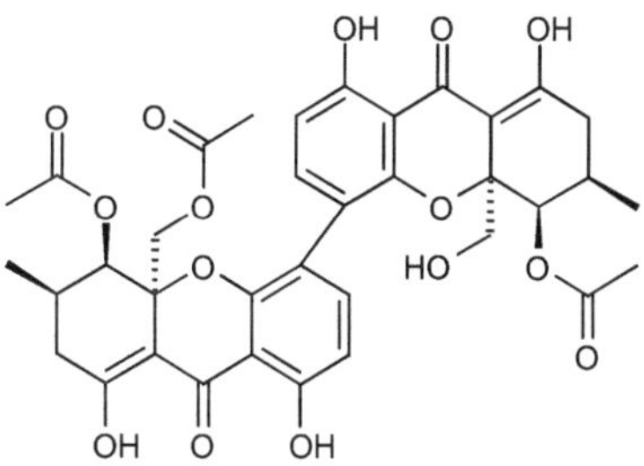

408 Deacetylphomoxanthone B

Type: Xanthone derivatives. $C_{34}H_{34}O_{14}$ Gum, $[\alpha]_D^{26}$ = +106.8° (*c* = 0.01, $CHCl_3$). Source: Mangrove-derived fungus *Phomopsis* sp. from mangrove *Acanthus ilicifolius* (branch, Hainan, China, first time isolated as MNP), terrestrial fungus *Phomopsis* sp. PSU-D15. Pharm: Cytotoxic (MDA-MB-435, IC_{50} = (14.40 ± 1.18)μmol/L, control Epirubicin, IC_{50} = (0.56 ± 0.06)μmol/L; HCT116, IC_{50} = (7.12 ± 0.70)μmol/L, Epirubicin, IC_{50} = (0.48 ± 0.03)μmol/L; Calu3, IC_{50} = (4.14 ± 0.02)μmol/L, Epirubicin, IC_{50} = (1.03 ± 0.10)μmol/L; Huh7, IC_{50} = (29.20 ± 0.19)μmol/L, Epirubicin, IC_{50} = (0.96 ± 0.01)μmol/L; MCF-10A, IC_{50} > 50 μmol/L, Epirubicin, IC_{50} = (0.48 ± 0.08)μmol/L). Ref: V. Rukachaisirikul, et al, Phytochemistry, 2008, 69, 783 | B. Ding, et al, Mar. Drugs, 2013, 11, 4961

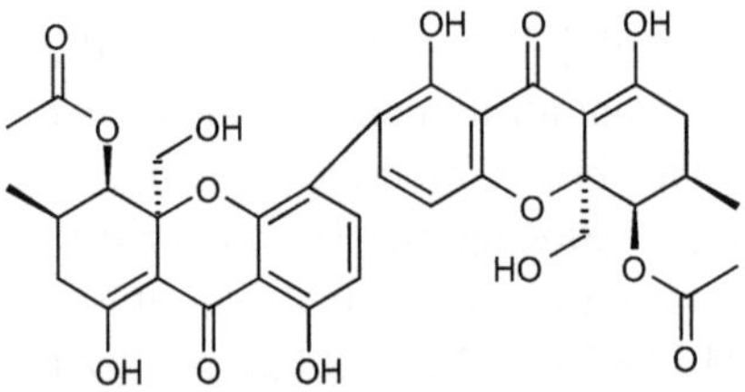

409 Deacetyl-phomoxanthone C

Type: Xanthone derivatives. $C_{34}H_{34}O_{14}$ Yellow powder, mp 145.5–146.7 °C, $[\alpha]_D^{25}$ = +101.6° (*c* = 0.44, MeOH). Source: Mangrove-derived fungus *Phomopsis* sp. from mangrove *Acanthus ilicifolius* (branch, Hainan, China). Pharm: Cytotoxic (MDA-MB-435, IC_{50} > 50 μmol/L; HCT116, IC_{50} = (44.06 ± 3.29)μmol/L, control Epirubicin, IC_{50} = (0.48 ± 0.03)μmol/L; Calu3, IC_{50} = (43.35 ± 2.09)μmol/L, Epirubicin, IC_{50} = (1.03 ± 0.10)μmol/L; Huh7, IC_{50} > 50 μmol/L; MCF-10A, IC_{50} > 50 μmol/L). Ref: B. Ding, et al, Mar. Drugs, 2013, 11, 4961

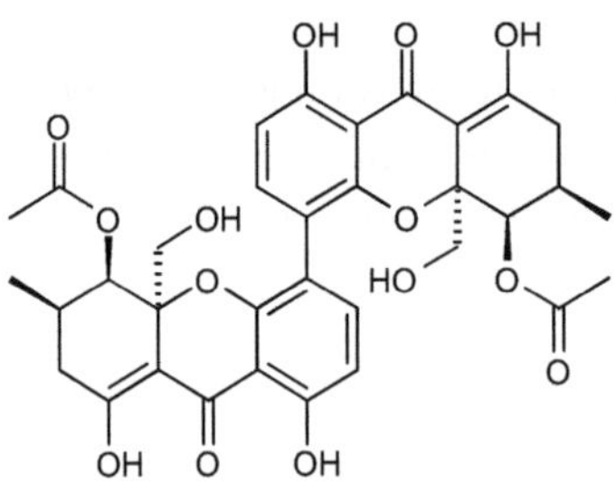

410 Dicerandrol A

Type: Xanthone derivatives. $C_{34}H_{34}O_{14}$ Amorph. yellow powder, $[\alpha]_D^{25}$ = −50.9° (*c* = 0.1, $CHCl_3$). Source: Mangrove-derived fungus *Phomopsis* sp. from mangrove *Acanthus ilicifolius* (branch, Hainan, China, first time isolated as MNP), terrestrial fungus *Phomopsis longicolla* from *Dicerandra frutescens* (endangered). Pharm: Cytotoxic

(MDA-MB-435, IC_{50} (3.03 ± 0.12)μmol/L, control Epirubicin, IC_{50} = (0.56 ± 0.06)μmol/L; HCT116, IC_{50} = (2.64 ± 0.03)μmol/L, Epirubicin, IC_{50} = (0.48 ± 0.03)μmol/L; Calu3, IC_{50} = (1.76 ± 0.02)μmol/L, Epirubicin, IC_{50} = (1.03 ± 0.10)μmol/L; Huh7, IC_{50} = (4.19 ± 0.08)μmol/L, Epirubicin, IC_{50} = (0.96 ± 0.01)μmol/L; MCF-10A, IC_{50} = (28.32 ± 3.57)μmol/L, Epirubicin, IC_{50} = (0.48 ± 0.08)μmol/L); antibacterial. Ref: M. M. Wagenaar et al, JNP, 2001, 64, 1006 | B. Ding, et al, Mar. Drugs, 2013, 11, 4961

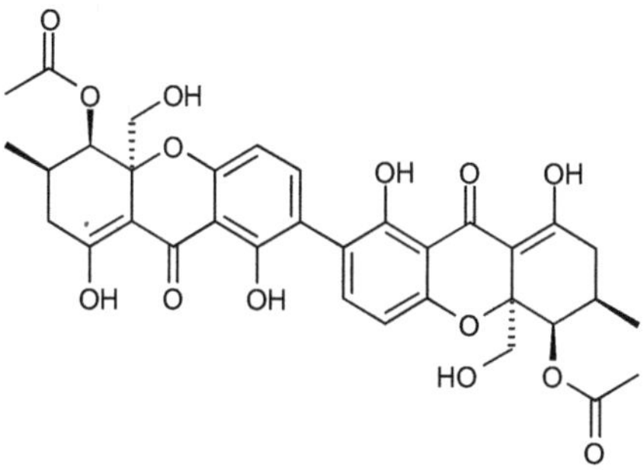

411 Dicerandrol B

Type: Xanthone derivatives. $C_{36}H_{36}O_{15}$ $[\alpha]_D^{25}$ = −6.5° (*c* = 0.1, $CHCl_3$). Source: Mangrove-derived fungus *Phomopsis* sp. from mangrove *Acanthus ilicifolius* (branch, Hainan, China first time isolated as MNP), terrestrial fungus *Phomopsis longicolla* from *Dicerandra frutescens* (endangered). Pharm: Cytotoxic (MDA-MB-435, IC_{50} = (8.65 ± 0.06)μmol/L, control Epirubicin, IC_{50} = (0.56 ± 0.06)μmol/L; HCT116, IC_{50} = (3.94 ± 0.39)μmol/L, Epirubicin, IC_{50} = (0.48 ± 0.03)μmol/L; Calu3, IC_{50} = (4.10 ± 0.08)μmol/L, Epirubicin, IC_{50} = (1.03 ± 0.10)μmol/L; Huh7, IC_{50} = (30.37 ± 1.10)μmol/L, Epirubicin, IC_{50} = (0.96 ± 0.01)μmol/L; MCF-10A, IC_{50} = (8.14 ± 1.27)μmol/L, Epirubicin, IC_{50} = (0.48 ± 0.08)μmol/L). Ref: M. M. Wagenaar et al, JNP, 2001, 64, 1006 | B. Ding, et al, Mar. Drugs, 2013, 11, 4961

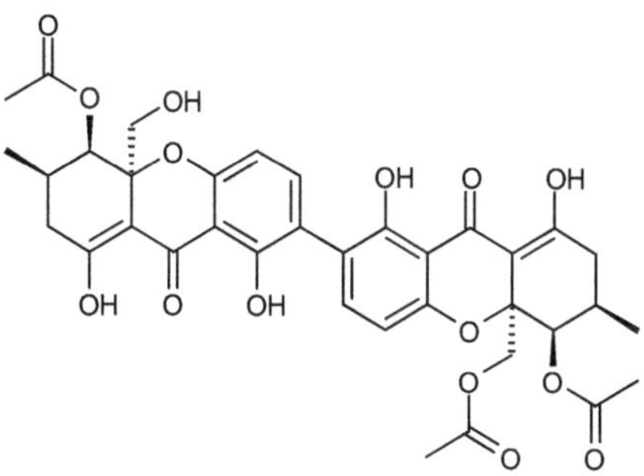

412 Dicerandrol C

Type: Xanthone derivatives. $C_{38}H_{38}O_{16}$ $[\alpha]_D$ = +44.3° (*c* = 0.003, $CHCl_3$). Source: Mangrove-derived fungus *Phomopsis* sp. from mangrove *Acanthus ilicifolius* (branch, Hainan, China first time isolated as MNP), terrestrial fungus *Phomopsis longicolla* from *Dicerandra frutescens* (endangered). Pharm: Cytotoxic (MDA-MB-435, IC_{50} = (44.10 ± 2.45)μmol/L, control Epirubicin, IC_{50} = (0.56 ± 0.06)μmol/L; HCT116, IC_{50} = (42.63 ± 2.90)μmol/L, Epirubicin, IC_{50} = (0.48 ± 0.03)μmol/L; Calu3, IC_{50} = (36.52 ± 0.32)μmol/L, Epirubicin, IC_{50} = (1.03 ± 0.10)μmol/L; Huh7, IC_{50} > 50 μmol/L, Epirubicin, IC_{50} = (0.96 ± 0.01)μmol/L; MCF-10A, IC_{50} = (33.05 ± 2.74)μmol/L, Epirubicin, IC_{50} = (0.48 ± 0.08)μmol/L). Ref: M. M. Wagenaar et al, JNP, 2001, 64, 1006 | B. Ding, et al, Mar. Drugs, 2013, 11, 4961

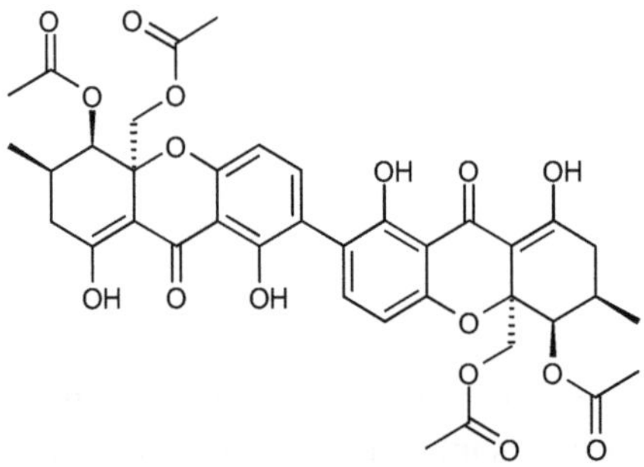

413 1,7-Dihydroxy-2-methoxy-3-(3-methylbut-2-enyl)-9*H*-xanthen- 9-one

Type: Xanthone derivatives. $C_{20}H_{18}O_5$ Yellow amorph. solid; mp 142–143 °C. Source: Mangrove-derived fungus *Phomopsis* sp. from mangrove *Avicennia marina* (Dong Zai, Hannan, China). Pharm: Cytotoxic (KB cell lines, IC_{50} = 20 μmol/L; KBV200 cell lines, IC_{50} = 30 μmol/L). Ref: Z. J. Huang, et al, Magn. Reson. Chem., 2010, 48, 80

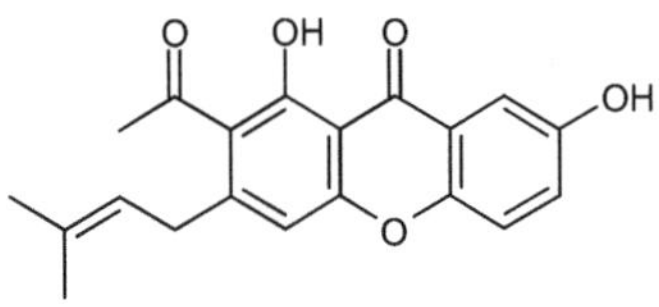

414 Dimethyl 8-methoxy-9-oxo-9*H*-xanthene-1,6-dicarboxylate

Type: Xanthone derivatives. $C_{18}H_{14}O_7$ Cryst., mp 188–189 °C. Source: Mangrove-derived fungus *Penicillium* sp. F 32 (culture broth) from mangrove *Acanthus ilicifolius* (bark, China). Pharm: Antifungal (*Fusarium oxysporum* f. *cubense*, MIC = 12.5 μg/mL). Ref: C. Shao, et al, Magn. Reson. Chem., 2008, 46, 1066

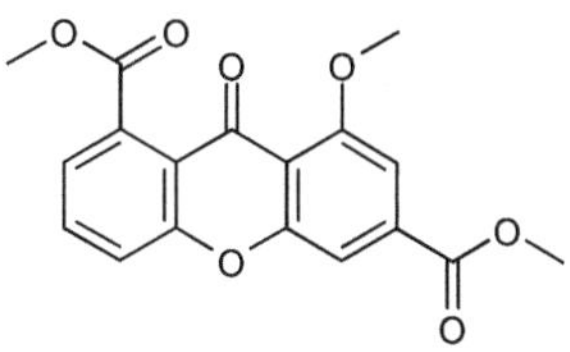

415 Emerixanthone A

Type: Xanthone derivatives. $C_{25}H_{27}ClO_6$ Yellow needle crystals, [α] = −78.1° (*c* = 0.0031, $CHCl_3$). Source: Deep-sea fungus *Emericella* sp. SCSIO 05240 (sediment sample, E120°0.975′ N19°0.664′, depth of 3258 m, open voyage to South China Sea, in August 2007). Pharm: Antibacterial (*Escherichia coli* ATCC 29922, *Klebsiella pneumonia* ATCC 13883, *Staphylococcus aureus* ATCC 29213, *Enterococcus faecalis* ATCC 29212, *Acinetobacter baumannii* ATCC 19606, and *Aeromonas hydrophila* ATCC 7966, all IZD = 4–6 mm, weak, control Ciprofloxacin, IZD = 35–40 mm). Ref: M. Fredimoses, et al, Mar. Drugs 2014, 12, 3190

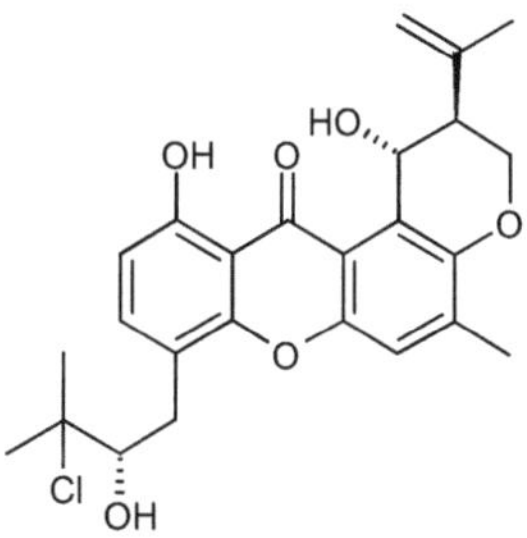

416 Emerixanthone C

Type: Xanthone derivatives. $C_{27}H_{32}O_7$ Yellow needle crystals, [α] = −95.5° (*c* = 0.0018, $CHCl_3$). Source: Deep-sea fungus *Emericella* sp. SCSIO 05240 (sediment sample, E120°0.975′ N19°0.664′, depth of 3258 m, open voyage to South China Sea, in August 2007). Pharm: Antibacterial (*Escherichia coli* ATCC 29922, *Klebsiella pneumonia* ATCC 13883, *Staphylococcus aureus* ATCC 29213, *Enterococcus faecalis* ATCC 29212, *Acinetobacter baumannii* ATCC 19606, and *Aeromonas hydrophila* ATCC 7966, all IZD = 4–6 mm, weak, control Ciprofloxacin, IZD = 35–40 mm). Ref: M. Fredimoses, et al, Mar. Drugs 2014, 12, 3190

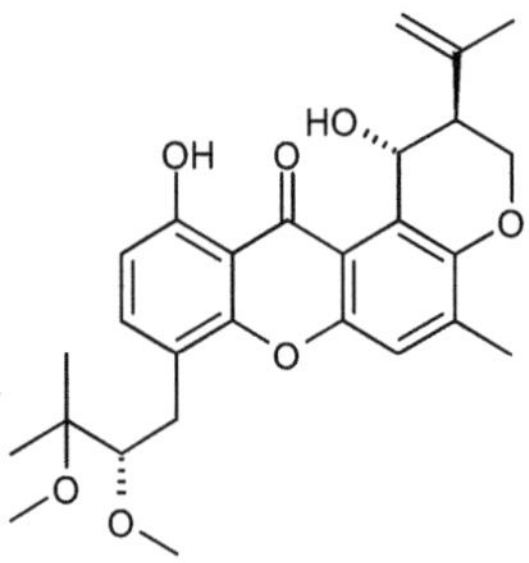

417 Emerixanthone D

Type: Xanthone derivatives. $C_{27}H_{30}O_8$ Yellow needle crystals, [α] = −93.3° (*c* = 0.009, $CHCl_3$). Source: Deep-sea fungus *Emericella* sp. SCSIO 05240 (sediment sample, E120°0.975′ N19°0.664′, depth of 3258 m, open voyage to South China Sea, in August 2007). Pharm: Antifungal (agricultural pathogens *Fusarium* sp., *Penicillium* sp., *Aspergillus niger, Rhizoctonia solani, Fusariumoxy sporium* f. sp. *niveum, Fusariumoxy sporium* f. sp. *cucumeris*, all IZD = 3–4 mm, mild, control Carbendazim, IZD = 40–45 mm). Ref: M. Fredimoses, et al, Mar. Drugs 2014, 12, 3190

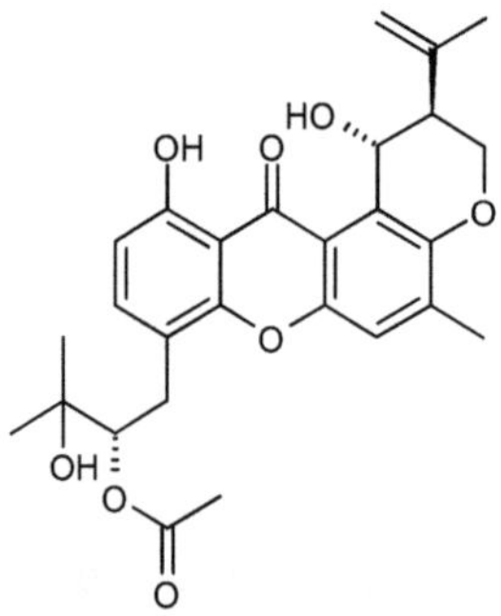

418 Engyodontiumone B

Type: Xanthone derivatives. $C_{16}H_{11}ClO_6$ Yellow needles. Source: Deep-sea fungus *Engyodontium album* DFFSCS021. Pharm: Cytotoxic (U937, IC_{50} = 55.5 μmol/L, control DOX, IC_{50} = 0.06 μmol/L; HeLa, IC_{50} = 96.1 μmol/L, DOX, IC_{50} = 0.8 μmol/L; MCF7, IC_{50} = 172.3 μmol/L, DOX, IC_{50} = 23.1 μmol/L; HepG2, IC_{50} = 73.8 μmol/L, DOX, IC_{50} = 3.3 μmol/L; Huh7, IC_{50} > 300 μmol/L, DOX, IC_{50} = 1.2 μmol/L). Ref: Q. Yao, et al, Mar. Drugs, 2014, 12, 5902

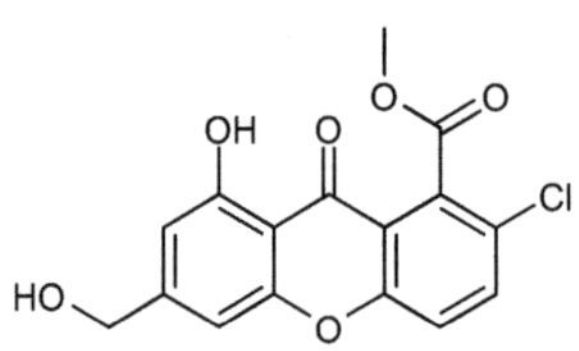

419 Engyodontiumone C

Type: Xanthone derivatives. $C_{16}H_{14}O_7$ Brown needles, $[\alpha]_D^{25}$ = +5° (*c* = 0.5, MeOH). Source: Deep-sea fungus *Engyodontium album* DFFSCS021. Pharm: Cytotoxic (U937, IC_{50} = 218.4 μmol/L, control DOX, IC_{50} = 0.06 μmol/L; HeLa, IC_{50} > 300 μmol/L, DOX, IC_{50} = 0.8 μmol/L; MCF7, IC_{50} > 300 μmol/L, DOX, IC_{50} = 23.1 μmol/L; HepG2, IC_{50} > 300 μmol/L, DOX, IC_{50} = 3.3 μmol/L; Huh7, IC_{50} > 300 μmol/L, DOX, IC_{50} = 1.2 μmol/L). Ref: Q. Yao, et al, Mar. Drugs, 2014, 12, 5902

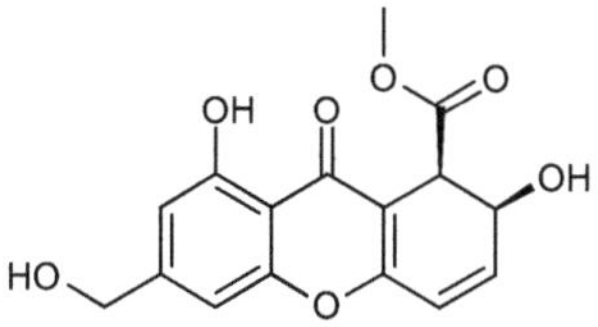

420 Engyodontiumone D

Type: Xanthone derivatives. $C_{16}H_{16}O_7$ Brown needles $[\alpha]_D^{25}$ = +14° (*c* = 0.3, MeOH). Source: Deep-sea fungus *Engyodontium album* DFFSCS021. Pharm: Cytotoxic (U937, IC_{50} = 208.6 μmol/L, control DOX, IC_{50} = 0.06 μmol/L; HeLa, IC_{50} > 300 μmol/L, DOX, IC_{50} = 0.8 μmol/L; MCF7, IC_{50} > 300 μmol/L, DOX, IC_{50} = 23.1 μmol/L; HepG2, IC_{50} > 300 μmol/L, DOX, IC_{50} = 3.3 μmol/L; Huh7, IC_{50} > 300 μmol/L, DOX, IC_{50} = 1.2 μmol/L). Ref: Q. Yao, et al, Mar. Drugs, 2014, 12, 5902

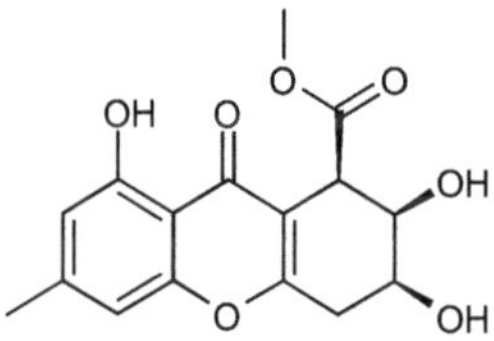

421 (±)-Engyodontiumone E

Type: Xanthone derivatives. $C_{16}H_{16}O_7$ Brown needles $[\alpha]_D^{25}$ = 0° (*c* = 0.4, MeOH). Source: Deep-sea fungus *Engyodontium album* DFFSCS021. Pharm: Cytotoxic (U937, IC_{50} = 15.9 μmol/L, control DOX, IC_{50} = 0.06 μmol/L; HeLa, IC_{50} = 205.2 μmol/L, DOX, IC_{50} = 0.8 μmol/L; MCF7, IC_{50} > 300 μmol/L, DOX, IC_{50} = 23.1 μmol/L; HepG2, IC_{50} > 300 μmol/L, DOX, IC_{50} = 3.3 μmol/L; Huh7, IC_{50} > 300 μmol/L, DOX, IC_{50} = 1.2 μmol/L). Ref: Q. Yao, et al, Mar. Drugs, 2014, 12, 5902

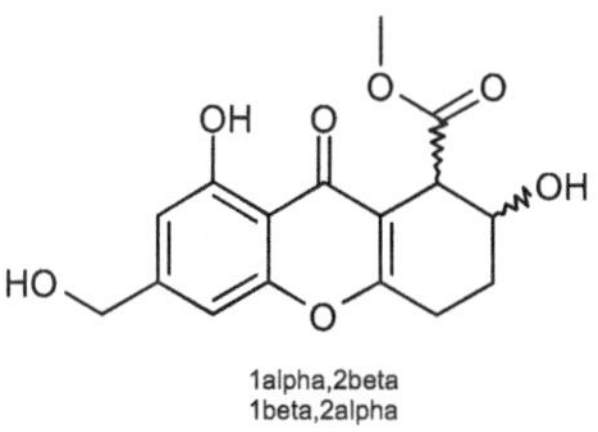

422 (±)-Engyodontiumone F

Type: Xanthone derivatives. $C_{16}H_{16}O_8$ Brown needles $[\alpha]_D^{25}$ = 0° (*c* = 0.3, MeOH). Source: Deep-sea fungus *Engyodontium album* DFFSCS021. Pharm: Cytotoxic (U937, IC_{50} = 192.7 µmol/L, control DOX, IC_{50} = 0.06 µmol/L; HeLa, IC_{50} > 300 µmol/L, DOX, IC_{50} = 0.8 µmol/L; MCF7, IC_{50} > 300 µmol/L, DOX, IC_{50} = 23.1 µmol/L; HepG2, IC_{50} > 300 µmol/L, DOX, IC_{50} = 3.3 µmol/L; Huh7, IC_{50} > 300 µmol/L, DOX, IC_{50} = 1.2 µmol/L). Ref: Q. Yao, et al, Mar. Drugs, 2014, 12, 5902

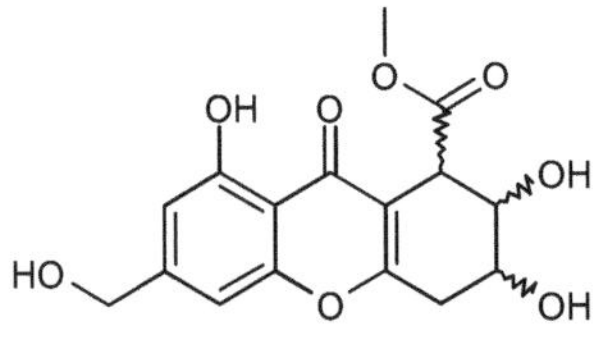

1beta,2beta,3beta
1alpha,2alpha,3alpha

423 (±)-Engyodontiumone G

Type: Xanthone derivatives. $C_{16}H_{16}O_8$ Brown needles $[\alpha]_D^{25}$ = 0° (*c* = 0.2, MeOH). Source: Deep-sea fungus *Engyodontium album* DFFSCS021. Pharm: Cytotoxic (U937, IC_{50} = 287.2 µmol/L, control DOX, IC_{50} = 0.06 µmol/L; HeLa, IC_{50} > 300 µmol/L, DOX, IC_{50} = 0.8 µmol/L; MCF7, IC_{50} > 300 µmol/L, DOX, IC_{50} = 23.1 µmol/L; HepG2, IC_{50} > 300 µmol/L, DOX, IC_{50} = 3.3 µmol/L; Huh7, IC_{50} > 300 µmol/L, DOX, IC_{50} = 1.2 µmol/L). Ref: Q. Yao, et al, Mar. Drugs, 2014, 12, 5902

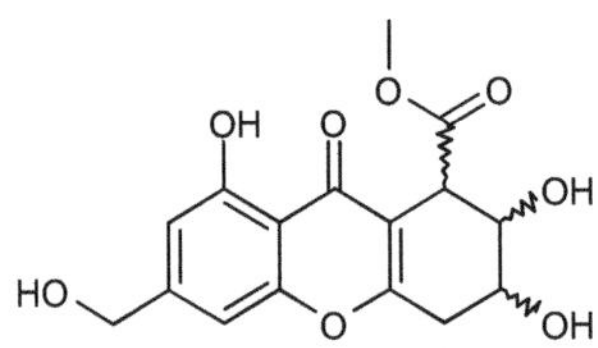

1beta,2beta,3alpha
1alpha,2alpha,3beta

424 Engyodontiumone H

Type: Xanthone derivatives. $C_{16}H_{14}O_7$ Brown needles, $[\alpha]_D^{25}$ = −56° (*c* = 1, MeOH). Source: Deep-sea fungus *Engyodontium album* DFFSCS021. Pharm: Cytotoxic (U937, IC_{50} = 4.9 µmol/L, control DOX, IC_{50} = 0.06 µmol/L; HeLa, IC_{50} = 24.8 µmol/L, DOX, IC_{50} = 0.8 µmol/L; MCF7, IC_{50} = 38.5 µmol/L, DOX, IC_{50} = 23.1 µmol/L; HepG2, IC_{50} = 60.5 µmol/L, DOX, IC_{50} = 3.3 µmol/L; Huh7, IC_{50} = 53.3 µmol/L, DOX, IC_{50} = 1.2 µmol/L); antibacterial (*Escherichia coli*, 25 µg/disc, IZ = 13.8 mm, control PG, 10 µg/disc, IZ = 31.8 mm, MIC = 64.0 µg/mL, PG, MIC = 2.0 µg/mL; *Bacillus subtilis*, 25 µg/disc, IZ = 16.5 mm, control PG, 10 µg/disc, IZ = 43.3 mm, MIC = 32.0 µg/mL, PG, MIC = 2.0 µg/mL). Ref: Q. Yao, et al, Mar. Drugs, 2014, 12, 5902

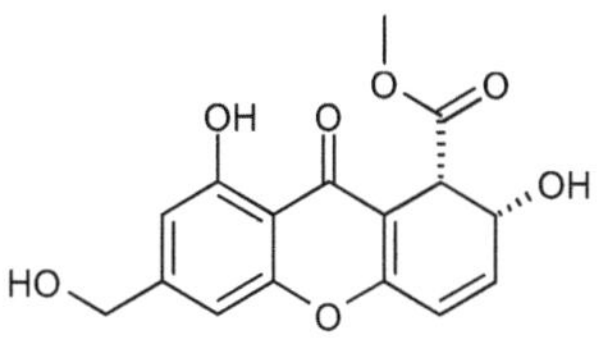

425 (2β)-Engyodontiumone H

AGI-B4 Type: Xanthone derivatives. $C_{16}H_{14}O_7$ Source: Deep-sea fungus *Engyodontium album* DFFSCS021. Pharm: Cytotoxic (U937, IC_{50} = 8.8 μmol/L, control DOX, IC_{50} = 0.06 μmol/L; HeLa, IC_{50} = 60.0 μmol/L, DOX, IC_{50} = 0.8 μmol/L; MCF7, IC_{50} = 102.2 μmol/L, DOX, IC_{50} = 23.1 μmol/L; HepG2, IC_{50} = 52.7 μmol/L, DOX, IC_{50} = 3.3 μmol/L; Huh7, IC_{50} = 133.3 μmol/L, DOX, IC_{50} = 1.2 μmol/L); antibacterial (*Escherichia coli*, 25 μg/disc, IZ = 15.8 mm, control PG, 10 μg/disc, IZ = 31.8 mm, MIC = 64 μg/mL, PG, MIC = 2.0 μg/mL; *Bacillus subtilis*, 25 μg/disc, IZ = 17.5 mm, control PG, 10 μg/disc, IZ = 43.3 mm, MIC = 64 μg/mL, PG, MIC = 2.0 μg/mL). Ref: Q. Yao, et al, Mar. Drugs, 2014, 12, 5902

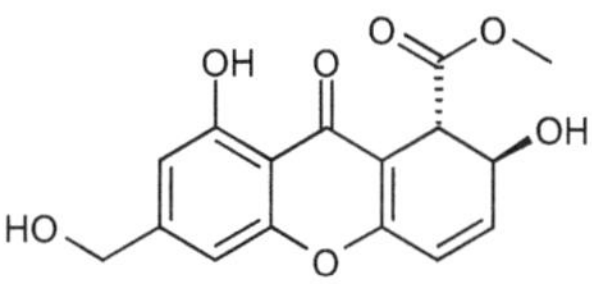

426 globosuxanthone A

Type: Xanthone derivatives. $C_{15}H_{12}O_7$ Yellow cryst., mp 152 °C, $[\alpha]_D^{25}$ = −50° (*c* = 1.2, MeOH/$CHCl_3$), $[\alpha]_D^{25}$ = −29° (*c* = 0.2, DMSO). Source: Marine-derived fungus *Beauveria bassiana* from an unidentified sponge (Iriomote I., Okinawa, Japan), an unidentified terrestrial plant. Pharm: Antineoplastic. Ref: H. Yamazaki, et al, Mar. Drugs, 2012, 10, 2691 | E. M. K. Wijeratne, et al, BoMC, 2006, 14, 7917

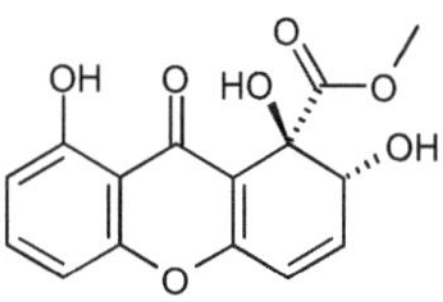

427 Grisephenone A

Type: Xanthone derivatives. $C_{17}H_{17}ClO_6$ Source: Sponge-derived fungus *Stachybotrys* sp. HH1 ZSDS1F1-2. Pharm: Antiviral (Enterovirus-71, IC_{50} = 50.0 μmol/L). Ref: C. Qin, et al, J. Antibiot., 2014, 68, 121

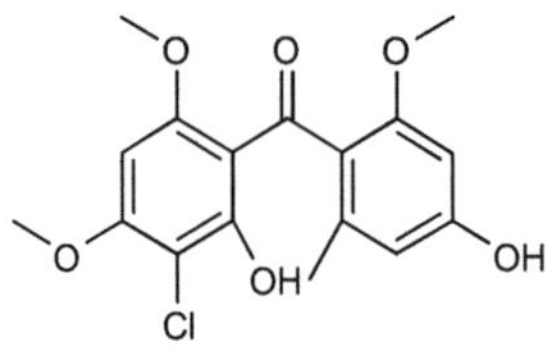

428 1-Hydroxy-4,7-dimethoxy-6-(3-oxobutyl)-9*H*-xanthen-9-one

Type: Xanthone derivatives. $C_{20}H_{18}O_6$ Yellow semisolid. Source: Mangrove-derived fungus *Phomopsis* sp. from mangrove *Avicennia marina* (Dong Sai of South China Sea coast). Pharm: Cytotoxic (KB cell lines, IC_{50} = 35 µmol/L; KBV200 cell lines, IC_{50} = 41 µmol/L). Ref: Z. J. Huang, et al, Magn. Reson. Chem., 2010, 48, 80

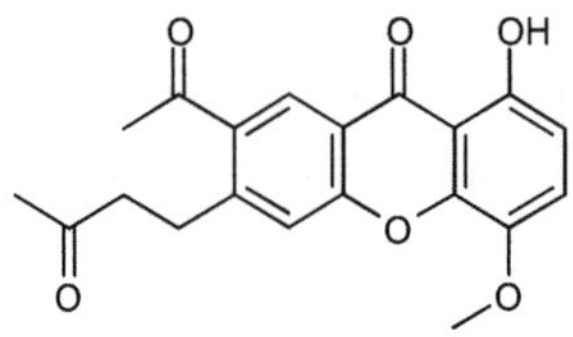

429 8-Hydroxy-3-methyl-9-oxo-9*H*-xanthene-1-carboxylic acid methyl ether

Type: Xanthone derivatives. $C_{16}H_{12}O_5$ Yellow needles. Source: Two unidentified marine fungi (co-culture broth, E33 and K38). Pharm: Antifungal (five flimantous fungal strains, 100 µg/mL: *Gloeasporium musae*, InRt = 53%, *Blumeria graminearum*, InRt = 4.6%, *Fusarium oxysporum*, InRt = 9.5%, *Perononphthora cichoralearum*, InRt = 48%, *Colletotrichum glocosporioides*, InRt = 28%). Ref: C. Y. Li, et al, Chem. Nat. Compd., 2011, 47, 382 | L. Xu, et al, Mar. Drugs, 2015, 13, 3479 (rev)

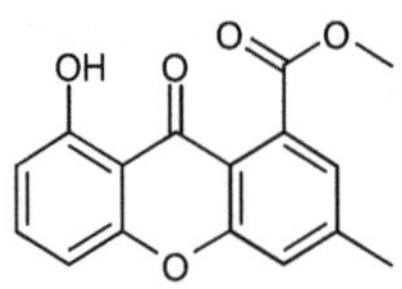

430 JBIR 97

Type: Xanthone derivatives. $C_{33}H_{28}O_{12}$ Yellow amorph. solid, $[\alpha]_D^{25}$ = +316° (*c* = 0.1, MeOH). Source: Marine-derived fungus *Tritirachium* sp. from sponge *Pseudoceratina purpurea* (Sakuraguchi, Ishigaki I., Okinawa). Pharm: Cytotoxic (WST-8colorimetric assay, hmn cervical carcinomaHeLa, IC_{50} = 11 µmol/L; ACC-MESO-1, IC_{50} = 31 µmol/L). Ref: J. Ueda, et al, J. Antibiot., 2010, 63, 615

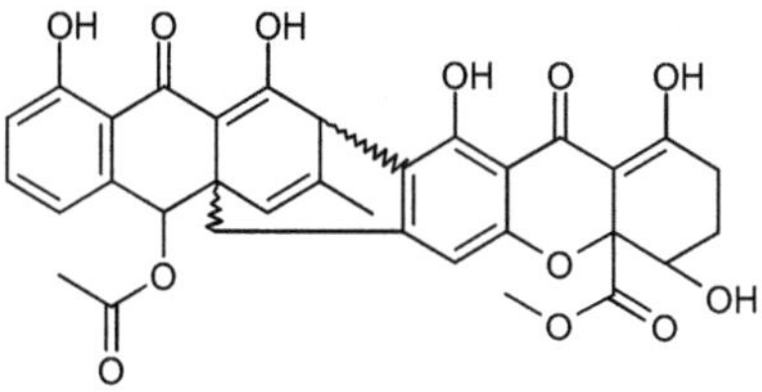

431 JBIR 98

Type: Xanthone derivatives. $C_{33}H_{28}O_{12}$ Yellow amorph. solid, $[\alpha]_D^{25} = +229°$ (c = 0.1, MeOH). Source: Marine-derived fungus *Tritirachium* sp. from sponge *Pseudoceratina purpurea* (Sakuraguchi, Ishigaki I., Okinawa). Pharm: Cytotoxic (WST-8colorimetric assay, hmn cervical carcinoma HeLa, IC_{50} = 17 μmol/L; ACC-MESO-1, IC_{50} = 63 μmol/L). Ref: J. Ueda, et al, J. Antibiot., 2010, 63, 615

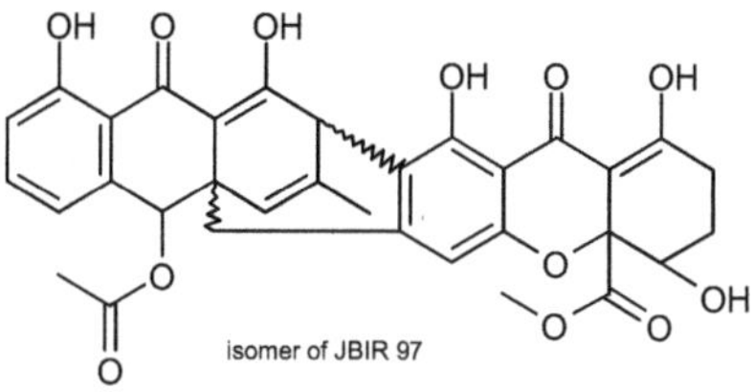

432 JBIR 99

Type: Xanthone derivatives. $C_{33}H_{28}O_{12}$ Yellow amorph. solid, $[\alpha]_D^{25} = +333°$ (c = 0.1, MeOH). Source: Marine-derived fungus *Tritirachium* sp. from sponge *Pseudoceratina purpurea* (Sakuraguchi, Ishigaki I., Okinawa). Pharm: Cytotoxic (WST-8colorimetric assay, hmn cervical carcinomaHeLa, IC_{50} = 17 μmol/L; ACC-MESO-1, IC_{50} = 59 μmol/L). Ref: J. Ueda, et al, J. Antibiot., 2010, 63, 615

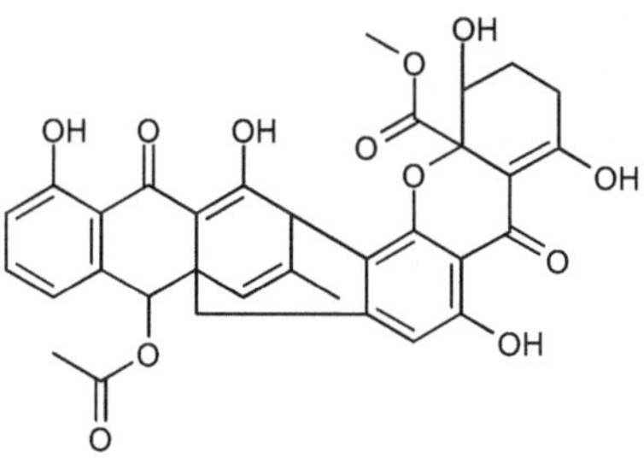

433 Monodictysin A

1,2,3,4,4*a*,9*a*-Hexahydro-1,3,4,8-tetrahydroxy-4*a*,6-dimethyl-9*H*-xanthen-9-one Type: Xanthone derivatives. $C_{15}H_{18}O_6$ Yellow cryst., $[\alpha]_D^{24} = +53°$ (c = 0.43, $CHCl_3$). Source: Marine-derived fungus *Monodictys putredinis* 187/195 15 I. Pharm: Cytochrome P450 1A (CYP1A) inhibitor (IC_{50} ≥ 50 μmol/L, control *α*-Naphthoflavone,

IC_{50} = 0.0035 μmol/L; 400 μmol/L without cytotoxicity); inducer of NAD(P)H quinone reductase (Hepa1c1c7, CD = 191.1 μmol/L, IC_{50} > 400 μmol/L); aromatase Cyp1A inhibitor (50 μmol/L, InRt = 32%). Ref: A. Krick, et al, JNP, 2007, 70, 353

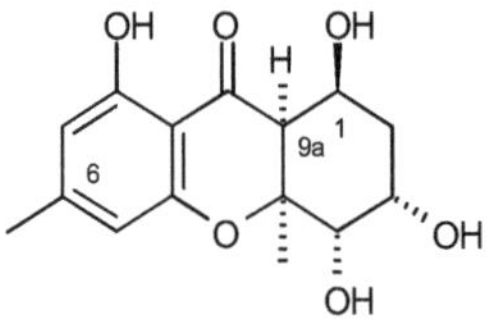

434 Monodictysin B

1,2,3,4,4*a*,9*a*-Hexahydro-1,4,8-trihydroxy-3,4*a*-dimethyl-9*H*-xanthen-9-one Type: Xanthone derivatives. $C_{15}H_{18}O_5$ Amorph. solid, $[\alpha]_D^{24}$ = +80.5° (*c* = 0.2, $CHCl_3$). Source: Marine-derived fungus *Monodictys putredinis* 187/195 15 I. Pharm: Cytochrome P450 1A inhibitor (IC_{50} = 23.3 μmol/L, control *α*-Naphthoflavone, IC_{50} = 0.0035 μmol/L); inducer of NAD(P)H quinone reductase (Hepa1c1c7, CD = 12.0 μmol/L, IC_{50} > 50 μmol/L); aromatase Cyp1A inhibitor (50 μmol/L, InRt = 9%). Ref: A. Krick, et al, JNP, 2007, 70, 353

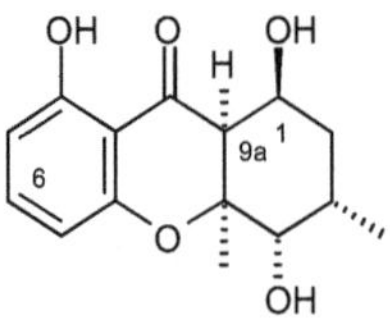

435 Monodictysin C

Type: Xanthone derivatives. $C_{16}H_{20}O_6$ Amorph. solid, $[\alpha]_D^{24}$ = +102° (*c* = 0.5, $CHCl_3$). Source: Marine-derived fungus *Monodictys putredinis.* Pharm: Cytochrome P450 1A inhibitor (IC_{50} = 3.0 μmol/L, control *α*-Naphthoflavone, IC_{50} = 0.0035 μmol/L); inducer of NAD(P)H quinone reductase (Hepa1c1c7, CD = 12.8 μmol/L, IC_{50} > 50 μmol/L); aromatase inhibitor (hmn recombinant aromatase (hmn Cyp 19 +P_{450} reductase supersomes) IC_{50} = 27 μmol/L, weak, control Ketokonazole, IC_{50} = 0.9 μmol/L); aromatase Cyp1A inhibitor (IC_{50} = 28.3 μmol/L, dose-dependent manner). Ref: A. Krick, et al, JNP, 2007, 70, 353

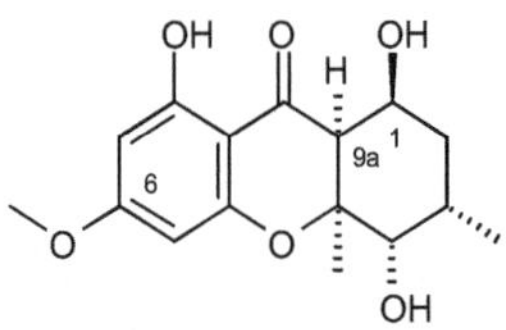

436 Monodictyxanthone

Type: Xanthone derivatives. $C_{15}H_{10}O_5$ Yellow solid. Source: Marine-derived fungus *Monodictys putredinis* 187/195 15 I. Pharm: Cytochrome P450 1A inhibitor (IC_{50} = 34.8 μmol/L); inducer of NAD(P)H quinone reductase (Hepa1c1c7, CD > 50 μmol/L, IC_{50} > 50 μmol/L); aromatase Cyp1A inhibitor (50 μmol/L, InRt = 37%). Ref: A. Krick, et al, JNP, 2007, 70, 353

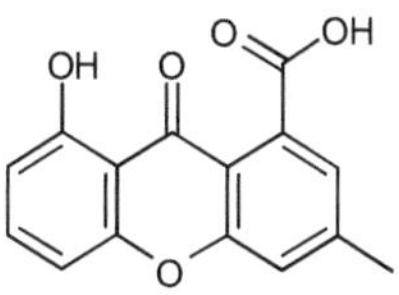

437 Norlichexanthone

3,6,8-Trihydroxy-1-methylxanthone Type: Xanthone derivatives. $C_{14}H_{10}O_5$ Yellow prisms (EtOH aq or Et_2O), mp 285–290 °C, mp 274–275 °C. Source: Marine-derived fungus *Arthrinium* sp. from sponge *Geodia cydonium* (Adriatic Sea, Italy), marine-derived fungus *Wardomyces anomalus*, sponge-derived fungus *Stachybotrys* sp. HH1 ZSDS1F1-2 (Qin, 2014). Pharm: Protein kinases inhibitors (16 different protein kinases: AKT1, IC_{50} > 100 μmol/L; ALK, IC_{50} = 41.3 μmol/L; ARK5, IC_{50} = 33.0 μmol/L; Aurora-B, IC_{50} = 3.0 μmol/L; AXL, IC_{50} > 100 μmol/L; FAK, IC_{50} > 100 μmol/L; IGF1-R, IC_{50} = 33.3 μmol/L; MEK1 wt, IC_{50} > 100 μmol/L; MET wt, IC_{50} > 100 μmol/L; NEK2, IC_{50} > 100 μmol/L; NEK6, IC_{50} > 100 μmol/L; PIM1, IC_{50} = 0.3 μmol/L; PLK1, IC_{50} = 75.7 μmol/L; PRK1, IC_{50} = 54.9 μmol/L; SRC, IC_{50} = 35.0 μmol/L; VEGF-R2, IC_{50} = 11.7 μmol/L); cytotoxic (L5178Y, IC_{50} = 1.16 μmol/L, control Kahalalide F, IC_{50} = 4.30 μmol/L; K562, IC_{50} = 253.50 μmol/L, control Cisplatin (CDDP), IC_{50} = 7.80 μmol/L; A2780, IC_{50} = 68.2 μmol/L, Cisplatin (CDDP), IC_{50} = 0.80 μmol/L; A2780CisR, IC_{50} = 74.0 μmol/L, Cisplatin (CDDP), IC_{50} = 8.40 μmol/L); active against *Clostridium welchii*; mycotoxin; antiviral (Enterovirus-71, IC_{50} = 40.3 μmol/L) (Qin, 2014). Ref: A. Abdel-Lateff, et al, JNP, 2003, 66, 706 | S. S. Ebada, et al, BoMC, 2011, 19, 4644 | C. Qin, et al, J. Antibiot., 2014, 68, 121

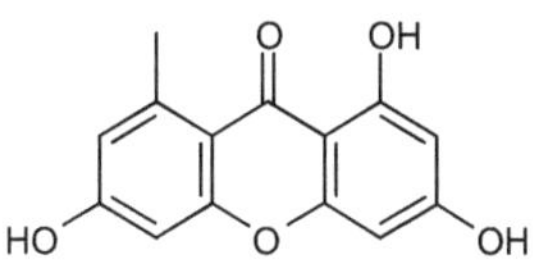

438 Oxisterigmatocystin A

Type: Xanthone derivatives. $C_{20}H_{18}O_8$ Source: Deep-sea fungus *Aspergillus versicolor*. Pharm: Cytotoxic (A549 and HL60). Ref: S. Cai, et al, J. Antibiot. 2011, 64, 193

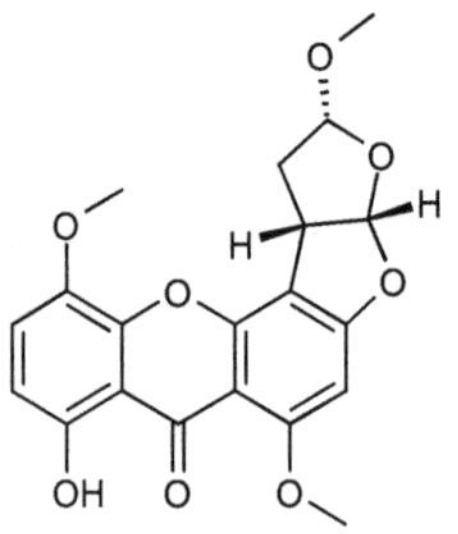

439 Oxisterigmatocystin B

Type: Xanthone derivatives. $C_{20}H_{18}O_8$ Source: Deep-sea fungus *Aspergillus versicolor*. Pharm: Cytotoxic (A549 and HL60). Ref: S. Cai, et al, J. Antibiot. 2011, 64, 193

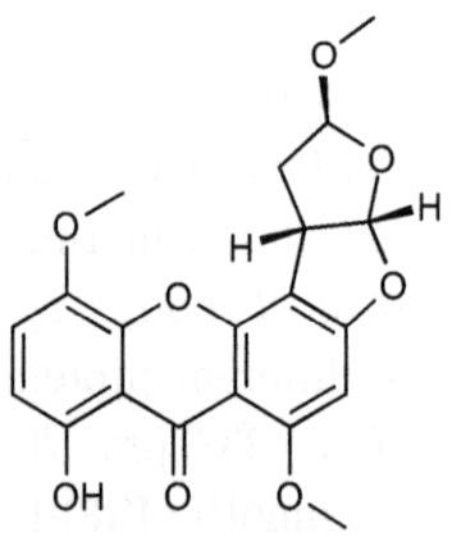

440 Oxisterigmatocystin C

Type: Xanthone derivatives. $C_{19}H_{16}O_7$ Source: Deep-sea fungus *Aspergillus versicolor*. Pharm: Cytotoxic (A549 and HL60). Ref: S. Cai, et al, J. Antibiot. 2011, 64, 193

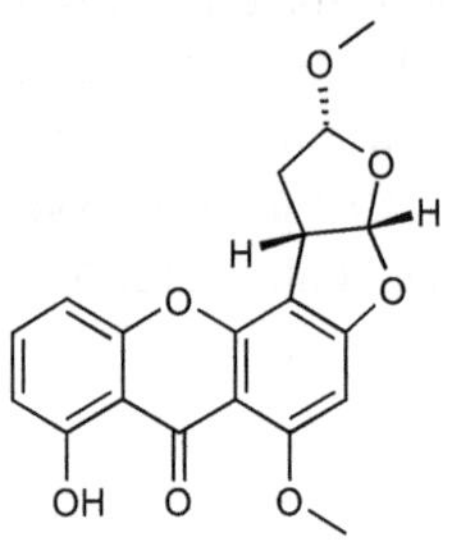

441 Paeciloxanthone

Type: Xanthone derivatives. $C_{20}H_{20}O_4$ Amorph. yellow solid, mp 137 °C. Source: Mangrove-derived fungus *Paecilomyces* sp. Tree1-7 from an unidentified mangrove (bark, Taiwan waters). Pharm: Cytotoxic (HepG2, IC_{50} = 1.08 µg/mL; antifungal (*Curvularia lunata* and *Candida albicans*); AChE inhibitor (IC_{50} = 2.25 µg/mL); antibacterial (standard disk assay, 40 µg/mL, *Escherichia coli*, IZD = 12 mm); antifungal (standard disk assay, 40 µg/mL, *Curvularia lunata*, IZD = 6 mm; *Candida albicans*, IZD = 10 mm). Ref: L. Wen, et al, J. Asian Nat. Prod. Res., 2008, 10, 133

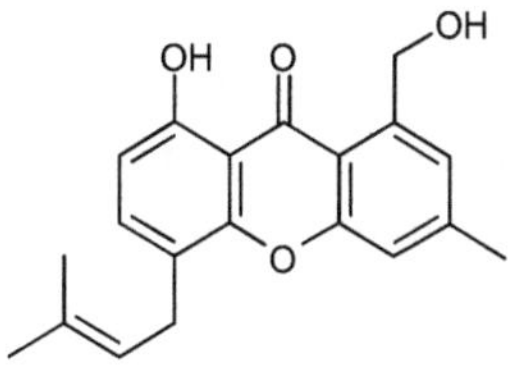

442 Penexanthone A

Type: Xanthone derivatives. $C_{36}H_{36}O_{15}$ Yellow powder, $[\alpha]_D^{23} = -36°$ (c = 0.1, MeOH). Source: Mangrove-derived fungus *Phomopsis* sp. from mangrove *Acanthus ilicifolius* (branch, Hainan, China, first time isolated as MNP), terrestrial fungus *Penicillium* sp. (Costa Rican rainforest). Pharm: Cytotoxic (Myeloma: Dox40, no stromal cell IC_{50} = 15 μmol/L, with stromal cell IC_{50} = 21.6 μmol/L; H929, no stromal IC_{50} = 54.5 μmol/L, with stromal cell IC_{50} = 35.2 μmol/L; KMS34, no stromal IC_{50} = 22.6 μmol/L, with stromal cell IC_{50} = 55.6 μmol/L; L363, no stromal IC_{50} = 66.0 μmol/L, with stromal cell IC_{50} = 79.4 μmol/L; MM1S, no stromal IC_{50} = 37.1 μmol/L, with stromal cell IC_{50} = 47.1 μmol/L; OCIMY5, no stromal IC_{50} = 18.8 μmol/L, with stromal cell IC_{50} = 47.7 μmol/L; OPM2, no stromal IC_{50} = 12.7 μmol/L, with stromal cell IC_{50} = 10.5 μmol/L; RPMI8226, no stromal IC_{50} = 12.9 μmol/L, with stromal cell IC_{50} = 8.9 μmol/L; Lymphoma: Farage, no stromal IC_{50} = 12.4 μmol/L, with stromal cell IC_{50} = 10.5 μmol/L; HT, no stromal IC_{50} = 10.8 μmol/L, with stromal cell IC_{50} = 9.9 μmol/L; OCILY17R, no stromal IC_{50} = 31.1 μmol/L, with stromal cell IC_{50} = 47.4 μmol/L; Leukemia: KU812F, no stromal IC_{50} = 14.6 μmol/L, with stromal cell IC_{50} = 24.2 μmol/L; Breast: MDA-MB-231, no stromal IC_{50} = 43.4 μmol/L, with stromal cell IC_{50} = 34.2 μmol/L; Prostate: PC3, no stromal IC_{50} = 122.3 μmol/L, with stromal cell IC_{50} = 66.3 μmol/L) (Cao, 2012); cytotoxic (MDA-MB-435, IC_{50} = (7.90 ± 0.58)μmol/L, control Epirubicin, IC_{50} = (0.56 ± 0.06)μmol/L; HCT116, IC_{50} = (6.92 ± 0.38)μmol/L, Epirubicin, IC_{50} = (0.48 ± 0.03)μmol/L; Calu3, IC_{50} = (6.44 ± 0.86)μmol/L, Epirubicin, IC_{50} = (1.03 ± 0.10)μmol/L; Huh7, IC_{50} = (42.84 ± 3.58)μmol/L, Epirubicin, IC_{50} = (0.96 ± 0.01)μmol/L; MCF-10A, IC_{50} = (16.13 ± 1.57)μmol/L, Epirubicin, IC_{50} = (0.48 ± 0.08)μmol/L); (Ding, 2013). Ref: S. Cao, et al, JNP, 2012, 75, 793 | B. Ding, et al, Mar. Drugs, 2013, 11, 4961

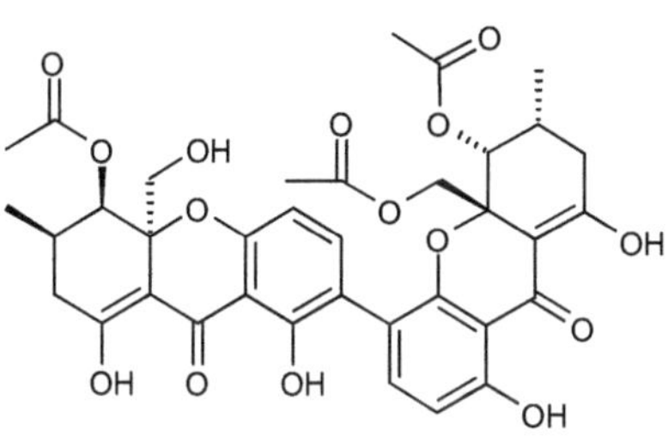

443 Phomolactonexanthone A

Type: Xanthone derivatives. $C_{34}H_{34}O_{14}$ Yellow powder, mp 102.1–103.2 °C, $[\alpha]_D^{25}$ = −75° (*c* = 0.2, MeOH). Source: Mangrove-derived fungus *Phomopsis* sp. from mangrove *Acanthus ilicifolius* (branch, Hainan, China). Pharm: Cytotoxic (MDA-MB-435, IC_{50} > 50 μmol/L; HCT116, IC_{50} > 50 μmol/L; Calu3, IC_{50} = (43.45 ± 2.51)μmol/L, control Epirubicin, IC_{50} = (1.03 ± 0.10)μmol/L; Huh7, IC_{50} > 50 μmol/L; MCF-10A, IC_{50} > 50 μmol/L). Ref: B. Ding, et al, Mar. Drugs, 2013, 11, 4961

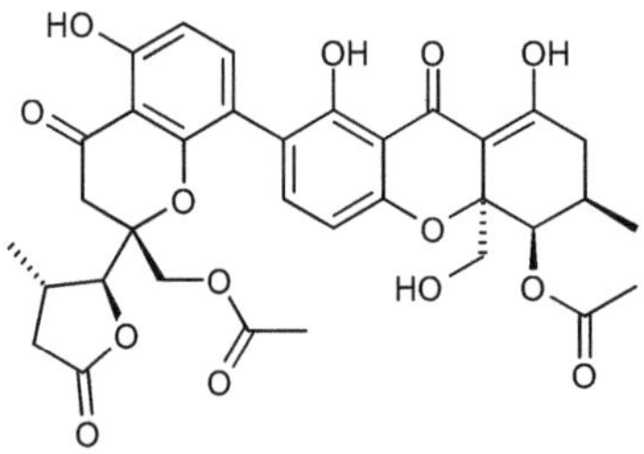

444 Phomolactonexanthone B

Type: Xanthone derivatives. $C_{34}H_{34}O_{14}$ Yellow powder, mp 101.6–102.4 °C, $[\alpha]_D^{25}$ = −21.7° (*c* = 0.6, MeOH). Source: Mangrove-derived fungus *Phomopsis* sp. from mangrove *Acanthus ilicifolius* (branch, Hainan, China). Pharm: Cytotoxic (MDA-MB-435, HCT116, Calu3, Huh7, and MCF-10A, all IC_{50} > 50 μmol/L). Ref: B. Ding, et al, Mar. Drugs, 2013, 11, 4961

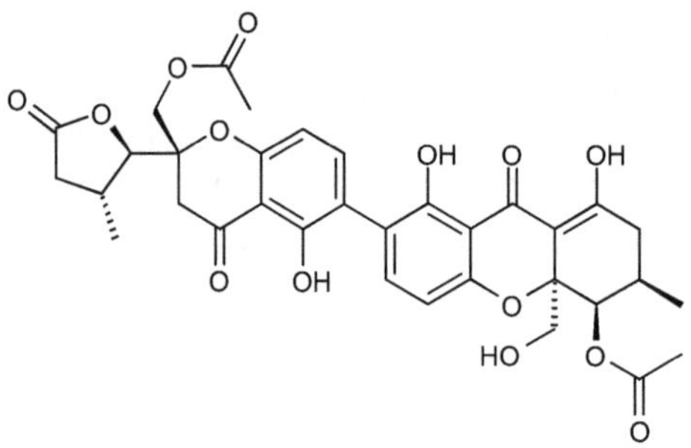

445 Stachybogrisephenone B

Type: Xanthone derivatives. $C_{16}H_{15}ClO_6$ Source: Sponge-derived fungus *Stachybotrys* sp. HH1 ZSDS1F1-2. Pharm: Antiviral (Enterovirus-71, IC_{50} = 30.1 μmol/L). Ref: C. Qin, et al, J. Antibiot., 2014, 68, 121

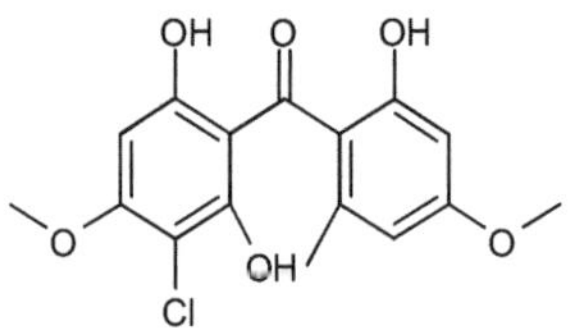

446 Sydowinin A

Type: Xanthone derivatives. $C_{16}H_{12}O_6$ Yellow needles (Me_2CO), mp 218–220 °C. Source: Deep-sea fungus *Engyodontium album* DFFSCS021, fungus *Aspergillus sydowi*. Pharm: Cytotoxic (U937, IC_{50} = 75.6 μmol/L, control DOX, IC_{50} = 0.06 μmol/L; HeLa, IC_{50} > 300 μmol/L, DOX, IC_{50} = 0.8 μmol/L; MCF7, IC_{50} > 300 μmol/L, DOX, IC_{50} = 23.1 μmol/L; HepG2, IC_{50} > 300 μmol/L, DOX, IC_{50} = 3.3 μmol/L; Huh7, IC_{50} > 300 μmol/L, DOX, IC_{50} = 1.2 μmol/L). Ref: T. Hamasaki, et al, Agric. Biol. Chem., 1975, 39, 2337; 2341 | Q. Yao, et al, Mar. Drugs, 2014, 12, 5902

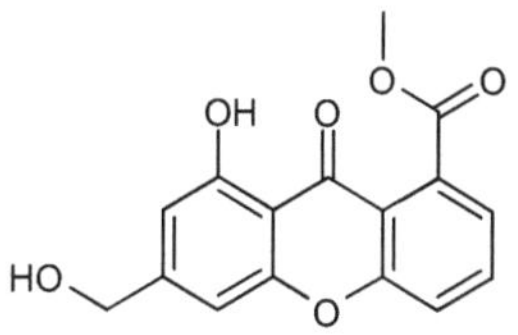

447 Sydowinin B

Antibiotic MS 347B Type: Xanthone derivatives. $C_{16}H_{12}O_7$ Yellow needles (Me2CO), mp 220 °C (dec). Source: Deep-sea fungus *Engyodontium album* DFFSCS021, fungi *Aspergillus sydowi* and *Aspergillus* sp. KY52178. Pharm: Cytotoxic (U937, IC_{50} = 127.0 μmol/L, control DOX, IC_{50} = 0.06 μmol/L; HeLa, IC_{50} > 300 μmol/L, DOX, IC_{50} = 0.8 μmol/L; MCF7, IC_{50} > 300 μmol/L, DOX, IC_{50} = 23.1 μmol/L; HepG2, IC_{50} > 300 μmol/L, DOX, IC_{50} = 3.3 μmol/L; Huh7, IC_{50} > 300 μmol/L, DOX, IC_{50} = 1.2 μmol/L). Ref: T. Hamasaki, et al, Agric. Biol. Chem., 1975, 39, 2337; 1973, 2341 | S. Nakanishi, et al, J. Antibiot., 1993, 46, 1775 | Q. Yao, et al, Mar. Drugs, 2014, 12, 5902

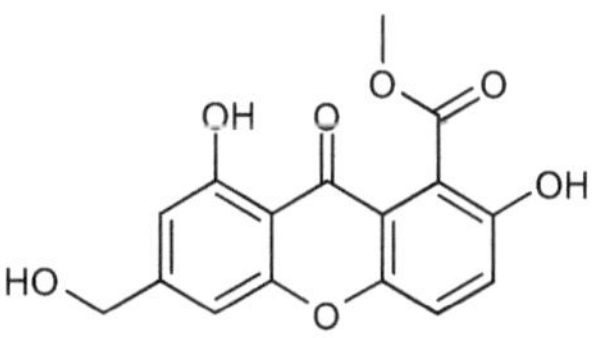

448 TAN-931

Type: Xanthone derivatives. $C_{15}H_{10}O_7$ Source: Acid-tolerant fungus *Penicillium purpurogenum* JS03-21 (region of red soil by Lujiang River, Yunnan, China). Pharm: Antiviral (IFV virus, IC_{50} = 58.6 μmol/L). Ref: H. Wang, et al, JNP, 2011, 74, 2014

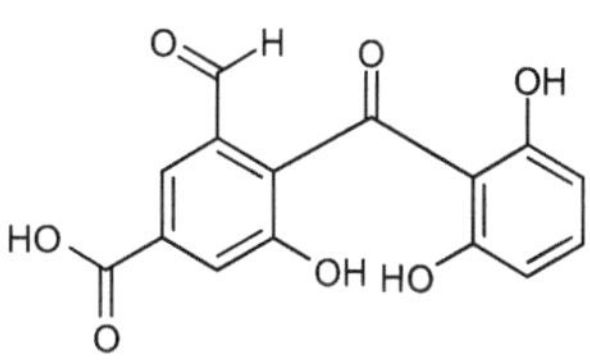

449 Varixanthone

Type: Xanthone derivatives. $C_{26}H_{28}O_8$ Yellow needles, mp 125–127 °C, $[\alpha]_D^{25}$ = +62.1° (c = 1.13, $CHCl_3$). Source: Marine-derived fungus *Emericella variecolor* M75-2. Pharm: Antibacterial (*Escherichia coli, Bacillus subtilis*, and *Staphylococcus aureus*, MIC = 12.5 μg/mL; *Enterococcus faecalis*, MIC = 50 μg/mL). Ref: J. Malmstrom, et al, JNP, 2002, 65, 364 | T. S. Bugni, et al, NPR, 2004, 21, 143 (rev) | M. Saleem, et al, NPR, 2007, 24, 1142 (rev)

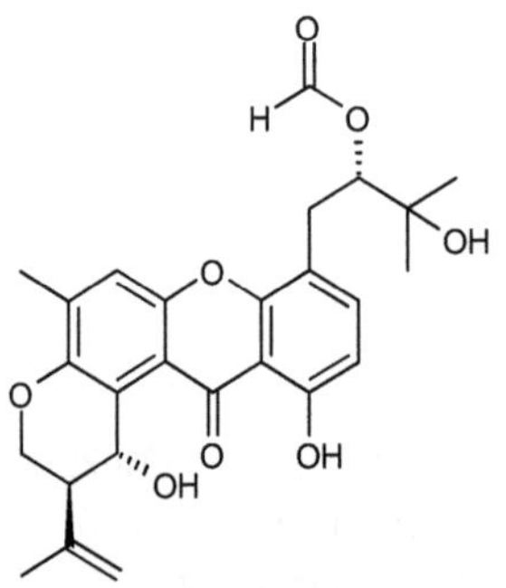

450 Vinaxanthone

Type: Xanthone derivatives. $C_{28}H_{16}O_{14}$ Pale yellow cryst. +1/2H_2O (EtOAc), mp 280 ° C. Source: Marine-derived fungus *Penicillium glabrum*. Pharm: Phospholipase C inhibitor; semaphorin inhibitor; CD-4-binding activity. Ref: K. Kumagai, et al, J. Antibiot., 2003, 56, 610

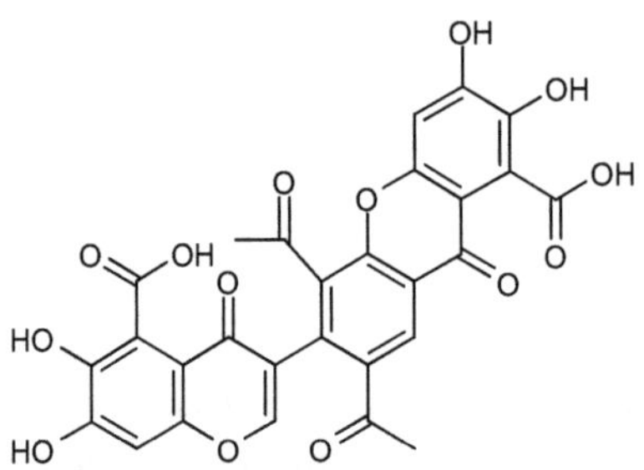

451 Yicathin B

Type: Xanthone derivatives. $C_{16}H_{12}O_6$ Source: Marine-derived fungus *Aspergillus wentii* (endophytic) from red alga *Gymnogongrus flabelliformis* (Pingtan I., Fujian, China). Pharm: Antimicrobial. Ref: R. -R. Sun, et al, Magn. Reson. Chem., 2013, 51, 65

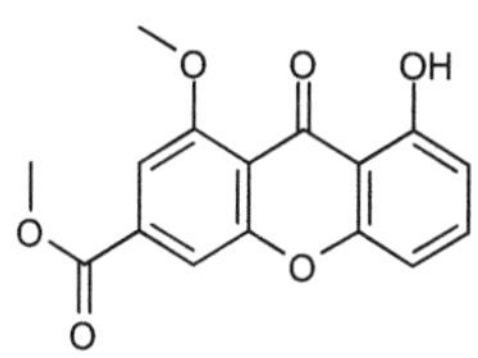

452 Yicathin C

Type: Xanthone derivatives. $C_{15}H_{10}O_6$ Source: Marine-derived fungus *Aspergillus wentii* (endophytic) from red alga *Gymnogongrus flabelliformis* (Pingtan I., Fujian, China). Pharm: Antimicrobial. Ref: R. -R. Sun, et al, Magn. Reson. Chem., 2013, 51, 65

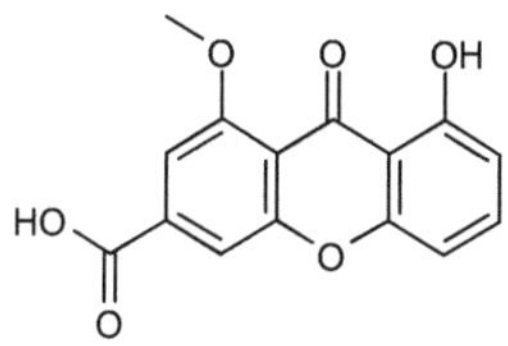

453 Aquastatin A

Type: Depsidones. $C_{36}H_{52}O_{12}$ Powder. Source: Marine-derived fungus *Cosmospora* sp. SF-5060. Pharm: Mammalian adenosine triphosphatases inhibitor; protein tyrosine phosphatase 1B inhibitor; enoyl-ACP reductases inhibitor; antibacterial. Ref: K. Hamano, et al, J. Antibiot., 1993, 46, 1648 | C. Seo, et al, BoMCL, 2009, 19, 6095 | Y. -J. Kwon, et al, Biol. Pharm. Bull., 2009, 32, 2061

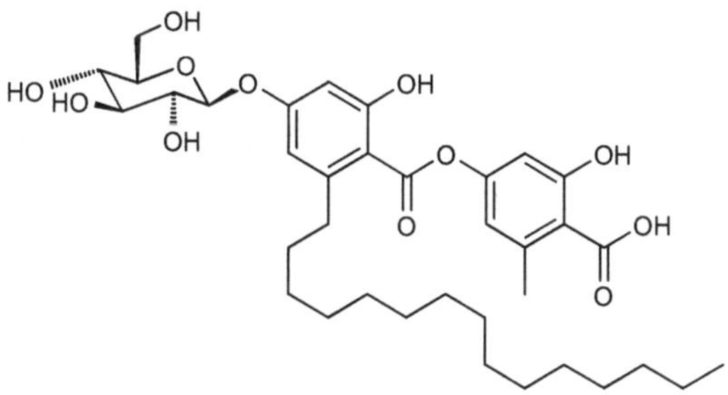

454 Arugosin A

Type: Depsidones. $C_{25}H_{28}O_6$ Yellow oil. Source: Marine-derived fungus *Emericella nidulans* var. *acristata* from an unidentified green alga (Mediterranean Sea). Pharm: Cytotoxic (*in vitro* survival and proliferation assay, panel of 36 hmn tumor cell lines, 10 μg/mL, 19% active in 7 out of 36 cell lines); antibacterial (*Bacillus megaterium*, mixture with Arugosin B, 50 μg/disk, IZD = 4 mm). Ref: N. Kawahara, et al, JCS Perkin I, 1988, 907 | Kralj, et al, JNP, 2006, 69, 995

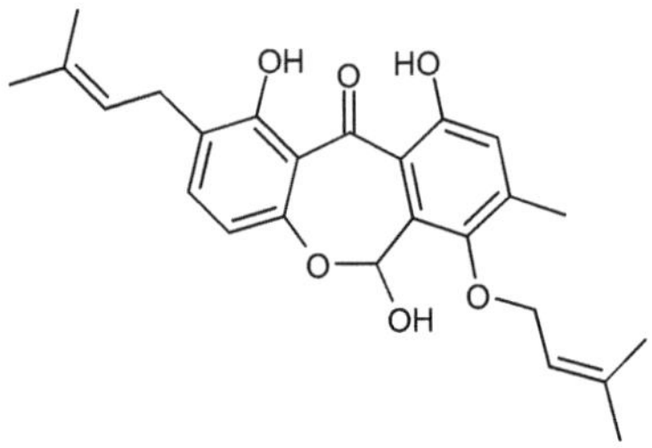

455 Arugosin B

Type: Depsidones. $C_{25}H_{28}O_6$ Yellow oil. Source: Marine-derived fungus *Emericella nidulans* var. *acristata* from an unidentified green alga (Mediterranean Sea). Pharm: Cytotoxic (*in vitro* survival and proliferation assay, panel of 36 hmn tumor cell lines, 10 µg/mL, 19% active in 7 out of 36 cell lines); antibacterial (*Bacillus megaterium*, mixture with Arugosin A, 50 µg/disk, IZD = 4 mm). Ref: N. Kawahara, et al, JCS Perkin I, 1988, 907 | Kralj, et al, JNP, 2006, 69, 995

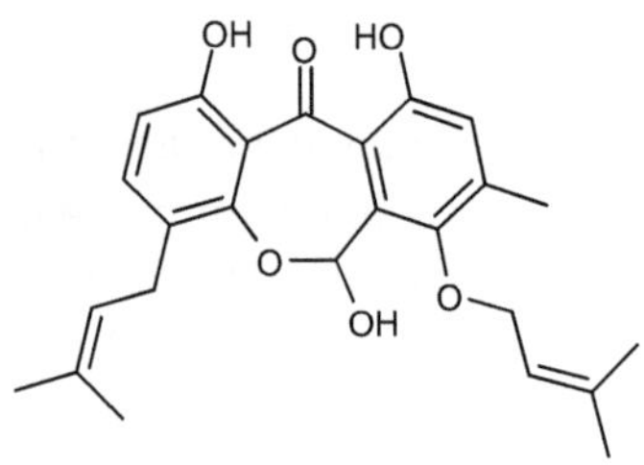

456 Corynesidone A

Type: Depsidones. $C_{15}H_{12}O_5$ Solid, mp 235–237 °C. Source: Marine-derived fungus *Corynespora cassiicola* L36. Pharm: Aromatase inhibitor. Ref: P. Chomcheon, et al, Phytochemistry, 2009, 70, 407

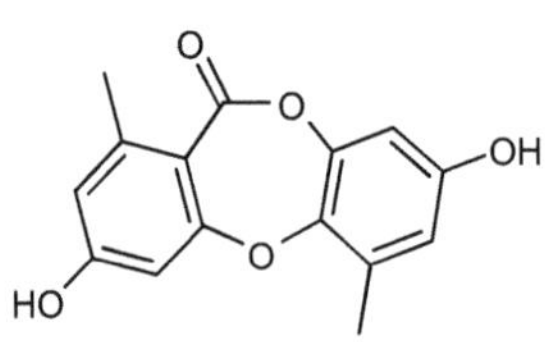

457 9-Dehydroxyeurotinone

2,4,7-Trihydroxy-9-methyldibenz[*b*,*e*]oxepin-6(11*H*)-one Type: Depsidones. $C_{15}H_{12}O_5$ White amorphous powder, $[\alpha]_D^{25}$ = −30.8° (*c* = 0.13, MeOH). Source: Mangrove-derived fungus *Eurotium rubrum* from semimangrove *Hibiscus tiliaceus* (Hainan, China). Pharm: Cytotoxic (SW1990, IC_{50} = 25 µg/mL). Ref: H. -J. Yan, et al, Helv. Chim. Acta, 2012, 95, 163

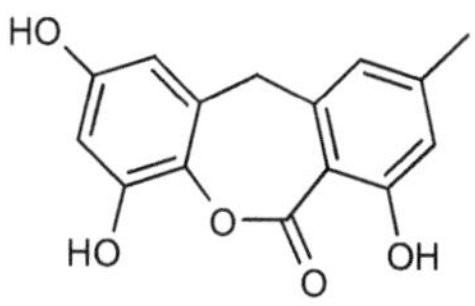

458 9-Dehydroxyeurotinone

2,4,7-Trihydroxy-9-methyldibenz[*b*,*e*]oxepin-6(11*H*)-one Type: Depsidones. $C_{15}H_{12}O_5$ White amorphous powder. $[\alpha]_D^{25} = -30.8°$ (c = 0.13, MeOH). Source: Marine-derived fungus *Eurotium rubrum* (endophytic). Pharm: Antibacterial (100 mg/disk, *Enterobacter coli*, IZ = 7 mm). Ref: H. J. Yan, et al, Helv. Chim. Acta, 2012, 95, 163

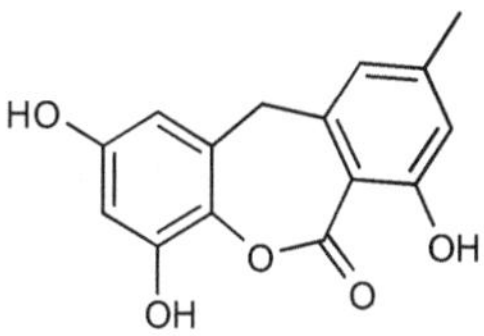

459 Guisinol

Type: Depsidones. $C_{23}H_{25}ClO_5$ Yellowish oil. Source: Marine-derived fungus *Emericella unguis* (teleomorph of fungus *Aspergillus unguis*) from an unidentified mollusc and an unidentified jellyfish (Venezuela). Pharm: Antibacterial (*Staphylococcus aureus*, 5 mg/mL DMSO, 15 µl added to a 4 mm well, mild). Ref: J. Nielsen, et al, Phytochemistry, 1999, 50, 263 | T. S. Bugni, et al, NPR, 2004, 21, 143 (rev)

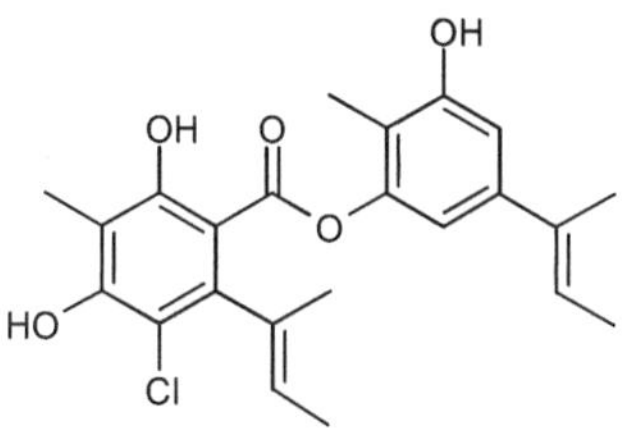

460 2-*O*-Methyl-4-*O*-(*α*-*D*-ribofuranosyl)-9-deoxyeurotinone

Type: Depsidones. $C_{21}H_{22}O_9$ Source: Mangrove-derived fungus *Eurotium rubrum* from semimangrove *Hibiscus tiliaceus* (China waters). Pharm: Antioxidant (DPPH scavenger, moderate). Ref: D. -L. Li, et al, J. Microbiol. Biotechnol., 2009, 19, 675

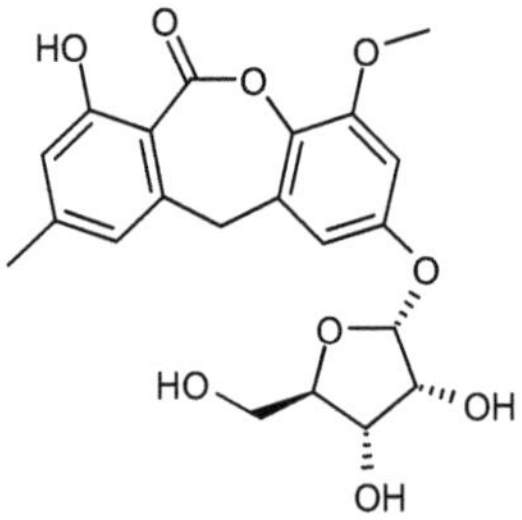

461 3′-*O*-Methyldehydroisopenicillide

Antibiotic MC 142 Type: Depsidones. $C_{22}H_{24}O_6$ Pawder, $[\alpha]_D$ = −25° ($CHCl_3$), $[\alpha]_D$ = +11.6° (*c* = 0.17, $CHCl_3$). Source: Mangrove-derived fungus *Penicillium* sp. from mangrove *Bruguiera gymnorrhiza* (rhizospheric soil, shaken culture, Hainan, China). fungi *Penicillium purpurogenum* and *Penicillium* sp. Pharm: Antihyperlipidaemic. Ref: H. Kawamura, et al, Nat. Prod. Lett., 2000, 14, 477 | Y. Zhang, et al, JNP, 2012, 75, 1888

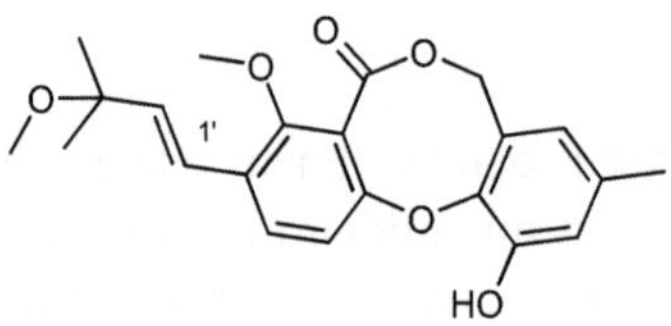

462 2-*O*-Methyl-9-deoxyeurotinone

Type: Depsidones. $C_{16}H_{14}O_5$ Amorph. powder. Source: Mangrove-derived fungus *Eurotium rubrum* from semimangrove *Hibiscus tiliaceus* (China waters). Pharm: Antioxidant (DPPH scavenger, moderate). Ref: D. -L. Li, et al, J. Microbiol. Biotechnol., 2009, 19, 675

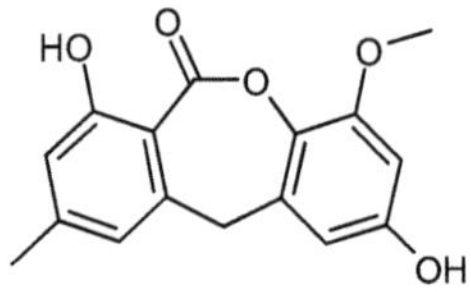

463 Penicillide

Type: Depsidones. $C_{21}H_{24}O_6$ Amorph. powder, $[\alpha]_D^{24}$ = +4.9° (*c* = 0.82, MeOH), $[\alpha]_D$ = +21° (*c* = 1, MeOH). Source: Mangrove-derived fungus *Penicillium* sp. from mangrove *Bruguiera gymnorrhiza* (rhizospheric soil, shaken culture, Hainan, China). Pharm: Plant growth inhibitor. Ref: T. Sassa, et al, Tet. Lett., 1974, 3941 | Y. Zhang, et al, JNP, 2012, 75, 1888

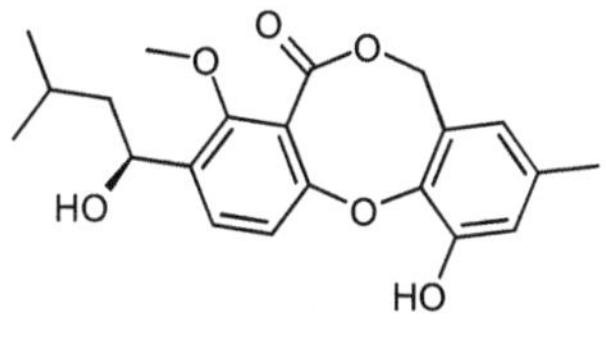

464 Spiromastixone A

Type: Depsidones. $C_{19}H_{20}O_5$ White powder, $[\alpha]_D^{25}$ = +36° (*c* = 0.05, MeOH). Source: Deep-sea fungus *Spiromastix* sp. (MCCC 3A00308, sediment, depth of 2869 m, South Atlantic Ocean GPS 13.7501W, 15.1668S). Pharm: Antibacterial (*Staphylococcus aureus* ATCC 29213, MIC = 4 μg/mL, control Penicillin G, MIC = 0.125 μg/mL; *Bacillus thuringiensis* SCSIO BT01, MIC = 4 μg/mL, Penicillin G, MIC = 128 μg/mL; *Bacillus subtilis* SCSIO BT01, MIC = 8 μg/mL, Penicillin G, MIC = 0.125 μg/mL; *Escherichia coli* ATCC 25922, MIC > 128 μg/mL, Penicillin G, MIC > 128 μg/mL). Ref: S. Niu, et al, JNP, 2014, 77, 1021

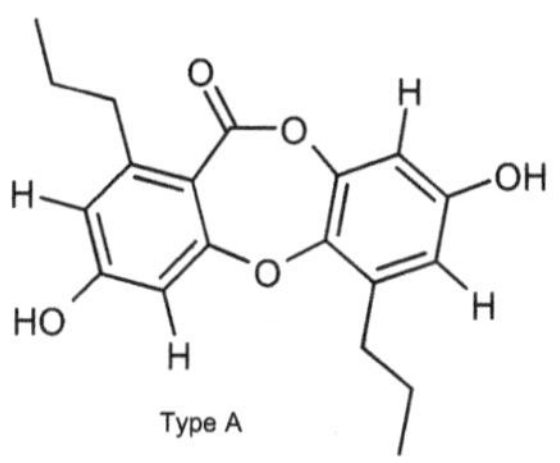

465 Spiromastixone B

Type: Depsidones. $C_{19}H_{19}ClO_5$ White powder, $[\alpha]_D^{25}$ = +18° (*c* = 0.05, MeOH). Source: Deep-sea fungus *Spiromastix* sp. (MCCC 3A00308, sediment, depth of 2869 m, South Atlantic Ocean GPS 13.7501W, 15.1668S). Pharm: Antibacterial (*Staphylococcus aureus* ATCC 29213, MIC = 8 μg/mL, control Penicillin G, MIC = 0.125 μg/mL; *Bacillus thuringiensis* SCSIO BT01, MIC = 4 μg/mL, Penicillin G, MIC = 128 μg/mL; *Bacillus subtilis* SCSIO BT01, MIC = 8 μg/mL, Penicillin G, MIC = 0.125 μg/mL; *Escherichia coli* ATCC 25922, MIC > 128 μg/mL, Penicillin G, MIC > 128 μg/mL). Ref: S. Niu, et al, JNP, 2014, 77, 1021

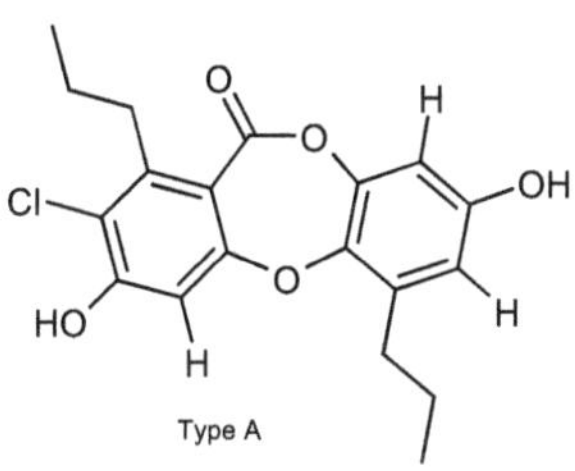

466 Spiromastixone C

Type: Depsidones. $C_{19}H_{19}ClO_5$ White powder, $[\alpha]_D^{25}$ = +22° (c = 0.05, MeOH). Source: Deep-sea fungus *Spiromastix* sp. (MCCC 3A00308, sediment, depth of 2869 m, South Atlantic Ocean GPS 13.7501W, 15.1668S). Pharm: Antibacterial (*Staphylococcus aureus* ATCC 29213, MIC = 8 μg/mL, control Penicillin G, MIC = 0.125 μg/mL; *Bacillus thuringiensis* SCSIO BT01, MIC = 8 μg/mL, Penicillin G, MIC = 128 μg/mL; *Bacillus subtilis* SCSIO BT01, MIC = 16 μg/mL, Penicillin G, MIC = 0.125 μg/mL; *Escherichia coli* ATCC 25922, MIC > 128 μg/mL, Penicillin G, MIC > 128 μg/mL); antibacterial (MSSA ATCC 29213, IC_{50} = 32 μmol/L, control Levofloxacin, IC_{50} = 0.25 μmol/L; MRSA ATCC 33591, IC_{50} = 64 μmol/L, Levofloxacin, IC_{50} = 0.25 μmol/L; MSSA 15, IC_{50} = 64 μmol/L, Levofloxacin, IC_{50} = 0.125 μmol/L; MSSA 12-28, IC_{50} = 64 μmol/L, Levofloxacin, IC_{50} = 0.25 μmol/L; MRSA 12-33, IC_{50} = 64 μmol/L, Levofloxacin, IC_{50} = 64 μmol/L; MSSE ATCC 12228, IC_{50} = 64 μmol/L, Levofloxacin, IC_{50} = 0.25 μmol/L; MSSE12-6, IC_{50} = 64 μmol/L, Levofloxacin, IC_{50} = 0.25 μmol/L; MRSE12-8, IC_{50} = 32 μmol/L, Levofloxacin, IC_{50} = 4 μmol/L; VSE ATCC 29212, IC_{50} = 256 μmol/L, Levofloxacin, IC_{50} = 1 μmol/L; VRE ATCC 51299, IC_{50} > 256 μmol/L, Levofloxacin, IC_{50} = 1 μmol/L; VSE 12-5, IC_{50} > 256 μmol/L, Levofloxacin, IC_{50} = 1 μmol/L; VRE 09-9, IC_{50} = 256 μmol/L, Levofloxacin, IC_{50} > 128 μmol/L; VRE ATCC 700221, IC_{50} > 256 μmol/L, Levofloxacin, IC_{50} = 64 μmol/L; VRE 12-1, IC_{50} = 256 μmol/L, Levofloxacin, IC_{50} = 64 μmol/L; VRE 12-3, IC_{50} > 256 μmol/L, Levofloxacin, IC_{50} = 64 μmol/L; ESBLs(−) ATCC 25922, IC_{50} > 256 μmol/L, Levofloxacin, IC_{50} ≤ 0.03 μmol/L; ESBLs(−) 1515, IC_{50} > 256 μmol/L, Levofloxacin, IC_{50} = 0.125 μmol/L; ESBLs(−) 12-14, IC_{50} > 256 μmol/L, Levofloxacin, IC_{50} = 1 μmol/L; ESBLs(+) 12-15, IC_{50} > 256 μmol/L, Levofloxacin, IC_{50} = 32 μmol/L; ESBLs(+) ATCC 700603, IC_{50} > 256 μmol/L, Levofloxacin, IC_{50} = 1 μmol/L; ESBLs(−) 7, IC_{50} > 256 μmol/L, Levofloxacin, IC_{50} = 0.25 μmol/L; NDM-1(+) ATCC BAA-2146, IC_{50} > 256 μmol/L, Levofloxacin, IC_{50} > 128 μmol/L; ESBLs(−) 12-4, IC_{50} > 256 μmol/L, Levofloxacin, IC_{50} = 0.125 μmol/L; ESBLs(+) 12-8, IC_{50} > 256 μmol/L, Levofloxacin, IC_{50} = 16 μmol/L). Ref: S. Niu, et al, JNP, 2014, 77, 1021

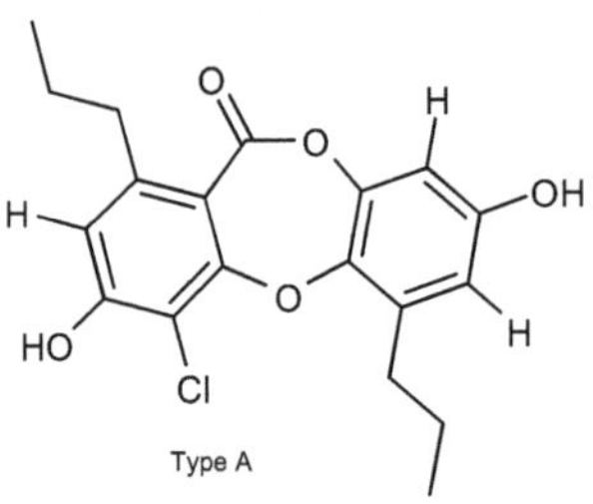

467 Spiromastixone D

Type: Depsidones. $C_{19}H_{18}Cl_2O_5$ White powder, $[\alpha]_D^{25}$ = +45° (c = 0.05, MeOH). Source: Deep-sea fungus *Spiromastix* sp. (MCCC 3A00308, sediment, depth of 2869 m, South Atlantic Ocean GPS 13.7501W, 15.1668S). Pharm: Antibacterial (*Staphylococcus aureus* ATCC 29213, MIC = 2 μg/mL, control Penicillin G, MIC = 0.125 μg/mL; *Bacillus thuringiensis* SCSIO BT01, MIC = 2 μg/mL, Penicillin G, MIC = 128 μg/mL; *Bacillus subtilis* SCSIO BT01, MIC = 2 μg/mL, Penicillin G, MIC = 0.125 μg/mL; *Escherichia coli* ATCC 25922, MIC > 128 μg/mL, Penicillin G, MIC > 128 μg/mL). Ref: S. Niu, et al, JNP, 2014, 77, 1021

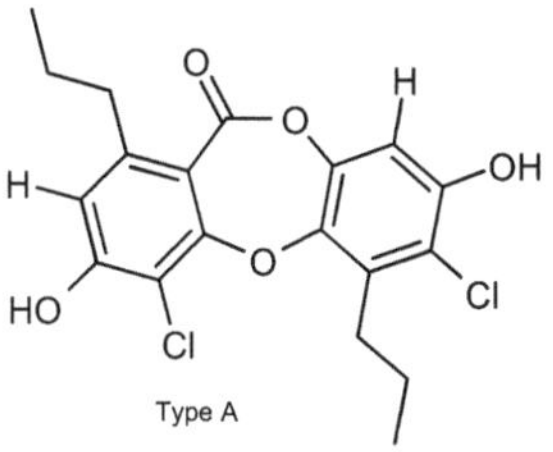

Type A

468 Spiromastixone E

Type: Depsidones. $C_{19}H_{18}Cl_2O_5$ White powder, $[\alpha]_D^{25} = +18°$ (c = 0.05, MeOH). Source: Deep-sea fungus *Spiromastix* sp. (MCCC 3A00308, sediment, depth of 2869 m, South Atlantic Ocean GPS 13.7501W, 15.1668S). Pharm: Antibacterial (*Staphylococcus aureus* ATCC 29213, MIC = 4 μg/mL, control Penicillin G, MIC = 0.125 μg/mL; *Bacillus thuringiensis* SCSIO BT01, MIC = 2 μg/mL, Penicillin G, MIC = 128 μg/mL; *Bacillus subtilis* SCSIO BT01, MIC = 4 μg/mL, Penicillin G, MIC = 0.125 μg/mL; *Escherichia coli* ATCC 25922, MIC > 128 μg/mL, Penicillin G, MIC > 128 μg/mL); antibacterial (MSSA ATCC 29213, IC_{50} = 64 μmol/L, control Levofloxacin, IC_{50} = 0.25 μmol/L; MRSA ATCC 33591, IC_{50} = 32 μmol/L, Levofloxacin, IC_{50} = 0.25 μmol/L; MSSA 15, IC_{50} = 64 μmol/L, Levofloxacin, IC_{50} = 0.125 μmol/L; MSSA 12-28, IC_{50} = 128 μmol/L, Levofloxacin, IC_{50} = 0.25 μmol/L; MRSA 12-33, IC_{50} = 128 μmol/L, Levofloxacin, IC_{50} = 64 μmol/L; MSSE ATCC 12228, IC_{50} = 32 μmol/L, Levofloxacin, IC_{50} = 0.25 μmol/L; MSSE12-6, IC_{50} = 32 μmol/L, Levofloxacin, IC_{50} = 0.25 μmol/L; MRSE12-8, IC_{50} = 64 μmol/L, Levofloxacin, IC_{50} = 4 μmol/L; VSE ATCC 29212, IC_{50} > 256 μmol/L, Levofloxacin, IC_{50} = 1 μmol/L; VRE ATCC 51299, IC_{50} > 256 μmol/L, Levofloxacin, IC_{50} = 1 μmol/L; VSE 12-5, IC_{50} > 256 μmol/L, Levofloxacin, IC_{50} = 1 μmol/L; VRE 09-9, IC_{50} > 256 μmol/L, Levofloxacin, IC_{50} > 128 μmol/L; VRE ATCC 700221, IC_{50} > 256 μmol/L, Levofloxacin, IC_{50} = 64 μmol/L; VRE 12-1, IC_{50} > 256 μmol/L, Levofloxacin, IC_{50} = 64 μmol/L; VRE 12-3, IC_{50} > 256 μmol/L, Levofloxacin, IC_{50} = 64 μmol/L; ESBLs (−) ATCC 25922, IC_{50} > 256 μmol/L, Levofloxacin, IC_{50} ≤ 0.03 μmol/L; ESBLs(−) 1515, IC_{50} > 256 μmol/L, Levofloxacin, IC_{50} = 0.125 μmol/L; ESBLs(−) 12-14, IC_{50} > 256 μmol/L, Levofloxacin, IC_{50} = 1 μmol/L; ESBLs(+) 12-15, IC_{50} > 256 μmol/L, Levofloxacin, IC_{50} = 32 μmol/L; ESBLs(+) ATCC 700603, IC_{50} > 256 μmol/L, Levofloxacin, IC_{50} = 1 μmol/L; ESBLs(−) 7, IC_{50} > 256 μmol/L, Levofloxacin, IC_{50} = 0.25 μmol/L; NDM-1(+) ATCC BAA-2146, IC_{50} > 256 μmol/L, Levofloxacin, IC_{50} > 128 μmol/L; ESBLs (−) 12-4, IC_{50} > 256 μmol/L, Levofloxacin, IC_{50} = 0.125 μmol/L; ESBLs(+) 12-8,

IC_{50} > 256 μmol/L, Levofloxacin, IC_{50} = 16 μmol/L). Ref: S. Niu, et al, JNP, 2014, 77, 1021

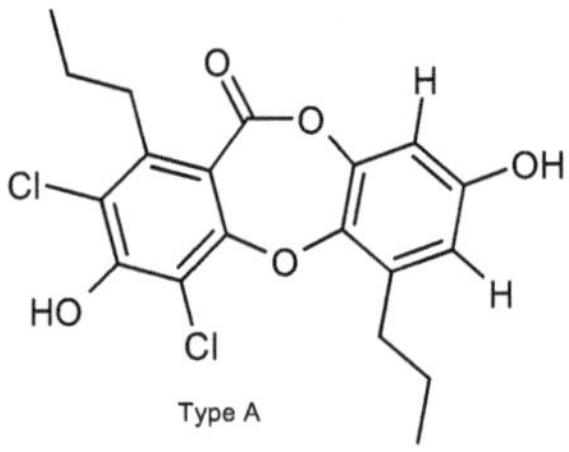

469 Spiromastixone F

Type: Depsidones. $C_{19}H_{17}Cl_3O_5$ White powder, $[\alpha]_D^{25}$ = +31° (*c* = 0.05, MeOH). Source: Deep-sea fungus *Spiromastix* sp. (MCCC 3A00308, sediment, depth of 2869 m, South Atlantic Ocean GPS 13.7501W, 15.1668S). Pharm: Antibacterial (*Staphylococcus aureus* ATCC 29213, MIC = 2 μg/mL, control Penicillin G, MIC = 0.125 μg/mL; *Bacillus thuringiensis* SCSIO BT01, MIC = 1 μg/mL, Penicillin G, MIC = 128 μg/mL; *Bacillus subtilis* SCSIO BT01, MIC = 0.5 μg/mL, Penicillin G, MIC = 0.125 μg/mL; *Escherichia coli* ATCC 25922, MIC > 128 μg/mL, Penicillin G, MIC > 128 μg/mL); antibacterial (MSSA ATCC 29213, IC_{50} = 4 μmol/L, control Levofloxacin, IC_{50} = 0.25 μmol/L; MRSA ATCC 33591, IC_{50} = 4 μmol/L, Levofloxacin, IC_{50} = 0.25 μmol/L; MSSA 15, IC_{50} = 4 μmol/L, Levofloxacin, IC_{50} = 0.125 μmol/L; MSSA 12-28, IC_{50} = 4 μmol/L, Levofloxacin, IC_{50} = 0.25 μmol/L; MRSA 12-33, IC_{50} = 4 μmol/L, Levofloxacin, IC_{50} = 64 μmol/L; MSSE ATCC 12228, IC_{50} = 2 μmol/L, Levofloxacin, IC_{50} = 0.25 μmol/L; MSSE12-6, IC_{50} = 4 μmol/L, Levofloxacin, IC_{50} = 0.25 μmol/L; MRSE12-8, IC_{50} = 4 μmol/L, Levofloxacin, IC_{50} = 4 μmol/L; VSE ATCC 29212, IC_{50} = 32 μmol/L, Levofloxacin, IC_{50} = 1 μmol/L; VRE ATCC 51299, IC_{50} = 32 μmol/L, Levofloxacin, IC_{50} = 1 μmol/L; VSE 12-5, IC_{50} = 32 μmol/L, Levofloxacin, IC_{50} = 1 μmol/L; VRE 09-9, IC_{50} = 32 μmol/L, Levofloxacin, IC_{50} > 128 μmol/L; VRE ATCC 700221, IC_{50} = 32 μmol/L, Levofloxacin, IC_{50} = 64 μmol/L; VRE 12-1, IC_{50} = 16 μmol/L, Levofloxacin, IC_{50} = 64 μmol/L; VRE 12-3, IC_{50} = 32 μmol/L, Levofloxacin, IC_{50} = 64 μmol/L; ESBLs(−) ATCC 25922, IC_{50} > 256 μmol/L, Levofloxacin, IC_{50} ≤ 0.03 μmol/L; ESBLs (−) 1515, IC_{50} > 256 μmol/L, Levofloxacin, IC_{50} = 0.125 μmol/L; ESBLs(−) 12-14, IC_{50} > 256 μmol/L, Levofloxacin, IC_{50} = 1 μmol/L; ESBLs(+) 12-15, IC_{50} > 256 μmol/L, Levofloxacin, IC_{50} = 32 μmol/L; ESBLs(+) ATCC 700603, IC_{50} > 256 μmol/L, Levofloxacin, IC_{50} = 1 μmol/L; ESBLs(−) 7, IC_{50} > 256 μmol/L, Levofloxacin, IC_{50} = 0.25 μmol/L; NDM-1(+) ATCC BAA-2146, IC_{50} > 256 μmol/L, Levofloxacin, IC_{50} > 128 μmol/L; ESBLs(−) 12-4, IC_{50} > 256 μmol/L, Levofloxacin, IC_{50} = 0.125 μmol/L; ESBLs(+) 12-8, IC_{50} > 256 μmol/L, Levofloxacin, IC_{50} = 16 μmol/L). Ref: S. Niu, et al, JNP, 2014, 77, 1021

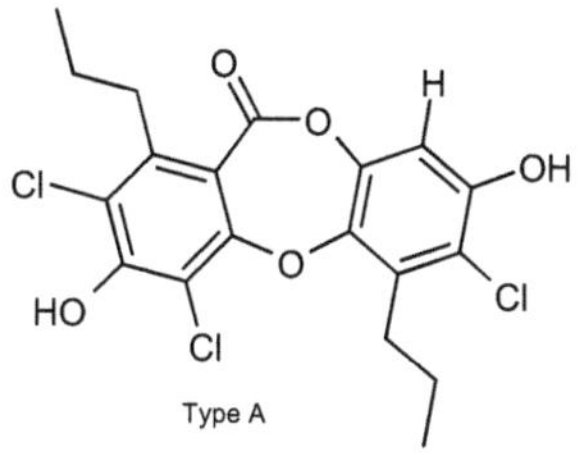

470 Spiromastixone G

Type: Depsidones. $C_{20}H_{19}Cl_3O_5$ White powder, $[\alpha]_D^{25}$ = +46° (*c* = 0.05, MeOH). Source: Deep-sea fungus *Spiromastix* sp. (MCCC 3A00308, sediment, depth of 2869 m, South Atlantic Ocean GPS 13.7501W, 15.1668S). Pharm: Antibacterial (*Staphylococcus aureus* ATCC 29213, MIC = 0.5 μg/mL, control Penicillin G, MIC = 0.125 μg/mL; *Bacillus thuringiensis* SCSIO BT01, MIC = 0.5 μg/mL, Penicillin G, MIC = 128 μg/mL; *Bacillus subtilis* SCSIO BT01, MIC = 0.25 μg/mL, Penicillin G, MIC = 0.125 μg/mL; *Escherichia coli* ATCC 25922, MIC > 128 μg/mL, Penicillin G, MIC > 128 μg/mL); antibacterial (MSSA ATCC 29213, IC_{50} = 8 μmol/L, control Levofloxacin, IC_{50} = 0.25 μmol/L; MRSA ATCC 33591, IC_{50} = 4 μmol/L, Levofloxacin, IC_{50} = 0.25 μmol/L; MSSA 15, IC_{50} = 4 μmol/L, Levofloxacin, IC_{50} = 0.125 μmol/L; MSSA 12-28, IC_{50} = 4 μmol/L, Levofloxacin, IC_{50} = 0.25 μmol/L; MRSA 12-33, IC_{50} = 8 μmol/L, Levofloxacin, IC_{50} = 64 μmol/L; MSSE ATCC 12228, IC_{50} = 4 μmol/L, Levofloxacin, IC_{50} = 0.25 μmol/L; MSSE12-6, IC_{50} = 4 μmol/L, Levofloxacin, IC_{50} = 0.25 μmol/L; MRSE12-8, IC_{50} = 2 μmol/L, Levofloxacin, IC_{50} = 4 μmol/L; VSE ATCC 29212, IC_{50} = 128 μmol/L, Levofloxacin, IC_{50} = 1 μmol/L; VRE ATCC 51299, IC_{50} > 256 μmol/L, Levofloxacin, IC_{50} = 1 μmol/L; VSE 12-5, IC_{50} > 256 μmol/L, Levofloxacin, IC_{50} = 1 μmol/L; VRE 09-9, IC_{50} = 256 μmol/L, Levofloxacin, IC_{50} > 128 μmol/L; VRE ATCC 700221, IC_{50} = 256 μmol/L, Levofloxacin, IC_{50} = 64 μmol/L; VRE 12-1, IC_{50} = 16 μmol/L, Levofloxacin, IC_{50} = 64 μmol/L; VRE 12-3, IC_{50} = 256 μmol/L, Levofloxacin, IC_{50} = 64 μmol/L; ESBLs(−) ATCC 25922, IC_{50} > 256 μmol/L, Levofloxacin, IC_{50} ≤ 0.03 μmol/L; ESBLs(−) 1515, IC_{50} > 256 μmol/L, Levofloxacin, IC_{50} = 0.125 μmol/L; ESBLs(−) 12-14, IC_{50} > 256 μmol/L, Levofloxacin, IC_{50} = 1 μmol/L; ESBLs(+) 12-15, IC_{50} > 256 μmol/L, Levofloxacin, IC_{50} = 32 μmol/L; ESBLs(+) ATCC 700603, IC_{50} > 256 μmol/L, Levofloxacin, IC_{50} = 1 μmol/L; ESBLs(−) 7, IC_{50} > 256 μmol/L, Levofloxacin, IC_{50} = 0.25 μmol/L; NDM-1(+) ATCC BAA-2146, IC_{50} > 256 μmol/L, Levofloxacin, IC_{50} > 128 μmol/L; ESBLs(−) 12-4, IC_{50} > 256 μmol/L, Levofloxacin, IC_{50} = 0.125 μmol/L; ESBLs(+) 12-8, IC_{50} > 256 μmol/L, Levofloxacin, IC_{50} = 16 μmol/L). Ref: S. Niu, et al, JNP, 2014, 77, 1021

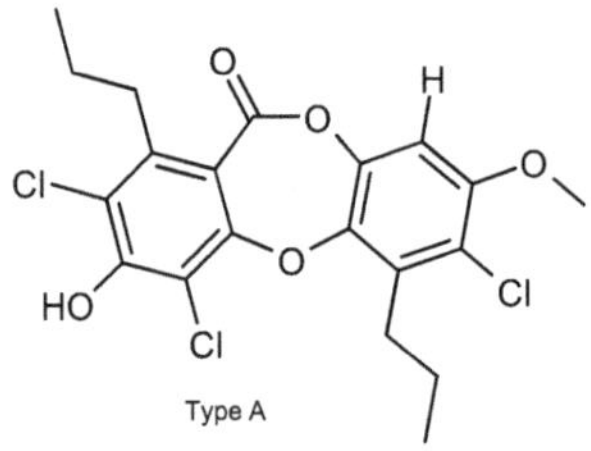

471 Spiromastixone H

Type: Depsidones. $C_{19}H_{17}Cl_3O_5$ White powder, $[\alpha]_D^{25}$ = +43° (*c* = 0.05, MeOH). Source: Deep-sea fungus *Spiromastix* sp. (MCCC 3A00308, sediment, depth of 2869 m, South Atlantic Ocean GPS 13.7501W, 15.1668S). Pharm: Antibacterial (*Staphylococcus aureus* ATCC 29213, MIC = 4 μg/mL, control Penicillin G, MIC = 0.125 μg/mL; *Bacillus thuringiensis* SCSIO BT01, MIC = 1 μg/mL, Penicillin G, MIC = 128 μg/mL; *Bacillus subtilis* SCSIO BT01, MIC = 1 μg/mL, Penicillin G, MIC = 0.125 μg/mL; *Escherichia coli* ATCC 25922, MIC > 128 μg/mL, Penicillin G, MIC > 128 μg/mL); antibacterial (MSSA ATCC 29213, IC_{50} = 4 μmol/L, control Levofloxacin, IC_{50} = 0.25 μmol/L; MRSA ATCC 33591, IC_{50} = 4 μmol/L, Levofloxacin, IC_{50} = 0.25 μmol/L; MSSA 15, IC_{50} = 4 μmol/L, Levofloxacin, IC_{50} = 0.125 μmol/L; MSSA 12-28, IC_{50} = 8 μmol/L, Levofloxacin, IC_{50} = 0.25 μmol/L; MRSA 12-33, IC_{50} = 8 μmol/L, Levofloxacin, IC_{50} = 64 μmol/L; MSSE ATCC 12228, IC_{50} = 16 μmol/L, Levofloxacin, IC_{50} = 0.25 μmol/L; MSSE12-6, IC_{50} = 16 μmol/L, Levofloxacin, IC_{50} = 0.25 μmol/L; MRSE12-8, IC_{50} = 4 μmol/L, Levofloxacin, IC_{50} = 4 μmol/L; VSE ATCC 29212, IC_{50} = 16 μmol/L, Levofloxacin, IC_{50} = 1 μmol/L; VRE ATCC 51299, IC_{50} = 16 μmol/L, Levofloxacin, IC_{50} = 1 μmol/L; VSE 12-5, IC_{50} = 16 μmol/L, Levofloxacin, IC_{50} = 1 μmol/L; VRE 09-9, IC_{50} = 16 μmol/L, Levofloxacin, IC_{50} > 128 μmol/L; VRE ATCC 700221, IC_{50} = 16 μmol/L, Levofloxacin, IC_{50} = 64 μmol/L; VRE 12-1, IC_{50} = 16 μmol/L, Levofloxacin, IC_{50} = 64 μmol/L; VRE 12-3, IC_{50} = 32 μmol/L, Levofloxacin, IC_{50} = 64 μmol/L; ESBLs(–) ATCC 25922, IC_{50} = 64 μmol/L, Levofloxacin, IC_{50} ≤ 0.03 μmol/L; ESBLs(–) 1515, IC_{50} = 64 μmol/L, Levofloxacin, IC_{50} = 0.125 μmol/L; ESBLs(–) 12-14, IC_{50} = 64 μmol/L, Levofloxacin, IC_{50} = 1 μmol/L; ESBLs(+) 12-15, IC_{50} = 64 μmol/L, Levofloxacin, IC_{50} = 32 μmol/L; ESBLs(+) ATCC 700603, IC_{50} = 256 μmol/L, Levofloxacin, IC_{50} = 1 μmol/L; ESBLs(–) 7, IC_{50} = 256 μmol/L, Levofloxacin, IC_{50} = 0.25 μmol/L; NDM-1(+) ATCC BAA-2146, IC_{50} = 256 μmol/L, Levofloxacin, IC_{50} > 128 μmol/L; ESBLs(–) 12-4, IC_{50} = 128 μmol/L, Levofloxacin, IC_{50} = 0.125 μmol/L; ESBLs(+) 12-8, IC_{50} = 128 μmol/L, Levofloxacin, IC_{50} = 16 μmol/L). Ref: S. Niu, et al, JNP, 2014, 77, 1021

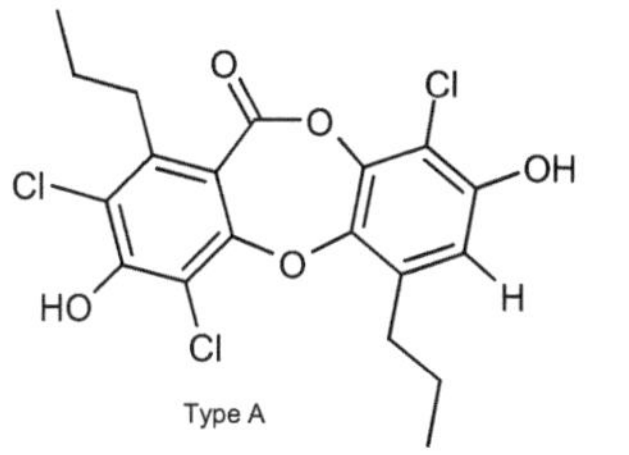

Type A

472 Spiromastixone I

Type: Depsidones. $C_{19}H_{16}Cl_4O_5$ White powder, $[\alpha]_D^{25}$ = +23° (*c* = 0.05, MeOH). Source: Deep-sea fungus *Spiromastix* sp. (MCCC 3A00308, sediment, depth of 2869 m, South Atlantic Ocean GPS 13.7501W, 15.1668S). Pharm: Antibacterial (*Staphylococcus aureus* ATCC 29213, MIC = 4 μg/mL, control Penicillin G, MIC = 0.125 μg/mL; *Bacillus thuringiensis* SCSIO BT01, MIC = 2 μg/mL, Penicillin G, MIC = 128 μg/mL; *Bacillus subtilis* SCSIO BT01, MIC = 2 μg/mL, Penicillin G, MIC = 0.125 μg/mL; *Escherichia coli* ATCC 25922, MIC > 128 μg/mL, Penicillin G, MIC > 128 μg/mL); antibacterial (MSSA ATCC 29213, IC_{50} = 16 μmol/L, control Levofloxacin, IC_{50} = 0.25 μmol/L; MRSA ATCC 33591, IC_{50} = 16 μmol/L, Levofloxacin, IC_{50} = 0.25 μmol/L; MSSA 15, IC_{50} = 16 μmol/L, Levofloxacin, IC_{50} = 0.125 μmol/L; MSSA 12-28, IC_{50} = 16 μmol/L, Levofloxacin, IC_{50} = 0.25 μmol/L; MRSA 12-33, IC_{50} = 32 μmol/L, Levofloxacin, IC_{50} = 64 μmol/L; MSSE ATCC 12228, IC_{50} = 4 μmol/L, Levofloxacin, IC_{50} = 0.25 μmol/L; MSSE12-6, IC_{50} = 4 μmol/L, Levofloxacin, IC_{50} = 0.25 μmol/L; MRSE12-8, IC_{50} = 8 μmol/L, Levofloxacin, IC_{50} = 4 μmol/L; VSE ATCC 29212, IC_{50} = 64 μmol/L, Levofloxacin, IC_{50} = 1 μmol/L; VRE ATCC 51299, IC_{50} = 64 μmol/L, Levofloxacin, IC_{50} = 1 μmol/L; VSE 12-5, IC_{50} = 64 μmol/L, Levofloxacin, IC_{50} = 1 μmol/L; VRE 09-9, IC_{50} = 64 μmol/L, Levofloxacin, IC_{50} > 128 μmol/L; VRE ATCC 700221, IC_{50} = 64 μmol/L, Levofloxacin, IC_{50} = 64 μmol/L; VRE 12-1, IC_{50} = 64 μmol/L, Levofloxacin, IC_{50} = 64 μmol/L; VRE 12-3, IC_{50} = 64 μmol/L, Levofloxacin, IC_{50} = 64 μmol/L; ESBLs(–) ATCC 25922, IC_{50} > 256 μmol/L, Levofloxacin, IC_{50} ≤ 0.03 μmol/L; ESBLs(–) 1515, IC_{50} > 256 μmol/L, Levofloxacin, IC_{50} = 0.125 μmol/L; ESBLs(–) 12-14, IC_{50} > 256 μmol/L, Levofloxacin, IC_{50} = 1 μmol/L; ESBLs(+) 12-15, IC_{50} > 256 μmol/L, Levofloxacin, IC_{50} = 32 μmol/L; ESBLs(+) ATCC 700603, IC_{50} > 256 μmol/L, Levofloxacin, IC_{50} = 1 μmol/L; ESBLs(–) 7, IC_{50} > 256 μmol/L, Levofloxacin, IC_{50} = 0.25 μmol/L; NDM-1(+) ATCC BAA-2146, IC_{50} > 256 μmol/L, Levofloxacin, IC_{50} > 128 μmol/L; ESBLs(–) 12-4, IC_{50} > 256 μmol/L, Levofloxacin, IC_{50} = 0.125 μmol/L; ESBLs(+) 12-8, IC_{50} > 256 μmol/L, Levofloxacin, IC_{50} = 16 μmol/L). Ref: S. Niu, et al, JNP, 2014, 77, 1021

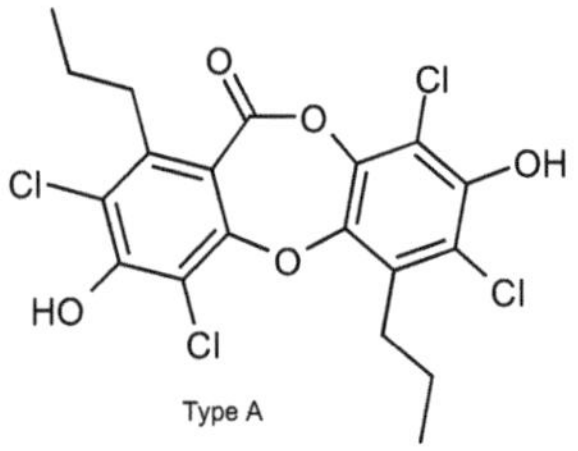

473 Spiromastixone J

Type: Depsidones. $C_{20}H_{18}Cl_4O_5$ White powder, $[\alpha]_D^{25}$ = +33° (*c* = 0.05, MeOH). Source: Deep-sea fungus *Spiromastix* sp. (MCCC 3A00308, sediment, depth of 2869 m, South Atlantic Ocean GPS 13.7501W, 15.1668S). Pharm: Antibacterial (*Staphylococcus aureus* ATCC 29213, MIC = 0.125 μg/mL, control Penicillin G, MIC = 0.125 μg/mL; *Bacillus thuringiensis* SCSIO BT01, MIC = 0.25 μg/mL, Penicillin G, MIC = 128 μg/mL; *Bacillus subtilis* SCSIO BT01, MIC = 0.125 μg/mL, Penicillin G, MIC = 0.125 μg/mL; *Escherichia coli* ATCC 25922, MIC > 128 μg/mL, Penicillin G, MIC > 128 μg/mL); antibacterial (MSSA ATCC 29213, IC_{50} = 4 μmol/L, control Levofloxacin, IC_{50} = 0.25 μmol/L; MRSA ATCC 33591, IC_{50} = 2 μmol/L, Levofloxacin, IC_{50} = 0.25 μmol/L; MSSA 15, IC_{50} = 2 μmol/L, Levofloxacin, IC_{50} = 0.125 μmol/L; MSSA 12-28, IC_{50} = 4 μmol/L, Levofloxacin, IC_{50} = 0.25 μmol/L; MRSA 12-33, IC_{50} = 4 μmol/L, Levofloxacin, IC_{50} = 64 μmol/L; MSSE ATCC 12228, IC_{50} = 4 μmol/L, Levofloxacin, IC_{50} = 0.25 μmol/L; MSSE12-6, IC_{50} = 2 μmol/L, Levofloxacin, IC_{50} = 0.25 μmol/L; MRSE12-8, IC_{50} = 1 μmol/L, Levofloxacin, IC_{50} = 4 μmol/L; VSE ATCC 29212, IC_{50} = 4 μmol/L, Levofloxacin, IC_{50} = 1 μmol/L; VRE ATCC 51299, IC_{50} = 4 μmol/L, Levofloxacin, IC_{50} = 1 μmol/L; VSE 12-5, IC_{50} = 4 μmol/L, Levofloxacin, IC_{50} = 1 μmol/L; VRE 09-9, IC_{50} = 4 μmol/L, Levofloxacin, IC_{50} > 128 μmol/L; VRE ATCC 700221, IC_{50} = 4 μmol/L, Levofloxacin, IC_{50} = 64 μmol/L; VRE 12-1, IC_{50} = 4 μmol/L, Levofloxacin, IC_{50} = 64 μmol/L; VRE 12-3, IC_{50} = 4 μmol/L, Levofloxacin, IC_{50} = 64 μmol/L; ESBLs(–) ATCC 25922, IC_{50} > 256 μmol/L, Levofloxacin, IC_{50} ≤ 0.03 μmol/L; ESBLs(–) 1515, IC_{50} > 256 μmol/L, Levofloxacin, IC_{50} = 0.125 μmol/L; ESBLs(–) 12-14, IC_{50} > 256 μmol/L, Levofloxacin, IC_{50} = 1 μmol/L; ESBLs(+) 12-15, IC_{50} > 256 μmol/L, Levofloxacin, IC_{50} = 32 μmol/L; ESBLs(+) ATCC 700603, IC_{50} > 256 μmol/L, Levofloxacin, IC_{50} = 1 μmol/L; ESBLs(–) 7, IC_{50} > 256 μmol/L, Levofloxacin, IC_{50} = 0.25 μmol/L; NDM-1(+) ATCC BAA-2146, IC_{50} > 256 μmol/L, Levofloxacin, IC_{50} > 128 μmol/L; ESBLs(–) 12-4, IC_{50} > 256 μmol/L, Levofloxacin, IC_{50} = 0.125 μmol/L; ESBLs(+) 12-8, IC_{50} > 256 μmol/L, Levofloxacin, IC_{50} = 16 μmol/L). Ref: S. Niu, et al, JNP, 2014, 77, 1021

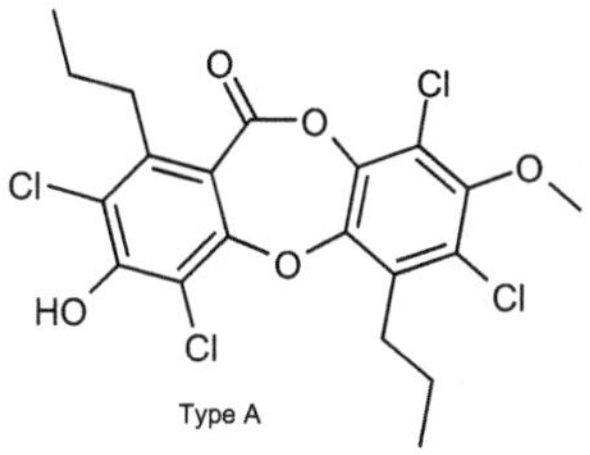

474 Spiromastixone K

Type: Depsidones. $C_{20}H_{19}Cl_3O_5$ White powder, $[\alpha]_D^{25}$ = +34° (c = 0.05, MeOH). Source: Deep-sea fungus *Spiromastix* sp. (MCCC 3A00308, sediment, depth of 2869 m, South Atlantic Ocean GPS 13.7501W, 15.1668S). Pharm: Antibacterial (*Staphylococcus aureus* ATCC 29213, MIC = 0.5 μg/mL, control Penicillin G, MIC = 0.125 μg/mL; *Bacillus thuringiensis* SCSIO BT01, MIC = 0.5 μg/mL, Penicillin G, MIC = 128 μg/mL; *Bacillus subtilis* SCSIO BT01, MIC = 0.5 μg/mL, Penicillin G, MIC = 0.125 μg/mL; *Escherichia coli* ATCC 25922, MIC > 128 μg/mL, Penicillin G, MIC > 128 μg/mL). Ref: S. Niu, et al, JNP, 2014, 77, 1021

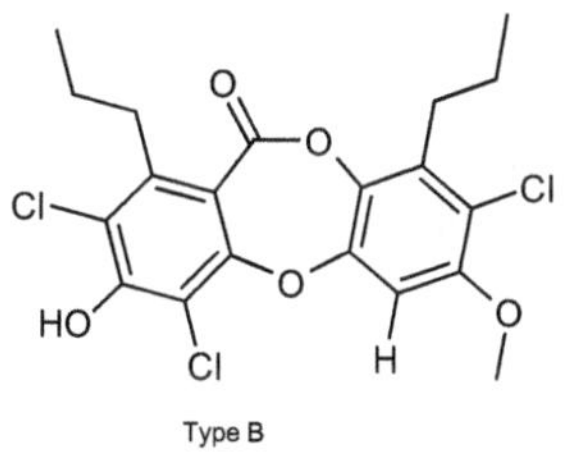

475 Spiromastixone L

Type: Depsidones. $C_{20}H_{18}Cl_4O_5$ White powder, $[\alpha]_D^{25}$ = +36° (c = 0.05, MeOH). Source: Deep-sea fungus *Spiromastix* sp. (MCCC 3A00308, sediment, depth of 2869 m, South Atlantic Ocean GPS 13.7501W, 15.1668S). Pharm: Antibacterial (*Staphylococcus aureus* ATCC 29213, MIC = 0.25 μg/mL, control Penicillin G, MIC = 0.125 μg/mL; *Bacillus thuringiensis* SCSIO BT01, MIC = 0.5 μg/mL, Penicillin G, MIC = 128 μg/mL; *Bacillus subtilis* SCSIO BT01, MIC = 0.25 μg/mL, Penicillin G, MIC = 0.125 μg/mL; *Escherichia coli* ATCC 25922, MIC > 128 μg/mL, Penicillin G, MIC > 128 μg/mL). Ref: S. Niu, et al, JNP, 2014, 77, 1021

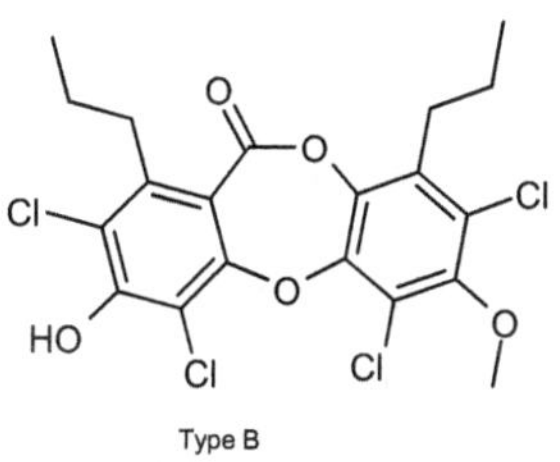

476 Spiromastixone M

Type: Depsidones. $C_{19}H_{18}Cl_2O_5$ White powder, $[\alpha]_D^{25}$ = +43° (*c* = 0.05, MeOH). Source: Deep-sea fungus *Spiromastix* sp. (MCCC 3A00308, sediment, depth of 2869 m, South Atlantic Ocean GPS 13.7501W, 15.1668S). Pharm: Antibacterial (*Staphylococcus aureus* ATCC 29213, MIC = 4 μg/mL, control Penicillin G, MIC = 0.125 μg/mL; *Bacillus thuringiensis* SCSIO BT01, MIC = 2 μg/mL, Penicillin G, MIC = 128 μg/mL; *Bacillus subtilis* SCSIO BT01, MIC = 2 μg/mL, Penicillin G, MIC = 0.125 μg/mL; *Escherichia coli* ATCC 25922, MIC > 128 μg/mL, Penicillin G, MIC > 128 μg/mL). Ref: S. Niu, et al, JNP, 2014, 77, 1021

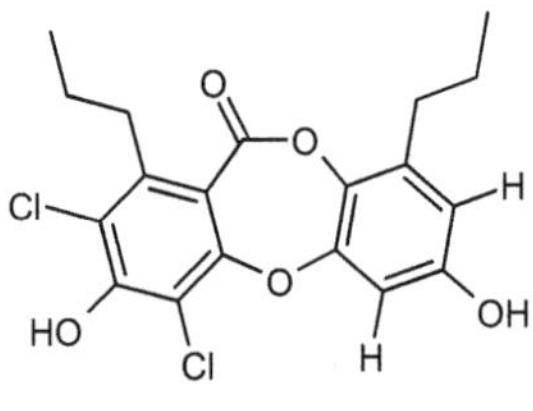

477 Spiromastixone N

Type: Depsidones. $C_{19}H_{17}Cl_3O_5$ White powder, $[\alpha]_D^{25}$ = +48° (*c* = 0.05, MeOH). Source: Deep-sea fungus *Spiromastix* sp. (MCCC 3A00308, sediment, depth of 2869 m, South Atlantic Ocean GPS 13.7501W, 15.1668S). Pharm: Antibacterial (*Staphylococcus aureus* ATCC 29213, MIC = 1 μg/mL, control Penicillin G, MIC = 0.125 μg/mL; *Bacillus thuringiensis* SCSIO BT01, MIC = 1 μg/mL, Penicillin G, MIC = 128 μg/mL; *Bacillus subtilis* SCSIO BT01, MIC = 0.5 μg/mL, Penicillin G, MIC = 0.125 μg/mL; *Escherichia coli* ATCC 25922, MIC > 128 μg/mL, Penicillin G, MIC > 128 μg/mL). Ref: S. Niu, et al, JNP, 2014, 77, 1021

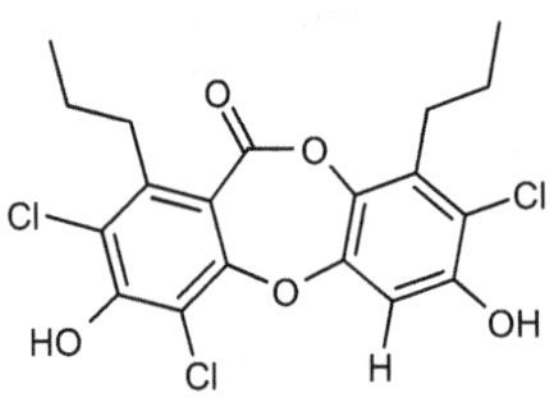

478 Spiromastixone O

Type: Depsidones. $C_{19}H_{16}Cl_4O_5$ White powder, $[\alpha]_D^{25}$ = +36° (*c* = 0.05, MeOH). Source: Deep-sea fungus *Spiromastix* sp. (MCCC 3A00308, sediment, depth of 2869 m, South Atlantic Ocean GPS 13.7501W, 15.1668S). Pharm: Antibacterial (*Staphylococcus aureus* ATCC 29213, MIC = 4 μg/mL, control Penicillin G, MIC = 0.125 μg/mL; *Bacillus thuringiensis* SCSIO BT01, MIC = 2 μg/mL, Penicillin G, MIC = 128 μg/mL; *Bacillus subtilis* SCSIO BT01, MIC = 2 μg/mL, Penicillin G,

MIC = 0.125 μg/mL; *Escherichia coli* ATCC 25922, MIC > 128 μg/mL, Penicillin G, MIC > 128 μg/mL). Ref: S. Niu, et al, JNP, 2014, 77, 1021

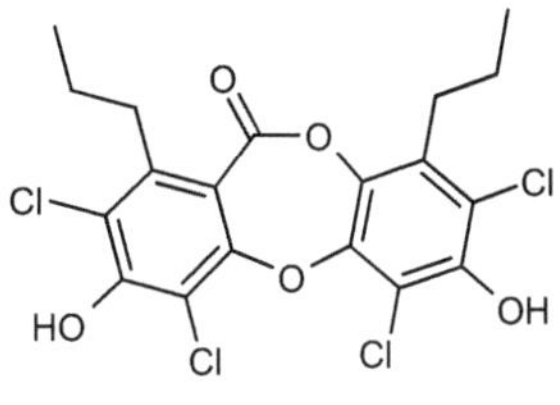

Type B

479 Stromemycin

Type: Depsidones. $C_{38}H_{48}O_{12}$ Amorph. red powder, mp 139 °C, $[\alpha]_D^{20}$ = +29° (c = 0.1, EtOH). Source: Marine-derived fungus *Emericella variecolor* from sponge *Haliclona valliculata*. Pharm: Metalloproteinase inhibitor. Ref: Pat. Coop. Treaty (WIPO), 2001, 01 44 264; CA, 135, 60263d | G. Bringmann, et al, Phytochemistry, 2003, 63, 437

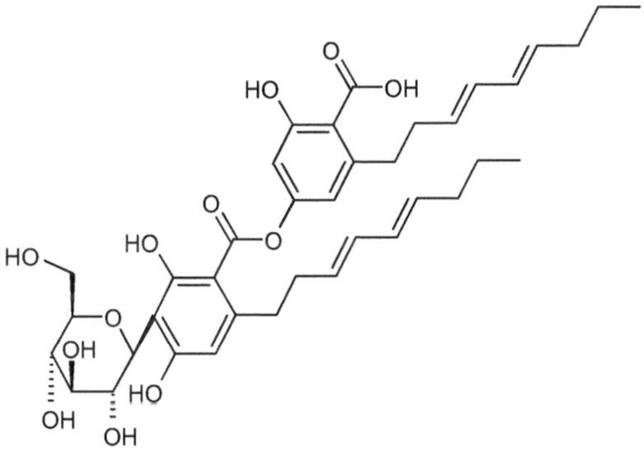

480 7-*O*-Acetylsecopeni-cillide C

Type: Simple biphenyl ethers. $C_{22}H_{24}O_7$ Source: Mangrove-derived fungus *Penicillium* sp. MA-37 from mangrove rhizospheric soil. Pharm: Antibacterial (*Micrococcus luteus*, MIC = 64 μg/mL; *Enterobacter coli*, MIC = 16 μg/mL). Ref: Y. Zhang, et al, JNP, 2012, 75, 1888

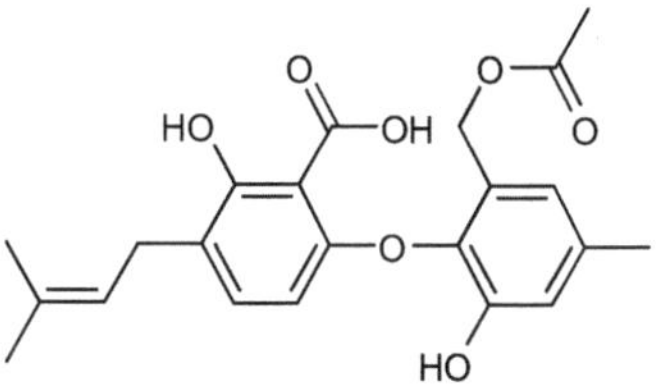

481 Ambigol A

3′,4,5′,6-Tetrachloro-3-(2,4-dichlorophenoxy)-2,2′-biphenyldiol. Type: Simple biphenyl ethers. $C_{18}H_8Cl_6O_3$ Powder, mp 181.5–183.5 °C, $[\alpha]_D = 0°$. Source: Cyanobacterium *Fischerella ambigua.* Pharm: Molluscacidal (*Biomphalaria glabrata*); antibacterial (*Micrococcus luteus* ATCC 9341, *Bacillus subtilis* ATCC 6633, and *Escherichia coli* ATCC 25922.18); antifungal (*P. oralicum* CBS 219.30); cytotoxic (KB cells); toxic (brine shrimp *Artemia salina*). Ref: B. S. Falch, et al, JOC, 1993, 58, 6570 | B. S. Falch, et al, Planta Med., 1995, 61, 321

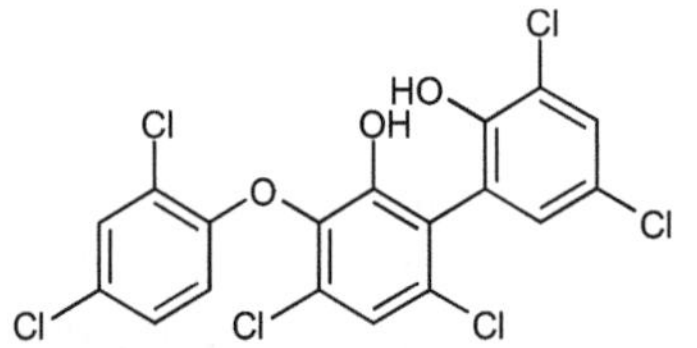

482 Ambigol B

Type: Simple biphenyl ethers. $C_{18}H_8Cl_6O_3$ Pale yellow powder. Source: Cyanobacterium *Fischerella ambigua.* Pharm: Molluscacidal (*Biomphalaria glabrata*); antibacterial (*Micrococcus luteus* ATCC 9341, *Bacillus subtilis* ATCC 6633, and *Escherichia coli* ATCC 25922.18); antifungal (*P. oralicum* CBS 219.30); cytotoxic (KB cells); toxic (brine shrimp *Artemia salina*). Ref: B. S. Falch, et al, JOC, 1993, 58, 6570 | B. S. Falch, et al, Planta Med., 1995, 61, 321

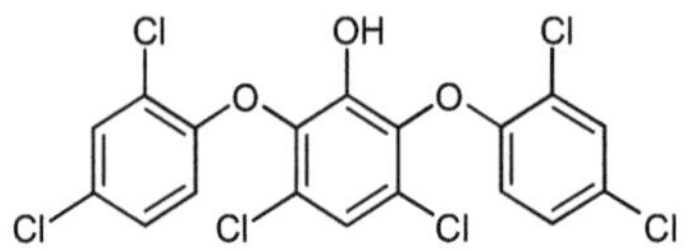

483 Ambigol C

Type: Simple biphenyl ethers. $C_{18}H_8Cl_6O_3$ Amorph. powder, mp 197 °C. Source: Cyanobacterium *Fischerella ambigua.* Pharm: Antiplasmodial (*Plasmodium falciparum*); antitrypanosomal (*Trypanosoma brucei rhodesiense*). Ref: G. M. König, et al, JNP, 2005, 68, 459

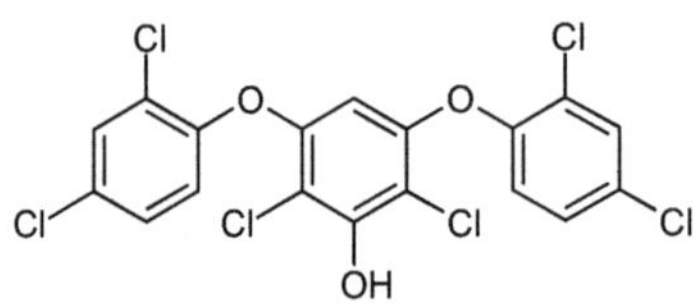

484 Awajanoran

Type: Simple biphenyl ethers. $C_{19}H_{22}O_4$ Oil, $[\alpha]_D^{25} = -21°$ (c = 0.1, MeOH). Source: Marine-derived fungus *Acremonium* sp. AWA16-1 (from sea mud sample, Japan waters). Pharm: Cytotoxic (A549, moderate); antibacterial (five different bacterial strains); antifungal (yeast *Candida albicans*). Ref: J. -H. Jang, et al, J. Antibiot., 2006, 59, 428

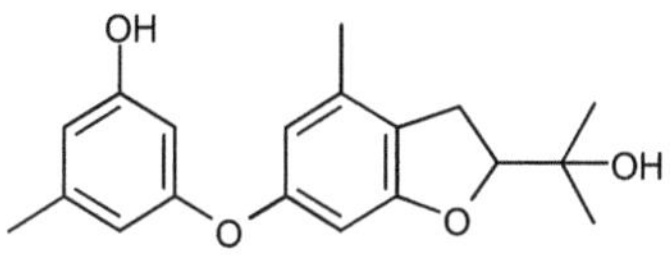

485 2-(5-Bromo-2-methoxy-phenoxy)-3,5-dibromophenol

Type: Simple biphenyl ethers. $C_{13}H_9Br_3O_3$ Source: Sponge *Phyllospongia dendyi* (Palau, Oceania). Pharm: Toxic (variety of micro- and macro-algae). Ref: T. Hattori, et al, Fish. Sci., 2001, 67, 899

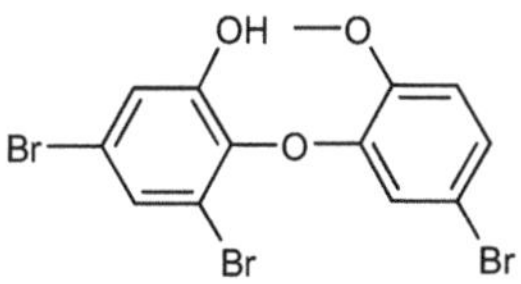

486 Chrysophaentin A

Type: Simple biphenyl ethers. $C_{32}H_{24}Cl_4O_8$ Amorph. powder. Source: Brown alga *Chrysophaeum taylorii*. Pharm: Antibacterial (MRSA, MIC = 1.5 μg/mL, VREF, MIC = 2.9 μg/mL MMOA: inhibits GTPase activity of bacterial cell division protein FtsZ, FtsZ is a structural homolog of eukaryotic tubulin and, similar to tubulin, is a GTPase that polymerizes in a GTP-regulated manner). Ref: A. Plaza, et al, JACS, 2010, 132, 9069

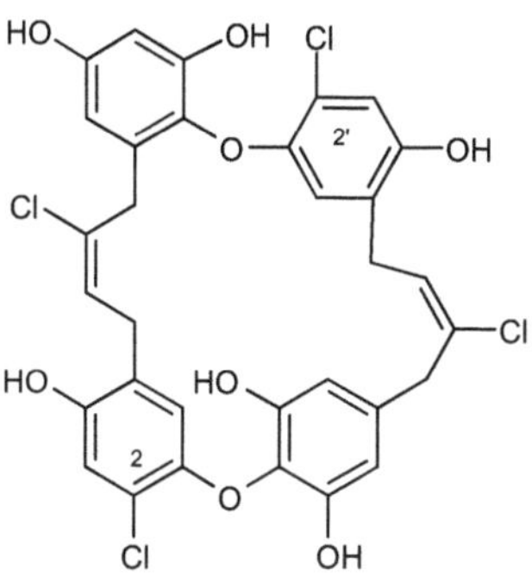

487 Chrysophaentin E

Type: Simple biphenyl ethers. $C_{32}H_{26}Cl_4O_6$ Amorph. powder. Source: Brown alga *Chrysophaeum taylorii*. Pharm: Antibacterial. Ref: A. Plaza, et al, JACS, 2010, 132, 9069

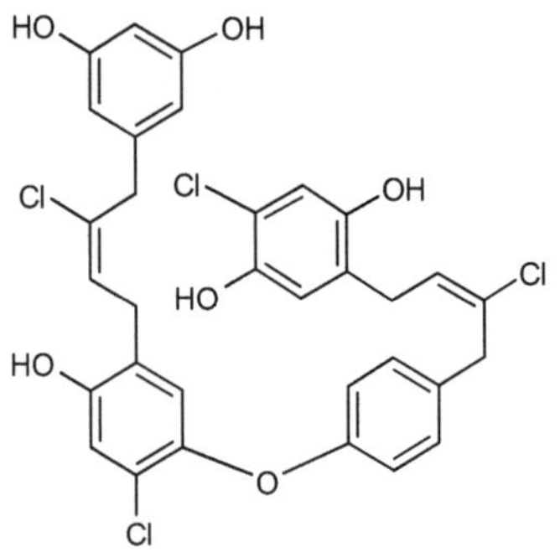

488 Chrysophaentin F

Type: Simple biphenyl ethers. $C_{32}H_{24}Cl_4O_8$ Amorph. powder. Source: Brown alga *Chrysophaeum taylorii*. Pharm: Antibacterial. Ref: A. Plaza, et al, JACS, 2010, 132, 9069

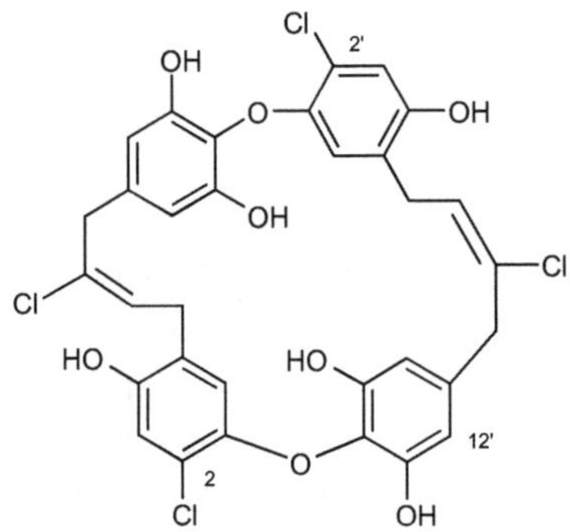

489 Cordyol C

Type: Simple biphenyl ethers. $C_{12}H_{10}O_4$ Source: Marine-derived fungus *Aspergillus sydowii* ZSDS1-F6. Pharm: Antiviral (H3N2 virus, IC_{50} = 78.5 μmol/L). Ref: J. -F. Wang, et al, J. Antibiot., 2014, 67, 581

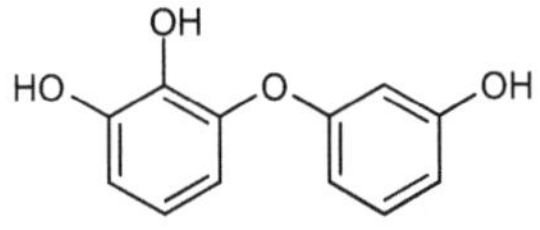

490 Cordyol C′

2,3,3′-Trihydroxy-5,5′-dimethyldiphenyl ether Type: Simple biphenyl ethers. $C_{14}H_{14}O_4$ Amorph. powder. Source: Sponge-derived fungus *Aspergillus sydowii* ZSDS1-F6, deep-sea fungus *Engyodontium album* DFFSCS021, fungus *Cordyceps* sp. BCC1861. Pharm: Antiviral (IFV virus, IC_{50} = 78.5 μmol/L); antiviral (HSV-1 virus);

antibacterial. Ref: T. Bunyapaiboonsri, et al, CPB, 2007, 55, 304 | J. -F. Wang, et al, J. Antibiot., 2014, 67, 581 | Q. Yao, et al, Mar. Drugs, 2014, 12, 5902

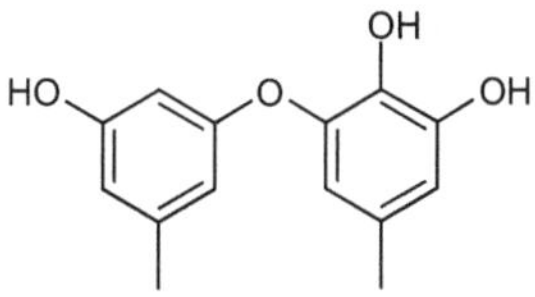

491 Cordyol E

3-*O*-Methyldiorcinol Type: Simple biphenyl ethers. $C_{15}H_{16}O_3$ Source: Mangrove-derived fungi *Penicillium expansum* from mangrove *Excoecaria agallocha* and *Aspergillus* sp. XS-20090066. Pharm: Antibacterial (*Staphylococcus epidermidis, Staphylococcus aureus, Vibrio anguillarum, Vibrio parahemolyticus* and *Pseudomonas putida*, MIC = 25.6–51.2 μmol/L). Ref: J. F. Wang, et al, Planta Med., 2012, 78, 1861 | M. Chen, et al,. JNP, 2013, 76, 547

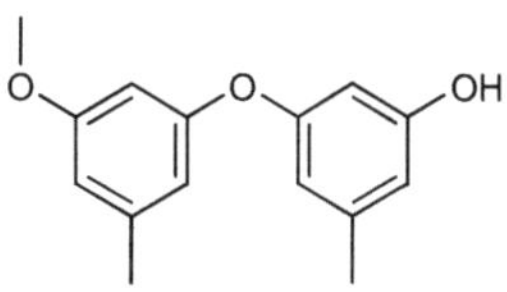

492 2,3-Dibromo-5-chloro-6-(2,4-dibromophenoxy)phenol

Type: Simple biphenyl ethers. $C_{12}H_5Br_4ClO_2$ Powder. Source: Sponge *Dysidea granulosa* (Milne Bay, Papua New Guinea). Pharm: Bcl-2 protein inhibitor (mild). Ref: L. Calcul, et al, JNP, 2009, 72, 443

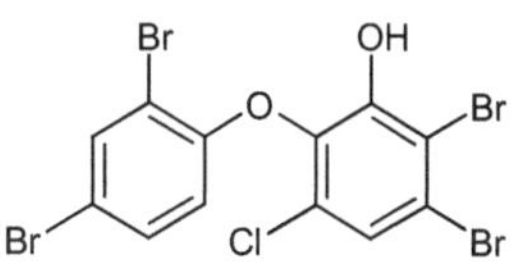

493 3,5-Dibromo-2-(2,4-dibromophenoxy)-4-methoxyphenol

Type: Simple biphenyl ethers. $C_{13}H_8Br_4O_3$ Green gum. Source: Sponge *Dysidea granulosa* (Milne Bay, Papua New Guinea). Pharm: Bcl-2 protein inhibitor (mild). Ref: L. Calcul, et al, JNP, 2009, 72, 443

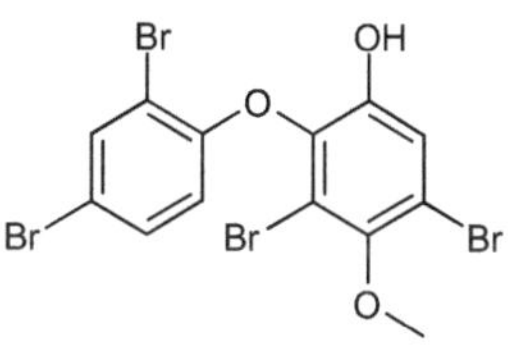

494 4,6-Dibromo-2-(2′-methoxy-4′,6′-dibromophenoxy)phenol

Type: Simple biphenyl ethers. $C_{13}H_8Br_4O_3$ Source: Sponge *Dysidea* sp. (four sponge specimans, Indo-Pacific). Pharm: Inosine monophosphate dehydrogenase (IMPDH) inhibitor; guanosine monophosphate synthetase inhibitor; 15-lipoxygenase inhibitor. Ref: X. Fu, et al, JNP, 1995, 58, 1384

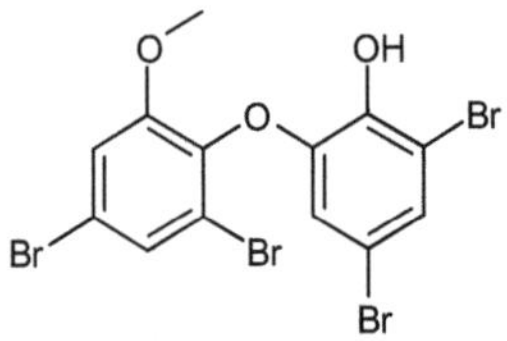

495 2-(3,5-Dibromo-2-methoxy-phenoxy)-3-bromophenol

Type: Simple biphenyl ethers. $C_{13}H_9Br_3O_3$ Source: Sponge *Phyllospongia dendyi* (Palau, Oceania). Pharm: Toxic (variety of micro- and macro-algae). Ref: T. Hattori, et al, Fish. Sci., 2001, 67, 899

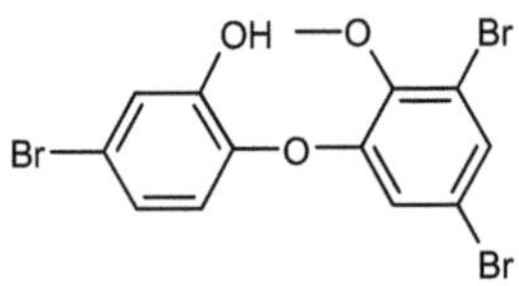

496 2-(4,6-Dibromo-2-methoxy-phenoxy)-3,5-dibromophenol

Type: Simple biphenyl ethers. $C_{13}H_8Br_4O_3$ Source: Sponge *Phyllospongia dendyi* (Palau, Oceania). Pharm: Toxic (variety of micro- and macro-algae). Ref: T. Hattori, et al, Fish. Sci., 2001, 67, 899

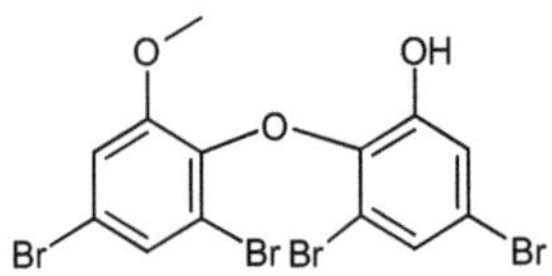

497 2-(3,5-Dibromo-2-methoxy-phenoxy)-3,4,5-tribromoanisole

Type: Simple biphenyl ethers. $C_{14}H_9Br_5O_2$ Cryst. ($CHCl_3$), mp 142–144 °C. Source: Sponge *Dysidea herbacea* (Australia). Pharm: Antimicrobial (*in vitro*). Ref: R. S. Norton, et al, Tetrahedron, 1981, 37, 2341

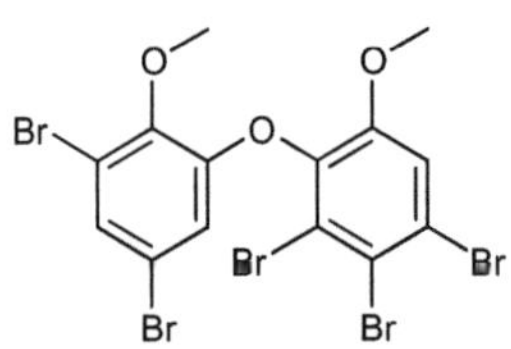

498 2-(2,4-Dibromo-phenoxy)-3,4,5-tribromophenol

Type: Simple biphenyl ethers. $C_{12}H_5Br_5O_2$ Cryst. (hexane), mp 195–196 °C, mp 185–186 °C. Source: Sponge *Dysidea herbacea* (Palau, Oceania). Pharm: Antimicrobial (*in vitro*). Ref: G. M. Sharma, et al, Tet. Lett., 1969, 1715 | B. Carté, et al, Tetrahedron, 1981, 37, 2335

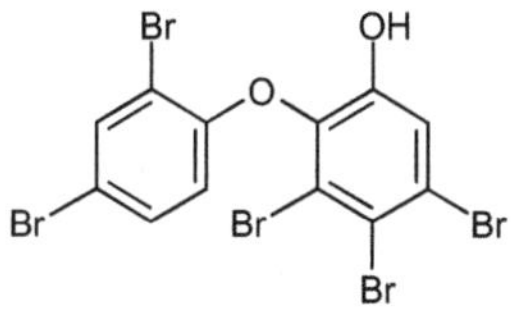

499 3,5-Dibromo-2-(4′,5′,6′-tribromo-2′-hydroxyphenoxy) phenol

Type: Simple biphenyl ethers. $C_{12}H_5Br_5O_3$ Fine needles (MeOH), mp 132–133 °C. Source: Sponge *Dysidea* sp. (four sponge specimans, Indo-Pacific). Pharm: Inosine monophosphate dehydrogenase (IMPDH) inhibitor; guanosine monophosphate synthetase inhibitor; 15-lipoxygenase inhibitor. Ref: X. Fu, et al, JNP, 1995, 58, 1384

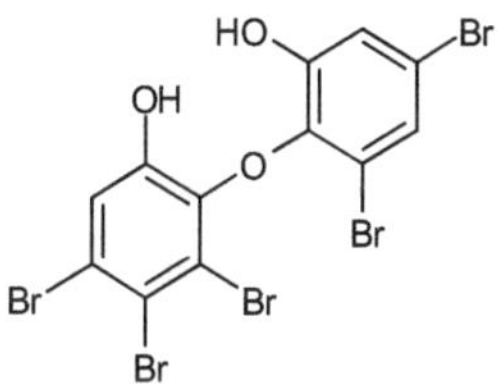

500 4,6-Dibromo-2-(4′,5′,6′-tribromo-2′-hydroxyphenoxy) phenol

Type: Simple biphenyl ethers. $C_{12}H_5Br_5O_3$ mp 132–133 °C. Source: Sponge *Dysidea* sp. (four sponge specimans, Indo-Pacific). Pharm: Inosine monophosphate dehydrogenase (IMPDH) inhibitor; guanosine monophosphate synthetase inhibitor; 15-lipoxygenase inhibitor. Ref: X. Fu, et al, JNP, 1995, 58, 1384

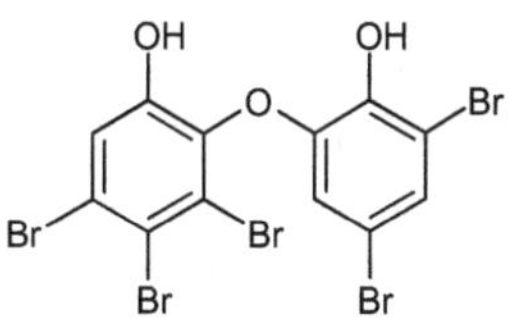

501 4,6-Dibromo-2-(4′,5′,6′-tribromo-2′-hydroxyphenoxy) phenol-1-methylether

Type: Simple biphenyl ethers. $C_{13}H_7Br_5O_3$ Viscous oil. Source: Sponge *Dysidea* sp. (four sponge specimans, Indo-Pacific). Pharm: Inosine monophosphate dehydrogenase (IMPDH) inhibitor; guanosine monophosphate synthetase inhibitor; 15-lipoxygenase inhibitor. Ref: X. Fu, et al, JNP, 1995, 58, 1384

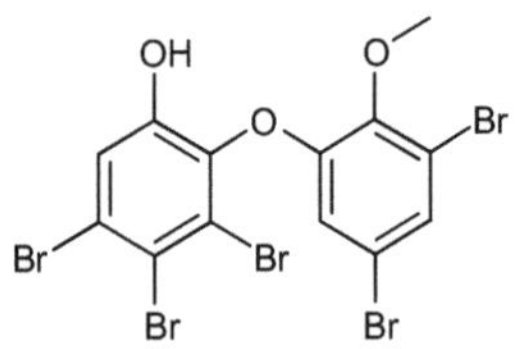

502 Dimethyl-2,3′-dimethylosoate

Type: Simple biphenyl ethers. $C_{19}H_{20}O_8$ Needles, mp 159–161 °C. Source: Marine-derived fungus *Aspergillus* sp. B-F-2 (from sediment sample, China). Pharm: Cytotoxic (K562, IC_{50} = 76.5 μmol/L, might due to arresting cell cycle in S phase). Ref: R. Liu, et al, J. Antibiot., 2006, 59, 362

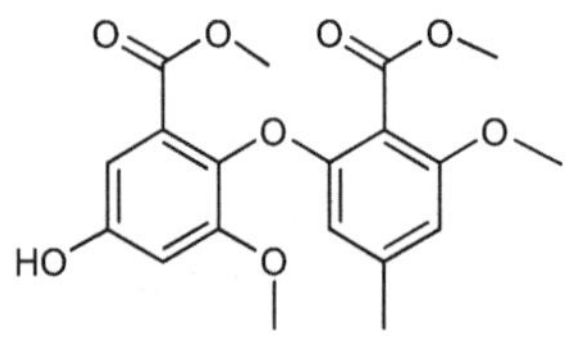

503 Diorcinol

Type: Simple biphenyl ethers. $C_{12}H_{10}O_3$ Source: Marine-derived fungus *Aspergillus sydowii* ZSDS1-F6. Pharm: Antiviral (H3N2 virus, IC_{50} = 66.5 μmol/L); antibacterial (*Klebsiella pneumonia*, MIC = 21.7 μmol/L). Ref: J. -F. Wang, et al, J. Antibiot., 2014, 67, 581

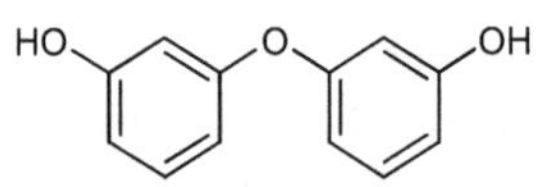

504 Diorcinol

Type: Simple biphenyl ethers. $C_{14}H_{14}O_3$ Viscous oil. Source: Marine-derived fungus *Aspergillus versicolor* (sediment, Sakhalin Bay, Sea of Okhotsk, Russia). Pharm: Antibacterial (gram-positive bacteria *Staphylococcus aureus* and *Bacillus subtilis*, MIC = 4.35 mmol/L); antifungal (yeast *Candida albicans*, MIC = 3.45 mmol/L); hemolytic (EC = 1.96 mmol/L); inhibits sperm fertilizing ability (IC_{50} = 0.078 mmol/L, at 10.87 mmol/L, has pronounced cytostatic effect and increases number of fertilized but undivided cells by 74%); cytotoxic (mouse splenic cells, MIC = 0.11 mmol/L). Ref: A. N. Yurchenko, et al, Russ. Chem. Bull., 2010, 59, 852

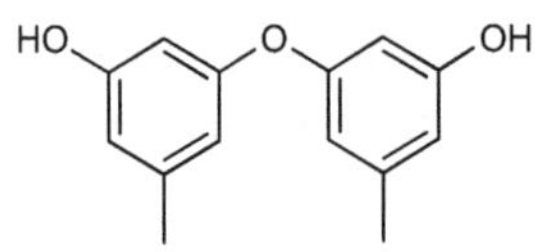

505 Diorcinol

3,3′-Dihydroxy-5,5′-dimethyldiphenyl ether Type: Simple biphenyl ethers. $C_{14}H_{14}O_3$ Viscous oil. Source: Sponge-derived fungus *Aspergillus sydowii* ZSDS1-F6, marine-derived *Aspergillus versicolor*, deep-sea fungus *Engyodontium album* DFFSCS021. Pharm: Anti-HIV (IC_{50} = 66.5 μmol/L); antibacterial. Ref: L. J. Fremlin, et al, JNP, 2009, 72, 666 | J. -F. Wang, et al, J. Antibiot., 2014, 67, 581 | Q. Yao, et al, Mar. Drugs, 2014, 12, 5902

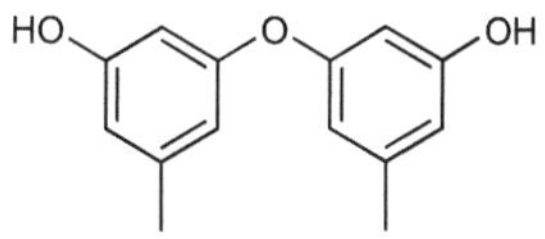

506 Diorcinol D

Type: Simple biphenyl ethers. $C_{19}H_{22}O_3$ Source: Marine-derived fungus *Aspergillus versicolor* (mud, Yellow Sea). Pharm: Cytotoxic (HTCLs). Ref: L. -N. Zhou, et al, Helv. Chim. Acta, 2011, 94, 1065 | H. Gao, et al, J. Antibiot., 2013, 66, 539

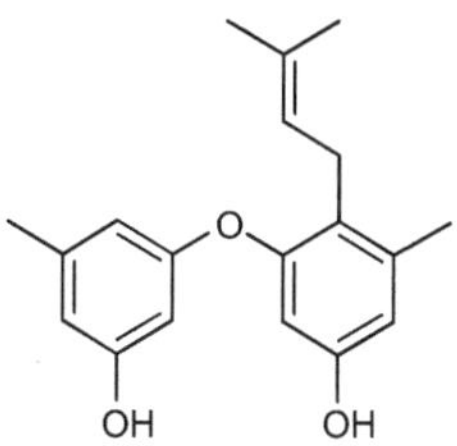

507 Diorcinol E

Type: Simple biphenyl ethers. $C_{19}H_{22}O_4$ Source: Marine-derived fungus *Aspergillus versicolor* (mud, Yellow Sea). Pharm: Cytotoxic (HTCLs). Ref: L. -N. Zhou, et al, Helv. Chim. Acta, 2011, 94, 1065 | H. Gao, et al, J. Antibiot., 2013, 66, 539

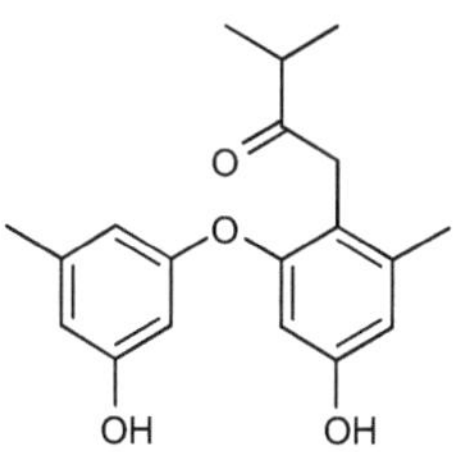

508 2,3,3′,4,5,5′-Hexabromo-2′,6-dihydroxydiphenyl ether

Type: Simple biphenyl ethers. $C_{12}H_4Br_6O_3$ Source: Sponges *Carteriospongia foliascens* [Syn. *Phyllospongia foliascens*] (Palau, Oceania) and *Dysidea herbacea* (Australia).

Pharm: Antimicrobial (*in vitro*). Ref: B. Carté, et al, Tetrahedron, 1981, 37, 2335 | R. S. Norton, et al, Tetrahedron, 1981, 37, 2341

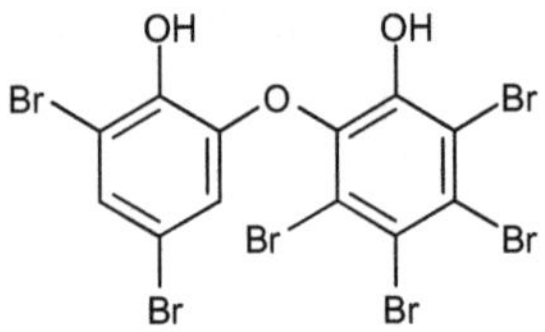

509 Lithothamnin A

Type: Simple biphenyl ethers. $C_{34}H_{27}Br_5N_4O_{10}$ Light yellow solid. Source: Red alga *Lithothamnion fragilissimum* (Lighthouse Reef, Palau Is.). Pharm: Cytotoxic (LOX, IC_{50} = 9.5 μmol/L; SNB19, IC_{50} = 7.6 μmol/L; OVCAR-3, IC_{50} = 7.6 μmol/L; Colon205, IC_{50} = 19.0 μmol/L; Molt4, IC_{50} = 19.0 μmol/L). Ref: A. W. W. Van Wyk, et al, JNP, 2011, 74, 1275

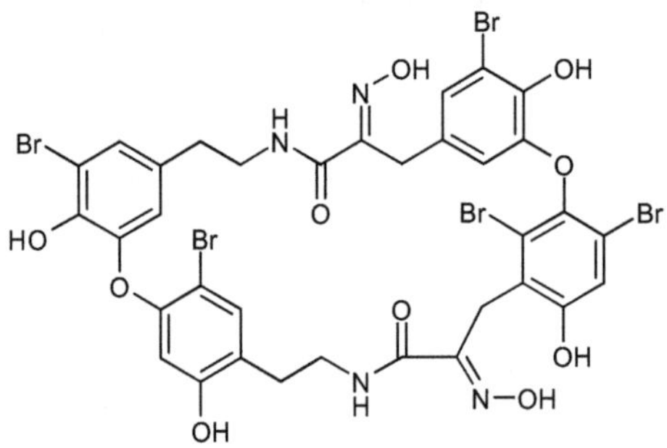

510 O-Methylspongiadioxin A

Type: Simple biphenyl ethers. $C_{13}H_6Br_4O_3$ Source: Sponge *Dysidea dendyi*. Pharm: Cell division inhibitor (fertilized eggs of sea urchin *Strongylocentrotus intermedius*, IC_{50} = 141 μmol/L). Ref: N. K. Utkina, et al, JNP, 2001, 64, 151; 2002, 65, 1213

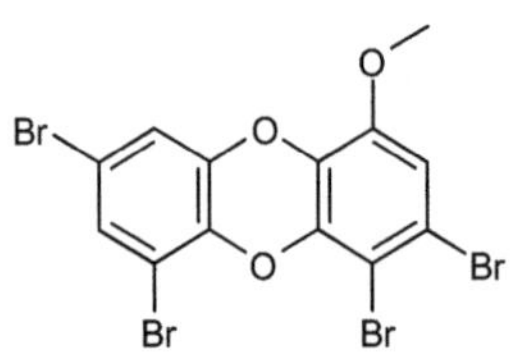

511 *O*-Methylspongiadioxin B

Type: Simple biphenyl ethers. $C_{13}H_6Br_4O_3$ Needles ($CHCl_3$) (semisynthetic), mp 204–206 °C (semisynthetic). Source: Sponge *Dysidea dendyi*. Pharm: Cell division inhibitor (fertilized eggs of sea urchin *Strongylocentrotus intermedius*, IC_{50} – 94 μmol/L). Ref: N. K. Utkina, et al, JNP, 2001, 64, 151; 2002, 65, 1213

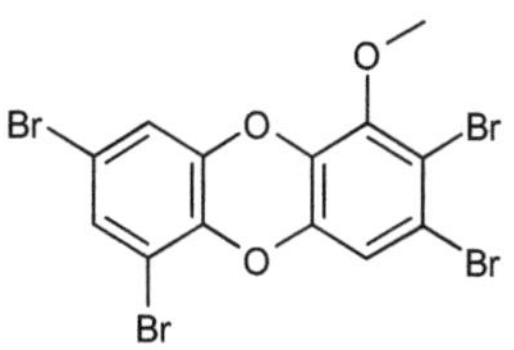

512 *O*-Methylspongiadioxin C

Type: Simple biphenyl ethers. $C_{13}H_7Br_3O_3$ Needles ($CHCl_3$), mp 205–207 °C. Source: Sponge *Dysidea dendyi*. Pharm: Cell division inhibitor (fertilized eggs of sea urchin *Strongylocentrotus intermedius*, IC_{50} = 166 µmol/L). Ref: N. K. Utkina, et al, JNP, 2001, 64, 151; 2002, 65, 1213

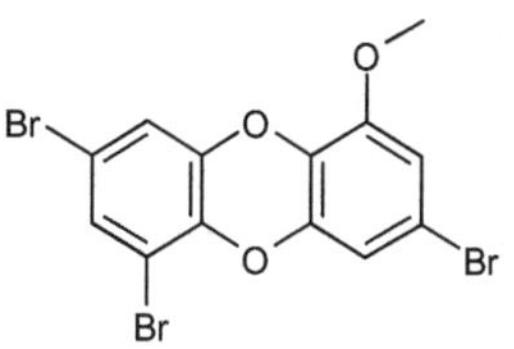

513 Microsphaerol

Type: Simple biphenyl ethers. $C_{23}H_{21}Cl_3O_5$ Source: Marine-derived fungus *Microsphaeropsis* sp. 7820 (endophytic). Pharm: Antibacterial (50 µg/disk: *Enterobacter coli*, IZ = 8 mm, *Bacillus megaterium*, IZ = 9 mm, *Microbotryum violaceum* IZ = 9 mm). Ref: H. Hussain, et al, Chem. Biodivers., 2015, 12, 289

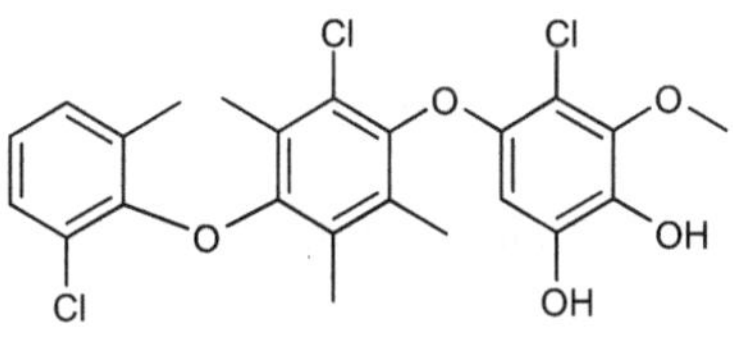

514 2,3′,4,5,5′-Pentabromo-2′,6-dihydroxydiphenyl ether

Type: Simple biphenyl ethers. $C_{12}H_5Br_5O_3$ Cryst. ($CHCl_3$), mp 183–184 °C. Source: Sponges *Carteriospongia* [Syn. *Phyllospongia*] *foliascens* (Palau, Oceania) and *Dysidea herbacea* (Australia). Pharm: Antimicrobial (*in vitro*). Ref: B. Carté, et al, Tetrahedron, 1981, 37, 2335 | R. S. Norton, et al, Tetrahedron, 1981, 37, 2341

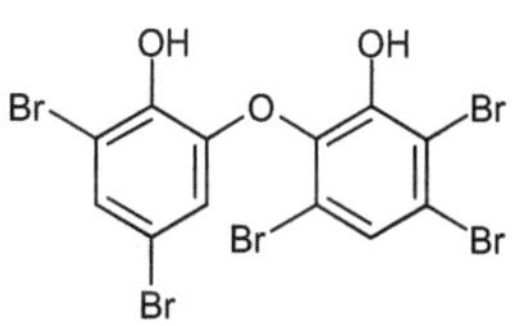

515 2,3′,4,4′,5′-Pentabromo-2′-hydroxydiphenyl ether

Type: Simple biphenyl ethers. $C_{12}H_5Br_5O_2$ mp 138–140 °C. Source: Sponge *Dysidea herbacea* (Palau, Oceania). Pharm: Antimicrobial (*in vitro*). Ref: G. M. Sharma, et al, Tet. Lett., 1969, 1715 | B. Carté, et al, Tetrahedron, 1981, 37, 2335

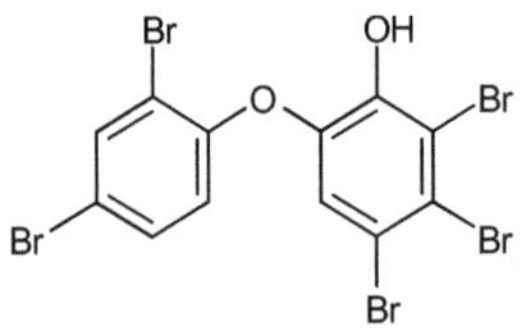

516 2,3′,4,5,5′-Pentabromo-2′-hydroxy-6-methoxydiphenyl ether

Type: Simple biphenyl ethers. $C_{13}H_7Br_5O_3$ Cryst. ($CHCl_3$), mp 202–204 °C. Source: Sponge *Dysidea* sp. (four sponge specimans, Indo-Pacific). Pharm: Inosine monophosphate dehydrogenase (IMPDH) inhibitor; guanosine monophosphate synthetase inhibitor; 15-lipoxygenase inhibitor. Ref: X. Fu, et al, JNP, 1995, 58, 1384

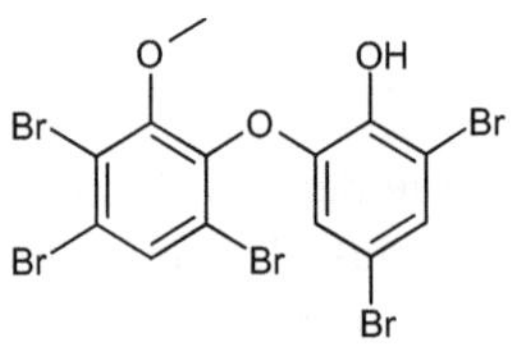

517 Pestalotether A

Type: Simple biphenyl ethers. $C_{18}H_{17}ClO_8$ Colorless gum. Source: Mangrove-derived fungus *Pestalotiopsis* sp. from mangrove *Rhizophora apiculata* (branch, Satun, Thailand). Pharm: Antifungal (*Cryptococcus neoformans* ATCC 90112, MIC = 200 μg/mL, *Candida albicans* NCPF3153, inactive). Ref: S. Klaiklay, et al, Tetrahedron, 2012, 68, 2299

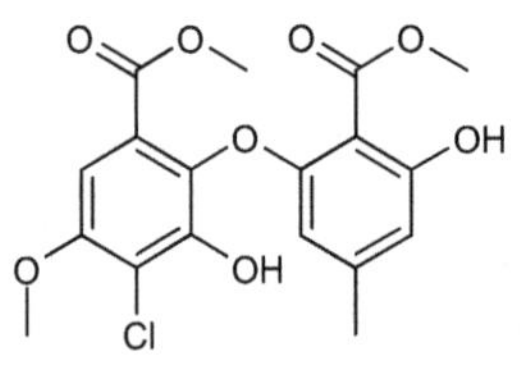

518 Pestalotether B

Type: Simple biphenyl ethers. $C_{16}H_{15}ClO_6$ Colorless gum. Source: Mangrove-derived fungus *Pestalotiopsis* sp. from mangrove *Rhizophora apiculata* (branch, Satun, Thailand). Pharm: Antifungal (*Cryptococcus neoformans* ATCC 90112, MIC = 200 μg/mL, *Candida albicans* NCPF3153, inactive). Ref: S. Klaiklay, et al, Tetrahedron, 2012, 68, 2299

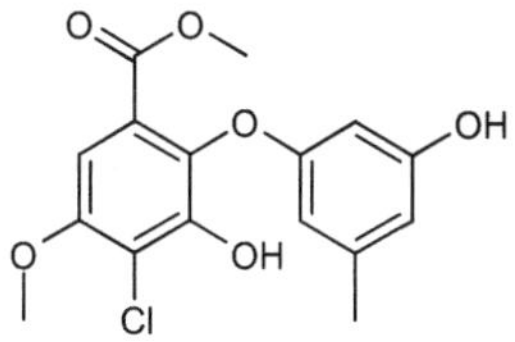

519 Spongiadioxin A

Type: Simple biphenyl ethers. $C_{12}H_4Br_4O_3$ Needles ($CHCl_3$), mp 241–242 °C. Source: Sponge *Dysidea dendyi* (North West coast of Australia). Pharm: Cell division inhibitor (fertilized eggs of sea urchin *Strongylocentrotus intermedius*, IC_{50} = 4.8 µmol/L); cytotoxic (EAC, ED_{50} = 29 µg/mL); LD_{50} (mouse) > 150 mg/kg. Ref: N. K. Utkina, et al, JNP, 2001, 64, 151; 2002, 65, 1213

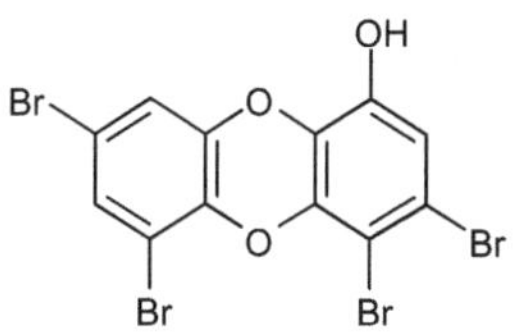

520 Spongiadioxin B

Type: Simple biphenyl ethers. $C_{12}H_4Br_4O_3$ Needles ($CHCl_3$), mp 245–247 °C. Source: Sponge *Dysidea dendyi*. Pharm: Cell division inhibitor (fertilized eggs of sea urchin *Strongylocentrotus intermedius*, IC_{50} = 1.1 µmol/L); cytotoxic (EAC, ED_{50} = 15.5 µg/mL); LD_{50} (mouse) > 150 mg/kg. Ref: N. K. Utkina, et al, JNP, 2001, 64, 151; 2002, 65, 1213

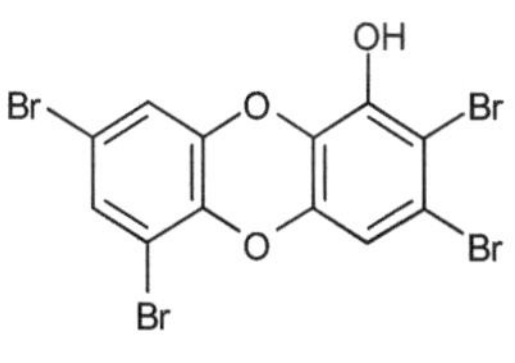

521 Spongiadioxin C

Type: Simple biphenyl ethers. $C_{12}H_5Br_3O_3$ Needles ($CHCl_3$), mp 203–205 °C. Source: Sponge *Dysidea dendyi*. Pharm: Cell division inhibitor (fertilized eggs of sea urchin *Strongylocentrotus intermedius*, IC_{50} = 5.7 µmol/L). Ref: N. K. Utkina, et al, JNP, 2001, 64, 151; 2002, 65, 1213

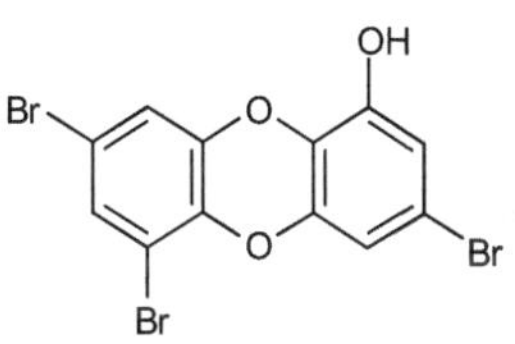

522 Tedarene A

Type: Simple biphenyl ethers. $C_{19}H_{18}O_2$ Source: Sponge *Tedania ignis* (Sweetings Cay, Bahamas). Pharm: Atropisomeric; NO production inhibitor (LPS-induced macrophages). Ref: V. Costantino, et al, JOC, 2012, 77, 6377

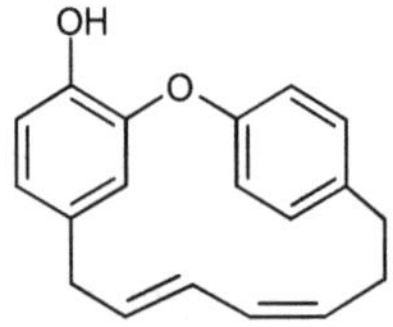

523 2,3′,4,5′-Tetrabromo-2′,6-dimethoxydiphenyl ether

Type: Simple biphenyl ethers. $C_{14}H_{10}Br_4O_3$ mp 86–88 °C. Source: Sponges *Dysidea herbacea* (Fiji, Australia) and *Carteriospongia* [Syn. *Phyllospongia*] *foliascens* (Palau, Oceania). Pharm: Antimicrobial (*in vitro*). Ref: R. S. Norton, et al, Tet. Lett., 1980, 21, 3801 | B. Carté, et al, Tetrahedron, 1981, 37, 2335 | 115 R. S. Norton, et al, Tetrahedron, 1981, 37, 2341

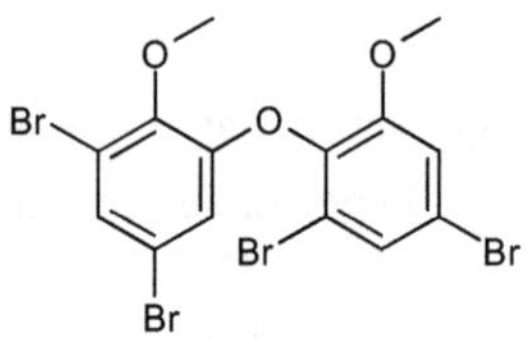

524 3′,5′,6′,6-Tetrabromo-2,4-dimethyldiphenyl ether

Type: Simple biphenyl ethers. $C_{14}H_{10}Br_4O$ Colorless oil. Source: Red alga *Laurencia similis*. Pharm: Hypoglycemic (PTP1B inhibitor) (*in vitro* assay, IC_{50} = 2.97 μg/mL, control HD, IC_{50} = 0.80 μg/mL). Ref: J. Qin, et al, Bioorganic Med. Chem. Lett., 2010, 20, 7152

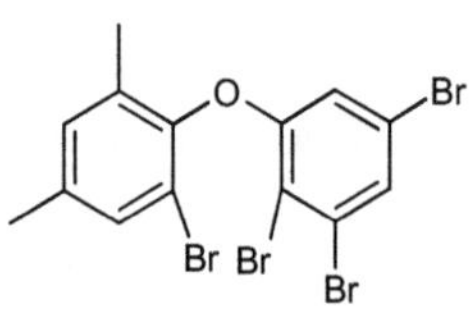

525 3,3′,5,5′-Tetrabromo-2-hydroxy-2′,6-dimethoxydiphenyl ether

Type: Simple biphenyl ethers. $C_{14}H_{10}Br_4O_4$ Source: Sponge *Phyllospongia dendyi* (Palau, Oceania). Pharm: Toxic (variety of micro- and macro-algae). Ref: T. Hattori, et al, Fish. Sci., 2001, 67, 899

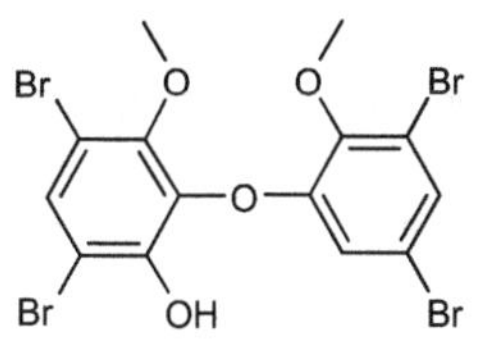

526 2,2′,4,4′-Tetrabromo-6-hydroxydiphenyl ether

Type: Simple biphenyl ethers. $C_{12}H_6Br_4O_2$ mp 172.5–173 °C. Source: Sponge *Dysidea herbacea* (Palau, Oceania). Pharm: Antimicrobial. Ref: B. Carté, et al, Tetrahedron, 1981, 37, 2335

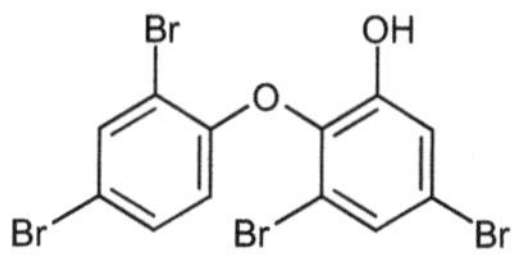

527 2,3′,4,5′-Tetrabromo-2′-hydroxydiphenyl ether

Type: Simple biphenyl ethers. $C_{12}H_6Br_4O_2$ Cryst. (hexane/petrol), mp 89–91 °C. Source: Sponges *Dysidea herbacea* and *Dysidea chlorea* (Palau, Oceania), cyanobacterium *Oscillatoria spongeliae*. Pharm: Antimicrobial. Ref: G. M. Sharma, et al, Tet. Lett., 1969, 1715 | B. Carté, et al, Tetrahedron, 1981, 37, 2335 | M. D. Unson, et al, Mar. Biol. 1994, 119, 1

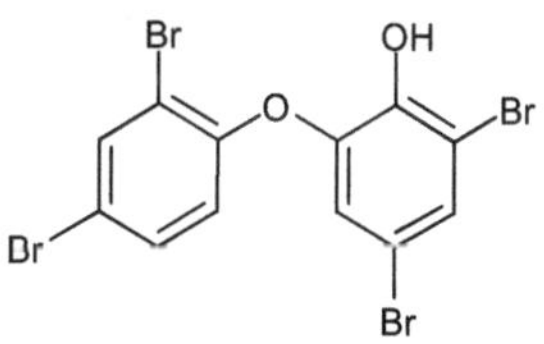

528 2,3′,4,5′-Tetrabromo-6-hydroxy-2′-methoxydiphenyl ether

Type: Simple biphenyl ethers. $C_{13}H_8Br_4O_3$ Pale green wax, mp 32–33 °C. Source: Sponges *Dysidea herbacea* (Australia) and *Dysidea* spp. Pharm: Antimicrobial. Ref: R. S. Norton, et al, Tetrhedron, 1981, 37, 2341 | N. K. Utkina, et al, Chem. Nat. Comp. 1994, 29, 291

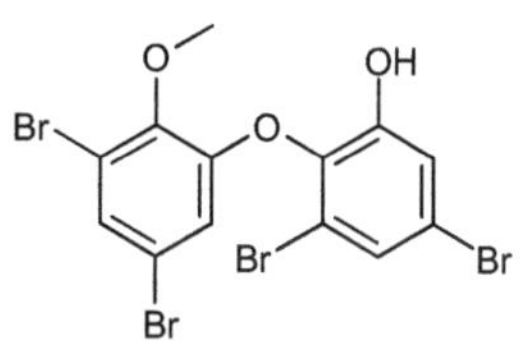

529 Tetraorcinol A

Type: Simple biphenyl ethers. $C_{28}H_{26}O_5$ Source: Marine-derived fungus *Aspergillus versicolor* from soft coral *Cladiella* sp. (Lingao, Hainan Province, China). Pharm: Antioxidant (DPPH scavenger, weak). Ref: Y. Zhuang, et al, Tetrahedron, 2011, 67, 7085

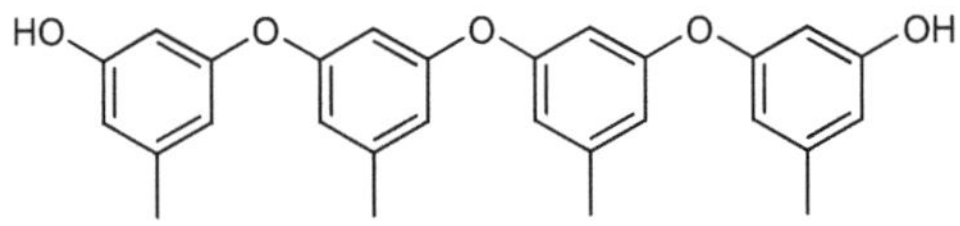

530 3,5,6-Tribromo-1-(2′-bromophenoxy)-2-benzene methyl ether

Type: Simple biphenyl ethers. $C_{13}H_8Br_4O_2$ Viscous oil. Source: Sponge *Dysidea herbacea* (West Sumatra, Indonesia). Pharm: Toxic (brine shrimp lethality test, IC_{50} = (26.25 ± 0.42)μg/mL)). Ref: D. Handayani, et al, JNP, 1997, 60, 1313

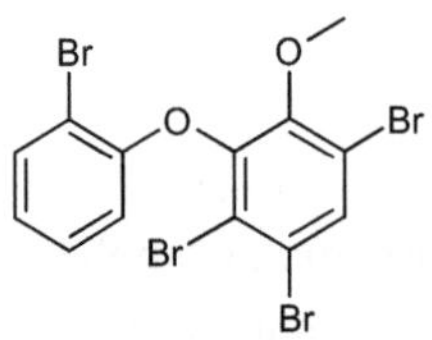

531 3,5,6-Tribromo-2-(2′-bromophenoxy) phenol

Type: Simple biphenyl ethers. $C_{12}H_6Br_4O_2$ Viscous oil. Source: Sponge *Dysidea herbacea* (West Sumatra, Indonesia). Pharm: Toxic (brine shrimp lethality test, IC_{50} = (3.20 ± 0.35)μg/mL)). Ref: D. Handayani, et al, JNP, 1997, 60, 1313

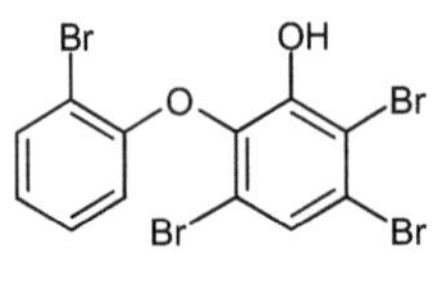

532 3,4,6-Tribromo-2-(2′-bromophenoxy) phenol

Type: Simple biphenyl ethers. $C_{12}H_6Br_4O_2$ Viscous oil. Source: Sponge *Dysidea herbacea* (West Sumatra, Indonesia). Pharm: Toxic (brine shrimp lethality test, IC_{50} = (3.20 ± 0.35)μg/mL). Ref: D. Handayani, et al, JNP, 1997, 60, 1313

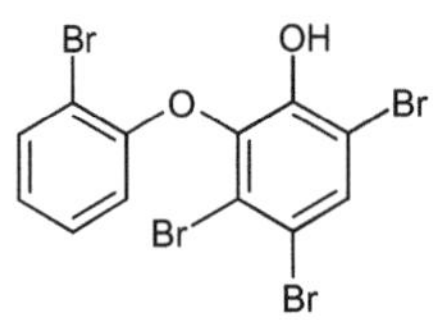

533 3,4,5-Tribromo-2-(2′-bromophenoxy)phenol

Type: Simple biphenyl ethers. $C_{12}H_6Br_4O_2$ Viscous oil. Source: Sponge *Dysidea herbacea* (West Sumatra, Indonesia). Pharm: Toxic (brine shrimp lethality test, IC_{50} = (3.30 ± 0.51)μg/mL). Ref: D. Handayani, et al, JNP, 1997, 60, 1313

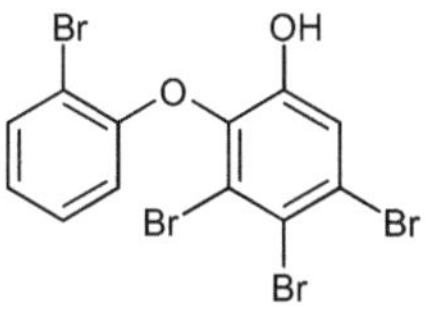

534 2,3,4-Tribromo-6-(3,5-dibromo-2-hydroxyphenoxy)phenol

Type: Simple biphenyl ethers. $C_{12}H_5Br_5O_3$ Needles ($CHCl_3$/DMSO). Source: Sponge *Lamellodysidea herbacea* (Vim Levu, Fiji). Pharm: Bcl-2 protein inhibitor (mild). Ref: L. Calcul, et al, JNP, 2009, 72, 443

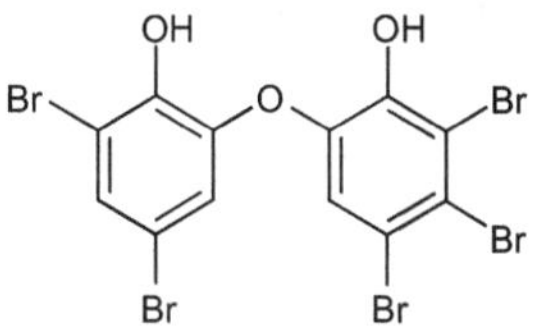

535 3,4,6-Tribromo-2-(3,5-dibromo-2-hydroxyphenoxy)phenol

Type: Simple biphenyl ethers. $C_{12}H_5Br_5O_3$ Needles (MeCN). Source: Sponge *Lamellodysidea herbacea* (Vim Levu, Fiji). Pharm: Bcl-2 protein inhibitor (mild). Ref: L. Calcul, et al, JNP, 2009, 72, 443

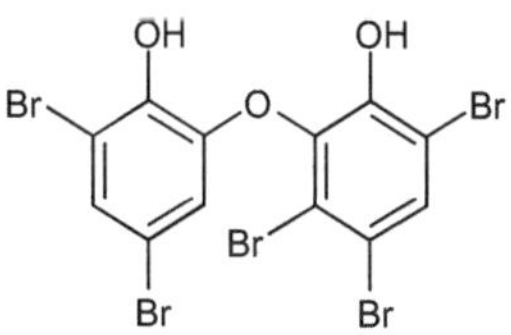

536 2,4′,5-Tribromo-2′,3′,6-trihydroxy-3,6′-bis(hydroxymethyl)diphenyl ether

Type: Simple biphenyl ethers. $C_{14}H_{11}Br_3O_6$ Pale yellow solid. Source: Red alga *Odonthalia corymbifera*. Pharm: α-Glucosidase inhibitor (IC_{50} = 25 μmol/L). Ref: H. Kurihara, et al, JNP, 1999, 62, 882

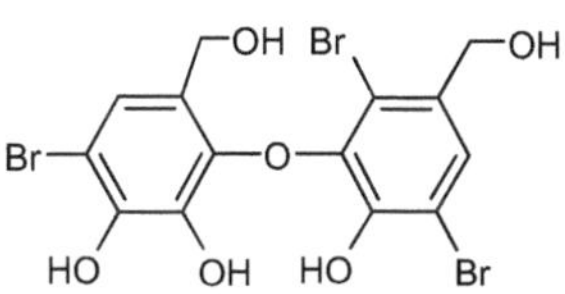

537 2,4,4′-Trichloro-2′-hydroxydiphenyl ether

Type: Simple biphenyl ethers. $C_{12}H_7Cl_3O_2$ Off-white cryst. powder or soft agglomerates, mp 54–57 °C, Insol. H_2O, sol. org. solvs. Source: Marine-derived bacterium *Micrococcus luteus* from sponge *Xestospongia* sp. (surface, Indo-Pacific). Pharm: Bacteriostatic; low acute toxicity; Fab I inhibitor. Ref: R. Zinkernagle, et al, Seifen-oele-fette wachse, 1967, 93, 670 | CRC Press, DNP on DVD, 2012, Version 20.2

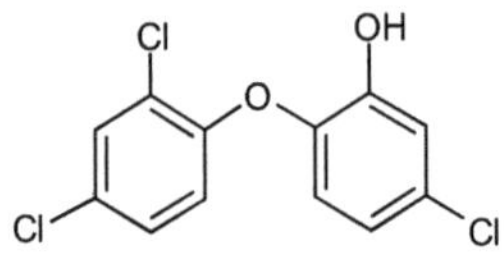

538 Altenusin

Type: Simple biphenyls. $C_{15}H_{14}O_6$ Source: Mangrove-derived fungus *Alternaria* sp. JCM9.2. Pharm: Antibacterial (MRSA, MIC = 31.25 μg/mL; *Enterococcus faecalis*, MIC = 62.5 μg/mL; *Enterobacter cloacae*, MIC = 125 μg/mL; *Candida albicans*, MIC = 125 μg/mL). Ref: J. Kjer, et al, JNP, 2009, 72, 2053

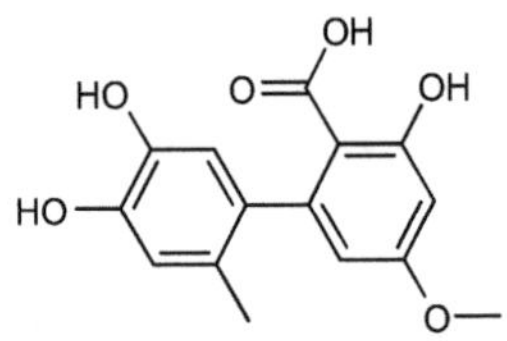

539 (2*E*,2′*Z*)-3,3′-(6,6′-Dihydroxybiphenyl-3,3′-diyl)diacrylic acid

Type: Simple biphenyls. $C_{18}H_{14}O_6$ White amorphous powder. Source: Marine-derived fungus *Spicaria elegans* KLA-03. Pharm: Antibacterial (*Enterobacter aerogenes*, MIC = 0.15 μmol/L, *Enterobacter coli*, MIC = 0.04 μmol/L, *Pseudomonas aeruginosa*, MIC = 1.53 μmol/L, *Staphylococcus aureus*, MIC = 0.77 μmol/L); antifungal (yeast *Candida albicans*, MIC = 0.38 μmol/L). Ref: Y. Wang, et al, Mar. Drugs, 2011, 9, 535

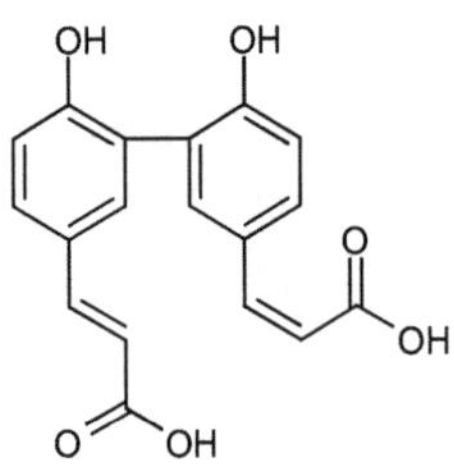

540 4-(3-Hydroxypropyl)-5,6-dimethoxybiphenyl-3,4′-diol

Type: Simple biphenyls. $C_{17}H_{20}O_5$ Needles, mp 205–207 °C. Source: Mangrove-derived fungus *Penicillium thomi* from mangrove *Bruguiera gymnorrhiza* (root, China).

Pharm: Cytotoxic (A549, HepG2 and HT29, moderate). Ref: G. Chen, et al, J. Asian Nat. Prod. Res., 2007, 9, 159

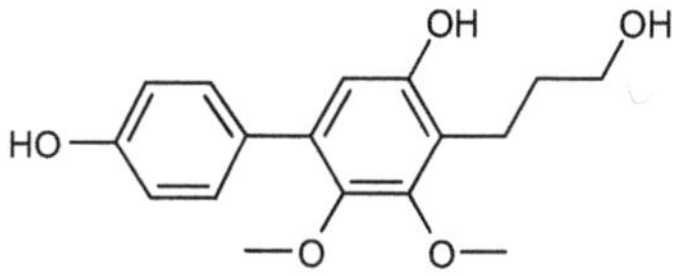

541 Tedarene B

Type: Simple biphenyls. $C_{19}H_{17}O_5S^{1-}$ Source: Sponge *Tedania ignis* (Sweetings Cay, Bahamas). Pharm: Atropisomeric. Ref: V. Costantino, et al, JOC, 2012, 77, 6377

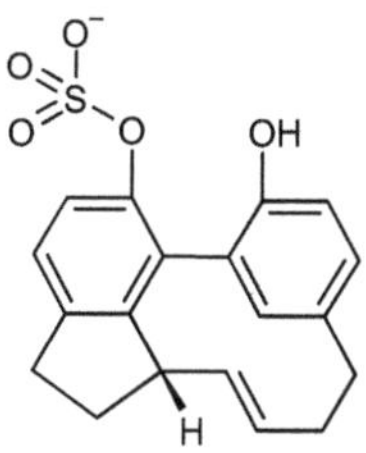

542 3′,4,4″-Trihydroxy-2′,6′-dimethoxy-*p*-terphenyl

Type: Simple biphenyls. $C_{20}H_{18}O_5$ Source: Deep-sea fungus *Aspergillus candidus*. Pharm: Antifungal (*Candida albicans*, great growth inhibition). Ref: F. Liu, et al, J. Xiamen Univ. 2013, 52, 670 (in Chinese)

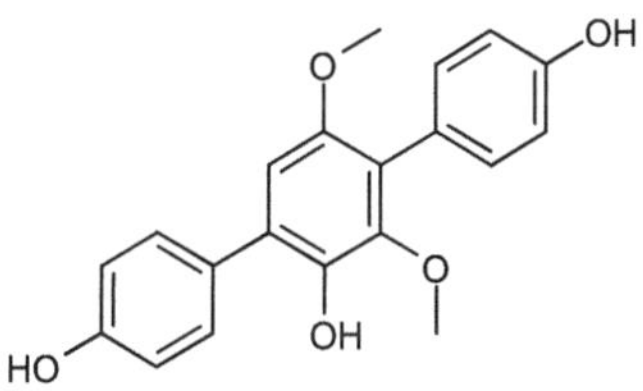

543 Ulocladol

Type: Simple biphenyls. $C_{16}H_{14}O_7$ Powder, mp 110–111 °C. Source: Marine-derived fungus *Ulocladium botrytis* (culture) from sponge *Myxilla incrustans*, marine-derived fungus *Microsphaeropsis olivacea*. Pharm: Tyrosine kinase p56lck inhibitor (ELISA-based assay, 0.02 μg/mL, reduced enzyme activity to 7%, since p56lck is necessary for T-cell activation, inhibitors of kinase could be useful for treating autoimmune diseases). Ref: U. Holler, et al, EurJOC, 1999, 2949 | D. R. Goldberg, et al, JMC, 2003, 46, 1337 | E. Hormazabal, et al, Z. Naturforsch., C, 2005, 60, 11

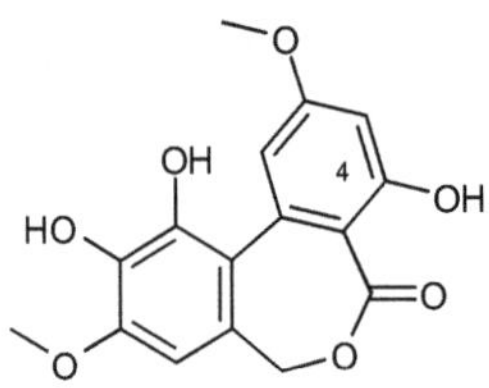

544 6,6'-Bieckol

8,8'-Bieckol Type: Phlorotannins. $C_{36}H_{22}O_{18}$ Prisms (H_2O), mp 300 °C. Source: Brown algae *Ecklonia kurome, Ecklonia cava.* and *Eisenia arborea.* Pharm: Hypoglycemic (*α*-glucosidase inhibitor) (*in vitro* assay, IC_{50} = 22.2 μmol/L); *hypoglycemic (α-amylase inhibitor)* (*in vitro* assay, IC_{50} > 500 μmol/L); antiplasmin inhibitor; *α*2-macroglobulin inhibitor; antiviral (inhibits HIV-1 infection IC_{50} = 1.07–1.72 μmol/L, MMOA: Viralp24 antigen production and reverse transcriptase inhibition). Ref: Y. Fukuyama, et al, CPB, 1989, 37, 2438 | M. Artan, et al, BoMC, 2008, 16, 7921 | S. H. Lee, et al, J. Sci. Food Agric. 2009, 89, 1552

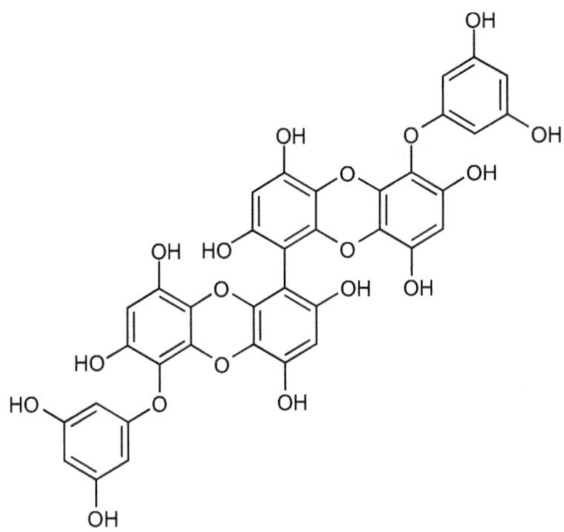

545 Crossbyanol B

Type: Phlorotannins. $C_{30}H_{15}Br_7O_{12}S_2$ Pale yellow solid. Source: Cyanobacterium *Leptolyngbya crossbyana* (Honaunau Reef, Hawaii). Pharm: Antibacterial (MRSA); toxic (brine shrimp, relatively potent). Ref: H. Choi, et al, JNP, 2010, 73, 517

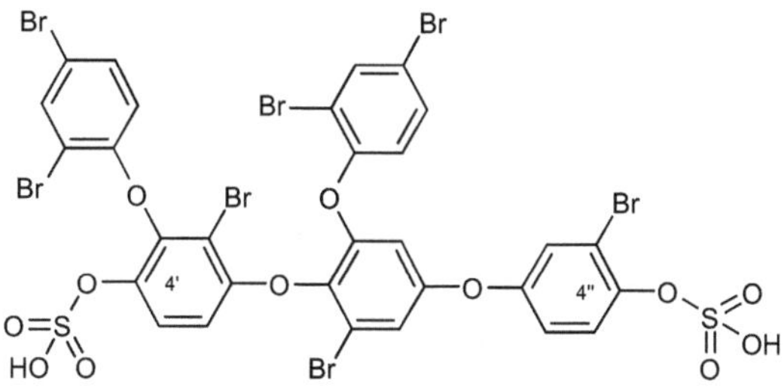

546 Dibenzo [1,4]dioxine-2,4,7,9-tetraol

Type: Phlorotannins. $C_{12}H_8O_6$ Source: Brown alga *Ecklonia maxima*. Pharm: Hypoglycemic (α-glucosidase inhibitor) (*in vitro* assay, IC_{50} = 33.69 μmol/L, control Acarbose, IC_{50} = 1013 μmol/L); antioxidant (DPPH scavenger, EC_{50} = 0.012 μmol/L, AAI (antioxidant activity index = final DPPH concentration/EC_{50}) = 4037, control Ascorbic acid, EC_{50} = 0.011 μmol/L, AAI = 4356). Ref: K. R. Rengasamy, et al, Food Chem., 2013, 141, 1412

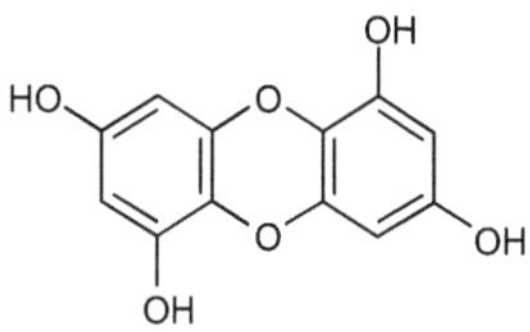

547 Dieckol

4‴-*O*-7-Bieckol Type: Phlorotannins. $C_{36}H_{22}O_{18}$ Cryst., mp 300 °C. Source: Brown algae *Ecklonia kurome, Ecklonia stolonifera Eisenia bicyclis* and *Ecklonia cava*. Pharm: Hypoglycemic (α-amylase inhibitor) (*in vitro* assay, 1 mmol/L, InRt = 97.5%); hypoglycemic (α-glucosidase inhibitor) (*in vitro* assay, IC_{50} = 1.61 μmol/L); hypoglycemic (α-glucosidase inhibitor) (IC_{50} = 10.8 μmol/L); hypoglycemic (α-amylase inhibitor) (IC_{50} = 124.9 μmol/L); hypoglycemic (α-glucosidase inhibitor) (*in vitro* assay, IC_{50} = 0.24 mmol/L); hypoglycemic (α-amylase inhibitor) (*in vitro* assay, IC_{50} = 0.66 mmol/L); hypoglycemic (AGE formation inhibition, *in vitro* assay, 1 mmol/L, InRt = 86.7%); Hypoglycemic (aldose reductase inhibitor) (rat lens aldose reductase *in vitro* assay, IC_{50} = 42.39 μmol/L); hypoglycemic (PTP1B inhibitor) (*in vitro* assay, IC_{50} = 1.18 μmol/L); hypoglycemic (C57BL/KsJ-db/db diabetic mice *in vivo*, reduction of blood glucose, glycosylated hemoglobin levels; reduction of hepatic lipids concentration and improvement of impaired glucose tolerance; reduction of glucose-6-phosphatase and phosphoenolpyruvate carboxykinase enzymes activities; and increased glucokinase activity); hypoglycemic (Streptozotocin-induced diabetic mice *in vivo*, reduction of post-prandial blood glucose level and delayed absorption of dietary carbohydrates); hypoglycemic (Pancreatic lipase inhibitor) (*in vitro* assay, IC_{50} = 0.26mg/mL); hypoglycemic (Pancreatic lipase inhibitor) (*in vitro* assay, IC_{50} = 99.3 μmol/L); antioxidant (10 μg/mL or 100 μg/mL); reduction of ROS over-production; anti-inflammatory (reduction of pro-inflammatory NF-κB, COX-2 and iNOS expressions); attenuates type II diabetes (db/db mouse model); reduction of oxidative stress (*in vivo*, increasing activities of hepatic antioxidant enzymes including glutathione peroxidase and superoxide dismutase); inihibits actions of plasma α_2-macroglobulin and α_2-plasmin inhibitor; histamine release inhibitor (hmn basophilic granulocyte KU-812, IC_{50} = 27.8 μmol/L). Ref: K. -W. Glombitza, et al, Phytochemistry, 1985, 24, 543 | Y. Okada, et al, JNP, 2004, 67, 103 | H. A. Jung, et al, Fish. Sci., 2008, 74, 1363 | S. H. Lee, et al, J. Sci. Food Agric., 2009, 89, 1552 | Q. T. Le, et al, Process Biochem, 2009, 44, 168 | S. H. Lee, et al, Food Chem. Toxicol. 2010, 48, 2633 | S. H. Lee, et al, Toxicol. Vitro, 2010, 24, 375 | H. E. Moon, et al, Biosci. Biotechnol. Biochem., 2011, 75, 1472 | S. H. Lee, et al, Food Chem. Toxicol., 2012, 50, 575 | K. B. W. R. Kim, et al, Biotechnol. Bioprocess Eng., 2012, 17, 739 | S. H. Eom, et al, Phytother. Res., 2013, 27, 148 | M. C. Kang, et al, Food Chem. Toxicol., 2013, 53, 294

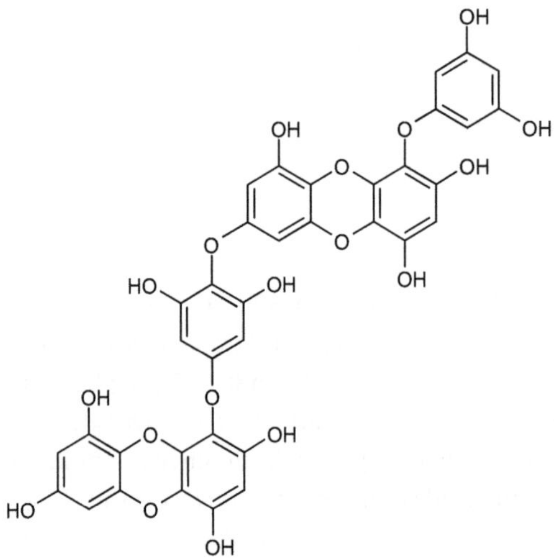

548 Dioxinodehydroeckol

Type: Phlorotannins. $C_{18}H_{10}O_9$ Powder. Source: Brown algae *Ecklonia stolonifera, Eisenia arborea, Ecklonia maxima* and *Eisenia bicyclis*. Pharm: Hypoglycemic (aldose reductase inhibitor) (rat lens aldose reductase *in vitro* assay, IC_{50} = 21.95 μmol/L); hypoglycemic (α-glucosidase inhibitor) (*in vitro* assay, IC_{50} = 34.60 μmol/L) (Moon, 2011); hypoglycemic (PTP1B inhibitor) (*in vitro* assay, IC_{50} = 29.97 μmol/L) (Moon, 2011); hypoglycemic (α-glucosidase inhibitor) (*in vitro* assay, IC_{50} = 97.33 nmol/L) (Eom, 2012); Hypoglycemic (α-amylase inhibitor) (*in vitro* assay, IC_{50} = 472.70 μmol/L) (Eom, 2012). Ref: K. -W. Glombitza, et al, Phytochemistry, 1985, 24, 543 | K. -W. Glombitza, et al, Planta Med., 1985, 51, 308 | H. S. Kang, et al, CPB,

2003, 51, 1012 H. A. Jung, et al, Fish. Sci., 2008, 74, 1363 | H. E. Moon, et al, Biosci. Biotechnol. Biochem., 2011, 75, 1472 | S. H. Eom, et al, J. Sci. Food Agric., 2012, 92, 2084

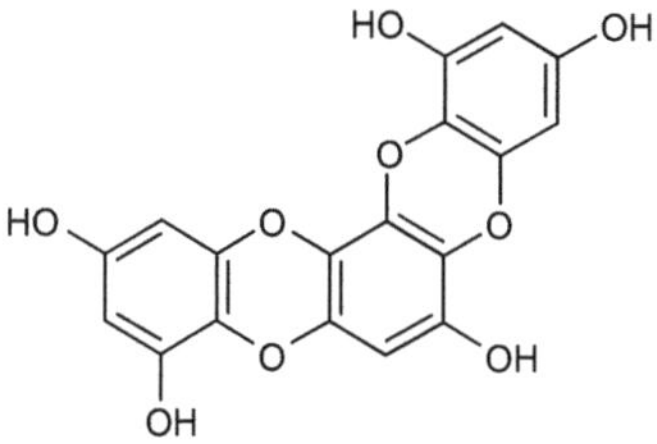

549 Diphlorethohydroxycarmalol

DPHC Type: Phlorotannins. $C_{24}H_{16}O_{13}$ Source: Brown algae *Carpophyllum maschalocarpum* and *Ishige okamurae.* Pharm: Hypoglycemic (*α*-glucosidase inhibitor) (*in vitro* assay, IC_{50} = 0.16 μmol/L); hypoglycemic (*α*-amylase inhibitor) (*in vitro* assay, IC_{50} = 0.53 μmol/L); hypoglycemic (Streptozotocin-induced diabetic mice *in vivo*, reduction of post-prandial blood glucose level and delayed absorption of dietary carbohydrates). Ref: S. -M. Li, et al, Phytochemistry, 1991, 30, 3417 | S. J. Heo, et al, J. Microbiol. Biotechnol., 2008, 18, 676 | S. J. Heo, et al, Eur. J. Pharmacol., 2009, 615, 252.

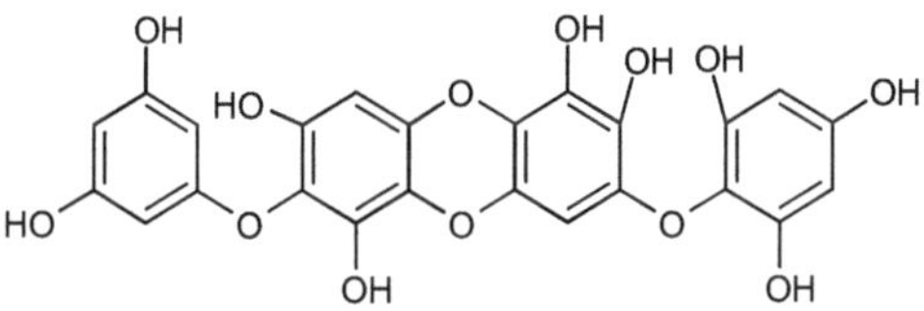

550 Eckol

Hexahydroxyphenoxydibenzo[1,4]dioxine Type: Phlorotannins. $C_{18}H_{12}O_9$ Cryst., mp 243–244 °C. Source: Brown algae *Ecklonia stolonifera, Ecklonia kurome, Ecklonia maxima, Eisenia arborea* and *Eisenia bicyclis.* Pharm: Hypoglycemic (*α*-amylase inhibitor) (*in vitro* assay, 1 mmol/L, InRt = 87.5%); hypoglycemic (*α*-glucosidase inhibitor) (*in vitro* assay, IC_{50} = 22.78 μmol/L); hypoglycemic (*α*-glucosidase inhibitor) (*in vitro* assay, IC_{50} = 11.16 μmol/L, control Acarbose, IC_{50} = 1013 μmol/L); hypoglycemic (AGE formation inhibition, *in vitro* assay, 1 mmol/L, InRt = 96.2%); hypoglycemic (aldose reductase inhibitor) (rat lens aldose reductase *in vitro* assay, IC_{50} = 54.68 μmol/L); hypoglycemic (PTP1B inhibitor) (*in vitro* assay, IC_{50} = 2.64 μmol/L); hypoglycemic (pancreatic lipase inhibitor) (*in vitro* assay, IC_{50} = 76.6 μmol/L); antioxidant (DPPH scavenger, EC_{50} = 0.008 μmol/L, AAI (antioxidant activity index = final DPPH concentration/EC_{50}) = 6096, control Ascorbic acid, EC_{50} = 0.011 μmol/L, AAI = 4356); *α*2-macroglobulin inhibitor; glycation inhibitor. Ref: K. -W. Glombitza, et al, Phytochemistry, 1985, 24, 543 | Y. Fukuyama, et al,

CPB, 1989, 37, 349 | Y. Okada, et al, JNP, 2004, 67, 103 | H. A. Jung, et al, Fish. Sci., 2008, 74, 1363 | H. E. Moon, et al, Biosci. Biotechnol. Biochem., 2011, 75, 1472 | K. R. Rengasamy, et al, Food Chem., 2013, 141, 1412 | S. H. Eom, et al, Phytother. Res. 2013, 27, 148

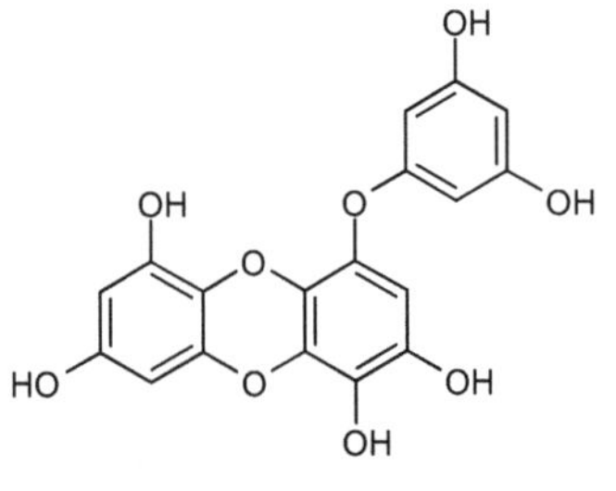

551 Fucodiphlorethol G

Type: Phlorotannins. $C_{24}H_{18}O_{12}$ Source: Brown alga *Ecklonia cava.* Pharm: Hypoglycemic (α-glucosidase inhibitor) (*in vitro* assay, IC_{50} = 19.5 μmol/L); hypoglycemic (α-amylase inhibitor) (*in vitro* assay, IC_{50} > 500 μmol/L). Ref: S. H. Lee, et al, J. Sci. Food Agric. 2009, 89, 1552

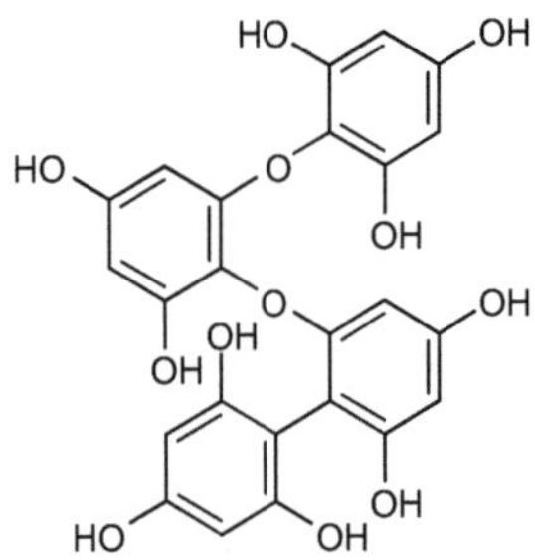

552 Fucofuroeckol A

Type: Phlorotannins. $C_{24}H_{14}O_{11}$ Source: Brown alga *Eisenia bicyclis.* Pharm: Hypoglycemic (α-amylase inhibitor) (*in vitro* assay, IC_{50} = 42.91 μmol/L); hypoglycemic (α-glucosidase inhibitor) (*in vitro* assay, IC_{50} = 131.34 nmol/L); hypoglycemic (Pancreatic lipase inhibitor) (*in vitro* assay, IC_{50} = 37.2 μmol/L). Ref: S. H. Eom, et al, et al, J. Sci. Food Agric. 2012, 92, 2084 | S. H. Eom, et al, Phytother. Res., 2013, 27, 148

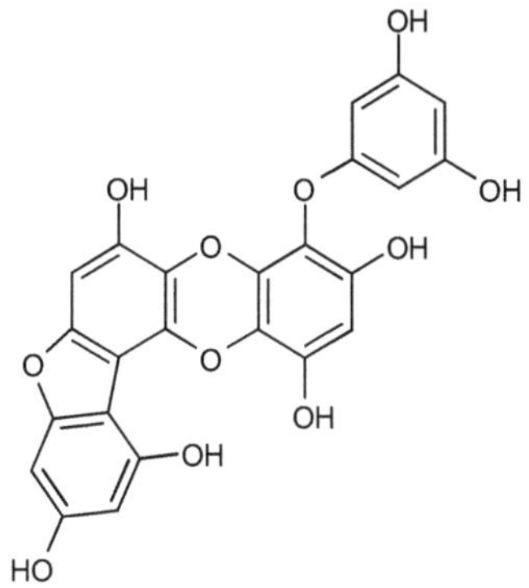

553 Octaphlorethol A

Type: Phlorotannins. $C_{48}H_{34}O_{24}$ Source: Brown alga *Ishige foliacea* (Jeju I., R. O. Korea). Pharm: Hypoglycemic (increased glucose uptake by GLUT4 via PI3-K/Akt and AMPK signalling pathways, L-6 rat myoblast cells *in vitro*); antioxidant (increased anti-oxidative glutathione peroxidase (GSH-px), catalase (CAT) and superoxide dismutase (SOD) activities); antioxidant (reduction of ROS over-production); cytoprotection of pancreatic *β*-cells (protects pancreatic *β*-cells damage caused by hyperglycaemia due to streptozotocin treatment); antidiabetic. Ref: S. H. Lee, et al, Biochem. Biophys. Res. Commun. 2012, 420, 576 | S. H. Lee, et al, Food Chem. Toxicol., 2013, 59, 643

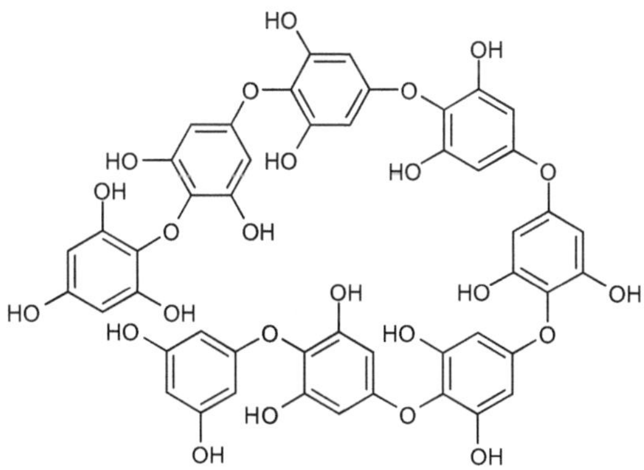

554 2-Phloro-6,6′-bieckol

2-O-(2,4,6-Trihydroxyphenyl)-6,6′-bieckol Type: Phlorotannins. $C_{42}H_{26}O_{21}$ Amorph. solid. Source: Brown alga *Ecklonia kurome*. Pharm: Antiplasmin inhibitor; *α*2-macroglobulin inhibitor. Ref: Y. Fukuyama, et al, CPB, 1989, 37, 2438

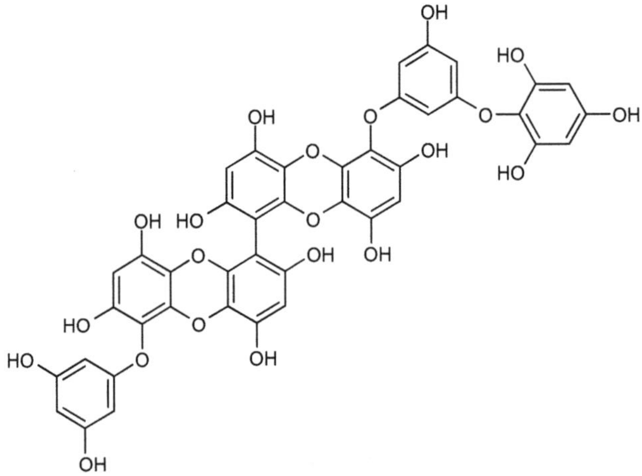

555 2-Phloroeckol

Type: Phlorotannins. $C_{24}H_{16}O_{12}$ Source: Brown alga Ecklonia kurome. Pharm: Inihibits action of plasma α2-macroglobulin; α2-plasmin inhibitor. Ref: K. -W. Glombitza, et al, Phytochemistry, 1985, 24, 543

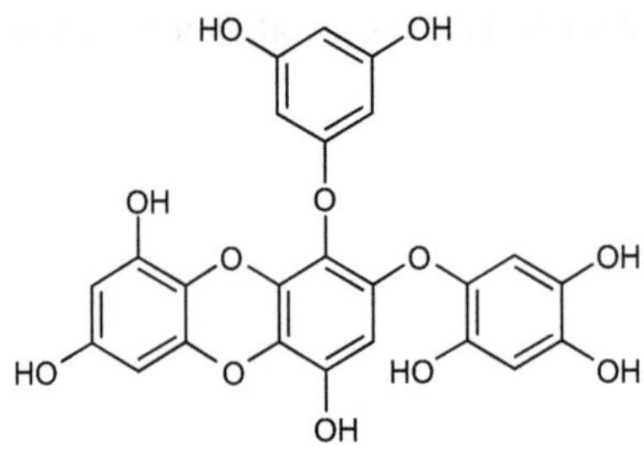

556 7-Phloroeckol

1-(3′,5′-Dihydroxyphenoxy)-7-(2″,4″,6″-trihydroxyphenoxy)-2,4,9-trihydroxydibenzo-1,4-dioxin Type: Phlorotannins. $C_{24}H_{16}O_{12}$ Source: Brown algae *Ecklonia stolonifera*, *Eisenia bicyclis*, and *Ecklonia cava*. Pharm: Hypoglycemic (AGE formation inhibition, *in vitro* assay, 1 mmol/L, InRt = 91.1%) (Okada, 2004); hypoglycemic (α-glucosidase inhibitor) (*in vitro* assay, IC_{50} = 6.13 μmol/L); hypoglycemic (α-glucosidase inhibitor) (*in vitro* assay, IC_{50} = 49.5 μmol/L); hypoglycemic (α-amylase inhibitor) (*in vitro* assay, IC_{50} = 250 μmol/L); hypoglycemic (aldose reductase inhibitor) (rat lens aldose reductase *in vitro* assay, IC_{50} = 27.54 μmol/L); hypoglycemic (PTP1B inhibitor) (*in vitro* assay, IC_{50} = 2.09 μmol/L); hypoglycemic (Pancreatic lipase inhibitor) (*in vitro* assay, IC_{50} = 12.7 μmol/L). Ref: Y. Okada, et al, JNP, 2004, 67, 103 | H. A. Jung, et al, Fish. Sci., 2008, 74, 1363 | S. H. Lee, et al, J. Sci. Food Agric. 2009, 89, 1552 | H. E. Moon, et al, Biosci. Biotechnol. Biochem., 2011, 75, 1472 | S. H. Eom, et al, Phytother. Res., 2013, 27, 148

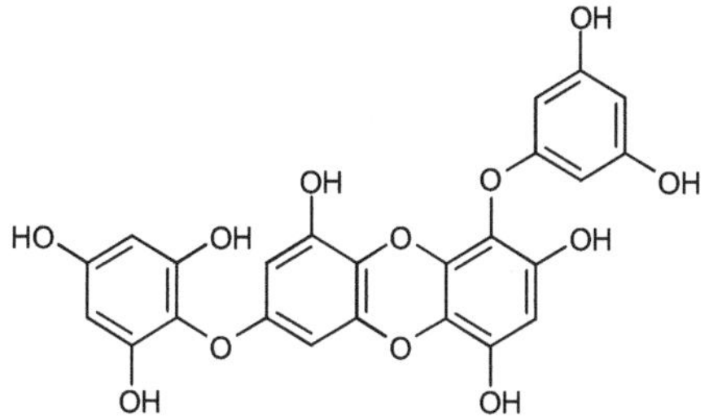

557 Phlorofucofuroeckol A

Type: Phlorotannins. $C_{30}H_{18}O_{14}$ Amorph. solid. Source: Brown algae (edible) *Ecklonia stolonifera, Eisenia bicyclis, Ecklonia kurome* and *Ecklonia cava*. Pharm: Hypoglycemic (α-glucosidase inhibitor) (*in vitro* assay, IC_{50} = 1.37 µmol/L); hypoglycemic (α-glucosidase inhibitor) (*in vitro* assay, IC_{50} = 19.7 µmol/L); hypoglycemic (α-amylase inhibitor) (*in vitro* assay, IC_{50} > 500 µmol/L); hypoglycemic (aldose reductase inhibitor) (rat lens aldose reductase *in vitro* assay, IC_{50} = 125.45 µmol/L) (Jung, 2008); hypoglycemic (PTP1B inhibitor) (*in vitro* assay, IC_{50} = 0.56 µmol/L) (Moon, 2011); antiplasmin inhibitor. Ref: Y. Fukuyama, et al, CPB, 1990, 38, 133 | H. A. Jung, et al, Fish. Sci., 2008, 74, 1363 | Y. Li, et al, BoMC, 2009, 17, 1963 | S. H. Lee, et al, J. Sci. Food Agric., 2009, 89, 1552 | H. E. Moon, et al, Biosci. Biotechnol. Biochem., 2011, 75, 1472

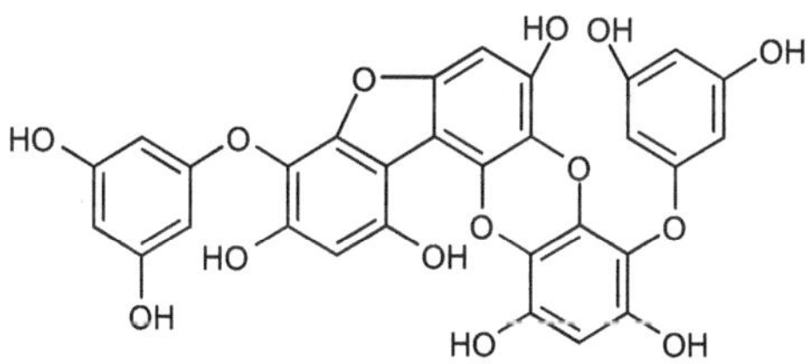

558 Phlorofucofuroeckol B

Type: Phlorotannins. $C_{30}H_{18}O_{14}$ Yellow-brown oil. Source: Brown alga *Eisenia arborea* (edible). Pharm: Anti-inflammatory (rat basophilic leukemia cell histamine release inhibition, IC50 = 7.8 µmol/L, MMOA: inhibition of β-hexosaminidase release); anti-allergic phlorotannins. Ref: Y. Sugiura, et al, Food Sci. Technol. Res. 2007, 13, 54

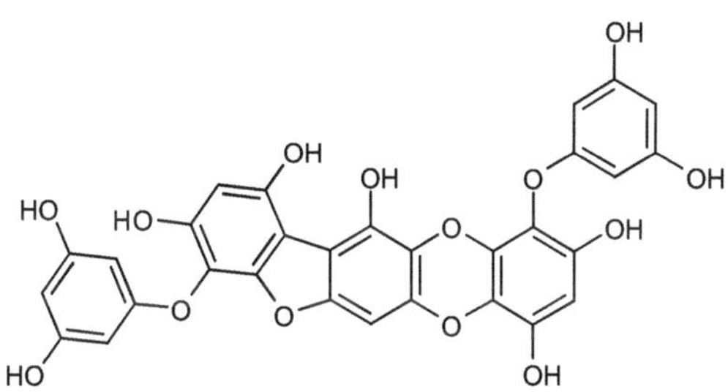

559 2,7″-Phloroglucinol-6,6′-bieckol

Type: Phlorotannins. $C_{48}H_{30}O_{23}$ Source: Brown alga *Ecklonia cava* (Jeju I., R. O. Korea). Pharm: Antioxidant. Ref: S. -M. Kang, et al, J. Funct. Foods, 2012, 4, 158

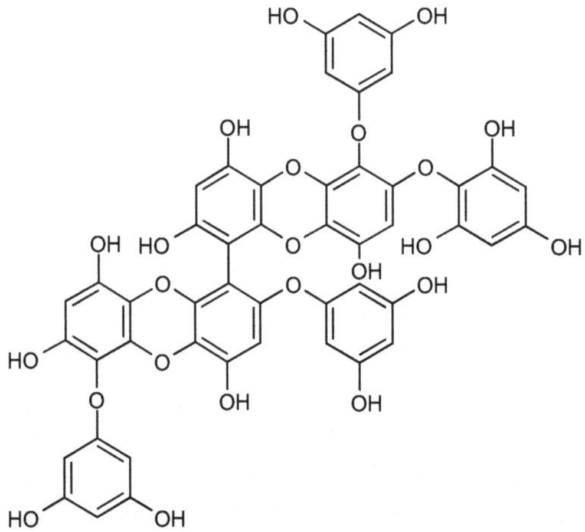

560 Phlorotannin DDBT

2-(4-(3,5-Dihydroxyphenoxy)-3,5-dihydroxyphenoxy) benzene-1,3,5-triol Type: Phlorotannins. $C_{18}H_{16}O_9$ Source: Brown alga *Sargassum patens*. Pharm: Hypoglycemic (α-amylase inhibitor) (*in vitro* assay, IC_{50} = 3.2 µg/mL); hypoglycemic (α-glucosidase inhibitor) (*in vitro* assay, rat intestinal maltase, IC_{50} = 114.0 µg/mL; rat instestinal sucrase, IC_{50} = 25.4 µg/mL). Ref: Y. Kawamura-Konishi, et al, J. Agric. Food Chem. 2012, 60, 5565

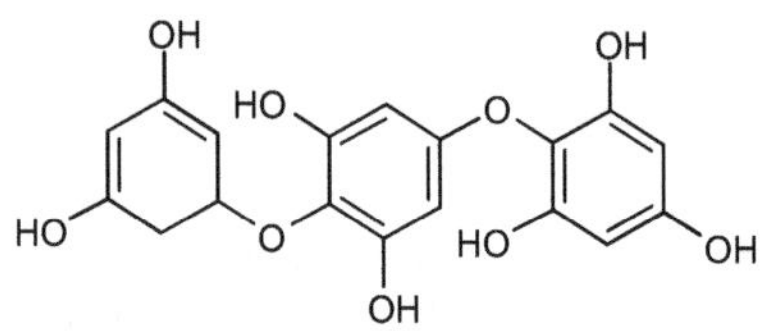

561 Phlorotannin representative trimer B

Type: Phlorotannins. $C_{18}H_{16}O_9$ Source: Brown alga *Fucus distichus* (subfraction FD-E-22, Alaska). Pharm: Hypoglycemic (α-glucosidase inhibitor) (*in vitro* assay, *Fucus distichus*' ethyl acetate subfraction 22 (FD-E-22), IC_{50} = 0.89 µmol/L); hypoglycemic (α-amylase inhibitor) (*in vitro* assay, *Fucus distichus*' ethyl acetate subfraction 22 (FD-E-22), IC_{50} = 13.9 µmol/L). Ref: J. Kellogg, et al, Mar. Drugs, 2014, 12, 5277

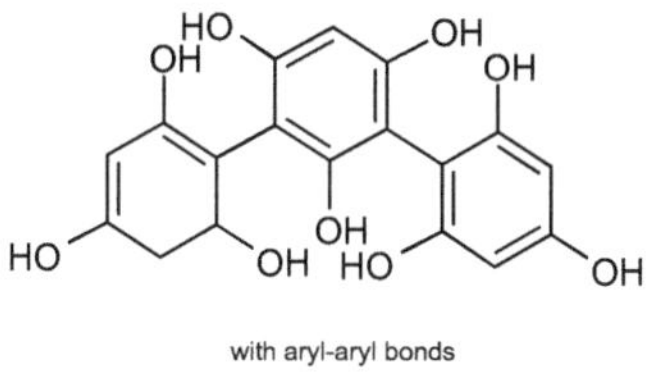

562 Phlorotannin representative trimer C

Type: Phlorotannins. $C_{18}H_{16}O_9$ Source: Brown alga *Fucus distichus* (subfraction FD-E-22, Alaska). Pharm: Hypoglycemic (α-glucosidase inhibitor) (*in vitro* assay, *Fucus distichus*' ethyl acetate subfraction 22 (FD-E-22), IC_{50} = 0.89 µmol/L); hypoglycemic (α-amylase inhibitor) (*in vitro* assay, *Fucus distichus*' ethyl acetate subfraction 22 (FD-E-22), IC_{50} = 13.9 µmol/L). Ref: J. Kellogg, et al, Mar. Drugs, 2014, 12, 5277

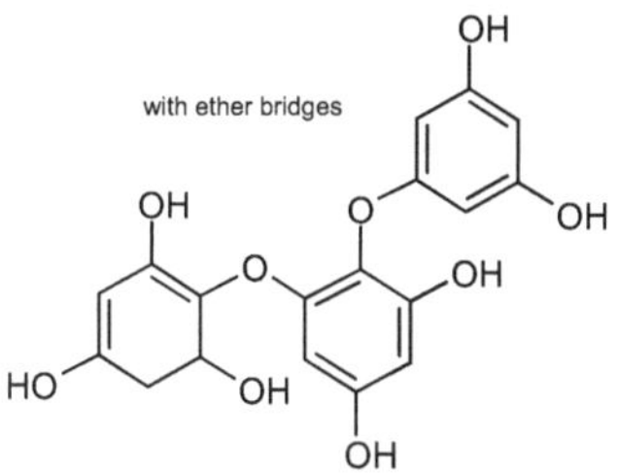

563 Phlorotannin representative trimer D

Type: Phlorotannins. $C_{18}H_{16}O_8$ Source: Brown alga *Fucus distichus* (subfraction FD-E-22, Alaska). Pharm: Hypoglycemic (α-glucosidase inhibitor) (*in vitro* assay, *Fucus distichus*' ethyl acetate subfraction 22 (FD-E-22), IC_{50} = 0.89 µmol/L); hypoglycemic (α-amylase inhibitor) (*in vitro* assay, *Fucus distichus*' ethyl acetate subfraction 22 (FD-E-22), IC_{50} = 13.9 µmol/L). Ref: J. Kellogg, et al, Mar. Drugs, 2014, 12, 5277

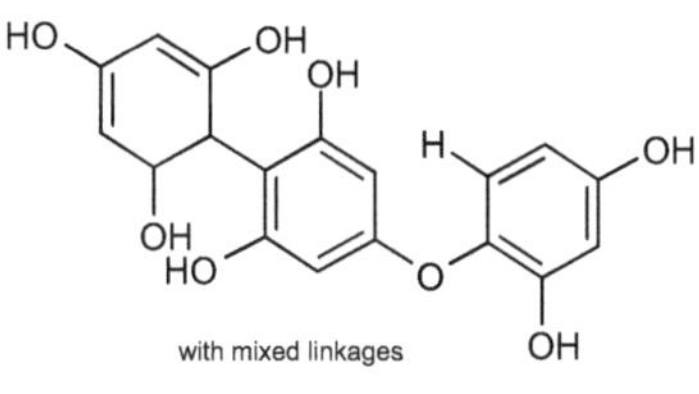

564 Aplysillin A

3-Bromo-5-[4-(3-bromo-4-hydroxyphenyl)-2,3-bis(sulfoxy)-1,3-butadienyl]-1,2-benzenediol Type: Diarylalkyls. $C_{16}H_{12}Br_2O_{11}S_2$ Source: Sponge *Aplysina fistularis fulva.* Pharm: Thrombin receptor antagonist (inhibits binding of [^{125}I]-thrombin to platelet membranes, IC_{50} = 20 µmol/L, weak). Ref: N. K. Gulavita, et al, JNP, 1995, 58, 954

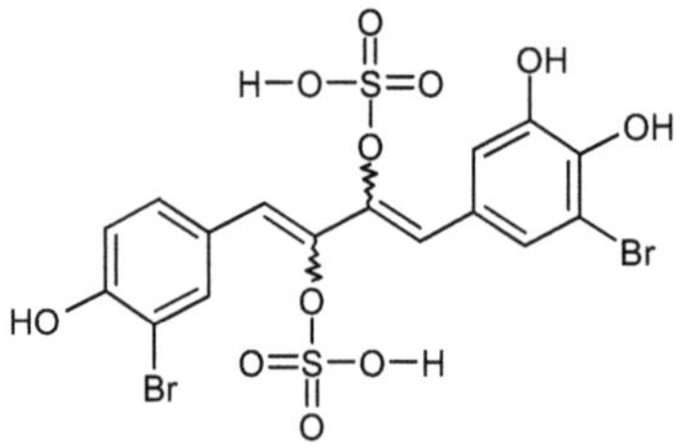

565 1,2-Bis(2,3,6-tribromo-4,5-dihydroxyphenyl)ethane

Type: Diarylalkyls. $C_{14}H_8Br_6O_4$ Amorph. solid, mp 230–232 °C. Source: Red alga *Symphyocladia latiuscula*. Pharm: Hypoglycemic (PTP1B inhibitor) (*in vitro* assay, IC_{50} = 3.5 μmol/L). Ref: X. -J. Duan, et al, JNP, 2007, 70, 1210 | X. Liu, et al, Chin. J. Oceanol. Limnol., 2011, 29, 686

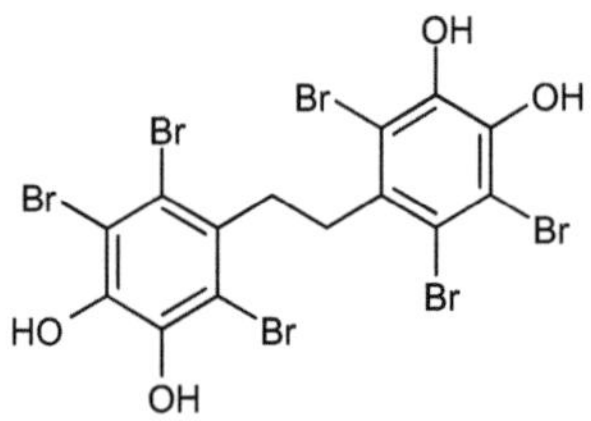

566 8-Carboxy-isoiantheran A

Type: Diarylalkyls. $C_{33}H_{18}Br_4O_{14}S_2$ Brownish amorphous solid. Source: Sponge *Ianthella quadrangulata* (Australia, polar extract). Pharm: Immune system activity (activation of Ca^{2+}-mobilization, IC_{50} = 0.48–1.3 μmol/L, MMOA: ionotropic $P2Y_{11}$ receptor activation); $P2Y_{11}$ receptor agonist (EC_{50} = 0.48 μmol/L, showed some selectivity for $P2Y_{11}$ over $P2Y_1$ and $P2Y_2$ receptors). Ref: H. Greve, et al, JMC, 2007, 50, 5600

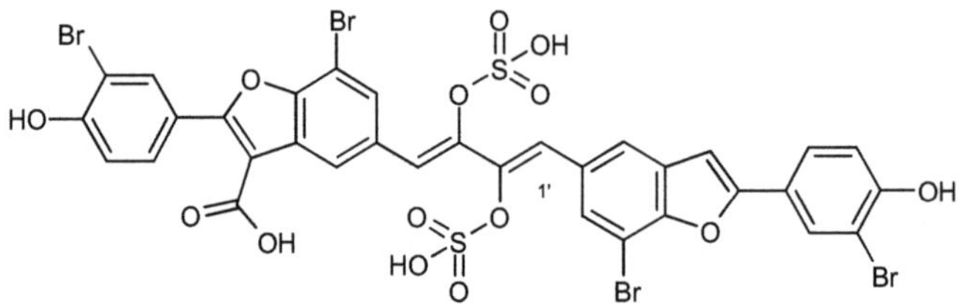

567 Colpol

2-Bromo-5-[4-(4-bromo-2-hydroxy-5-methoxyphenyl)-2-butenyl]-1,3-benzenediole
Type: Diarylalkyls. $C_{17}H_{16}Br_2O_4$ Source: Brown alga *Colpomenia sinuosa* (Red Sea). Pharm: Cytotoxic (P_{388}, A549, HT29 and CV-1, IC_{50} = 10 μg/mL). Ref: D. Green, et al, JNP, 1993, 56, 1201

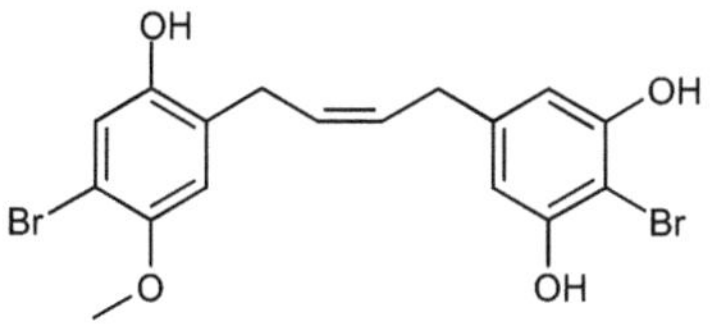

568 Iantheran A

Type: Diarylalkyls. $C_{32}H_{18}Br_4O_{12}S_2$ Pale yellow cryst. (di-Na salt), mp 172–173 °C (dec, di-Na salt). Source: Sponge *Ianthella* sp. (Australia). Pharm: Na/K-ATPase inhibitor (IC_{50} = 10 μmol/L). Ref: Y. Okamoto, et al, Tet. Lett., 1999, 40, 507 | Y. Okamoto, et al, BoMC, 2001, 9, 179

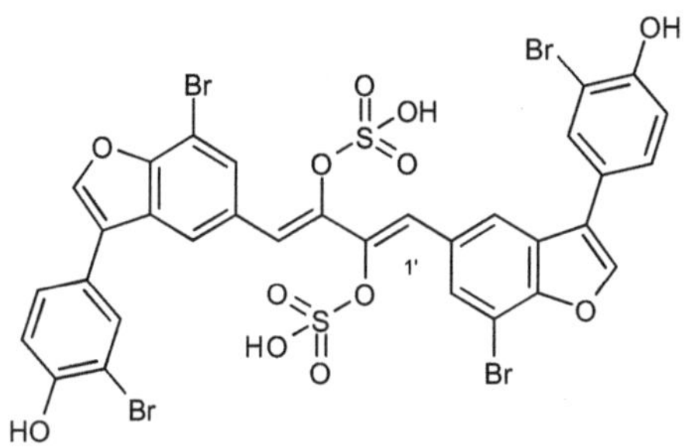

569 Iantheran B

8-Carboxy-isoiantheran A Type: Diarylalkyls. $C_{32}H_{22}Br_4O_{12}S_2$ mp 155–156 °C (dec) (di-Na salt). Source: Sponge *Ianthella* sp. (Great Barrier Reef, Australia). Pharm: Na/K-ATPase inhibitor. Ref: Y. Okamoto, et al, BoMC, 2001, 9, 179

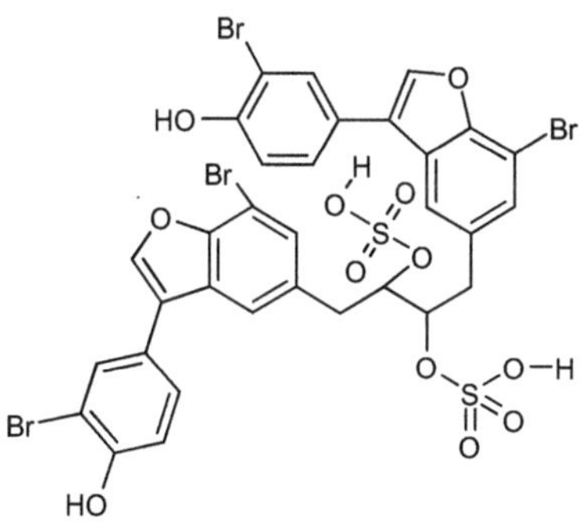

570 Isoiantheran A

Type: Diarylalkyls. $C_{32}H_{18}Br_4O_{12}S_2$ Brownish amorphous solid, $[\alpha]_D^{23}$ = −0.7° (*c* = 0.39, MeOH). Source: Sponge *Ianthella quadrangulata* (Australia, polar extract). Pharm: Immune system activity (activation of Ca^{2+}-mobilization, IC_{50} = 0.48–1.3 μmol/L,

MMOA: ionotropic $P2Y_{11}$ receptor activation); $P2Y_{11}$ receptor agonist (EC_{50} = 1.29 μmol/L). Ref: H. Greve, et al, JMC, 2007, 50, 5600

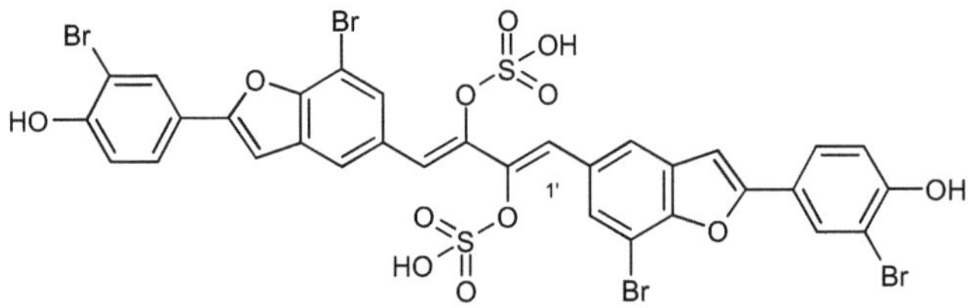

571 Panicein A_2

Type: Diarylalkyls. $C_{22}H_{26}O_3$ Oil. Source: Sponge *Reniera mucosa*. Pharm: Cytotoxic (P_{388}, A549, HT29 and MEL28, ED_{50} = 5 μg/mL). Ref: E. Zubía, et al, Tetrahedron, 1994, 50, 8153 | M. Gordaliza, et al, Mar. Drugs, 2010, 8, 2849 (rev)

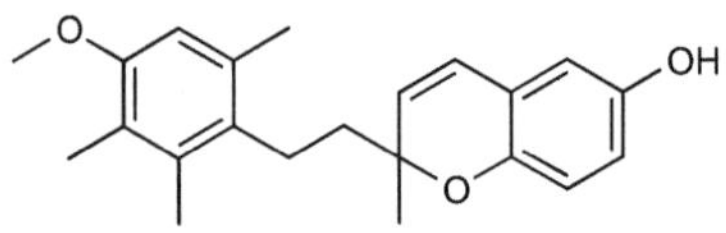

572 Panicein F_1

Type: Diarylalkyls. $C_{22}H_{28}O_4$ Orange oil. Source: Sponge *Reniera mucosa*. Pharm: Cytotoxic (P_{388}, A549 and MEL28, ED_{50} = 5 μg/mL); DHFR inhibitor (ED_{50} = 3 μg/mL). Ref: E. Zubia, et al, Tetrahedron, 1994, 50, 8153 | M. Gordaliza, et al, Mar. Drugs, 2010, 8, 2849 (rev)

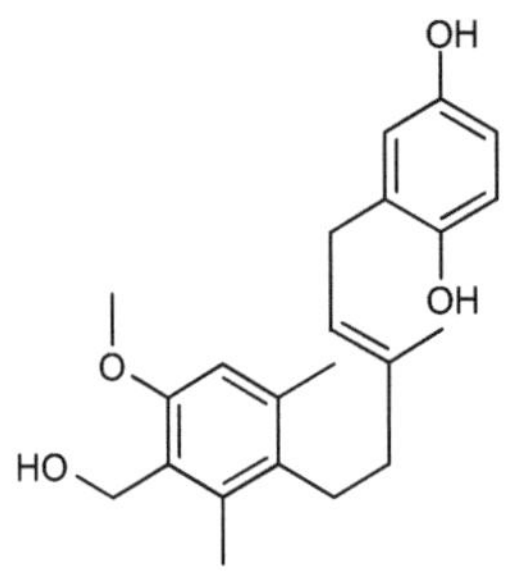

573 3″-Deoxy-6′-*O*-desmethylcandidusin B

Type: Terphenyls. $C_{19}H_{14}O_6$ Source: Mangrove-derived fungus *Penicillium chermesinum* from mangrove *Kandelia candel* (South China Sea, Guangdong, China). Pharm: Acetylcholinesterase inhibitor. Ref: H. Huang, et al, JNP, 2011, 74, 997

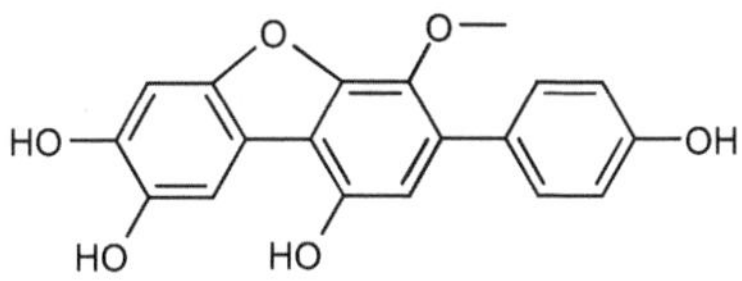

574 4''-Deoxyisoterprenin

Type: Terphenyls. $C_{25}H_{26}O_5$ Powder. Source: Marine-derived fungus *Aspergillus candidus* IF10 (from sediment sample, Japan waters). Pharm: Cytotoxic (KB-3-1, moderate). Ref: H. Wei, et al, J. Antibiot., 2007, 60, 586

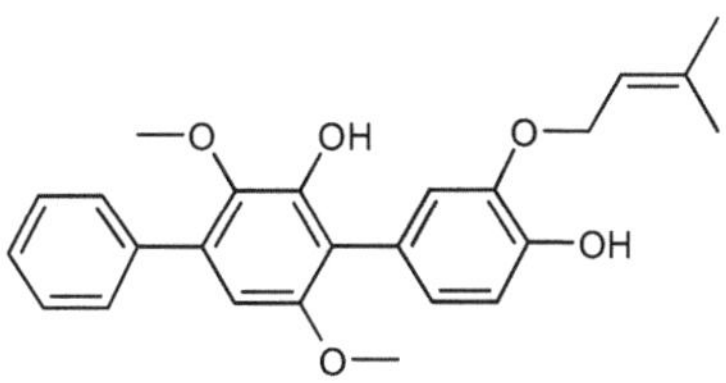

575 4''-Deoxyprenylterphenyllin

Type: Terphenyls. $C_{25}H_{26}O_4$ Powder. Source: Marine-derived fungus *Aspergillus candidus* IF10 (from sediment sample, Japan waters). Pharm: Cytotoxic (KB-3-1, moderate). Ref: H. Wei, et al, J. Antibiot., 2007, 60, 586

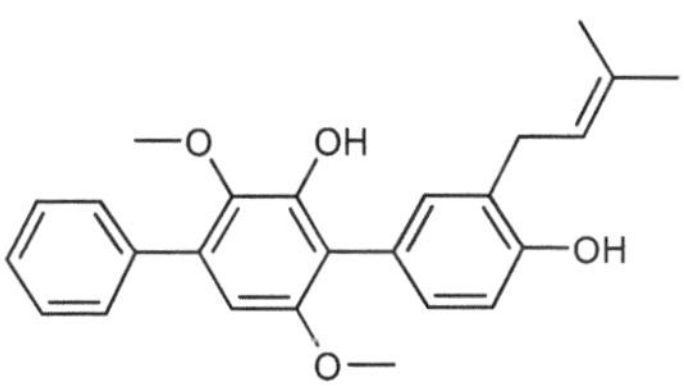

576 4''-Deoxyterprenin

Type: Terphenyls. $C_{25}H_{26}O_5$ Cryst. Source: Marine-derived fungus *Aspergillus candidus* (from sediment sample, Japan waters). Pharm: Cytotoxic (KB-3-1, moderate); Immunosuppressant; antimetabolite effect. Ref: H. Wei, et al, J. Antibiot., 2007, 60, 586

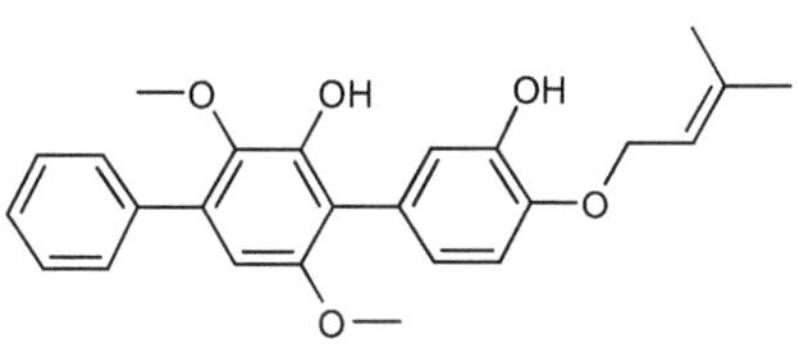

577 6′-*O*-Desmethylterphenyllin

Type: Terphenyls. $C_{19}H_{16}O_5$ Source: Mangrove-derived fungus *Penicillium chermesinum* from mangrove *Kandelia candel* (South China Sea, Guangdong, China). Pharm: α-Glucosidase inhibitor (strong). Ref: H. Huang, et al, JNP, 2011, 74, 997

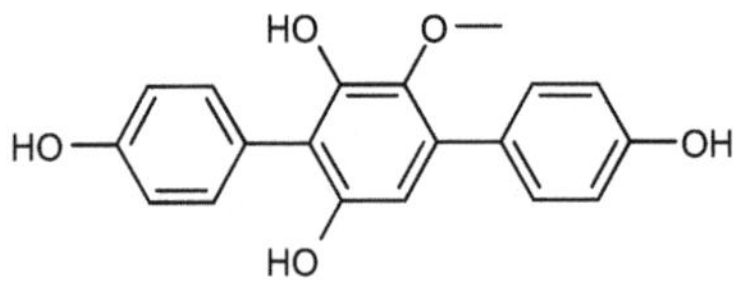

578 3-Hydroxy-6′-*O*-desmethylterphenyllin

Type: Terphenyls. $C_{19}H_{16}O_6$ Source: Mangrove-derived fungus *Penicillium chermesinum* from mangrove *Kandelia candel* (South China Sea, Guangdong, China). Pharm: α-Glucosidase inhibitor (strong). Ref: H. Huang, et al, JNP, 2011, 74, 997

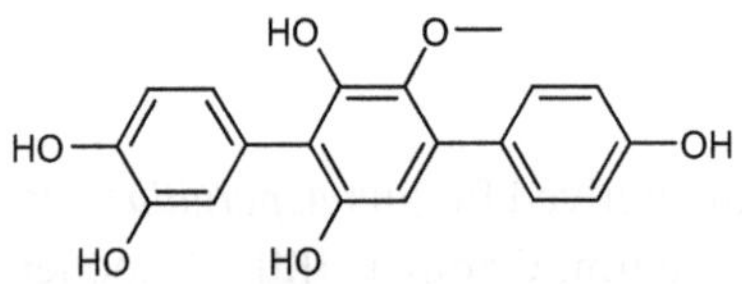

579 Prenylterphenyllin

Type: Terphenyls. $C_{25}H_{26}O_5$ Powder. Source: Marine-derived fungus *Aspergillus candidus* IF10 (from sediment sample, Japan waters). Pharm: Cytotoxic (KB-3-1, moderate). Ref: H. Wei, et al, J. Antibiot., 2007, 60, 586

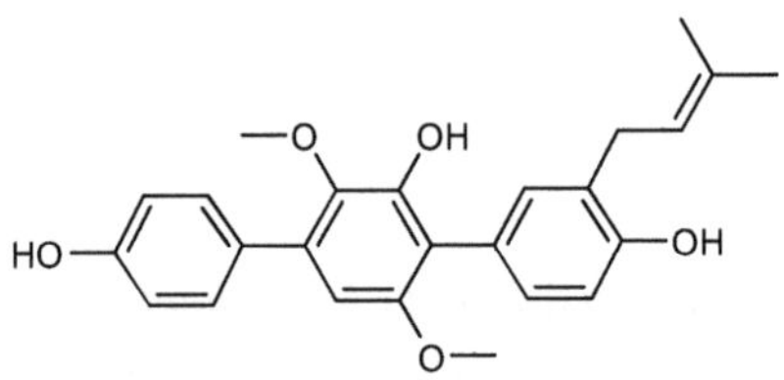

580 Terprenin

Type: Terphenyls. $C_{20}H_{26}O_6$ Prisms, mp 155.5–156 °C. Source: Terrestrial fungus *Aspergillus candidus*. Pharm: Immunosuppressant. Ref: T. Kamigauchi, et al, J. Antibiot., 1998, 51, 445

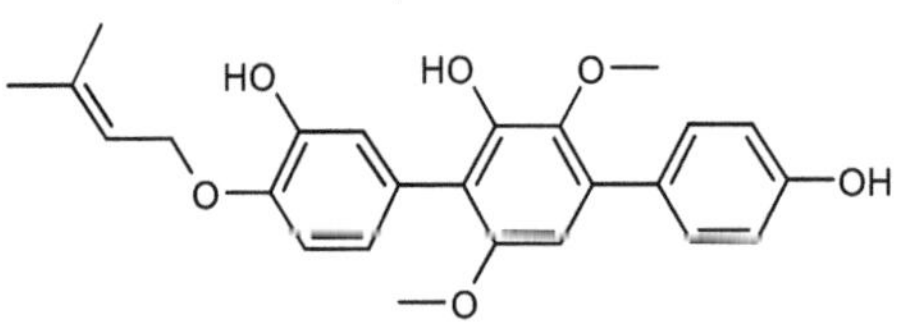

581 3′,4,4″-Trihydroxy-2′,5′-dimethoxy-*p*-terphenyl

Terphenyllin Type: Terphenyls. $C_{20}H_{18}O_5$ mp 239 °C (dec). Source: Deep-sea fungus *Aspergillus candidus*, terrestrial fungus *Aspergillus candidus*. Pharm: Plant growth inhibitor. Ref: R. Marchelli, et al, J. Antibiot., 1975, 28, 328 | F. Liu, et al, J. Xiamen Univ., 2013, 52, 670 (in Chinese)

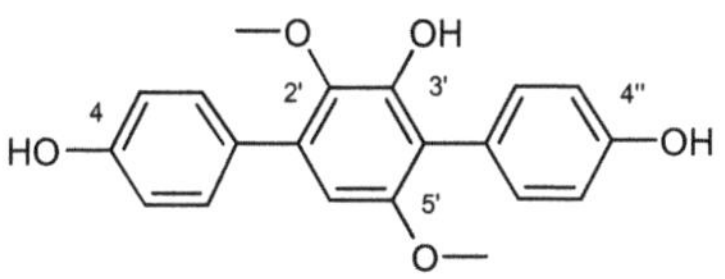

582 Aspulvinone E

Type: Pulvinone group. $C_{17}H_{12}O_5$ Yellow needles (MeOH), mp 282–284 °C, 275 °C. Source: Mangrove-derived fungus *Aspergillus terreus* GWQ-48 from unidentified mangrove (rhizospehere soil, Fujian Province, China), terrestrial fungus *Aspergillus terreus*. Pharm: Antiviral (IFV H1N1 virus inhibitor, IC_{50} = 192.05 μmol/L, significant); glucose-6-phosphatase inhibitor. Ref: N. Ojima, et al, Phytochemistry, 1973, 12, 2527 | Vertesy, L. et al, J. Antibiot., 2000, 53, 677 | H. Gao, et al, Bioorg. Med. Chem. Lett., 2013, 23, 1776 | S. Z. Moghadamtousi, et al, Mar. Drugs, 2015, 13, 4520 (rev)

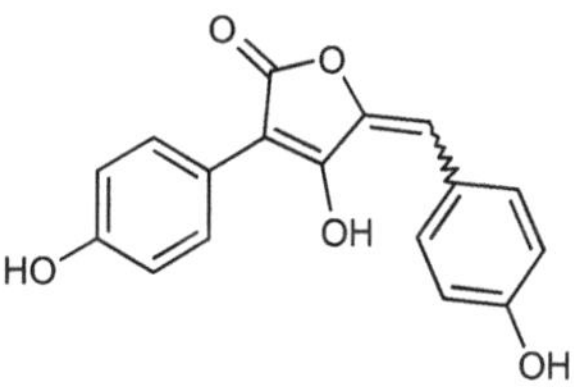

583 Cadiolide B

Type: Pulvinone group. $C_{24}H_{10}Br_6O_6$ Orange amorph. solid. Source: Ascidians *Pseudodistoma antinboja* (Tong-Yeong City, R. O. Korea) and *Botryllus* sp. Pharm: Antibacterial (MSSA and MRSA). Ref: C. J. Smith, et al, JOC, 1998, 63, 4147 | W. Wang, et al, JNP, 2012, 75, 2049

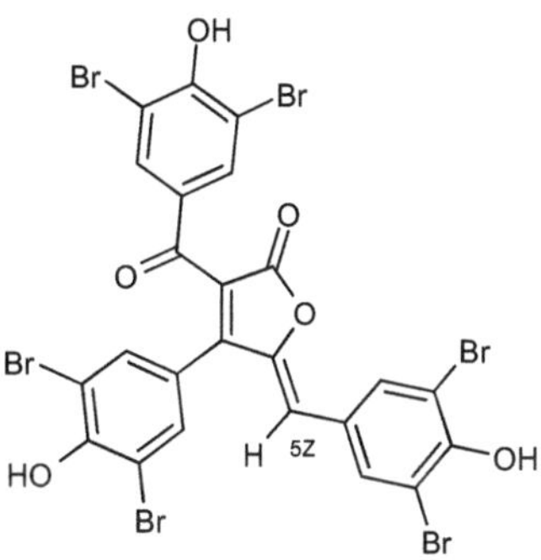

584 Cadiolide E

Type: Pulvinone group. $C_{24}H_{11}Br_5O_6$ Source: Ascidian *Pseudodistoma antinboja* (Tong-Yeong City, R. O. Korea). Pharm: Antibacterial (gram-positive and -negative bacteria). Ref: W. Wang, et al, JNP, 2012, 75, 2049

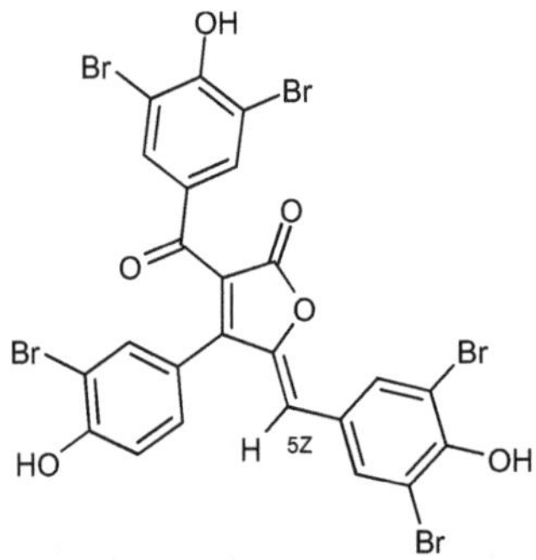

585 Cadiolide G

Type: Pulvinone group. $C_{24}H_{14}Br_4O_6$ Source: Ascidian *Synoicum* sp. (Chuja-do, R. O. Korea). Pharm: Antibacterial (gram-positive and -negative bacteria). Ref: T. H. Won, et al, JNP, 2012, 75, 2055

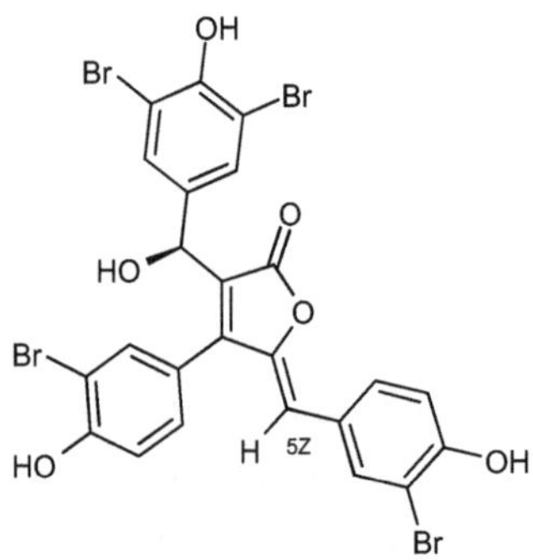

586 (5*E*)-Cadiolide H

Type: Pulvinone group. $C_{25}H_{16}Br_4O_6$ Source: Ascidian *Synoicum* sp. (Chuja-do, R. O. Korea). Pharm: Antibacterial (gram-positive and -negative bacteria). Ref: T. H. Won, et al, JNP, 2012, 75, 2055

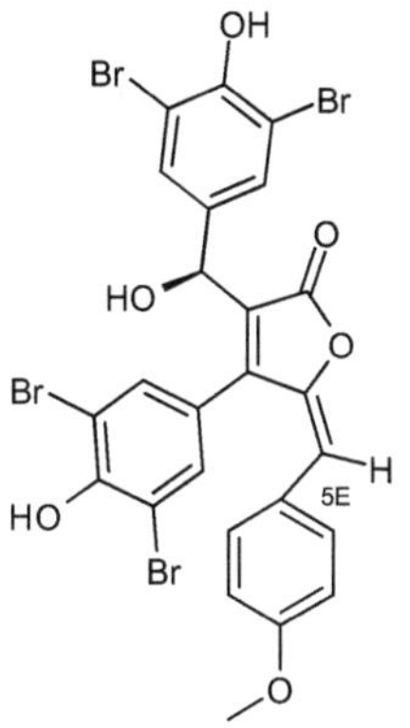

587 (5*Z*)-Cadiolide H

Type: Pulvinone group. $C_{25}H_{16}Br_4O_6$ Source: Ascidian *Synoicum* sp. (Chuja-do, R. O. Korea). Pharm: Antibacterial (gram-positive and -negative bacteria). Ref: T. H. Won, et al, JNP, 2012, 75, 2055

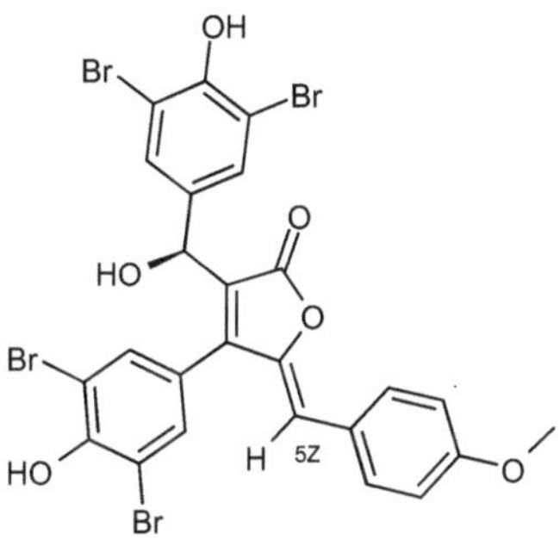

588 Cadiolide I

Type: Pulvinone group. $C_{25}H_{12}Br_6O_7$ Source: Ascidian *Synoicum* sp. (Chuja-do, R. O. Korea). Pharm: Antibacterial (gram-positive and -negative bacteria). Ref: T. H. Won, et al, JNP, 2012, 75, 2055

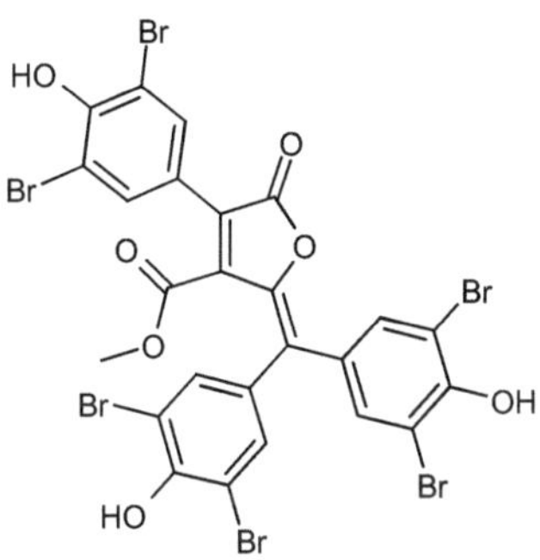

589 Isoaspulvinone E

Type: Pulvinone group. $C_{17}H_{12}O_5$ Source: Mangrove-derived fungus *Aspergillus terreus* from unidentified mangrove (rhizospehere soil, Fujian Province, China). Pharm: Antiviral (H1N1 virus inhibitor, significant, inhibits H1N1 viral neuraminidase). Ref: H. Gao, et al, Bioorg. Med. Chem. Lett., 2013, 23, 1776

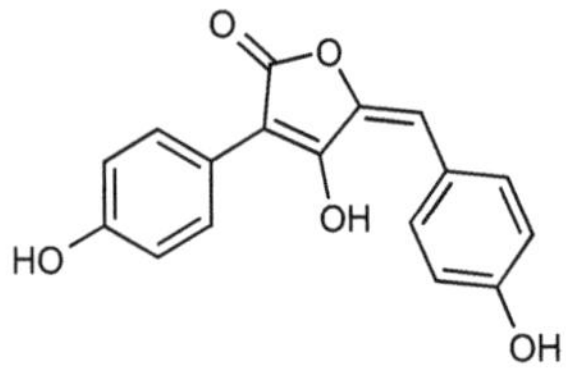

590 Pulvic acid

Pulvinic acid Type: Pulvinone group. $C_{12}H_8O_4$ Orange powder (Et_2O or $CHCl_3$), orange prisms (C_6H_6), mp 216–217 °C. Source: Mangrove-derived fungus *Aspergillus terreus* GWQ-48 from unidentified mangrove (rhizospehere soil, Fujian Province, China, first time MNP), various terrestrial lichen spp. Pharm: Antiviral (IFV H1N1 virus inhibitor, IC_{50} = 94.42 μmol/L, significant). Ref: P. Karrer,et al, Helv. Chim. Acta, 1926, 9, 446 | H. Gao, et al, Bioorg. Med. Chem. Lett., 2013, 23, 1776 | S. Z. Moghadamtousi, et al, Mar. Drugs, 2015, 13, 4520 (rev)

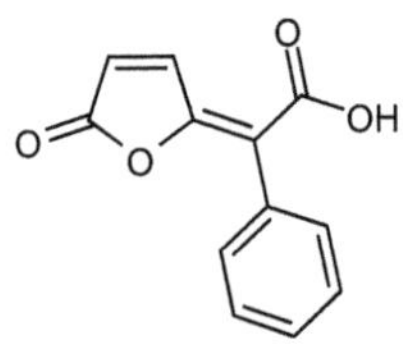

591 Rubrolide A

Type: Pulvinone group. $C_{17}H_8Br_4O_4$ Red amorph. solid (Me_2CO). Source: Ascidians *Ritterella rubra, Synoicum blochmanni* and *Synoicum prunum* (Australia). Pharm: PP1 inhibitor (moderate but selective); PP2A inhibitor (moderate but selective). Ref: S. Miao, et al, JOC, 1991, 56, 6275 | A. R. Carroll, et al, JOC, 1999, 64, 2680

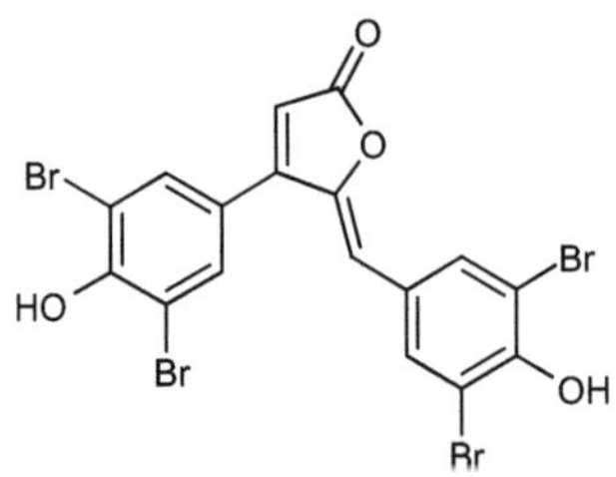

592 Rubrolide B

Type: Pulvinone group. $C_{17}H_7Br_4ClO_4$ Red amorph. solid (Me_2CO). Source: Ascidians *Ritterella rubra* and *Synoicum blochmanni*. Pharm: PP1 inhibitor (moderate but selective); PP2A inhibitor (moderate but selective). Ref: S. Miao, et al, JOC, 1991, 56, 6275

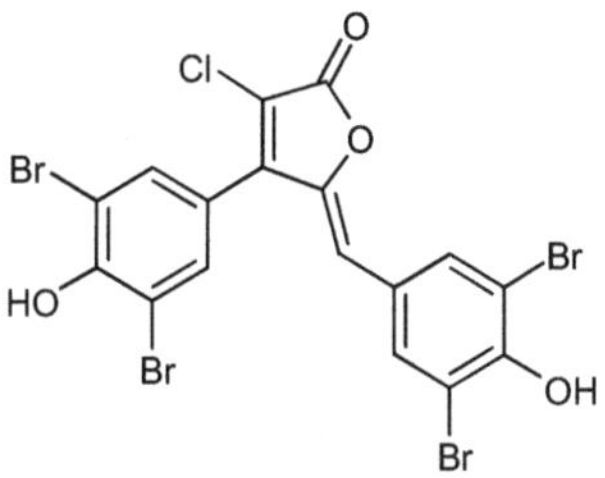

593 Rubrolide C

Type: Pulvinone group. $C_{17}H_{10}Br_2O_4$ Red amorph. solid (Me_2CO). Source: Ascidians *Ritterella rubra* and *Synoicum blochmanni*. Pharm: PP1 inhibitor (moderate but selective); PP2A inhibitor (moderate but selective). Ref: S. Miao, et al, JOC, 1991, 56, 6275 | J. Boukouvalas, et al, Tet. Lett., 1998, 39, 7665

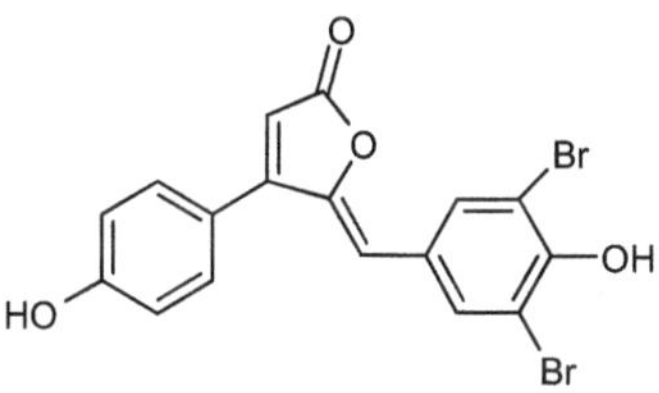

594 Rubrolide D

Type: Pulvinone group. $C_{17}H_{10}Br_2O_4$ Source: Ascidian *Ritterella rubra*. Pharm: PP1 inhibitor (moderate but selective); PP2A inhibitor (moderate but selective). Ref: S. Miao, et al, JOC, 1991, 56, 6275

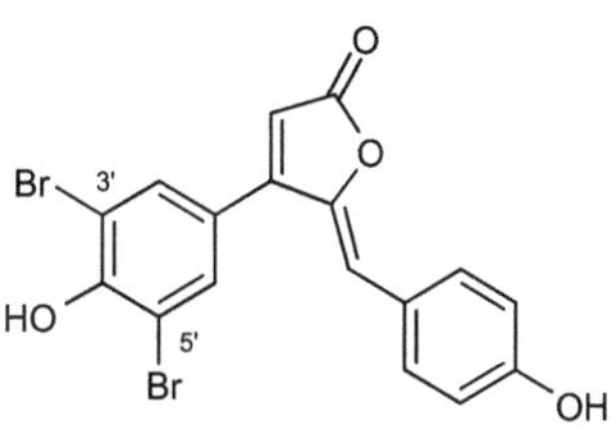

595 Rubrolide E

Type: Pulvinone group. $C_{17}H_{12}O_4$ Yellow solid, mp 278–281 °C. Source: Ascidian *Synoicum globosum* (Algoa Bay, South Africa), *Ritterella rubra* and *Synoicum blochmanni*. Pharm: PP1 inhibitor (moderate but selective); PP2A inhibitor (moderate but selective). Ref: S. Miao, et al, JOC, 1991, 56, 6275 | J. Boukouvalas, et al, Tet. Lett., 1998, 39, 7665 | J. Sikorska, et al, JNP, 2012, 75, 1824

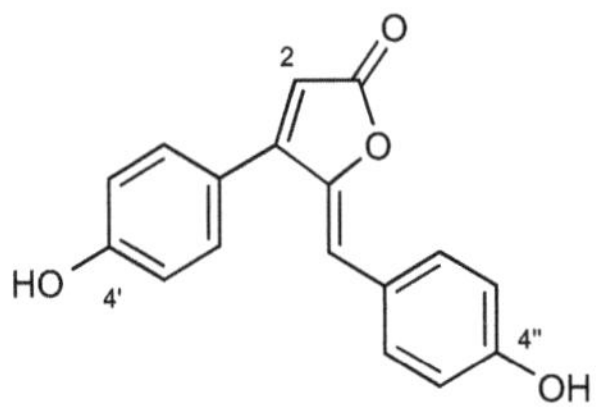

596 Rubrolide F

Type: Pulvinone group. $C_{18}H_{14}O_4$ Yellow solid. Source: Ascidians *Synoicum globosum* (Algoa Bay, South Africa) and *Ritterella rubra*. Pharm: PP1 inhibitor (moderate but selective); PP2A inhibitor (moderate but selective). Ref: S. Miao, et al, JOC, 1991, 56, 6275 | J. Sikorska, et al, JNP, 2012, 75, 1824

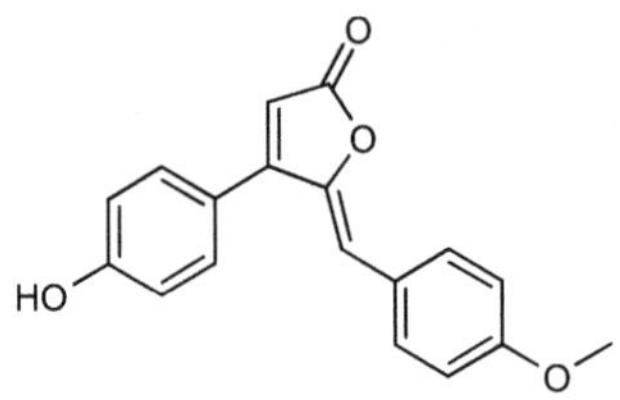

597 Rubrolide G

Type: Pulvinone group. $C_{17}H_{10}Br_4O_5$ Source: Ascidian *Ritterella rubra*. Pharm: PP1 inhibitor (moderate but selective); PP2A inhibitor (moderate but selective). Ref: S. Miao, et al, JOC, 1991, 56, 6275

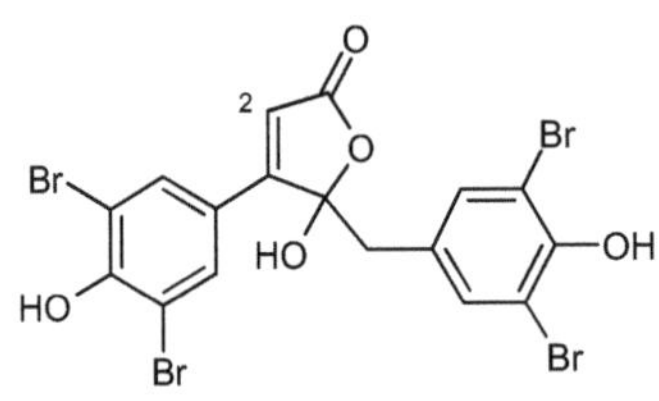

598 Rubrolide H

Type: Pulvinone group. $C_{17}H_9Br_4ClO_5$ Source: Ascidian *Ritterella rubra*. Pharm: PP1 inhibitor (moderate but selective); PP2A inhibitor (moderate but selective). Ref: S. Miao, et al, JOC, 1991, 56, 6275

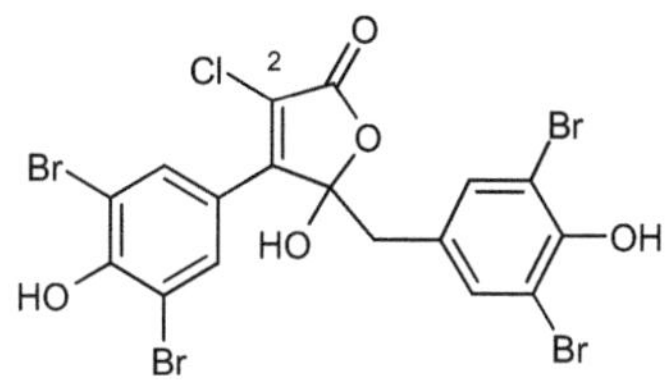

599 Rubrolide I

Type: Pulvinone group. $C_{17}H_8Br_3ClO_4$ Orange-yellow powder. Source: Ascidian *Synoicum blochmanni* (Spain). Pharm: Cytotoxic (HT29, ED_{50} = 5.0 μg/mL). Ref: M. J. Ortega, et al, Tetrahedron, 2000, 56, 3963

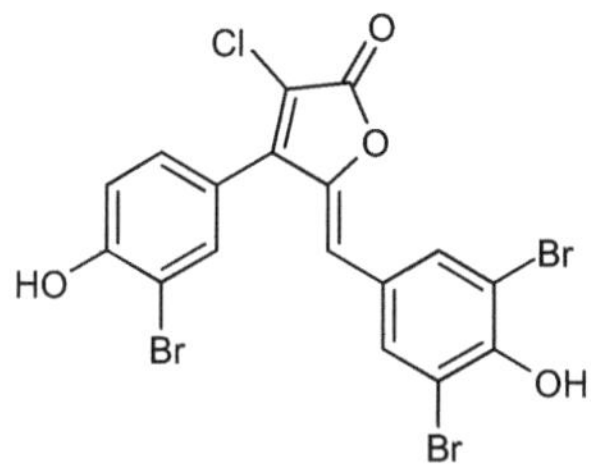

600 Rubrolide K

Type: Pulvinone group. $C_{17}H_9Br_2ClO_4$ Orange amorph. solid. Source: Ascidian *Synoicum blochmanni* (Spain). Pharm: Cytotoxic (P_{388}, ED_{50} = 2.5 μg/mL, A549, ED_{50} = 2.5 μg/mL, HT29, ED_{50} = 1.2 μg/mL). Ref: M. J. Ortega, et al, Tetrahedron, 2000, 56, 3963

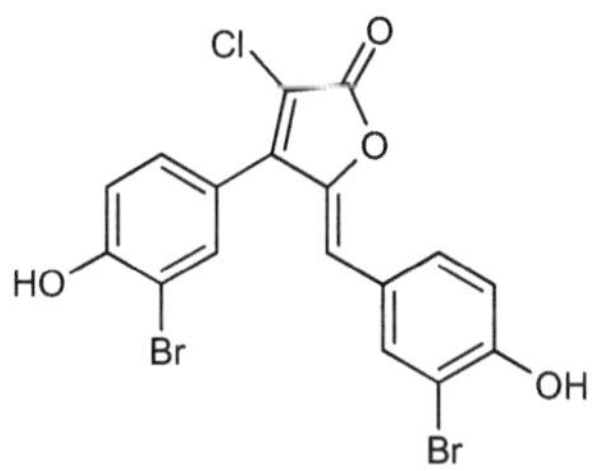

601 Rubrolide L

Type: Pulvinone group. $C_{17}H_9Br_2ClO_4$ Reddish-orange amorph. solid. Source: Ascidian *Synoicum blochmanni* (Spain). Pharm: Cytotoxic (P_{388}, ED_{50} = 5 μg/mL, A549, ED_{50} = 5 μg/mL, HT29, ED_{50} = 2.5 μg/mL, MEL28, ED_{50} = 5 μg/mL). Ref: M. J. Ortega, et al, Tetrahedron, 2000, 56, 3963

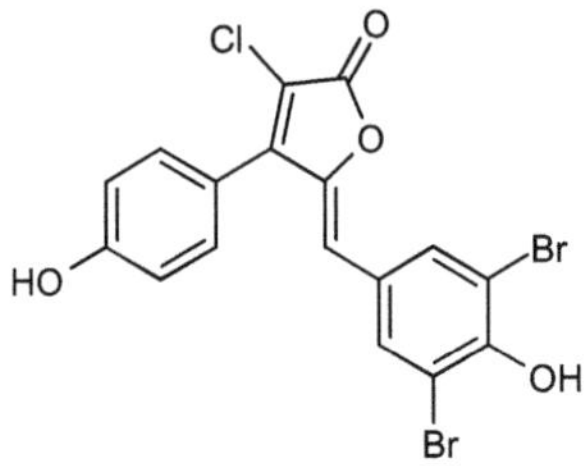

602 Rubrolide M

Type: Pulvinone group. $C_{17}H_{10}BrClO_4$ Orange amorph. solid. Source: Ascidian *Synoicum blochmanni* (Spain). Pharm: Cytotoxic (P_{388}, ED_{50} = 1.2 μg/mL, A549, ED_{50} = 1.2 μg/mL, HT29, ED_{50} = 1.2 μg/mL, MEL28, ED_{50} = 1.2 μg/mL). Ref: M. J. Ortega, et al, Tetrahedron, 2000, 56, 3963

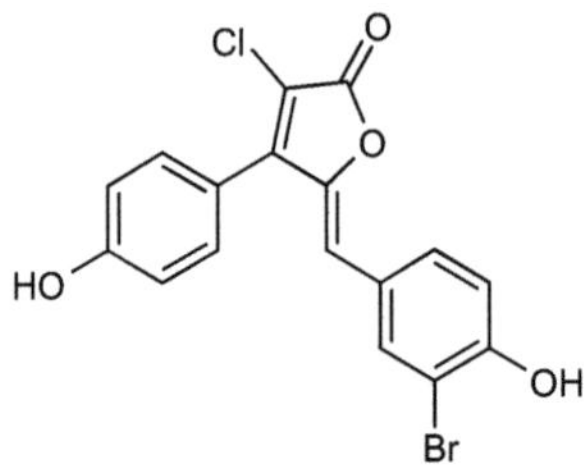

603 Rubrolide O

Type: Pulvinone group. $C_{17}H_9Br_2ClO_4$ Amorph. yellow solid. Source: Ascidian *Synoicum* sp. Pharm: Anti-inflammatory (9:1 ratio of *Z/E* isomers, inhibits superoxide production by hmn neutrophils *in vitro*, IC_{50} = 35 μmol/L, no short-term toxicity); anti-inflammatory (inhibits PMA-induced adhesion of neutrophils to microtitre plate); antiproliferative (48 h exposure MTT-detected assay, IC_{50} = 33 μmol/L). Ref: A. N. Pearce, et al, JNP, 2007, 70, 111

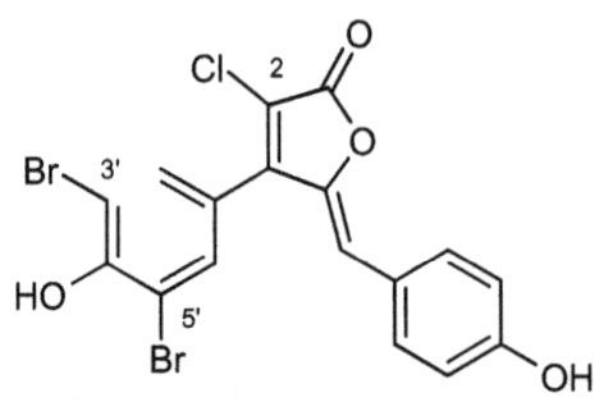

604 Carbamidocyclophane A

Type: Cylindrocyclophane derivatives. $C_{38}H_{54}Cl_4N_2O_8$ $[\alpha]_D^{22}$ = −1.2° (*c* = 0.4, MeOH). Source: Cyanobacterium *Nostoc* sp. CAVN 10. Pharm: Antibacterial (*Staphylococcus aureus* ATCC 6538, MIC = 0.1 mmol/L); cytotoxic (MCF7, IC_{50} = 0.86 μmol/L, hmn amniotic epithelial cell line Fl, IC_{50} = 3.3 μmol/L). Ref: H. T. N. Bui, et al, JNP, 2007, 70, 499

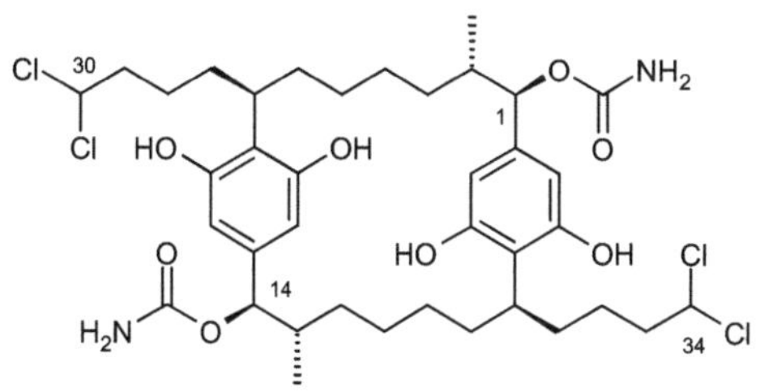

605 Carbamidocyclophane B

Type: Cylindrocyclophane derivatives. $C_{38}H_{55}Cl_3N_2O_8$ $[\alpha]_D^{22} = -0.3°$ (c = 0.34, MeOH). Source: Cyanobacterium *Nostoc* sp. CAVN 10. Pharm: Antibacterial (*Staphylococcus aureus* ATCC 6538, MIC = 0.04 mmol/L); cytotoxic (MCF7, IC_{50} = 1.8 μmol/L, hmn amniotic epithelial cell line Fl, IC_{50} = 4.2 μmol/L). Ref: H. T. N. Bui, et al, JNP, 2007, 70, 499

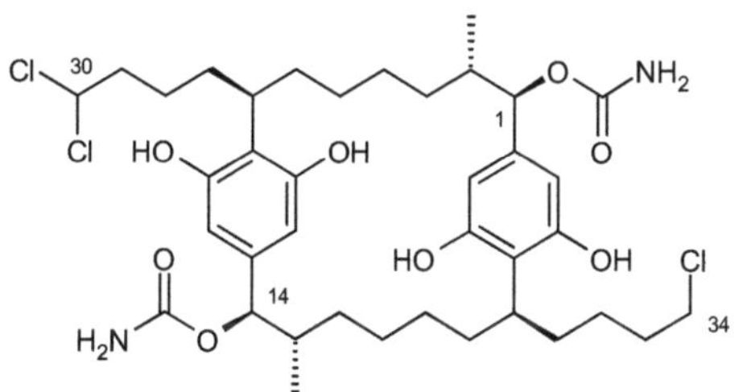

606 Carbamidocyclophane C

Type: Cylindrocyclophane derivatives. $C_{38}H_{56}Cl_2N_2O_8$ $[\alpha]_D^{22} = +8.5°$ (c = 0.56, MeOH). Source: Cyanobacterium *Nostoc* sp. CAVN 10. Pharm: Antibacterial (*Staphylococcus aureus* ATCC 6538, MIC = 0.06 mmol/L); cytotoxic (MCF7, IC_{50} = 2.2 μmol/L, hmn amniotic epithelial cell line Fl, IC_{50} = 5.1 μmol/L). Ref: H. T. N. Bui, et al, JNP, 2007, 70, 499

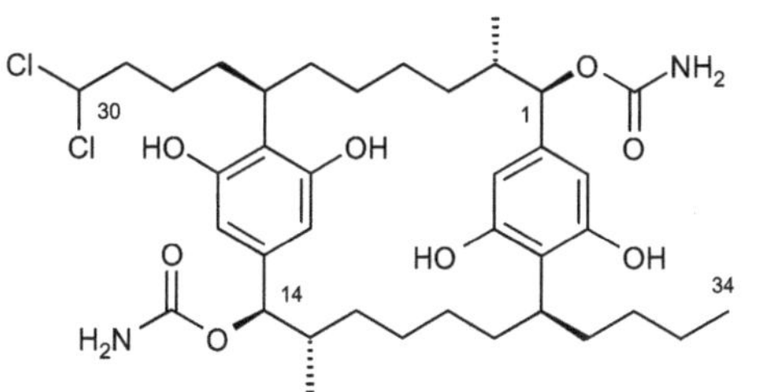

607 Carbamidocyclophane D

Type: Cylindrocyclophane derivatives. $C_{38}H_{57}ClN_2O_8$ $[\alpha]_D^{22} = +5.6°$ (c = 0.26, MeOH). Source: Cyanobacterium *Nostoc* sp. CAVN 10. Pharm: Antibacterial (*Staphylococcus aureus*); cytotoxic (MCF7 and hmn amniotic epithelial cell line Fl). Ref: H. T. N. Bui, et al, JNP, 2007, 70, 499

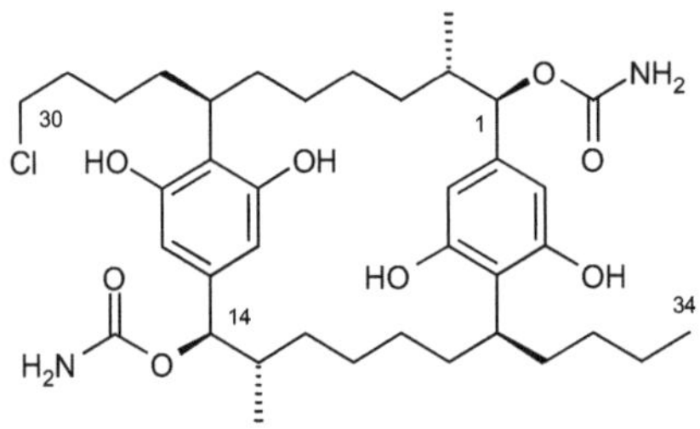

608 Carbamidocyclophane E

Type: Cylindrocyclophane derivatives. $C_{38}H_{58}N_2O_8$ $[\alpha]_D^{22}$ = +5.7° (*c* = 0.12, MeOH). Source: Cyanobacterium *Nostoc* sp. CAVN 10. Pharm: Antibacterial (*Staphylococcus aureus*); cytotoxic (MCF7 and han amniotic epithelial cell line Fl). Ref: H. T. N. Bui, et al, JNP, 2007, 70, 499

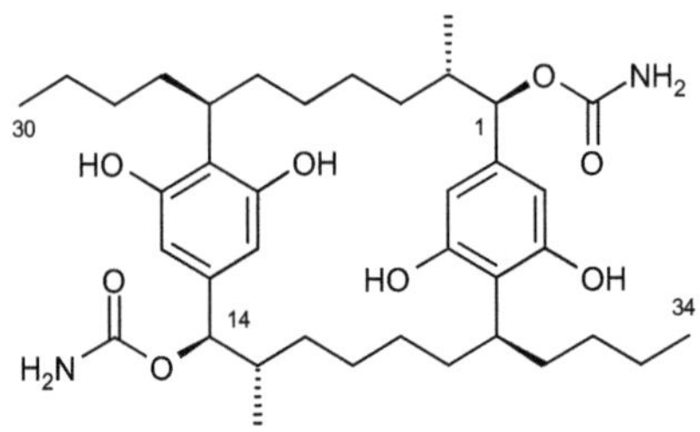

609 Cylindrocyclophane A

Type: Cylindrocyclophane derivatives. $C_{36}H_{56}O_6$ Cryst., mp 276–278 °C, $[\alpha]_D$ = –20° (*c* = 0.5, MeOH). Source: Cyanobacteria *Cylindrospermum licheniforme* ATCC 29204 and *Nostoc* sp. UIC 10022A. Pharm: 20*S*-proteasome inhibitor (IC_{50} = 33.9 µmol/L, control Bortezomib, IC_{50} = 2.5 nmol/L). Ref: B. S. Moore, et al, Tetrahedron, 1992, 48, 3001 | G. E. Chlipala, et al, JNP, 2010, 73, 1529

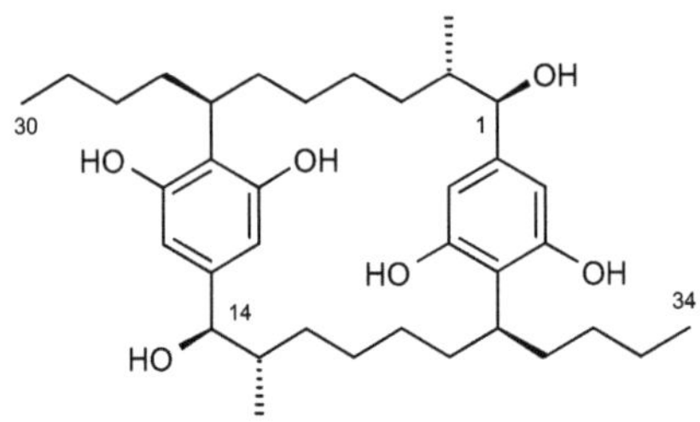

610 Cylindrocyclophane A_1

Type: Cylindrocyclophane derivatives. $C_{36}H_{55}ClO_6$ Amorph. powder. Source: Cyanobacterium *Nostoc* sp. UIC 10022A. Pharm: 20*S*-proteasome inhibitor (IC_{50} = 27.6 µmol/L, control Bortezomib, IC_{50} = 2.5 nmol/L). Ref: G. E. Chlipala, et al, JNP, 2010, 73, 1529

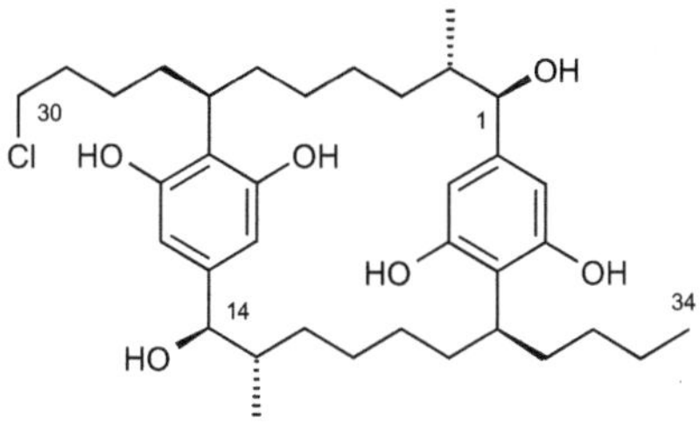

611 Cylindrocyclophane A_2

Type: Cylindrocyclophane derivatives. $C_{36}H_{54}Cl_2O_6$ Amorph. powder, $[\alpha]_D$ = −26° (*c* = 0.07, MeOH). Source: Cyanobacterium *Nostoc* sp. UIC 10022A. Pharm: 20*S*-proteasome inhibitor (IC_{50} = 2.55 μmol/L, control Bortezomib, IC_{50} = 2.5 nmol/L); cytotoxic (HT29, EC_{50} = 1.7 μmol/L, Positive control Camptothecin, EC_{50} = 60 nmol/L). Ref: G. E. Chlipala, et al, JNP, 2010, 73, 1529

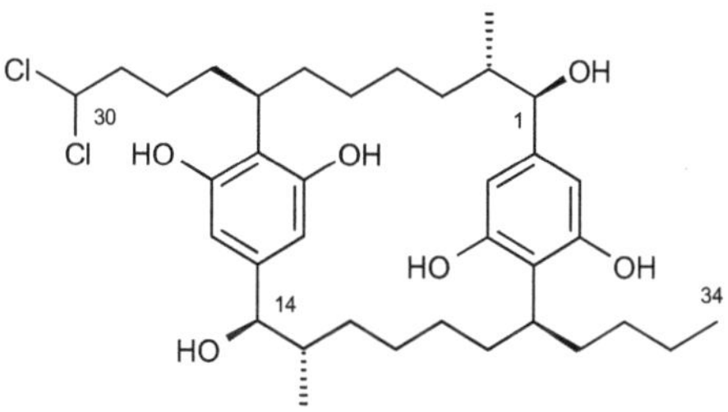

612 Cylindrocyclophane A_3

Type: Cylindrocyclophane derivatives. $C_{36}H_{53}Cl_3O_6$ Amorph. powder, $[\alpha]_D$ = −27° (*c* = 0.06, MeOH). Source: Cyanobacterium *Nostoc* sp. UIC 10022A. Pharm: 20*S*-proteasome inhibitor (IC_{50} = 2.75 μmol/L, control Bortezomib, IC_{50} = 2.5 nmol/L); cytotoxic (HT29, EC_{50} = 0.5 μmol/L, Positive control Camptothecin, EC_{50} = 60 nmol/L). Ref: G. E. Chlipala, et al, JNP, 2010, 73, 1529

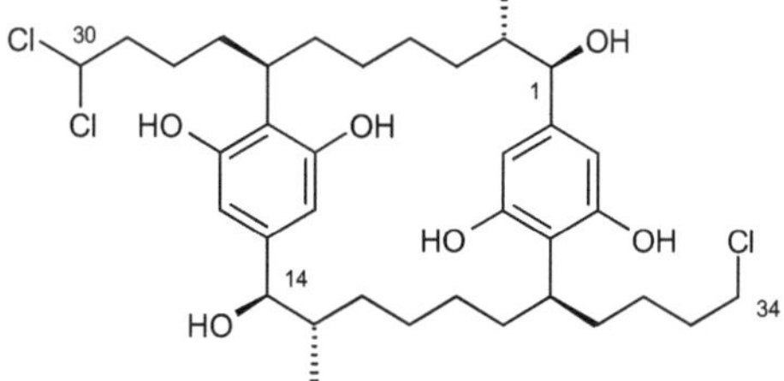

613 Cylindrocyclophane A_4

Type: Cylindrocyclophane derivatives. $C_{36}H_{52}Cl_4O_6$ Amorph. powder, $[\alpha]_D$ = −26° (*c* = 0.2, MeOH). Source: Cyanobacterium *Nostoc* sp. UIC 10022A. Pharm: 20*S*-proteasome inhibitor (IC_{50} = 3.93 μmol/L, control Bortezomib, IC_{50} = 2.5 nmol/L); cytotoxic

(HT29, EC_{50} = 2.0 µmol/L, Positive control Camptothecin, EC_{50} = 60 nmol/L). Ref: G. E. Chlipala, et al, JNP, 2010, 73, 1529

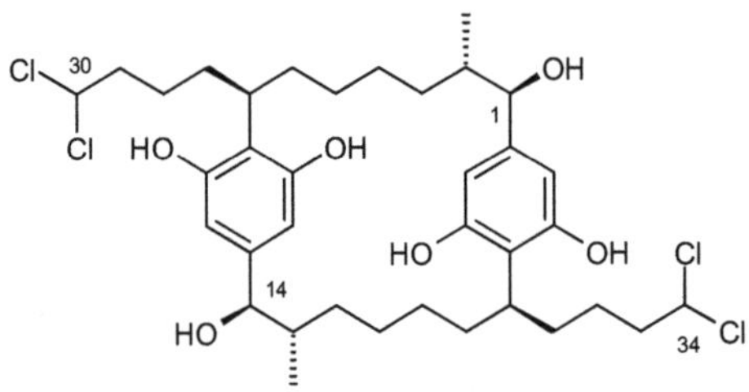

614 Cylindrocyclophane A_{B4}

Type: Cylindrocyclophane derivatives. $C_{36}H_{52}Br_4O_6$ Amorph. powder, $[\alpha]_D$ = −45° (*c* = 0.04, MeOH). Source: Cyanobacterium *Nostoc* sp. UIC 10022A. Pharm: 20*S*-proteasome inhibitor (IC_{50} = 2.23 µmol/L, control Bortezomib, IC_{50} = 2.5 nmol/L); cytotoxic (HT29, EC_{50} = 0.5 µmol/L, Positive control Camptothecin, EC_{50} = 60 nmol/L). Ref: G. E. Chlipala, et al, JNP, 2010, 73, 1529

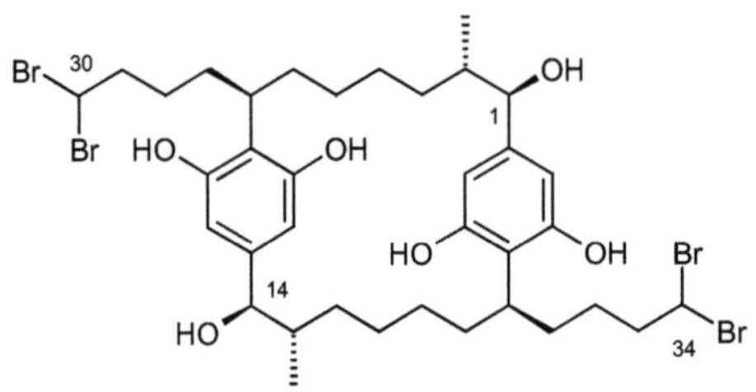

615 Cylindrocyclophane C

Type: Cylindrocyclophane derivatives. $C_{36}H_{56}O_5$ $[\alpha]_D$ = −40° (*c* = 0.1, MeOH). Source: Cyanobacteria *Cylindrospermum licheniforme* ATCC 29204 and *Nostoc* sp. UIC 10022A. Pharm: 20*S*-proteasome inhibitor (IC_{50} = 59.3 µmol/L, control Bortezomib, IC_{50} = 2.5 nmol/L). Ref: B. S. Moore, et al, Tetrahedron, 1992, 48, 3001 | G. E. Chlipala, et al, JNP, 2010, 73, 1529

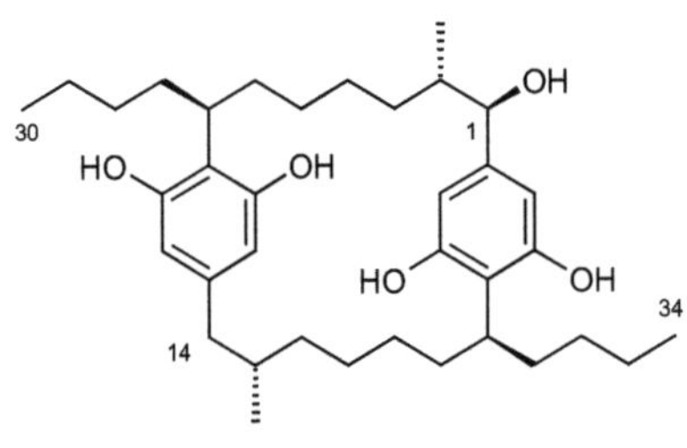

616 Cylindrocyclophane C_2

Type: Cylindrocyclophane derivatives. $C_{36}H_{54}Cl_2O_5$ Amorph. powder, $[\alpha]_D$ = −37° (*c* = 0.15, MeOH). Source: Cyanobacterium *Nostoc* sp. UIC 10022A. Pharm:

20*S*-proteasome inhibitor (IC_{50} = 22.8 µmol/L, control Bortezomib, IC_{50} = 2.5 nmol/L); cytotoxic (HT29, EC_{50} = 0.9 µmol/L, Positive control Camptothecin, EC_{50} = 60 nmol/L). Ref: G. E. Chlipala, et al, JNP, 2010, 73, 1529

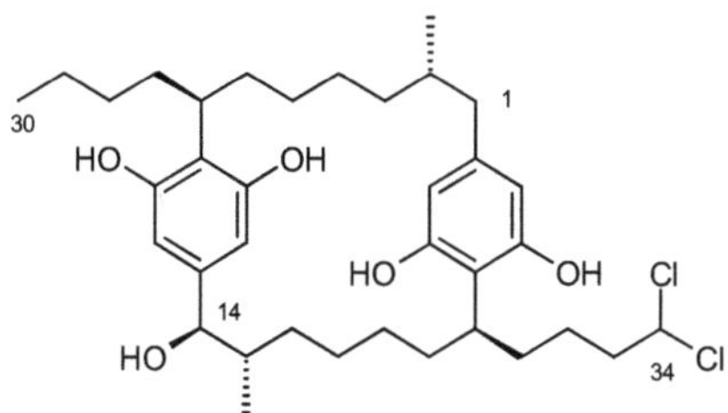

617 Cylindrocyclophane C$_4$

Type: Cylindrocyclophane derivatives. $C_{36}H_{52}Cl_4O_5$ Amorph. powder, $[\alpha]_D$ = −40° (*c* = 0.03, MeOH). Source: Cyanobacterium *Nostoc* sp. UIC 10022A. Pharm: 20*S*-proteasome inhibitor (IC_{50} = 11.2 µmol/L, control Bortezomib, IC_{50} = 2.5 nmol/L); cytotoxic (HT29, EC_{50} = 2.8 µmol/L, Positive control Camptothecin, EC_{50} = 60 nmol/L). Ref: G. E. Chlipala, et al, JNP, 2010, 73, 1529

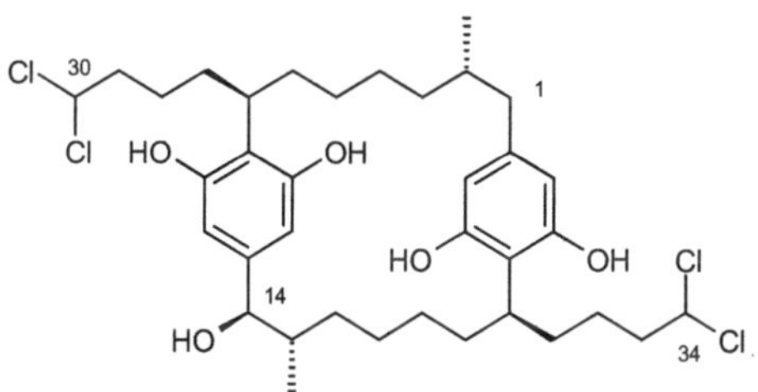

618 Cylindrocyclophane F

Type: Cylindrocyclophane derivatives. $C_{36}H_{56}O_4$ $[\alpha]_D$ = −72° (*c* = 0.9, MeOH). Source: Cyanobacteria *Cylindrospermum licheniforme* ATCC 29412 and *Nostoc* sp. UIC 10022A. Pharm: 20*S*-proteasome inhibitor (IC_{50} > 100 µmol/L, control Bortezomib, IC_{50} = 2.5 nmol/L). Ref: B. S. Moore, et al, Tetrahedron, 1992, 48, 3001 | G. E. Chlipala, et al, JNP, 2010, 73, 1529

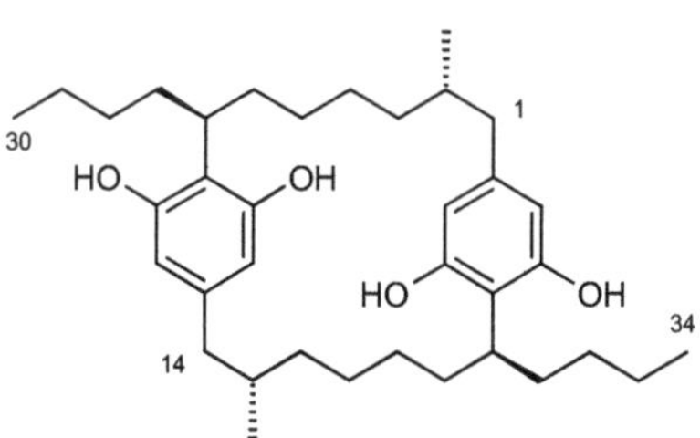

619 Cylindrocyclophane F_4

Type: Cylindrocyclophane derivatives. $C_{36}H_{52}Cl_4O_4$ Amorph. powder. Source: Cyanobacterium *Nostoc* sp. UIC 10022A. Pharm: 20*S*-proteasome inhibitor (IC_{50} = 44.8 μmol/L, control Bortezomib, IC_{50} = 2.5 nmol/L). Ref: G. E. Chlipala, et al, JNP, 2010, 73, 1529

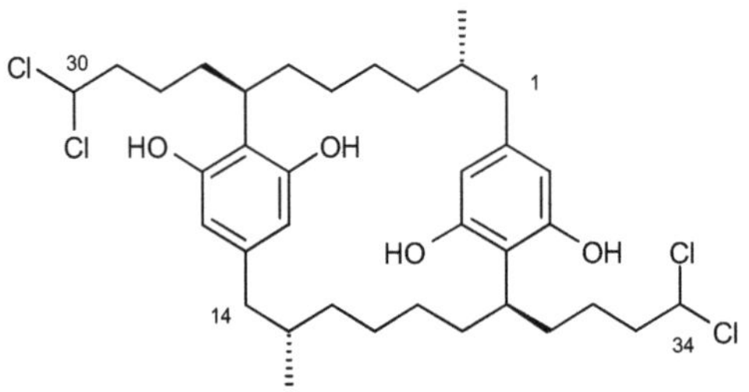

620 Nostocyclophane A

Type: Cylindrocyclophane derivatives. $C_{48}H_{74}Cl_2O_{16}$ Oil, $[\alpha]_D^{25}$ = −12° (*c* = 0.1, MeOH). Source: Cyanobacterium *Nostoc linckia* UTEX B1932. Pharm: Cytotoxic (KB, LoVo, IC_{50} = 1–2 μg/mL). Ref: B. S. Moore, et al, JACS, 1990, 112, 4061 | J. L. Chen, et al, JOC, 1991, 56, 4360

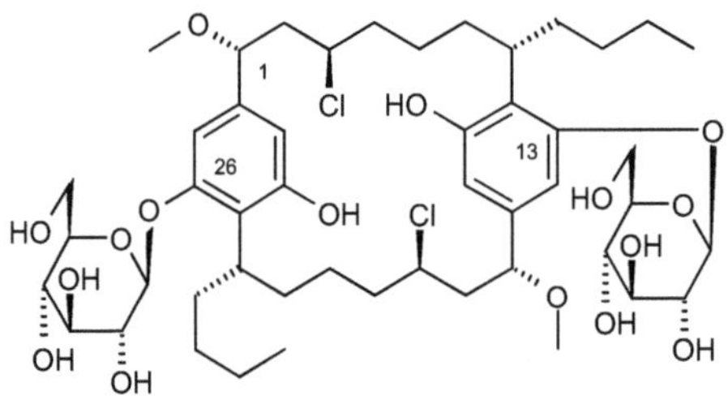

621 Nostocyclophane B

Type: Cylindrocyclophane derivatives. $C_{42}H_{64}Cl_2O_{11}$ Oil, $[\alpha]_D^{25}$ = −3.7° (*c* = 0.25, MeOH). Source: Cyanobacterium *Nostoc linckia* UTEX B1932. Pharm: Cytotoxic (KB, LoVo, IC_{50} = 1–2 μg/mL). Ref: B. S. Moore, et al, JACS, 1990, 112, 4061 | J. L. Chen, et al, JOC, 1991, 56, 4360

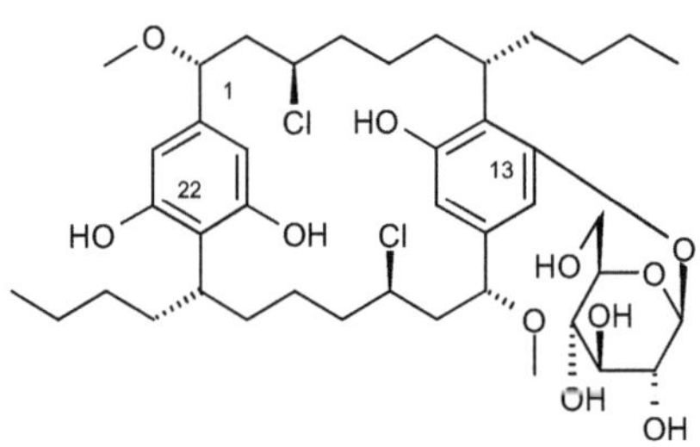

622 Nostocyclophane C

Type: Cylindrocyclophane derivatives. $C_{35}H_{52}Cl_2O_6$ Oil, $[\alpha]_D^{25}$ = −5.53° (*c* = 0.27, MeOH). Source: Cyanobacterium *Nostoc linckia* UTEX B1932. Pharm: Cytotoxic (KB, LoVo, IC_{50} = 1–2 μg/mL). Ref: B. S. Moore, et al, JACS, 1990, 112, 4061 | J. L. Chen, et al, JOC, 1991, 56, 4360

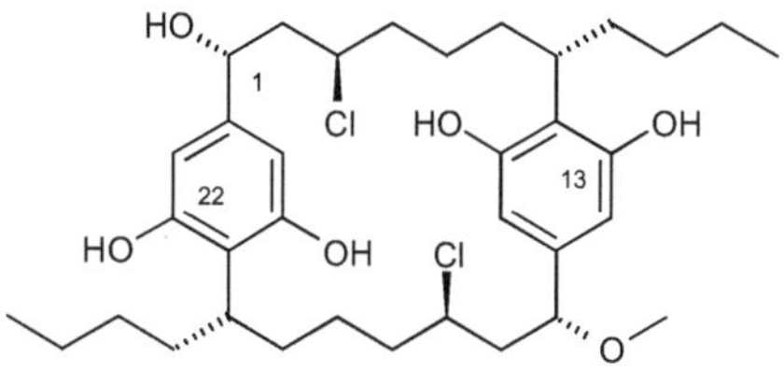

623 Nostocyclophane D

Type: Cylindrocyclophane derivatives. $C_{36}H_{54}Cl_2O_6$ Cryst. (EtOH aq), mp 242–243 °C, $[\alpha]_D^{25}$ = +10.8° (*c* = 0.4, MeOH). Source: Cyanobacterium *Nostoc linckia* UTEX B1932. Pharm: Cytotoxic (KB, LoVo, IC_{50} = 0.5 μg/mL). Ref: B. S. Moore, et al, JACS, 1990, 112, 4061 | J. L. Chen, et al, JOC, 1991, 56, 4360

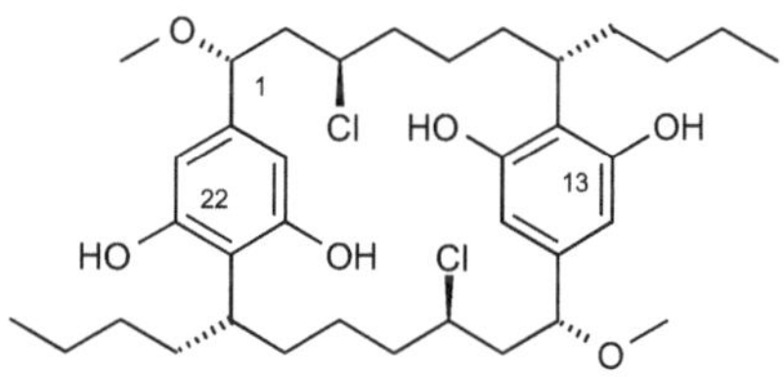

2.3 Tropolone Derivatives

624 Thiotropocin

4-Hydroxy-8-thioxocyclohept[*c*][1,2]oxathiol-3(8*H*)-one; Antibiotic CB 104 Type: Tropolone derivatives. $C_8H_4O_3S_2$ Fine orange needles, mp 222–225 °C (dec), mp 300 °C, Source: Marine-derived bacterium *Caulobacter* sp. PK654. Pharm: Antibacterial (gram-positive and -negative bacteria), antifungal; antimycoplasmas, LD_{50} (mus, orl) = 50–100 mg/kg. Ref: Y. Kawano, et al, J. Mar. Biotechnol., 1997, 5, 225; 1998, 6, 49

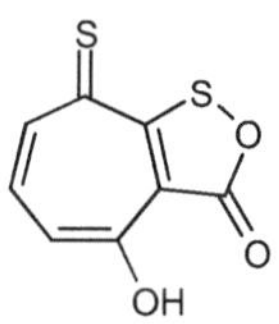

625 Tropolone

2-Hydroxy-2,4,6-cycloheptatrien-1-one Type: Tropolone derivatives. $C_7H_6O_2$ Needles (hexane or petrol), mp 49–50 °C. Source: Marine-derived bacterium *Phaeobacter gallaciensis* DSM17395. Pharm: LD_{50} (rat, ipr) = 190 mg/kg. Ref: V. Thiel, et al, Org. Biomol. Chem., 2010, 8, 234

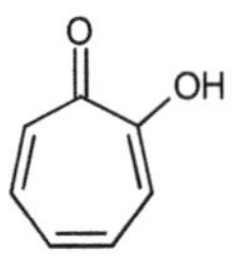

2.4 Benzofuranoids

626 6,7-Dihydro-3-(hydroxymethyl)-6-methyl-4(5*H*)-benzofuranone

Type: Benzofurans. $C_{10}H_{12}O_3$ Source: Marine-derived fungus *Phomopsis* sp. hzla01-1 (from unidentified marine origin). Pharm: Antibacterial (*Escherichia coli, Candida albicans, Saccharomyces cerevisiae, Bacillus subtilis*). Ref: X. P. Du, et al, J. Antibiot., 2008, 61, 250

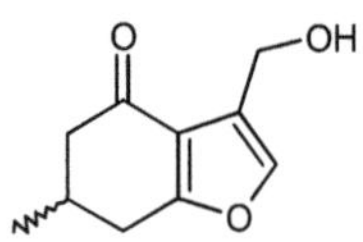

627 Felinone B

Type: Benzofurans. $C_{12}H_{14}O_3$ Source: Marine-derived fungus *Beauveria felina* EN-135 (entomopathogenic). Pharm: Antibacterial (*Pseudomonas aeruginosa*, MIC = 32 μg/mL). Ref: F. Y. Du, et al, Mar. Drugs, 2014, 12, 2816

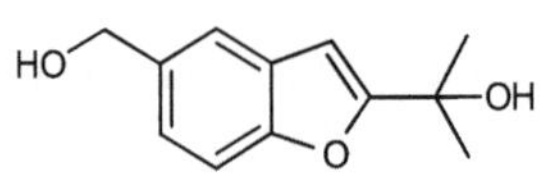

628 Isoacremine D

Type: Benzofurans. $C_{12}H_{14}O_3$ Source: Marine-derived fungus *Myceliophthora lutea*. Pharm: Antibacterial (*Staphylococcus aureus*, MIC = 200 μg/mL). Ref: O. F. Smetanina, et al, Chem. Nat. Compd., 2011, 47, 385

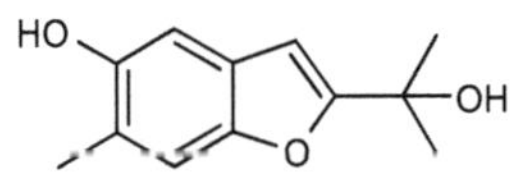

629 Phialofurone

Type: Benzofurans. $C_{14}H_{16}O_4$ Yellow powder, $[\alpha]_D^{20} = -4.9°$ (c = 0.16, MeOH). Source: Marine-derived fungus *Phialocephala* sp. (deep sea sediment, East Pacific Ocean). Pharm: Cytotoxic (P_{388}, IC_{50} = (0.2 ± 0.01)μmol/L, control *cis*-Diaminedichloroplatinum (CDDP), IC_{50} = (0.04 ± 0.03)μmol/L; K562, IC_{50} = (22.4 ± 0.9)μmol/L, *cis*-Diaminedichloroplatinum (CDDP), IC_{50} = (0.08 ± 0.05)μmol/L). Ref: D. -H. Li, et al, Chem. Biodiversity, 2011, 8, 895

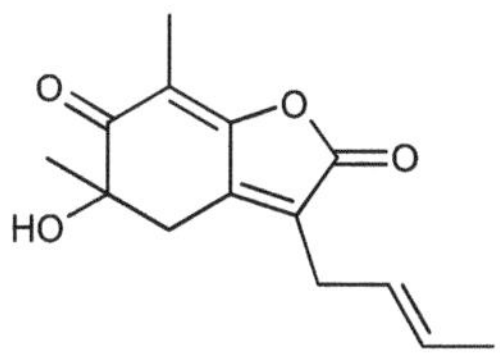

630 Thelepin

Type: Benzofurans. $C_{14}H_9Br_3O_3$ Pale yellow cryst. (MeOH), mp 202–203 °C. Source: Annelid worm *Thelepus setosus*. Pharm: Antifungal. Ref: T. Higa, et al, Tetrahedron, 1975, 31, 2379 | O. Tsuge, et al, Chem. Lett., 1984, 1415

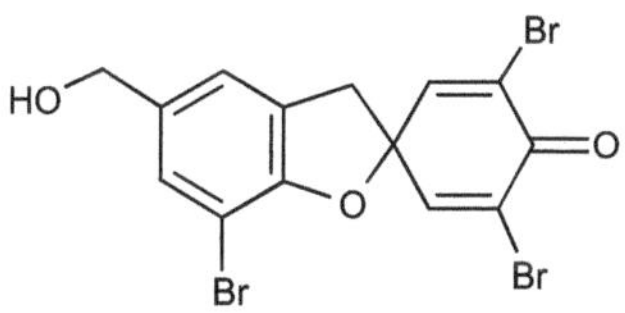

631 Ustusorane E

Type: Benzofurans. $C_{15}H_{22}O_4$ Amorph. solid, $[\alpha]_D^{20} = -11°$ (c = 0.1, MeOH). Source: Mangrove-derived fungus *Aspergillus ustus* 094102 from mangrove *Bruguiera gymnorrhiza* (rhizosphere soil, Wenchang, Hainan, China). Pharm: Cytotoxic (A549, IC_{50} > 100 μmol/L, control VP-16, IC_{50} = 0.63 μmol/L; HL60, IC_{50} = 0.13 μmol/L, VP-16, IC_{50} = 0.042 μmol/L). Ref: Z. Lu, et al, JNP, 2009, 72, 1761

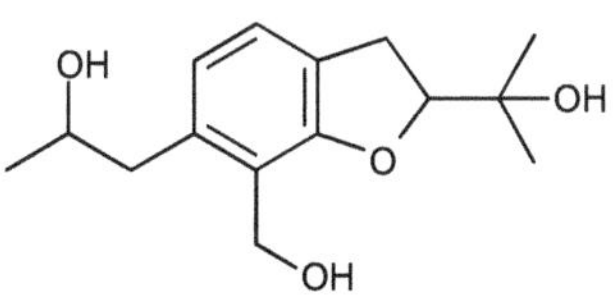

632 (±)-Acetophthalidin

3-Acetyl-5,7-dihydroxy-1(3*H*)-isobenzofuranone Type: Isobenzofurans. $C_{10}H_8O_5$ Powder, mp 213–2215 °C (dec), mp 195–205 °C Source: Marine-derived fungus *Aspergillus* sp. (from sediment). Pharm: Cytotoxic; mammalian cell cycle inhibitor (tsFT210, 6.25 μg/mL, caused complete arrest in G2/M phase). Ref: C. -B. Cui, et al, J. Antibiot.,

1996, 49, 216 | S. Nomoto, et al, Liebigs Ann., 1997, 721 | T. S. Bugni, et al, NPR, 2004, 21, 143 (rev)

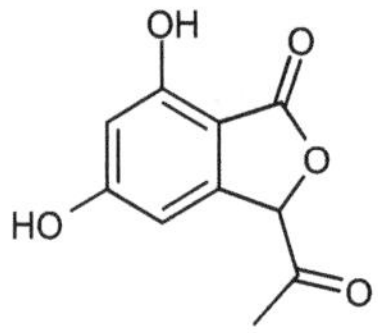

633 3-Acetyl-7-hydroxy-5-methoxy-3,4-dimethyl-1(3*H*)-isobenzofuranone

Type: Isobenzofurans. $C_{13}H_{14}O_5$ Needles, $[\alpha]_D^{29}$ = +200° (*c* = 0.05, EtOH). Source: Marine-derived fungi *Halorosellinia oceanica* BCC5149 and *Leptosphaeria* sp. KTC 727. Pharm: Antituberculosis (*Mycobacterium tuberculosis,* MIC = 200 μg/mL, weak). Ref: M. Chinworrungsee, et al, JCS Perkin I, 2002, 2473 | W. C. Tayone, et al, JNP, 2011, 74, 425

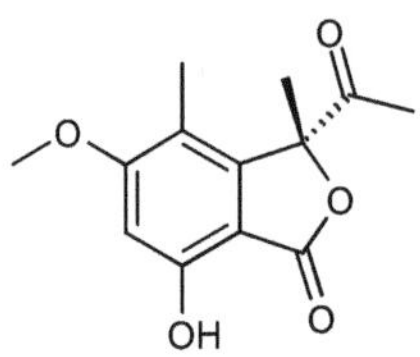

634 Ascolactone B

Type: Isobenzofurans. $C_{15}H_{16}O_7$ Amorph. solid, $[\alpha]_D^{25}$ = −169.2° (*c* = 0.5, MeOH). Source: Marine-derived fungus *Ascochyta salicorniae* from green alga *Ulva* sp. (North Sea coast, Germany). Pharm: PPs inhibitors (MPtpB (mycobacterial protein tyrosine phosphatase B), IC_{50} = 95.0 μmol/L; Cdc25a, PTP1B, VHR, MPtpA and VE-PTP, all inactive). Ref: S. F. Seibert, et al, Org. Biomol. Chem., 2006, 4, 2233

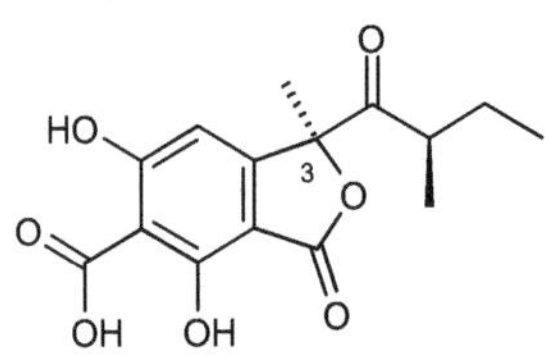

635 3-Butyl-7-methoxyphthalide

Type: Isobenzofurans. $C_{13}H_{16}O_3$ Viscous liquid, $[\alpha]_D^{27}$ = −49.1° (*c* = 1.8, $CHCl_3$). Source: An unidentified marine-derived fungus CRIF2 (order Pleosporales) from an unidentified sponge (Thailand). Pharm: Cytotoxic (several cell lines). Ref: V. Prachyawarakorn, et al, Planta Med., 2008, 74, 69

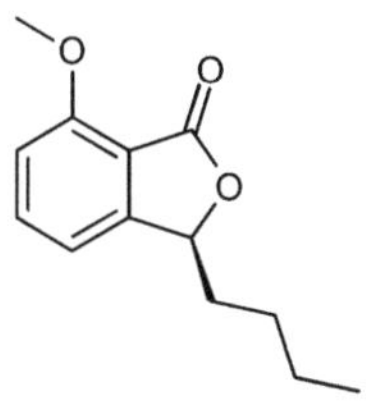

636 Corollosporin

3-Hexyl-3,7-dihydroxy-1(3*H*)-isobenzofuranone Type: Isobenzofurans. $C_{14}H_{18}O_4$ Cryst. (hexane/EtOAc), mp 79.5–81.5 °C. Source: Marine-derived fungus *Corollospora maritima*. Pharm: Antibacterial. Ref: K. Liberra, et al, Pharmazie, 1998, 53, 578

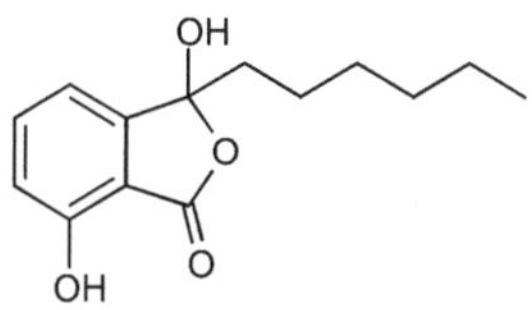

637 Cytosporone E

Type: Isobenzofurans. $C_{15}H_{22}O_5$ Source: Mangrove-derived fungus *Leucostoma persoonii* from mangrove *Rhizophora mangle* (branch, Everglades, Florida), *Cytospora* sp. Pharm: Antiplasmodial (*Plasmodium falciparum*, IC_{90} = 13 µmol/L, to A549 cytotoxic IC_{90} = 437 µmol/L, 90% inhibition therapeutic index TI_{90} = 33); antibacterial (MRSA, MIC = 72 µmol/L, minimum bactericidal concentration MBC_{90} = 45 µmol/L, minimum biofilm eradication counts $MBEC_{90}$ = 39 µmol/L; A549 IC_{50} = 280 µmol/L, IC_{90} = 440 µmol/L). Ref: M. P. Singh, et al, Mar. Drugs, 2007, 5, 71 | J. Beau, et al, Mar. Drugs, 2012, 10, 762

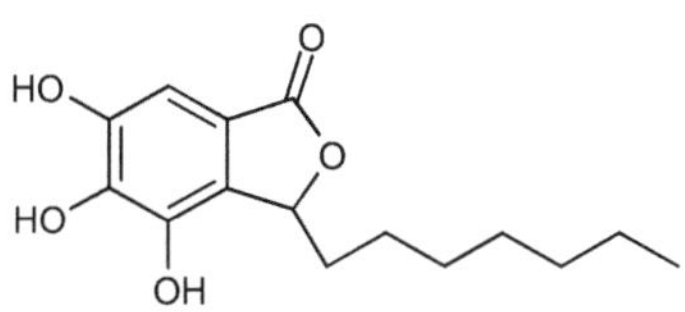

638 Diaporthelactone

Type: Isobenzofurans. $C_{11}H_{10}O_4$ Cryst. (EtOAc). Source: Marine-derived fungus *Diaporthe* sp. HLY2. Pharm: Cytotoxic (KB, IC_{50} = 6.25 µg/mL; Raji cell lines, IC_{50} = 5.51 µg/mL). Ref: X. Lin, et al, FEMS Microbiol. Lett., 2005, 251, 53

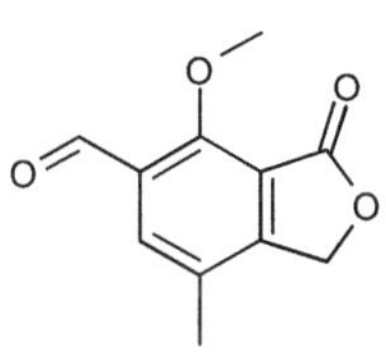

639 Epicoccone

Type: Isobenzofurans. $C_9H_8O_5$ Cryst., mp 249–250 °C. Source: Marine-derived fungus *Epicoccum* sp. from brown alga *Fucus vesiculosus,* terrestrial fungus (*Epicoccum purpurascens* MYC 1097). Pharm: Antioxidant (DPPH scavenger, 25 μg/mL, ScRt = 95%; inhibits peroxidation of linolenic acid, TBARS assay, 37 μg/mL, InRt = 62%). Ref: A. Abdel-Lateff, et al, Planta Med., 2003, 69, 831 | N. H. Lee, et al, Bull. Korean Chem. Soc., 2007, 28, 877 | M. Saleem, et al, NPR, 2007, 24, 1142 (rev)

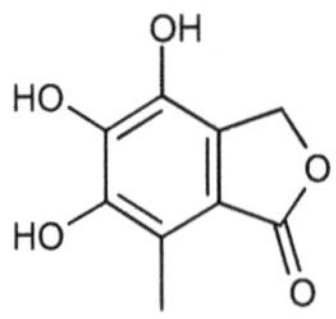

640 Marilone A

Type: Isobenzofurans. $C_{21}H_{28}O_4$ Transparent oil. Source: Marine-derived fungus *Stachylidium* sp. from sponge *Callyspongia flammea* (Bear I., Sydney, Australia). Pharm: Antiplasmodial (*Plasmodium berghei,* liver stages, IC_{50} = 12.1 μmol/L); cytotoxic (NCI-H460, MCF7 and SF268, average GI_{50} = 36.7 μmol/L, weak antiproliferative). Ref: C. Almeida, et al, Beilstein J. Org. Chem., 2011, 7, 1636

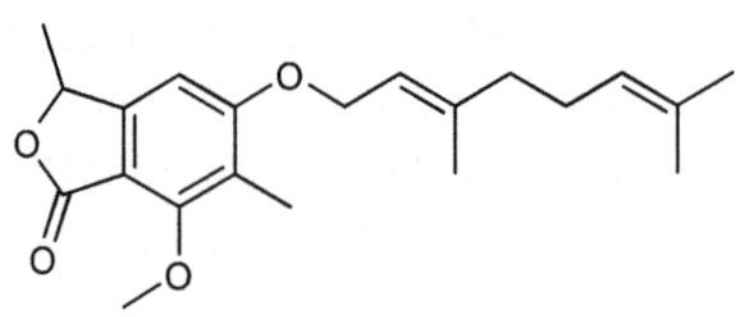

641 Marilone B

Type: Isobenzofurans. $C_{11}H_{12}O_4$ White amorph. solid. Source: Marine-derived fungus *Stachylidium* sp. from sponge *Callyspongia flammea* (Bear I., Sydney, Australia). Pharm: Selective antagonist of serotonin receptor 5-HT_{2B} (*K*i = 7.7 μmol/L). Ref: C. Almeida, et al, Beilstein J. Org. Chem., 2011, 7, 1636

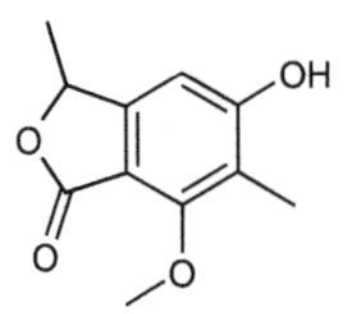

642 Marilone C

Type: Isobenzofurans. $C_{20}H_{26}O_4$ Transparent oil. Source: Marine-derived fungus *Stachylidium* sp. from sponge *Callyspongia flammea* (Bear I., Sydney, Australia).

Pharm: Cytotoxic (NCI-H460, MCF7 and SF268, average GI_{50} = 26.6 μmol/L, weak, antiproliferative). Ref: C. Almeida, et al, Beilstein J. Org. Chem., 2011, 7, 1636

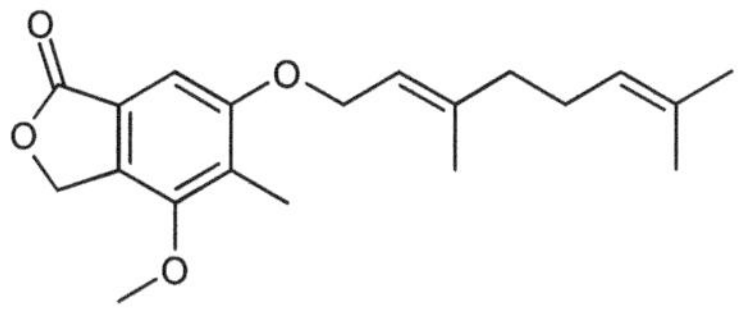

643 Mycophenolic acid

Type: Isobenzofurans. $C_{17}H_{20}O_6$ Needles (H_2O), mp 141 °C. Source: Terrestrial fungi *Penicillium* spp. Pharm: Antineoplastic; antiviral; immunosuppressant; used in treating psoriasis and leishmaniasis; used in prophylaxis of acute kidney rejection. Ref: CRC Press, DNP on DVD, 2012, version20.2

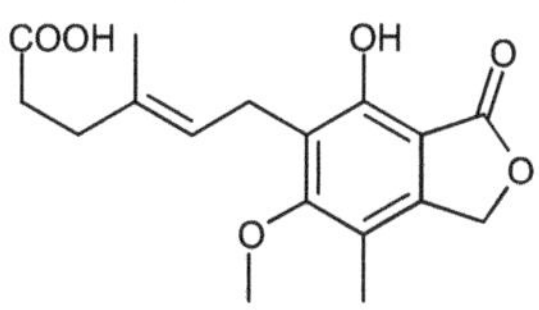

644 Paecilocin A

Type: Isobenzofurans. $C_{16}H_{22}O_3$ White amorph. powder, $[\alpha]_D^{25}$ = −8.6° (*c* = 0.25,$CHCl_3$). Source: Marine-derived fungus *Paecilomyces variotii* from scyphomedusa *Neopilema nomurai* (Southern coast of R. O. Korea). Pharm: Antibacterial (*Staphylococcus aureus* SG 511, MIC > 40 μg/mL, control Tetracycline, MIC = 1.25 μg/mL; MRSA 3089, MIC > 40 μg/mL, control Vancomycin, MIC = 0.625 μg/mL; MDR *Vibrio parahemolyticus* 7001, MIC > 40 μg/mL, control Levofloxacin, MIC = 0.156 μg/mL). Ref: J. Liu, et al, JNP, 2011,74, 1826

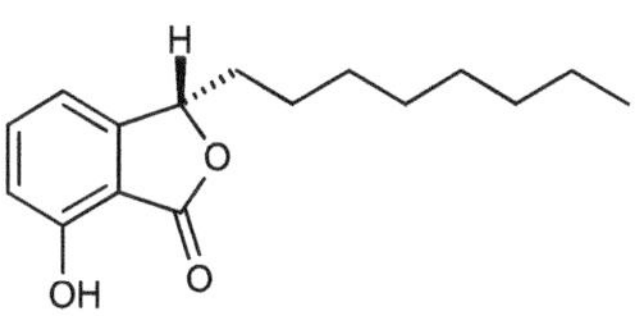

645 Paecilocin B

Type: Isobenzofurans. $C_{18}H_{28}O_4$ Yellow oil, $[\alpha]_D^{25}$ = −4.1° (*c* = 0.10, MeOH). Source: Marine-derived fungus *Paecilomyces variotii* from scyphomedusa *Neopilema nomurai* (Southern coast of R. O. Korea). Pharm: Antibacterial (*Staphylococcus aureus* SG 511, MIC = 5 μg/mL, control Tetracycline, MIC = 1.25 μg/mL; MRSA 3089, MIC = 20 μg/mL, control Vancomycin, MIC = 0.625 μg/mL; MDR *Vibrio parahemolyticus*

7001, MIC > 40 μg/mL, control Levofloxacin, MIC = 0.156 μg/mL). Ref: J. Liu, et al, JNP, 2011,74, 1826

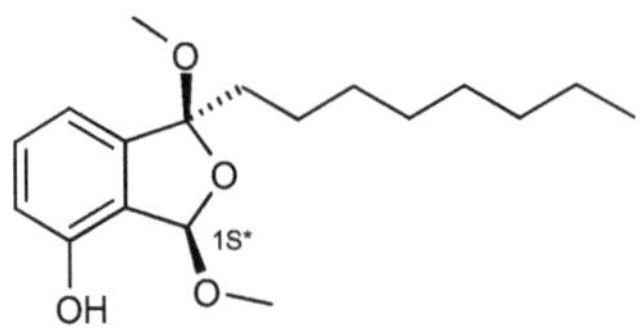

646 Paecilocin C

Type: Isobenzofurans. $C_{18}H_{28}O_4$ Yellow oil, $[\alpha]_D^{25} = -12.6°$ (c = 0.10, MeOH). Source: Marine-derived fungus *Paecilomyces variotii* from scyphomedusa *Neopilema nomurai* (Southern coast of R. O. Korea). Pharm: Antibacterial (*Staphylococcus aureus* SG 511, MIC = 20 μg/mL, control Tetracycline, MIC = 1.25 μg/mL; MRSA 3089, MIC = 40 μg/mL, control Vancomycin, MIC = 0.625 μg/mL; MDR *Vibrio parahemolyticus* 7001, MIC > 40 μg/mL, control Levofloxacin, MIC = 0.156 μg/mL). Ref: J. Liu, et al, JNP, 2011,74, 1826

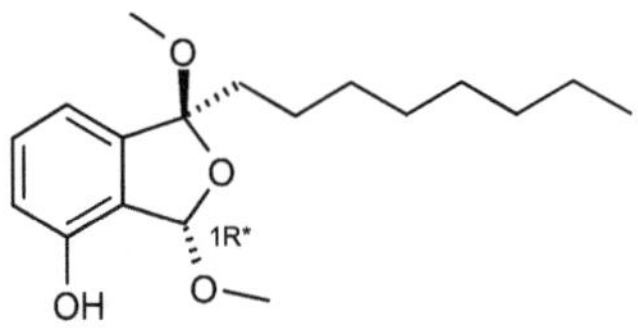

647 Pestaliopen A

Type: Isobenzofurans. $C_{27}H_{33}ClO_{10}$ Source: Mangrove-derived fungus *Pestalotiopsis* sp. (endophytic) from mangrove *Rhizophora mucronata* (leaves, Hainan, China). Pharm: Antibacterial (*Enterococcus faecalis,* modest). Ref: Y. Hemberger, et al, Chem. -Eur. J., 2013, 19, 15556

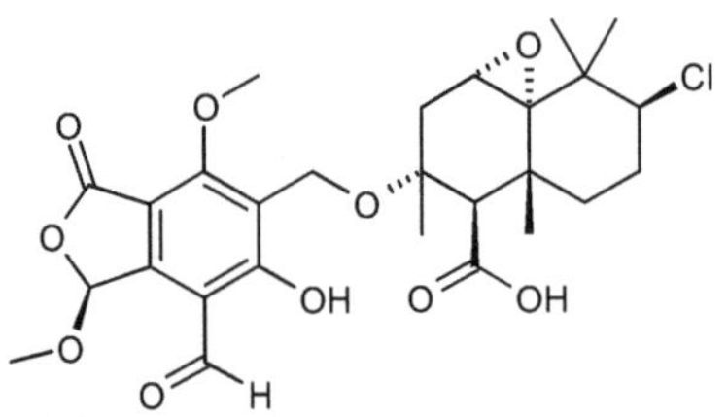

648 Purpurester A

Type: Isobenzofurans. $C_{13}H_{16}O_5$ Source: Acid-tolerant fungus *Penicillium purpurogenum* JS03-21 (region of red soil by Lujiang River, Yunnan, China). Pharm: Antiviral (IFV virus, IC_{50} = 85.3 μmol/L). Ref: H. Wang, et al, JNP, 2011, 74, 2014

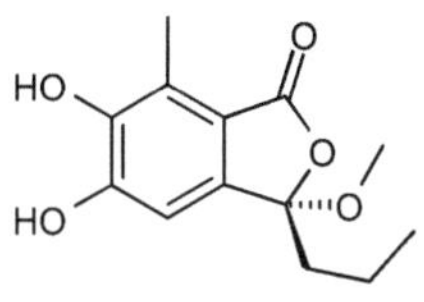

649 Talaroflavone

Type: Isobenzofurans. $C_{14}H_{12}O_6$ Flakes, mp 230 °C (dec), $[\alpha]_D$ = +181° (*c* = 0.74, MeOH). Source: Marine-derived unidentified fungus from unidentified algae (Japan waters), terrestrial fungus (*Talaromyces flavus*). Pharm: Eukaryotic DNA polymerases selective inhibitor (X and Y families); inhibitor inactive (other eukaryotic or prokaryotic polymerases, or other DNA metabolic enzymes). Ref: W. A. Ayer, et al, Can. J. Chem., 1990, 68, 2085 | M. Naganuma, et al, BoMC, 2008, 16, 2939

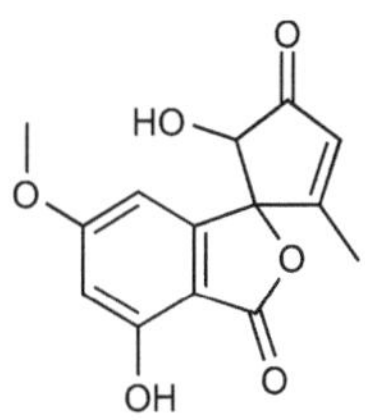

2.5 Benzopyranoids

650 Aflatoxin B2b

Type: 1-Benzopyrans. $C_{22}H_{20}O_{10}$ Source: Mangrove-derived fungus *Aspergillus flavus* 092008 from semimangrove *Hibiscus tiliaceus* (root, Wenchang, Hainan, China). Pharm: Antibacterial (*Enterobacter coli*, MIC = 22.5 μmol/L, *Bacillus subtilis*, MIC = 1.7 μmol/L, *Enterobacter aerogenes*, MIC = 1.1 μmol/L). Ref: H. Wang, et al, Arch. Pharmacal Res., 2012, 35, 1387

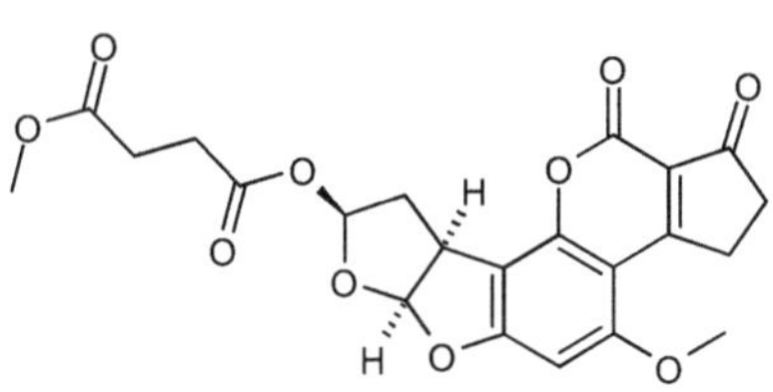

651 Ammonificin C

Type: 1-Benzopyrans. $C_{23}H_{20}BrNO_6$ $[\alpha]_D^{24}$ = −45° (*c* = 0.8, MeOH). Source: Marine-derived bacterium *Thermovibrio ammonificans* (hydrothermal vent, East Pacific Rise). Pharm: Apoptosis inducer (2 μmol/L, control Staurosporine, 0.1 μmol/L). Ref: E. H. Andrianasolo, et al, Mar. Drugs, 2012, 10, 2300

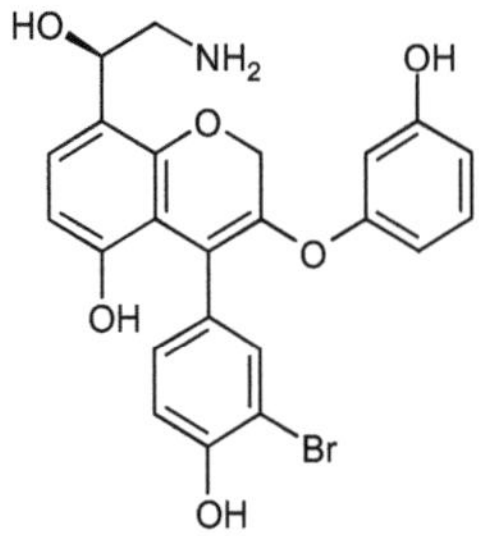

652 Ammonificin D

Type: 1-Benzopyrans. $C_{23}H_{19}Br_2NO_5$ $[\alpha]_D^{24} = -45°$ (c = 0.8, MeOH). Source: Marine-derived bacterium *Thermovibrio ammonificans* (hydrothermal vent, East Pacific Rise). Pharm: Apoptosis inducer (3 µmol/L, control Staurosporine, 0.1 µmol/L). Ref: E. H. Andrianasolo, et al, Mar. Drugs, 2012, 10, 2300

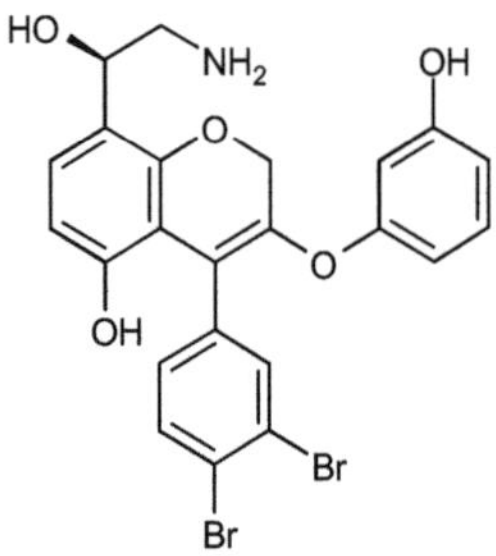

653 Antibiotic SB236049

Type: 1-Benzopyrans. $C_{14}H_9NO_6$ Yellow solid. Source: Fungus *Chaetomium funicola* TCF 6040. Pharm: Metallo-β-lactamase inhibitor (metallo-β-lactamase of *Bacteroides fragilis* CfiA, Ki = 15 µmol/L); metallo-β-lactamase inhibitor (metallo-β-lactamase of *Bacillus cereus* II, Zn^{2+} 1 µmol/L, IC_{50} = 0.3 µmol/L; Zn^{2+} 100 µmol/L, IC_{50} = 0.7 µmol/L; metallo-β-lactamase of *Bacteroides fragilis* CfiA, Zn^{2+} 100 µmol/L, IC_{50} = 2 µmol/L; metallo-β-lactamase of *Pseudomonas aeruginosa* IMP-1, Zn^{2+} 100 µmol/L, IC_{50} = 151 µmol/L; metallo-β-lactamase of L-1, Zn^{2+} 100 µmol/L, IC_{50} > 1000 µmol/L; angiotensin-converting enzyme ACE, Zn^{2+} 1 µmol/L, IC_{50} > 333 µmol/L). Ref: D. J. Payne, et al, Antimicrob. Agents Chemother., 2002, 46, 1880

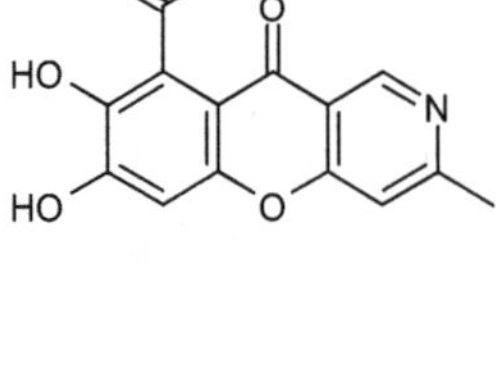

654 Apralactone A

Type: 1-Benzopyrans. $C_{18}H_{20}O_6$ Amorph. solid, $[\alpha]_D^{20}$ = +10° (*c* = 0.3, EtOH). Source: Marine-derived fungus *Curvularia* sp. 768 from red alga *Acanthophora spicifera* (Guam). Pharm: Cytotoxic (panel of 36 hmn tumour cell lines, mean IC_{50} = 9.87 μmol/L BXF-1218L, IC_{50} = 12.3 μmol/L; BXF-T24, IC_{50} = 6.36 μmol/L; CNXF-SF268, IC_{50} = 6.85 μmol/L; CXF-HCT116, IC_{50} = 9.59 μmol/L; CXF-HT29, IC_{50} = 10.26 μmol/L; GXF-251L, IC_{50} = 9.35 μmol/L; LXF-1121L, IC_{50} = 11.43 μmol/L; LXF-289L, IC_{50} = 15.77 μmol/L; LXF-526L, IC_{50} = 4.61 μmol/L; LXF-629L, IC_{50} = 9.71 μmol/L; LXF-H460, IC_{50} = 6.74 μmol/L; MAXF-401NL, IC_{50} = 9.17 μmol/L; MAXF-MCF7, IC_{50} = 10.11 μmol/L; MEXF-276L, IC_{50} = 9.03 μmol/L; MEXF-462NL, IC_{50} = 12.2 μmol/L; MEXF-520L, IC_{50} = 7.66 μmol/L; OVXF-899L, IC_{50} = 9.56 μmol/L; OVXF-OVCAR3, IC_{50} = 10.97 μmol/L; PAXF-1657L, IC_{50} = 13.25 μmol/L; PAXF-PANC1, IC_{50} = 11.43 μmol/L; PRXF-22RV1, IC_{50} = 7.66 μmol/L; PRXF-DU145, IC_{50} = 7.94 μmol/L; PRXF-LNCAP, IC_{50} = 9.52 μmol/L; PRXF-PC3M, IC_{50} = 10.05 μmol/L; PXF-1752L, IC_{50} = 16.78 μmol/L; RXF-1781L, IC_{50} = 16.57 μmol/L; RXF-486L, IC_{50} = 12.64 μmol/L; UXF-1138L, IC_{50} = 10.11 μmol/L). Ref: H. Greve, et al, EurJOC, 2008, 5085

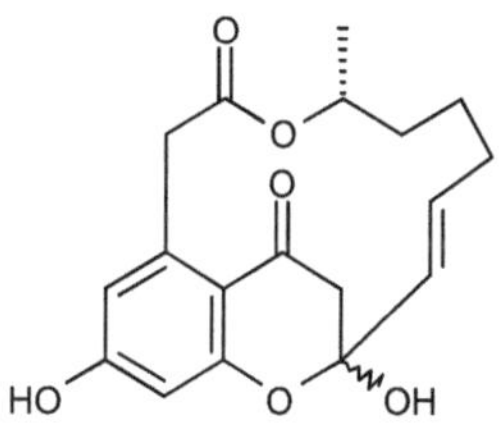

655 Aspergillitine

Type: 1-Benzopyrans. $C_{15}H_{13}NO_2$ Yellow powder (MeOH). Source: Marine-derived fungus *Aspergillus versicolor* from sponge *Xestospongia exigua* (Mengangan I., Bali, Indonesia, Sep 1997). Pharm: Antibacterial (*Bacillus subtilis*, moderate). Ref: W. Lin, et al, JNP, 2003, 66, 57

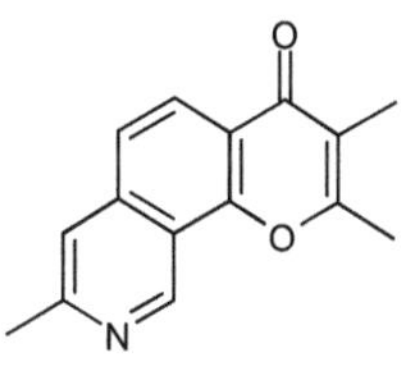

656 Caffeoylate ester

Type: 1-Benzopyrans. $C_{24}H_{18}O_8$ Source: Mangrove *Laguncularia racemosa*. Pharm: Antioxidant (significant); hmn kinases inhibitor (FLT3 and SAK, moderate). Ref: C. Shi, et al, Phytochemistry, 2010, 71, 435

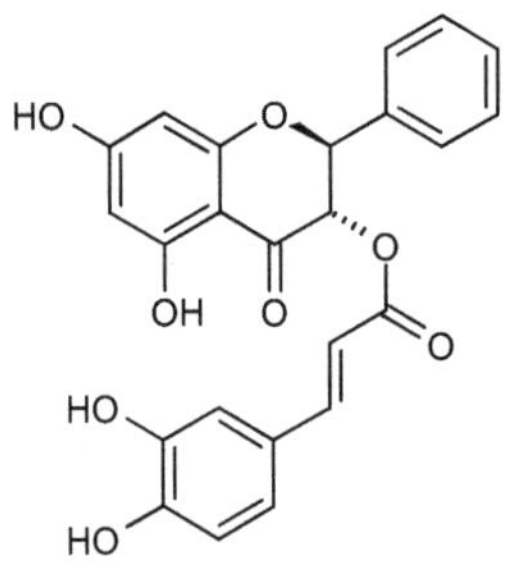

657 Chromanone A

Type: 1-Benzopyrans. $C_{12}H_{12}O_4$ Source: Marine-derived fungus *Penicillium* sp. from green alga *Ulva* sp. (Suez Canal, Egypt). Pharm: Antioxidant (hydroxyl radical scavenger, potent); inhibits cytochrome P450 1A (CYP1A); glutathione S-transferases (GST) inhibitor; enzymes inducer (epoxide hydrolase, glutathione S-transferase, enzymes involved in carcinogen metabolism). Ref: A. M. Gamal-Eldeen, et al, Environ. Toxicol. Pharmacol., 2009, 28, 317

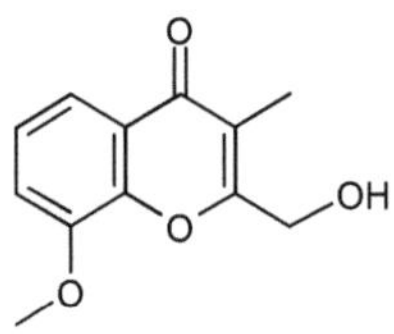

658 Chromenol

2-Methyl-2-(prenylmethyl)-2*H*-1-benzopyran-6-ol Type: 1-Benzopyrans. $C_{16}H_{20}O_2$ Light yellow oil. Source: Ascidian *Aplidium constellatum*. Pharm: Anti-inflammatory. Ref: N. M. Targett, et al, JNP, 1984, 47, 556

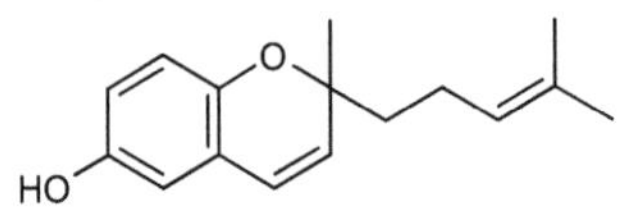

659 (±)-Dictyochromenol

Type: 1-Benzopyrans. $C_{21}H_{28}O_2$ Oil. Source: Brown alga *Dictyopteris undulata*. Pharm: Ichthyotoxin (27 ppm). Ref: M. -N, Dave, et al, Heterocycles, 1984, 22, 2301

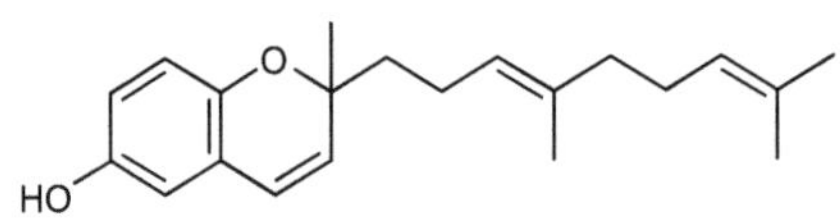

660 (2*R*)-2,3-Dihydro-7-hydroxy-6,8-dimethyl-2-[(*E*)-prop-l-enyl]chromen-4-one

Type: 1-Benzopyrans. $C_{14}H_{16}O_3$ Yellow meedles, $[\alpha]_D^{20}$ = +44.5° (*c* = 0.10, MeOH). Source: Deep-sea fungus *Aspergillus sydowi* YH11-2 (Guam, depth of 1000 m, E144°43′ N13°26′). Pharm: Cytotoxic (P_{388}, IC_{50} = 0.14 μmol/L, control CDDP, IC_{50} = 0.039 μmol/L). Ref: L. Tian, et al, Arch. Pharm. Res,. 2007, 30, 1051

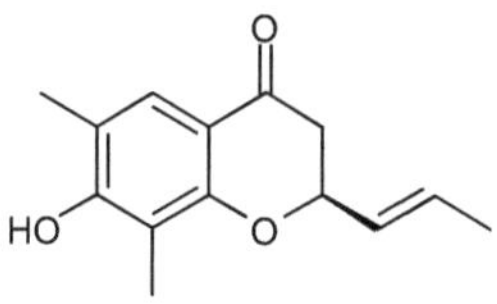

661 2,3-Dihydro-7-hydroxy-6-methyl-2-(1-propenyl)-4*H*-1-benzopyran-4-one

Type: 1-Benzopyrans. $C_{13}H_{14}O_3$ Source: Marine-derived fungus *Trichoderma* sp. from starfish *Acanthaster planci* (Hainan Sanya National Coral Reef Reserve, China). Pharm: Cytotoxic (several HTCLs cells, moderate). Ref: W. -J. Lan, et al, Nat. Prod. Commun., 2012, 7, 1337

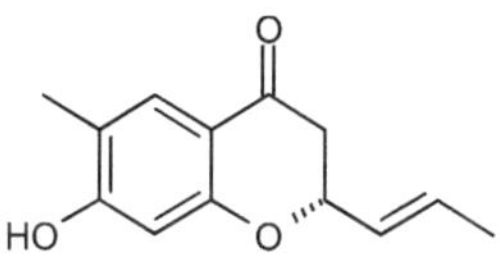

662 2,2-Dimethyl-2*H*-1-benzopyran-6-ol

Type: 1-Benzopyrans. $C_{11}H_{12}O_2$ Cryst. (Et_2O/pentane), mp 86–87 °C. Source: Ascidian *Aplidium californicum*. Pharm: Antimutagenic. Ref: B. M. Howard, et al, Tet. Lett., 1979, 4449

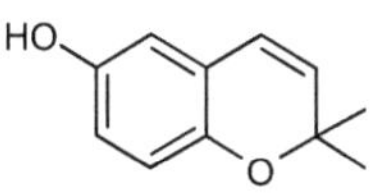

663 Monodictyochrome A

Type: 1-Benzopyrans. $C_{30}H_{30}O_{11}$ Yellow solid, $[\alpha]_D^{24}$ = −67° (*c* = 0.42, $CHCl_3$). Source: Marine-derived fungus *Monodictys putredinis*. Pharm: Cytochrome P450 1A inhibitor (IC_{50} = 5.3 μmol/L); inducer of NAD(P)H quinone reductase (cultured mouse Hepa1c1c7 cells, CD (concentration required to double the specific activity of quinone reductase) = 22.1 μmol/L, IC_{50} > 50 μmol/L); aromatase inhibitor (IC_{50} = 24.4 μmol/L, weak). Ref: A. Pontius, et al, JNP, 2008, 71, 1793

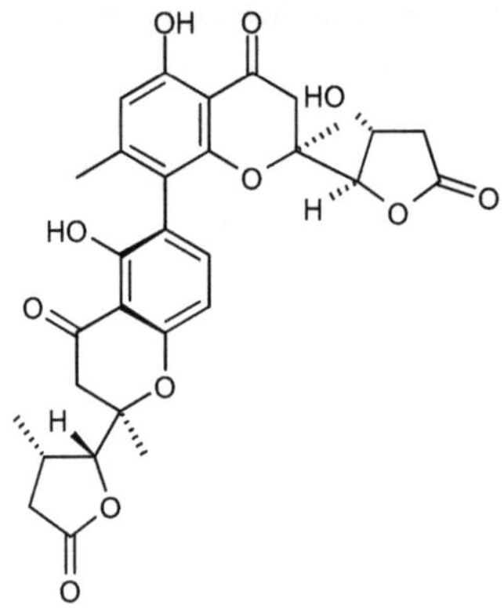

664 Monodictyochrome B

Type: 1-Benzopyrans. $C_{30}H_{30}O_{11}$ Yellow solid, $[\alpha]_D^{24} = +74°$ (c = 0.6, $CHCl_3$). Source: Marine-derived fungus *Monodictys putredinis*. Pharm: Cytochrome P450 1A inhibitor (IC_{50} = 7.5 μmol/L); inducer of NAD(P)H quinone reductase (cultured mouse Hepa1c1c7 cells, CD (concentration required to double the specific activity of QR) = 24.8 μmol/L, IC_{50} > 50 μmol/L); aromatase inhibitor (IC_{50} = 16.5 μmol/L, weak). Ref: A. Pontius, et al, JNP, 2008, 71, 1793

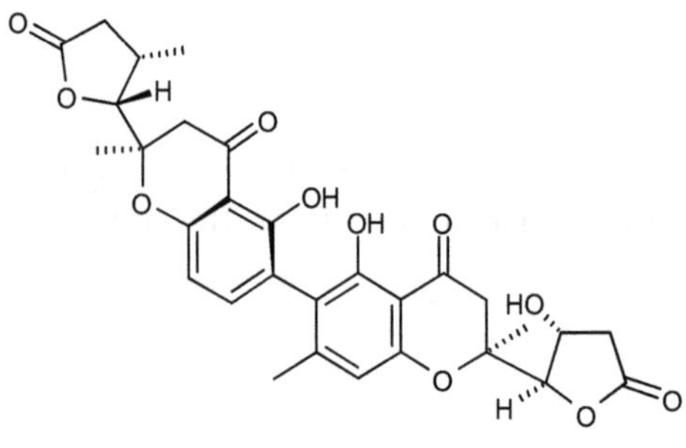

665 Oxirapentyne E

Type: 1-Benzopyrans. $C_{16}H_{20}O_5$ Source: Marine-derived fungus *Isaria felina* (sediment, Vietnam). Pharm: Growth stimulant (corn and barley rootlets). Ref: A. N. Yurchenko, et al, Chem. Nat. Compd., 2013, 49, 857

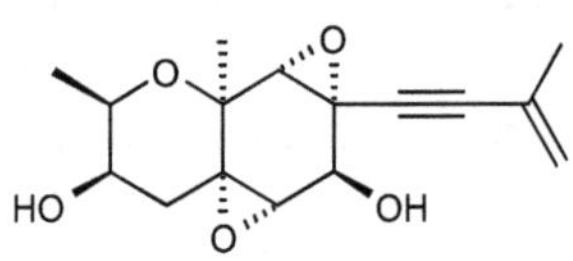

666 Penilactone A

Type: 1-Benzopyrans. $C_{25}H_{26}O_9$ Source: Deep-sea fungus *Penicillium crustosum* PRB-2 (Antarctic Ocean). Pharm: Antineoplastic (weak); cytotoxic (10 mmol/L, NF-κB, InRt = 40%). Ref: G. Wu, et al, Tetrahedron, 2012, 68, 9745

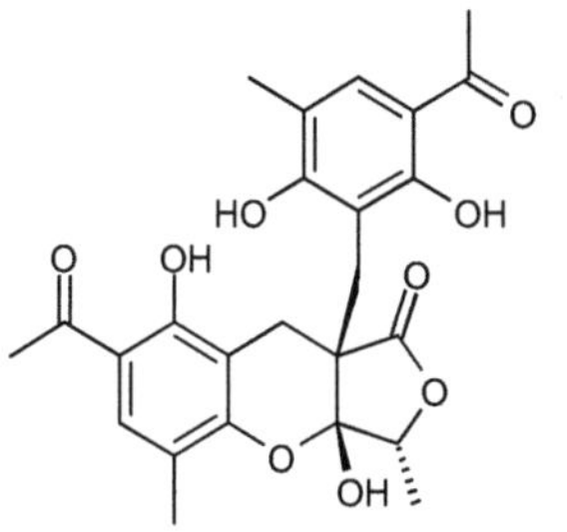

667 Penilactone B

Type: 1-Benzopyrans. $C_{26}H_{26}O_{11}$ Source: Deep-sea fungus *Penicillium crustosum* PRB-2 (Antarctic Ocean). Pharm: Antineoplastic (weak). Ref: G. Wu, et al, Tetrahedron, 2012, 68, 9745

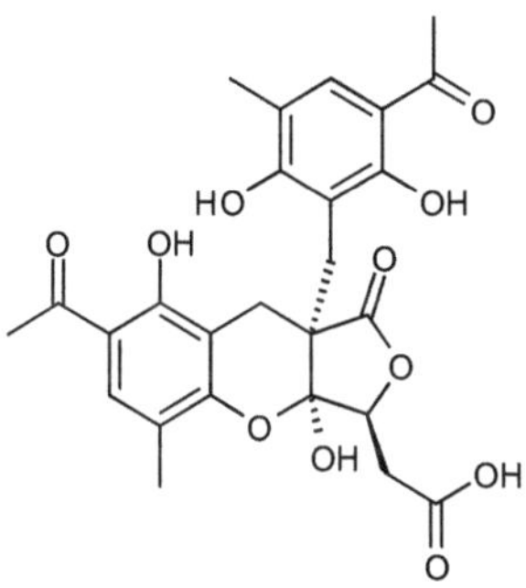

668 Penostatin A

Type: 1-Benzopyrans. $C_{22}H_{32}O_3$ Needles, mp 73–75 °C, $[\alpha]_D$ = +133° (c = 0.2, $CHCl_3$). Source: Marine-derived fungus *Penicillium* sp. OUPS-79 from green alga *Enteromorpha intestinalis*. Pharm: Cytotoxic (P_{388}, ED_{50} = 0.8 μg/mL). Ref: C. Takahashi, et al, Tet. Lett., 1996, 37, 655 | C. Iwamoto, et al, Tetrahedron, 1999, 55, 14353

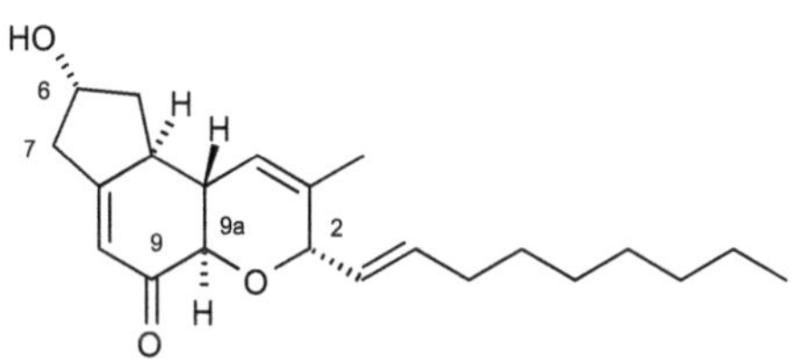

669 Penostatin B

Type: 1-Benzopyrans. $C_{22}H_{32}O_3$ Powder, mp 63–66 °C, $[\alpha]_D$ = −103° (*c* = 0.5, $CHCl_3$). Source: Marine-derived fungus *Penicillium* sp. OUPS-79 from green alga *Enteromorpha intestinalis*. Pharm: Cytotoxic (P_{388}, ED_{50} = 1.2 μg/mL). Ref: C. Takahashi, et al, Tet. Lett., 1996, 37, 655 | C. Iwamoto, et al, Tetrahedron, 1999, 55, 14353

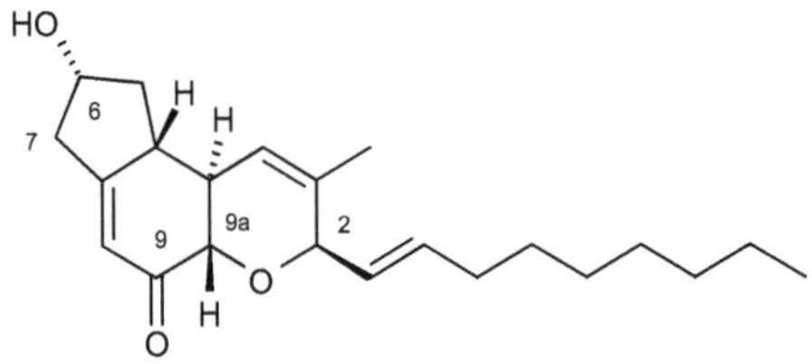

670 Penostatin C

Type: 1-Benzopyrans. $C_{22}H_{30}O_2$ Powder, mp 63–65 °C, $[\alpha]_D$ = +120° (*c* = 1, $CHCl_3$). Source: Marine-derived fungus *Penicillium* sp. OUPS-79 from green alga *Enteromorpha intestinalis*. Pharm: Cytotoxic (P_{388}, ED_{50} = 1.0 μg/mL (1.1 μg/mL); BSY1, ED_{50} = 2.0 μg/mL; MCF7, ED_{50} = 1.6 μg/mL; HCC2998, ED_{50} = 2.0 μg/mL; NCI-H522, ED_{50} = 2.5 μg/mL; DMS114, ED_{50} = 1.9 μg/mL; OVCAR-3, ED_{50} = 2.4 μg/mL; MKN1, ED_{50} = 1.7 μg/mL). Ref: C. Takahashi, et al, Tet. Lett., 1996, 37, 655 | C. Iwamoto, et al, Tetrahedron, 1999, 55, 14353

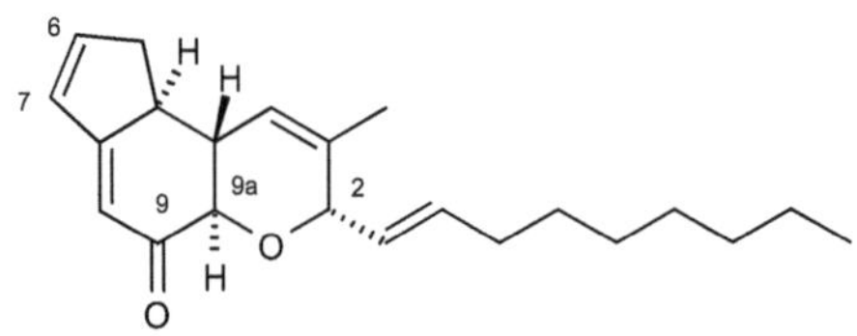

671 Penostatin D

Type: 1-Benzopyrans. $C_{22}H_{34}O_3$ Powder, mp 106–110 °C, $[\alpha]_D$ = −26.7° (*c* = 0.1, $CHCl_3$). Source: Marine-derived fungus *Penicillium* sp. OUPS-79 from green alga *Enteromorpha intestinalis*. Pharm: Cytotoxic (P_{388}, ED_{50} = 11.0 μg/mL (11.5 μg/mL)). Ref: C. Takahashi, et al, Tet. Lett., 1996, 37, 655 | C. Iwamoto, et al, Tetrahedron, 1999, 55, 14353

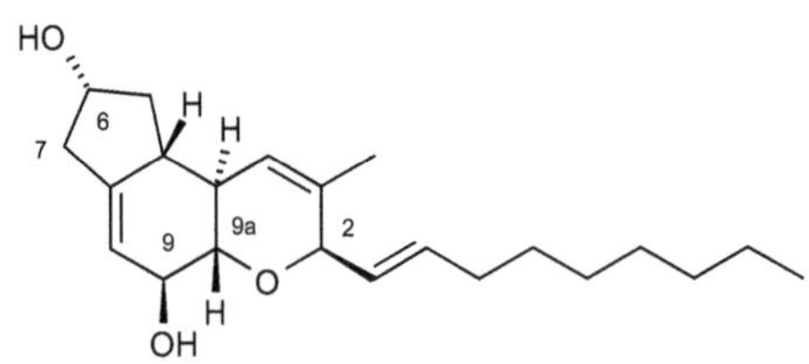

672 Pestalotiopsone F

Type: 1-Benzopyrans. $C_{17}H_{20}O_5$ Amorph. solid (MeOH). Source: Mangrove-derived fungus *Pestalotiopsis* sp. JCM2A4 from mangrove *Rhizophora mucronata* (leaves, Hainan, China). Pharm: Cytotoxic (murine cancer cell line L5178Y, EC_{50} = 8.93 μg/mL). Ref: J. Xu, et al, JNP, 2009, 72, 662

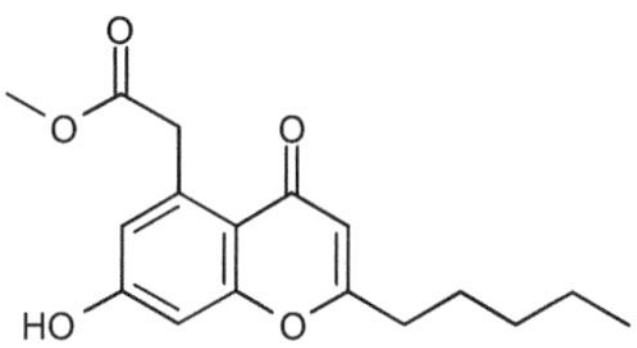

673 Sacrochromenol A sulfate

Type: 1-Benzopyrans. $C_{36}H_{52}O_5S$ Source: Sponges *Sarcotragus spinulosus* (deepwater) and *Ircinia spinosula*. Pharm: Na/K-ATPase inhibitor. Ref: V. A. Stonik, et al, JNP, 1992, 55, 1256 | L. A. Tziveleka, et al, Chem. Biodiversity, 2005, 2, 901

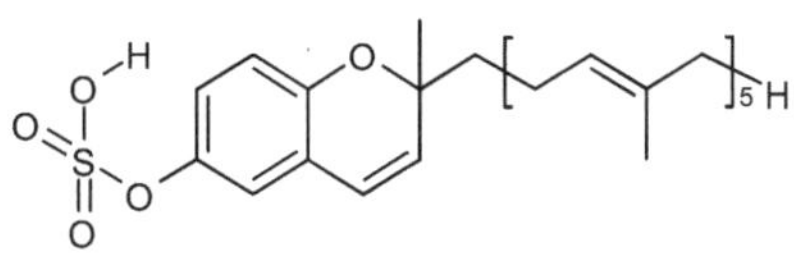

674 Sacrochromenol B sulfate

Type: 1-Benzopyrans. $C_{41}H_{60}O_5S$ Source: Sponge *Sarcotragus spinulosus* (deepwater). Pharm: Na/K-ATPase inhibitor. Ref: V. A. Stonik, et al, JNP, 1992, 55, 1256

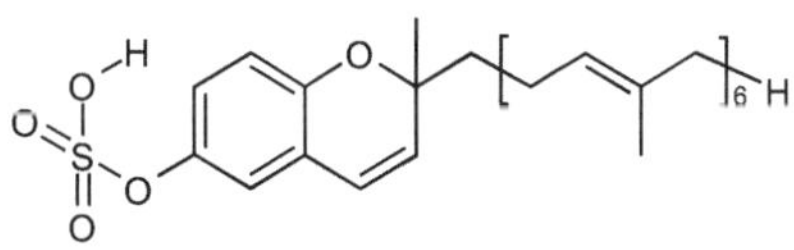

675 Sacrochromenol C sulfate

Type: 1-Benzopyrans. $C_{46}H_{68}O_5S$ Source: Sponge *Sarcotragus spinulosus* (deepwater). Pharm: Na/K-ATPase inhibitor. Ref: V. A. Stonik, et al, JNP, 1992, 55, 1256

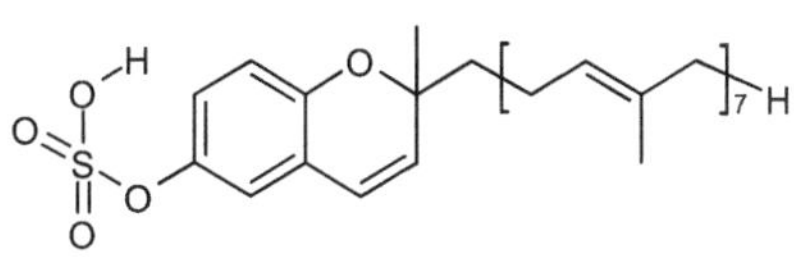

676 Trichodermatide B

Type: 1-Benzopyrans. $C_{16}H_{24}O_4$ Needles (MeOH), mp 65–68 °C, $[\alpha]_D^{20}$ = +64° (*c* = 0.05, MeOH). Source: Marine-derived fungus *Trichoderma reesei* (sediment sample, China waters). Pharm: Cytotoxic (MTT assay, hmn melanoma A375-S2 cell line, IC_{50} = 187.3 μg/mL). Ref: Y. Sun, et al, Org. Lett., 2008, 10, 393

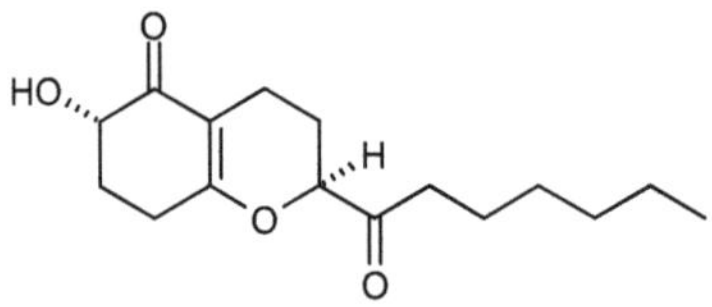

677 Trichodermatide C

2-(1-Heptenyl)-2,3,4,6,7,8-hexahydro-6-hydroxy-5*H*-1-benzopyran-5-one Type: 1-Benzopyrans. $C_{16}H_{24}O_3$ Oil, $[\alpha]_D^{20}$ = +80° (*c* = 0.02, MeOH). Source: Marine-derived fungus *Trichoderma reesei* (sediment sample, China waters). Pharm: Cytotoxic (MTT assay, hmn melanoma A375-S2 cell line, IC_{50} = 38.8 μg/mL). Ref: Y. Sun, et al, Org. Lett., 2008, 10, 393

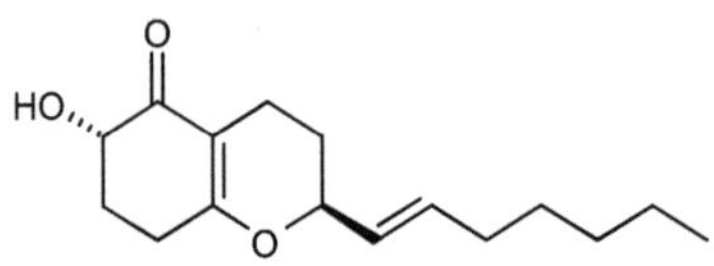

678 Trichodermatide D

Type: 1-Benzopyrans. $C_{16}H_{24}O_4$ Oil, $[\alpha]_D^{20}$ = +22.5° (*c* = 0.01, MeOH). Source: Marine-derived fungus *Trichoderma reesei* (sediment sample, China waters). Pharm: Cytotoxic (MTT assay, hmn melanoma A375-S2 cell line, IC_{50} = 222.2 μg/mL). Ref: Y. Sun, et al, Org. Lett., 2008, 10, 393

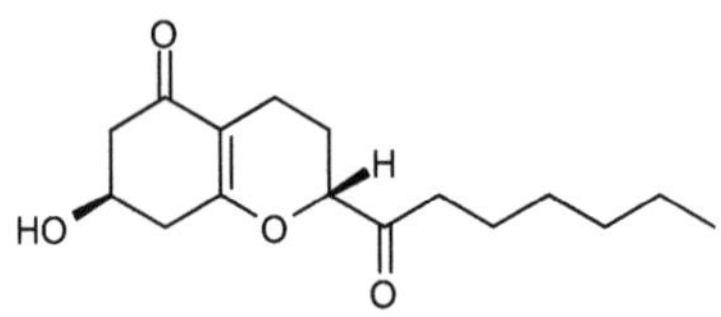

679 Xiamenmycin C

Type: 1-Benzopyrans. $C_{17}H_{23}NO_3$ Source: Marine-derived streptomycete *Streptomyces xiamenensis* (double mutant strain, sediment, East Pacific Ocean). Pharm: Cytotoxic (WI26, cell proliferation inhibitor). Ref: Z. -Y. You, et al, Mar. Drugs, 2013, 11, 4035

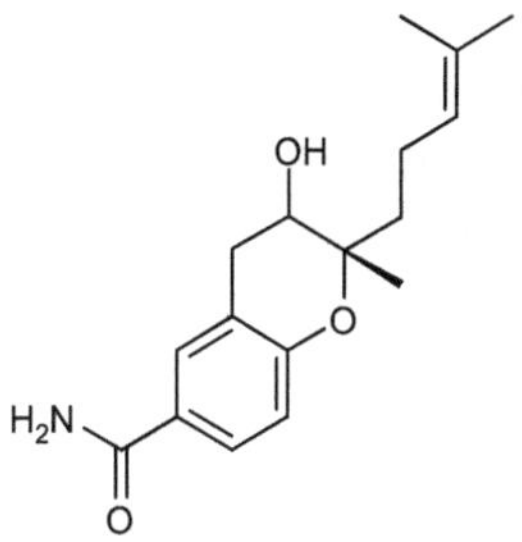

680 Xiamenmycin D

Type: 1-Benzopyrans. $C_{22}H_{31}NO_6$ Source: Marine-derived streptomycete *Streptomyces xiamenensis* (double mutant strain, sediment, East Pacific Ocean). Pharm: Cytotoxic (WI26, cell proliferation inhibitor). Ref: Z. -Y. You, et al, Mar. Drugs, 2013, 11, 4035

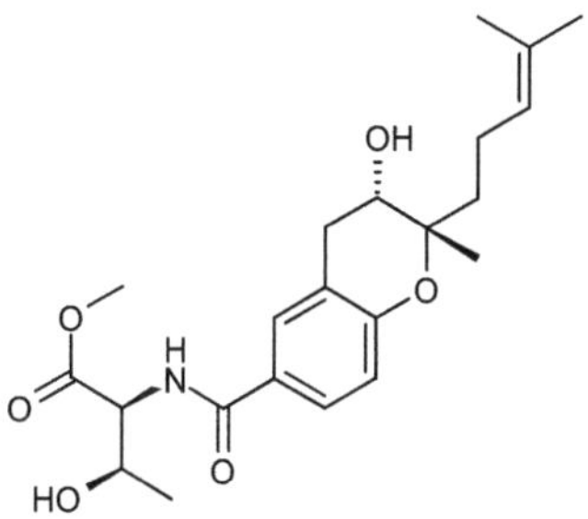

681 Anhydrofulvic acid

Type: Pyrano-1-benzopyrans. $C_{14}H_{10}O_7$ Yellow prisms (DMF aq), mp 237–242 °C. Source: Marine-derived fungus *Penicillium afacidum*. Pharm: Antifungal; collagenase inhibitor. Ref: K. -I. Fujita, et al, J. Antibiot., 1999, 52, 628

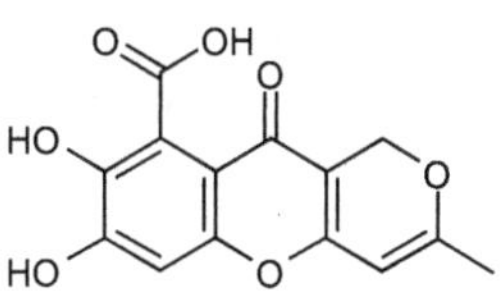

682 Antibiotic SB236050

Type: Pyrano-1-benzopyrans. $C_{15}H_{14}O_8$ Lignt yellow solid (> 95% pure). Source: Fungus *Chaetomium funicola* TCF 6040. Pharm: Metallo-*β*-lactamase inhibitor (bacterium *Bacillus cereus* II, *K*i = 88 µmol/L; bacterium *Bacteroides fragilis* CfiA, *K*i = 10 µmol/L; bacterium *Pseudomonas aeruginosa* IMP-1, *K*i = 32 µmol/L); metallo-*β*-lactamase inhibitor (metallo-*β*-lactamase of *Bacillus cereus* II, Zn^{2+} 1 µmol/L, IC_{50} = 389 µmol/L; Zn^{2+} 100 µmol/L, IC_{50} = 256 µmol/L; metallo-*β*-lactamase of *Bacteroides fragilis* CfiA, Zn^{2+} 100 µmol/L, IC_{50} = 29 µmol/L; metallo-*β*-lactamase of *Pseudomonas aeruginosa* IMP-1, Zn^{2+} 100 µmol/L, IC_{50} = 113 µmol/L;

metallo-β-lactamase of L-1, Zn^{2+} 100 μmol/L, IC_{50} > 1000 μmol/L; angiotensin-converting enzyme ACE, Zn^{2+} 1 μmol/L, IC_{50} > 1000 μmol/L); synergist of Meropenem antibacterial (*Bacteroides fragilis* 262 and *Bacteroides fragilis* 460, meropenem with no synergist, MIC = 16 μg/mL; meropenem with 8 μg/mL synergist, MIC = 2 μg/mL; meropenem with 32 μg/mL synergist, MIC = 0.5 μg/mL; *Bacteroides fragilis* 288, meropenem with no synergist, MIC = 256 μg/mL; meropenem with 8 μg/mL synergist, MIC = 32 μg/mL; meropenem with 32 μg/mL synergist, MIC = 8 μg/mL; *Stenotrophomonas maltophilia* 511, meropenem with no synergist, MIC = 128 μg/mL; meropenem with 32 μg/mL synergist, MIC = 128 μg/mL; *Pseudomonas aeruginosa* 101, meropenem with no synergist, MIC = 512 μg/mL; meropenem with 32 μg/mL synergist, MIC = 512 μg/mL). Ref: D. J. Payne, et al, Antimicrob. Agents Chemother., 2002, 46, 1880

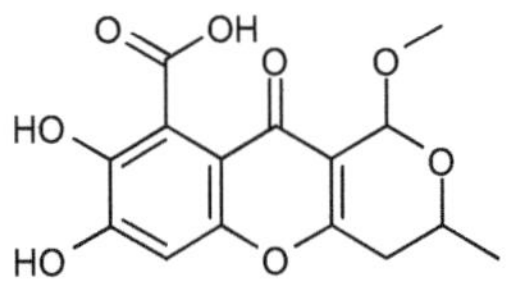

683 Antibiotic SB238569

Type: Pyrano-1-benzopyrans. $C_{15}H_{12}O_8$ Yellow solid (> 90% pure). Source: Fungus *Chaetomium funicola* TCF 6040. Pharm: Metallo-β-lactamase inhibitor (bacterium *Bacillus cereus* II, *Ki* = 79 μmol/L; bacterium *Bacteroides fragilis* CfiA, *Ki* = 3.4 μmol/L; bacterium *Pseudomonas aeruginosa* IMP-1, *Ki* = 17 μmol/L); metallo-β-lactamase inhibitor (metallo-β-lactamase of *Bacillus cereus*II, Zn^{2+} 1 μmol/L, IC_{50} = 13 μmol/L, Zn^{2+} 100 μmol/L, IC_{50} = 19 μmol/L; metallo-β-lactamase of *Bacteroides fragilis* CfiA, Zn^{2+} 100 μmol/L, IC_{50} = 7 μmol/L; metallo-β-lactamase of *Pseudomonas aeruginosa* IMP-1, Zn^{2+} 100 μmol/L, IC_{50} = 26 μmol/L; metallo-β-lactamase of L-1, Zn^{2+} 100 μmol/L, IC_{50} > 1000 μmol/L; angiotensin-converting enzyme ACE, Zn^{2+} 1 μmol/L, IC_{50} > 1000 μmol/L); synergist of Meropenem antibacterial (*Bacteroides fragilis* 262 and *Bacteroides fragilis* 460, meropenem with no synergist, MIC = 16 μg/mL; meropenem with 8 μg/mL synergist, MIC = 4 μg/mL; meropenem with 32 μg/mL synergist, MIC = 1 μg/mL; *Bacteroides fragilis* 288, meropenem with no synergist, MIC = 256 μg/mL; meropenem with 8 μg/mL synergist, MIC = 64 μg/mL; meropenem with 32 μg/mL synergist, MIC = 16 μg/mL; *Stenotrophomonas maltophilia* 511, meropenem with no synergist, MIC = 128 μg/mL; meropenem with 32 μg/mL synergist, MIC = 128 μg/mL; *Pseudomonas aeruginosa* 101, meropenem with no synergist, MIC = 512 μg/mL; meropenem with 32 μg/mL synergist, MIC = 512 μg/mL). Ref: D. J. Payne, et al, Antimicrob. Agents Chemother., 2002, 46, 1880

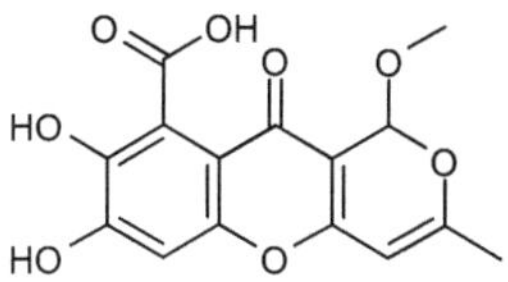

684 Chaetocyclinone A

Type: Pyrano-1-benzopyrans. $C_{17}H_{16}O_8$ Yellow solid. Source: Marine-derived fungus *Chaetomium* sp. Gö 100/2 from an unidentified marine alga. Pharm: Antifungal (inhibits phytopathogenic fungi *Phytophthora infestans* selectly, 31 ppm, growth InRt = 90%). Ref: S. Lösgen, et al, EurJOC, 2007, 2191

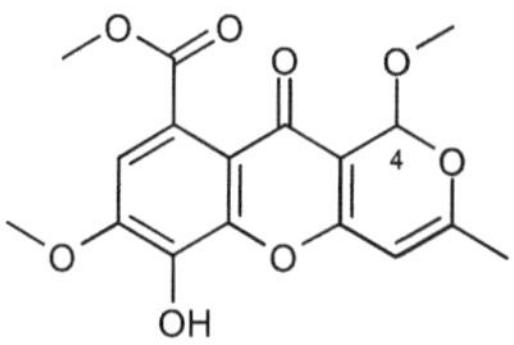

685 Bicoumanigrin

Type: Comarinoids. $C_{22}H_{18}O_8$ Amorph. powder. Source: Marine-derived fungus *Aspergillus niger* from sponge *Axinella damicornis*. Pharm: Cytotoxic (moderate). Ref: J. Hiort, et al, JNP, 2004, 67, 1532

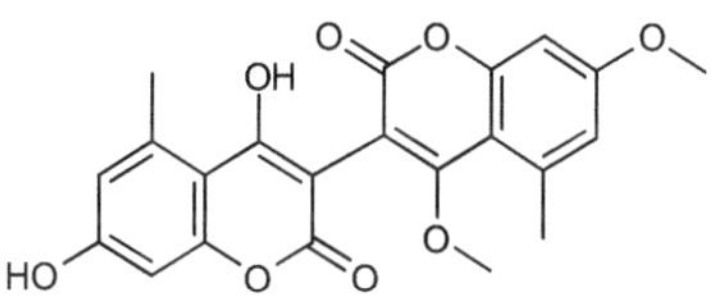

686 Esculetin-4-carboxylic acid ethyl ester

Type: Comarinoids. $C_{12}H_{10}O_6$ Amorph. green solid. Source: Sponge *Axinella* cf. *corrugata*. Pharm: Antiviral (SARS-Corona virus viral protease 3CL, IC_{50} = 46 μmol/L); SARS-coronovirus 3CL protease inhibitor. Ref: S. P. de Lira, et al, J. Braz. Chem. Soc., 007,18, 440

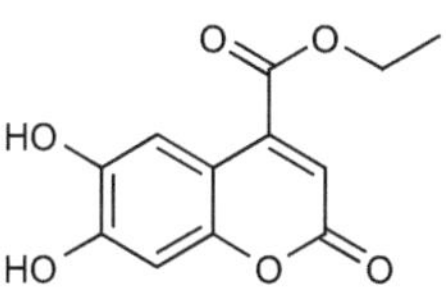

687 Ascochitin

Type: 2-Benzopyrans. $C_{15}H_{16}O_5$ Yellow cryst., mp 196–198 °C, $[\alpha]_D^{25}$ = −86° ($CHCl_3$). Source: Marine-derived fungus *Ascochyta salicorniae* from green alga *Ulva* sp. (North Sea coast, Germany), terrestrial fungi (*Ascochyta* spp.). Pharm: PPs inhibitors

(MPtpB (mycobacterial protein tyrosine phosphatase B), IC_{50} = 11.5 μmol/L; Cdc25a, IC_{50} = 69 μmol/L; PTP1B, IC_{50} = 38.5 μmol/L; VHR, MPtpA and VE-PTP, all inactive); antibacterial (gram-positive and -negative bacteria). Ref: L. Colombo, et al, JCS Perkin I, 1980, 675 | S. F. Seibert, et al, Org. Biomol. Chem., 2006, 4, 2233

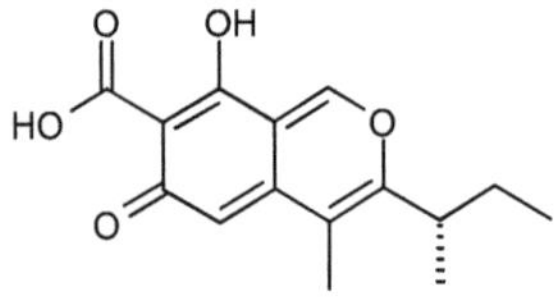

688 Asperbiphenyl

Type: 2-Benzopyrans. $C_{24}H_{30}O_6$ Amorph. brown powder, $[\alpha]_D^{21}$ = +0.1° (*c* = 0.26, MeOH). Source: Marine-derived fungus *Aspergillus* sp. MF-93 (seawater, Quang-Zhou Gulf, China). Pharm: Antiviral (inhibits multiplication of tobacco mosaic virus (TMV), 0.2mg/mL, InRt = 35.5%). Ref: Z. -J. Wu, et al, Pest Manag. Sci., 2009, 65, 60

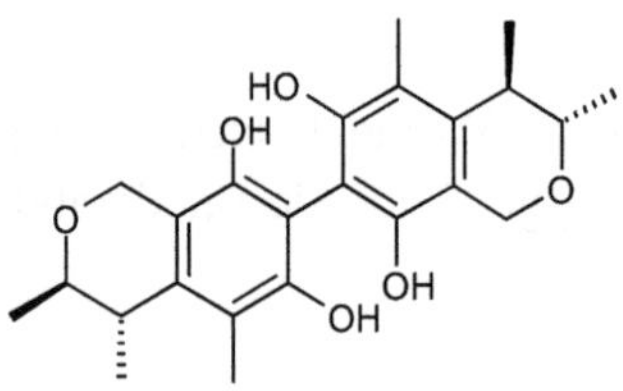

689 Aspergilone A

Type: 2-Benzopyrans. $C_{26}H_{26}O_3$ Colorless solid. Source: Marine-derived fungus *Aspergillus* sp. from gorgonian *Dichotella gemmacea* (South China Sea). Pharm: Cytotoxic (HL60, IC_{50} = 3.2 μg/mL; MCF7, IC_{50} = 25.0 μg/mL; A549, IC_{50} = 37.0 μg/mL); Antifoulant (barnacle *Balanus amphitrite,* at nontoxic concentration, EC_{50} = 7.68 μg/mL, settlement inhibitor, potent). Ref: C. -L. Shao, et al, BoMCL, 2011, 21, 690

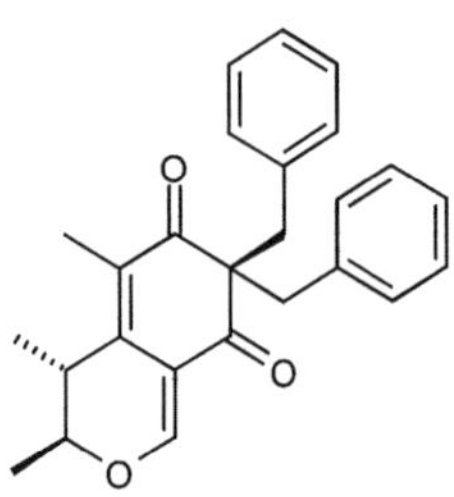

690 7-*epi*-Austdiol

Type: 2-Benzopyrans. $C_{12}H_{12}O_5$ Cryst. (EtOAc/MeOH), mp 174–176 °C, $[\alpha]_D^{20}$ = +116.1° (*c* = 0.1, MeOH). Source: Mangrove-derived fungus *Talaromyces* sp. from mangrove

Kandelia candel (stem bark, Qi'ao I., Zhuhai, China). Pharm: Cytotoxic (KB and KBV200, moderate). Ref: F. Liu, et al, Planta Med., 2010, 76, 185

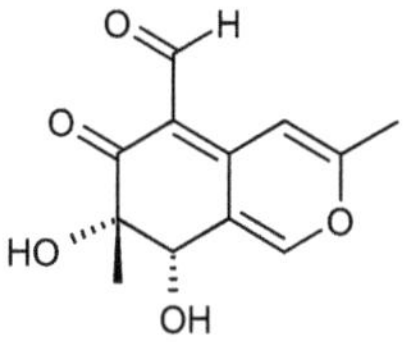

691 Citrinin

Monascidin A Type: 2-Benzopyrans. $C_{13}H_{14}O_5$ Lemon-yellow needles (EtOH or MeOH), mp 178–179 °C, $[\alpha]_D^{18}$ = −37.4° (EtOH). Source: Marine-derived fungus *Microsporum* sp. from red alga *Lomentaria catenata* (Korea, surface). Pharm: Antioxidant (DPPH scavenger, moderate); phytotoxic; cytotoxin; severe skin irritant; LD_{50} (mus, orl) = 112 mg/kg. Ref: Y. Li, et al, CPB, 2006, 54, 882

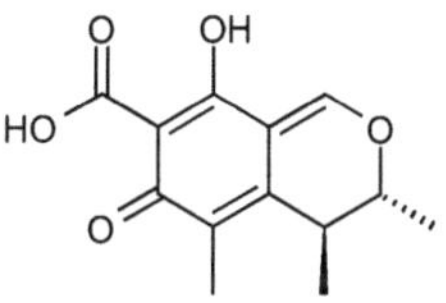

692 Cytosporone C

1-Heptyl-1,4-dihydro-6,8-dihydroxy-3*H*-2-benzopyran-3-one Type: 2-Benzopyrans. $C_{16}H_{22}O_4$ Cryst. (Me_2CO), $[\alpha]_D^{20}$ = +11° (*c* = 0.1, MeOH). Source: Mangrove-derived fungus *Leucostoma persoonii* from mangrove *Rhizophora mangle* (branch, Everglades, Florida). Pharm: Antibacterial (MRSA, MIC > 358 μmol/L; A549 IC_{50} = 690 μmol/L, IC_{90} = 840 μmol/L). Ref: S. F. Brady, et al, Org. Lett., 2000, 2, 4043 | M. P. Singh, et al, Mar. Drugs, 2007, 5, 71 | J. Xu, et al, BoMC, 2009, 17, 7362 | J. Beau, et al, Mar. Drugs, 2012, 10, 762

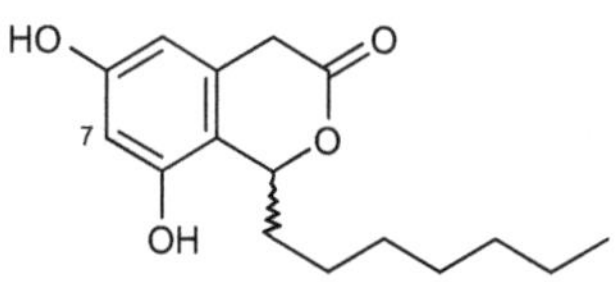

693 6-Desmethoxyhormothamnione

Type: 2-Benzopyrans. $C_{20}H_{18}O_7$ Light yellow solid. Source: Brown alga *Chrysophaeum taylorii*. Pharm: Cytotoxic (9KB, ID_{50} = 1 μg/mL). Ref: W. H. Gerwick, et al, Tet. Lett., 1986, 27, 1979 | W. H. Gerwick, JNP, 1989, 52, 252

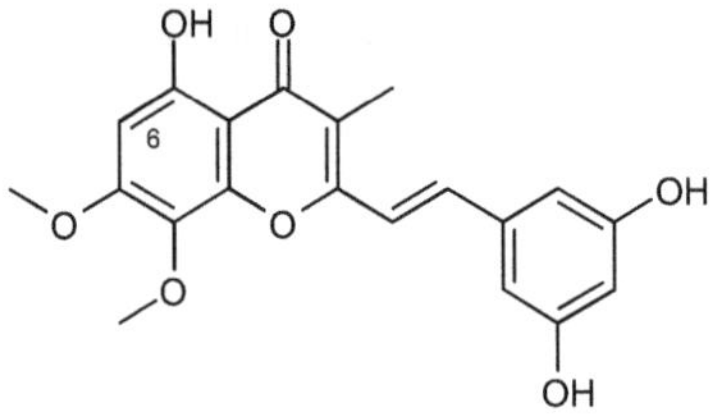

694 Dothiorelone D

Type: 2-Benzopyrans. $C_{15}H_{20}O_5$ Source: Mangrove-derived fungus *Dothiorella* sp. Pharm: Cytotoxic. Ref: Q. Yu, et al, CA, 2004, 142, 388784y

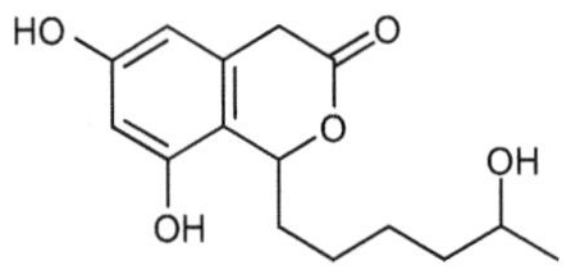

695 Hormothamnione

Type: 2-Benzopyrans. $C_{21}H_{20}O_8$ Yellow solid, mp 270 °C (dec). Source: Brown algae *Chrysophaeum taylorii* and *Hormothamnion enteromorphoides* (Puerto Rico). Pharm: Cytotoxic; RNA synthesis inhibitor. Ref: W. H. Gerwick, et al, Tet. Lett., 1986, 27, 1979 | W. H. Gerwick, JNP, 1989, 52, 252

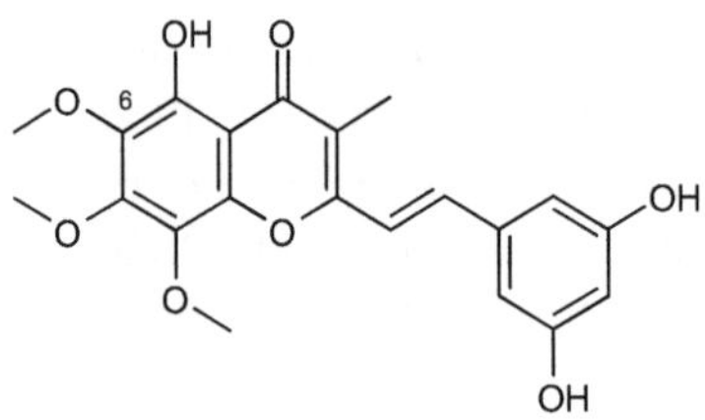

696 Isochromophilone XI

Type: 2-Benzopyrans. $C_{25}H_{29}ClO_8$ Source: Sponge *Tethya aurantium*, marine fungus *Bartalinia robillardoides* strain LF550. Pharm: Antibacterial (*Bacillus subtilis*, MIC = 55.6 μmol/L, *Staphylococcus lentus*, MIC = 78.4 μmol/L); antifungal (*Trichophyton rubrum*, MIC = 41.5 μmol/L). Ref: J. Wiese, et al, Mar. Drugs, 2011, 9, 561 | N. Jansen, et al, Mar. Drugs, 2013, 11, 800

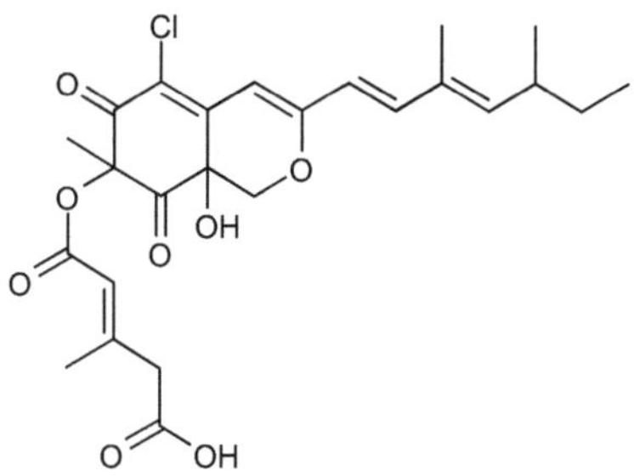

697 8-*O*-Methyl-*epi*-austdiol

Type: 2-Benzopyrans. $C_{13}H_{14}O_5$ Yellow powder, mp 160–161 °C, $[\alpha]_D^{20}$ = +115.5° (*c* = 0.1, MeOH). Source: Mangrove-derived fungus *Talaromyces* sp. ZH-154 from mangrove *Kandelia candel* (stem bark, Qi'ao I., Zhuhai, China). Pharm: Cytotoxic (KB and KBV200, moderate). Ref: F. Liu, et al, Planta Med., 2010, 76, 185

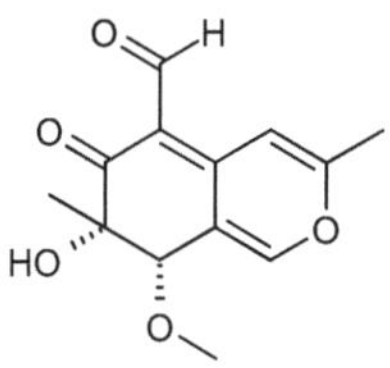

698 Peneciraistin A

Type: 2-Benzopyrans. $C_{15}H_{20}O_5$ Source: Marine-derived fungus *Penicillium raistrickii* (saline soil, Bohai Bay, Shandong, China). Pharm: Antioxidant (DPPH scavenger). Ref: L. -Y. Ma, et al, Tetrahedron, 2012, 68, 2276

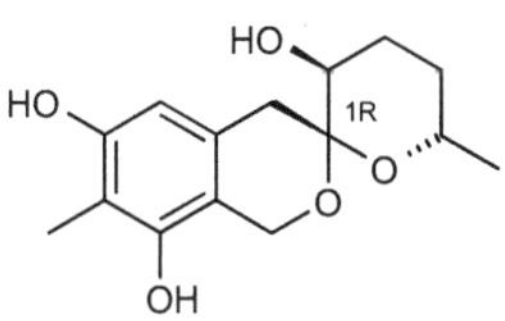

699 Peneciraistin B

Type: 2-Benzopyrans. $C_{15}H_{20}O_5$ Source: Marine-derived fungus *Penicillium raistrickii* (saline soil, Bohai Bay, Shandong, China). Pharm: Antioxidant (DPPH scavenger). Ref: L. -Y. Ma, et al, Tetrahedron, 2012, 68, 2276

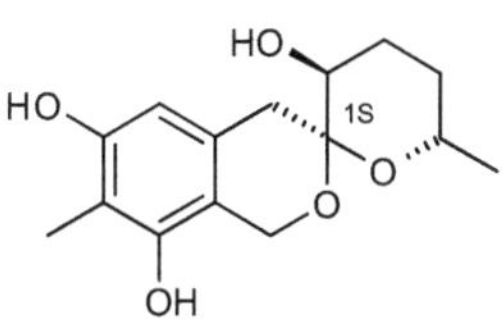

700 Peneciraistin C

Type: 2-Benzopyrans. $C_{19}H_{26}O_9$ Source: Marine-derived fungus *Penicillium raistrickii* (saline soil, Bohai Bay, Shandong, China). Pharm: Cytotoxic (A549 and MCF7, moderate). Ref: L. -Y. Ma, et al, Tetrahedron, 2012, 68, 2276

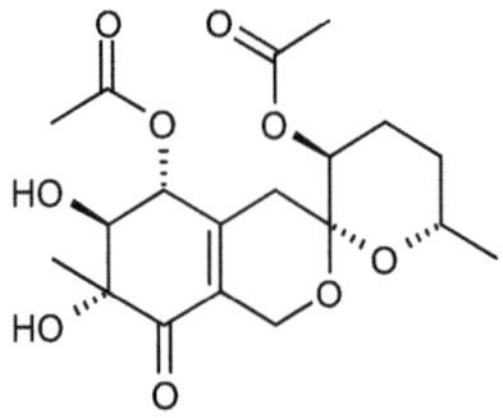

701 Peneciraistin D

Type: 2-Benzopyrans. $C_{15}H_{16}O_6$ Source: Marine-derived fungus *Penicillium raistrickii* (saline soil, Bohai Bay, Shandong, China). Pharm: Antioxidant (DPPH scavenger). Ref: L. -Y. Ma, et al, Tetrahedron, 2012, 68, 2276

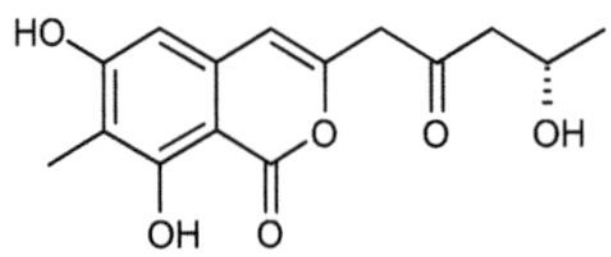

702 Penicillanthranin A

Type: 2-Benzopyrans. $C_{28}H_{24}O_{10}$ Source: Marine-derived fungus *Penicillium citrinum* from gorgonian sea fan *Annella* sp. (Similan Is., Phangnga, Thailand). Pharm: Antibacterial (*Staphylococcus aureus* and MRSA). Ref: N. Khamthong, et al, Tetrahedron, 2012, 68, 8245

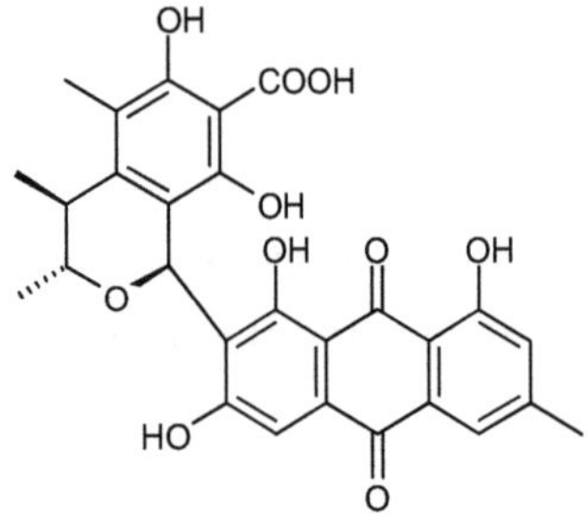

703 Penicitrinol C

Type: 2-Benzopyrans. $C_{14}H_{18}O_4$ Source: Marine-derived fungus *Penicillium citrinum* (sediment, Lanqi I., Fujian, China). Pharm: Cytotoxic (HL60, weak). Ref: L. Chen, et al, CPB, 2011, 59, 515

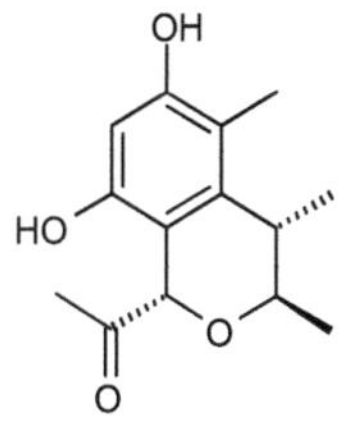

704 Pestalachloride C

Type: 2-Benzopyrans. $C_{21}H_{20}Cl_2O5$ Source: Marine-derived fungus *Pestalotiopsis* sp. ZJ-2009-7-6 from a soft coral. Pharm: Antibacterial (*Enterobacter coli*, MIC = 5.0 μmol/L, *Vibrio anguillarum*, MIC = 10.0 μmol/L, *Vibrio parahaemolyticus*, MIC = 20.0 μmol/L). Ref: M. Y. Wei, et al, Mar. Drugs, 2013, 11, 1050

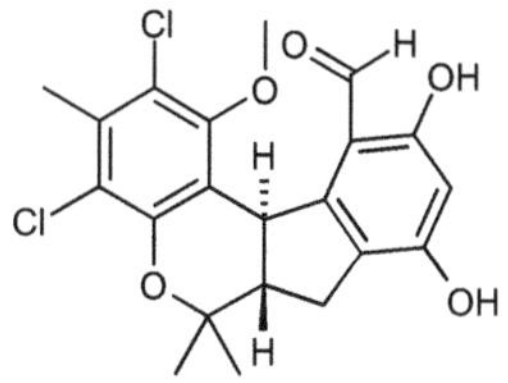

705 Pestalachloride D

Type: 2-Benzopyrans. $C_{21}H_{20}Cl_2O5$ Source: Marine-derived fungus *Pestalotiopsis* sp. ZJ-2009-7-6 from a soft coral. Pharm: Antibacterial (*Enterobacter coli*, MIC = 5.0 μmol/L, *Vibrio anguillarum*, MIC = 10.0 μmol/L, *Vibrio parahaemolyticus*, MIC = 20.0 μmol/L). Ref: M. Y. Wei, et al, Mar. Drugs, 2013, 11, 1050

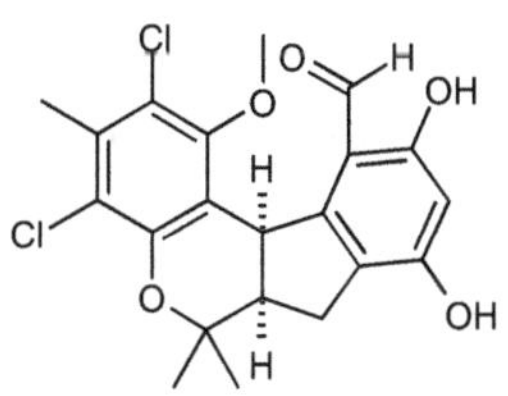

706 Pinophilin A

Type: 2-Benzopyrans. $C_{21}H_{22}O_7$ Source: Marine-derived fungus *Penicillium pinophilum* from green alga *Ulva fasciata* (Kasai, Tokyo, Japan). Pharm: Mammalian DNA polymerases inhibitor (A (pol *γ*), B (pols α, *δ*, and *ε*), and Y (pols *η*, *ι*, and *κ*) families, selective); cytotoxic (HTCLs cells, growth and proliferation inhibitor). Ref: Y. Myobatake, et al, JNP, 2012, 75, 135

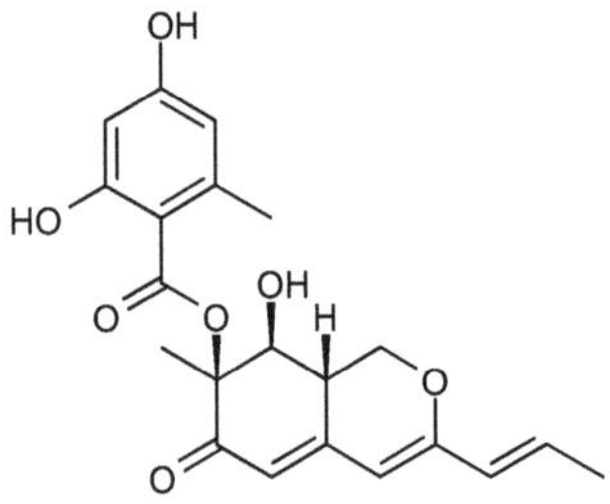

707 Pinophilin B

Type: 2-Benzopyrans. $C_{21}H_{22}O_8$ Source: Marine-derived fungus *Penicillium pinophilum* from green alga *Ulva fasciata* (Kasai, Tokyo, Japan). Pharm: Mammalian DNA polymerases inhibitor (A (pol *y*), B (pols α, *δ*, and *ε*), and Y (pols *η*, *ι*, and *κ*) families, selective); cytotoxic (HTCLs cells, growth and proliferation inhibitor). Ref: Y. Myobatake, et al, JNP, 2012, 75, 135

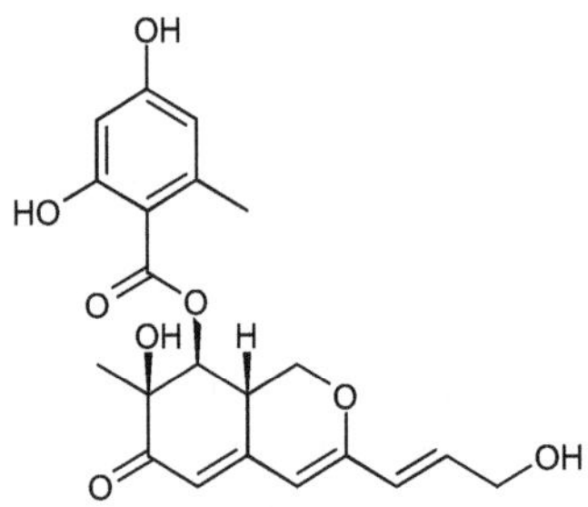

708 (3*R*,4*S*)-3,4,5-Trimethyisochroman-6,8-diol

Type: 2-Benzopyrans. $C_{12}H_{16}O_3$ Source: Deep-sea fungus *Aspergillus sydowi* YH11-2 (Guam, depth of 1000 m, E144°43′ N13°26′). Pharm: Cytotoxic (P_{388}, IC_{50} = 1.95 μmol/L, control CDDP, IC_{50} = 0.039 μmol/L). Ref: L. Tian, et al, Arch. Pharm. Res,. 2007, 30, 1051

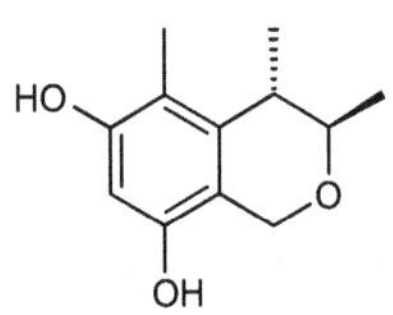

709 Chaetomugilin A

Type: Azaphilone 2-benzopyrans. $C_{23}H_{27}ClO_7$ Yellow powder ($CHCl_3$/MeOH), mp 149–151 °C, $[\alpha]_D$ = −50.1° (*c* = 0.16, EtOH). Source: Marine-derived fungus *Chaetomium globosum* OUPS-T106B-6 from fish *Mugil cephalus* (Katsuura Bay, Japan), terrestrial fungus (*Chaetomium globosum*). Pharm: Cytotoxic (P_{388}, IC_{50} = 8.7 μmol/L, control 5-FU, IC_{50} = 1.7 μmol/L; HL60, IC_{50} = 7.3 μmol/L, 5-FU, IC_{50} = 2.7 μmol/L);

cytotoxic (panel of 39 hmn cancer cell lines: HBC4, Log_{10} GI_{50} (mol/L) = −4.00; BSY1, Log_{10} GI_{50} (mol/L) = −4.00; HBC5, Log_{10} GI_{50} (mol/L) = −4.00; MCF7, Log_{10} GI_{50} (mol/L) = −4.43; MDA-MB-231, Log_{10} GI_{50} (mol/L) = −4.00; U251, Log_{10} GI_{50} (mol/L) = −4.00; SF268, Log_{10} GI_{50} (mol/L) = −4.00; SF295, Log_{10} GI_{50} (mol/L) = −4.00; SF539, Log_{10} GI_{50} (mol/L) = −4.00; SNB75, Log_{10} GI_{50} (mol/L) = −4.36; SNB78, Log_{10} GI_{50} (mol/L) = −4.00; HCC2998, Log_{10} GI_{50} (mol/L) = −4.00; KM12, Log_{10} GI_{50} (mol/L) = −4.00; HT29, Log_{10} GI_{50} (mol/L) = −4.00; HCT15, Log_{10} GI_{50} (mol/L) = −4.00; HCT116, Log_{10} GI_{50} (mol/L) = −4.00; NCI-H23, Log_{10} GI_{50} (mol/L) = −4.00; NCI-H226, Log_{10} GI_{50} (mol/L) = −4.18; NCI-H522, Log_{10} GI_{50} (mol/L) = −4.36; NCI-H460, Log_{10} GI_{50} (mol/L) = −4.00; A549, Log_{10} GI_{50} (mol/L) = −4.00; DMS273, Log_{10} GI_{50} (mol/L) = −4.00; DMS114, Log_{10} GI_{50} (mol/L) = −4.06; LOX-IMVI, Log_{10} GI_{50} (mol/L) = −4.15; OVCAR-3, Log_{10} GI_{50} (mol/L) = −4.00; OVCAR-4, Log_{10} GI_{50} (mol/L) = −4.00; OVCAR-5, Log_{10} GI_{50} (mol/L) = −4.00; OVCAR-8, Log_{10} GI_{50} (mol/L) = −4.00; SK-OV-3, Log_{10} GI_{50} (mol/L) = −4.00; RXF-631L, Log_{10} GI_{50} (mol/L) = −4.63; ACHN, Log_{10} GI_{50} (mol/L) = −5.51; St4, Log_{10} GI_{50} (mol/L) = −4.00; MKN1, Log_{10} GI_{50} (mol/L) = −4.00; MKN7, Log_{10} GI_{50} (mol/L) = −4.00; MKN28, Log_{10} GI_{50} (mol/L) = −4.73; MKN45, Log_{10} GI_{50} (mol/L) = −4.51; MKN74, Log_{10} GI_{50} (mol/L) = −5.24; DU145, Log_{10} GI_{50} (mol/L) = −4.00; PC3, Log_{10} GI_{50} (mol/L) = −4.00; MG-MID (mean value of Log_{10} GI_{50} (mol/L) for all cell lines tested) = −4.10; Δ (difference in Log_{10} GI_{50} (mol/L) value of most sensitive cell and MG-MID value) = 1.13; Range (difference in Log_{10} GI_{50} (mol/L) value of most sensitive cell and least sensitive cell) = 1.24 (effective value of selective cytotoxic activity: MG-MID <-5, Δ ≥ 0.5, and range ≥ 1.0; COMPARE program suggested a novel mechanism of action). Ref: T. Yamada, et al, Tet. Lett., 2008, 49, 4192 | Y. Muroga, et al, J. Antibiot., 2008, 61, 615

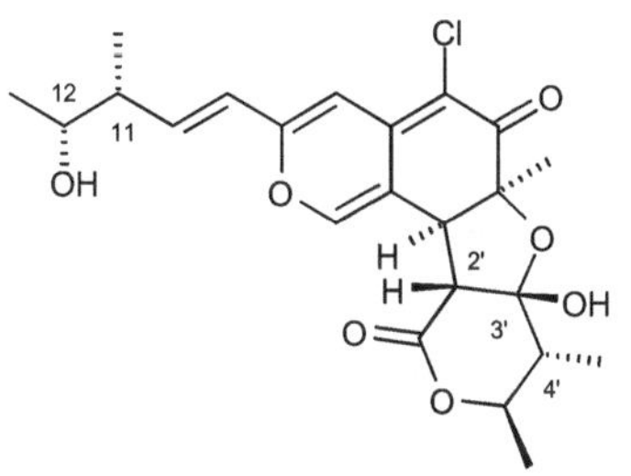

710 4′-*epi*-Chaetomugilin A

Type: Azaphilone 2-benzopyrans. $C_{23}H_{27}ClO_7$ Yellow powder, mp 135–137 °C, $[\alpha]_D^{22}$ = −262.5° (c = 0.12, EtOH). Source: Marine-derived fungus *Chaetomium globosum* OUPS-T106B-6 from fish *Mugil cephalus* (Katsuura Bay, Japan). Pharm: Cytotoxic (P_{388}, HL60, L_{1210} and KB, all IC_{50} > 100 µmol/L). Ref: Y. Muroga, et al, Helv. Chim. Acta, 2010, 93, 542

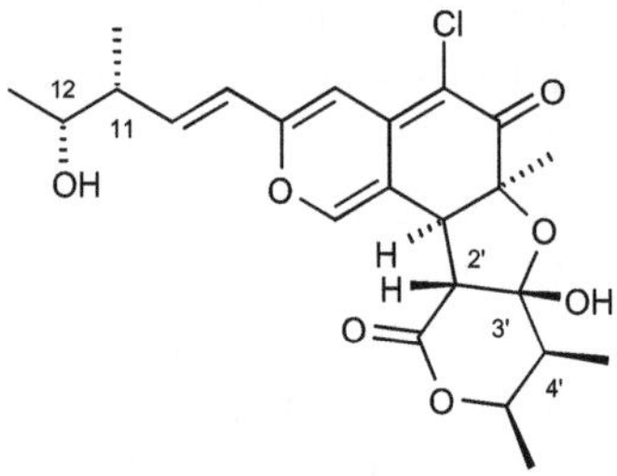

711 11-*epi*-Chaetomugilin A

Type: Azaphilone 2-benzopyrans. $C_{23}H_{27}ClO_7$ Yellow powder, mp 127–129 °C, $[\alpha]_D^{22}$ = −185.1° (*c* = 0.18, EtOH). Source: Marine-derived fungus *Chaetomium globosum* OUPS-T106B-6 from fish *Mugil cephalus* (Katsuura Bay, Japan). Pharm: Cytotoxic (P_{388}, IC_{50} = 88.9 µmol/L, control 5-FU, IC_{50} = 1.7 µmol/L; HL60, IC_{50} = 66.7 µmol/L, 5-FU, IC_{50} = 2.7 µmol/L; L_{1210}, IC_{50} > 100 µmol/L, 5-FU, IC_{50} = 1.1 µmol/L; KB, IC_{50} > 100 µmol/L, 5-FU, IC_{50} = 7.7 µmol/L). Ref: Y. Muroga, et al, Helv. Chim. Acta, 2010, 93, 542

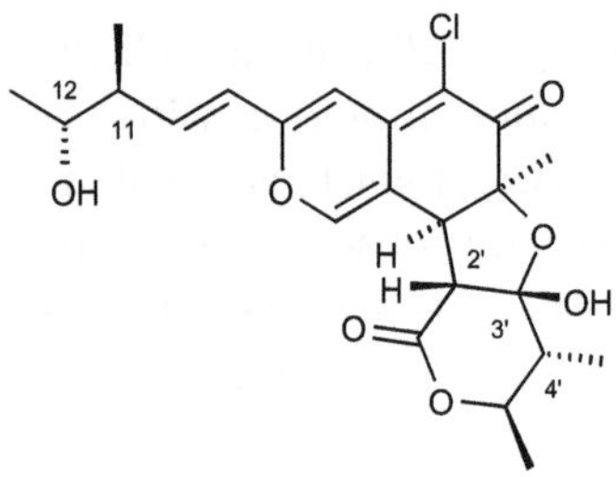

712 Chaetomugilin B

Type: Azaphilone 2-benzopyrans. $C_{24}H_{29}ClO_7$ Yellow prisms ($CHCl_3$/MeOH), mp 215–217 °C, $[\alpha]_D$ = −40.2° (*c* = 0.03, EtOH). Source: Marine-derived fungus *Chaetomium globosum* OUPS-T106B-6 from fish *Mugil cephalus* (Katsuura Bay, Japan). Pharm: Cytotoxic (P_{388}, IC_{50} = 18.7 µmol/L, control 5-FU, IC_{50} = 1.7 µmol/L; HL60, IC_{50} = 16.5 µmol/L, 5-FU, IC_{50} = 2.7 µmol/L); cytotoxic selective (panel of 39 hmn cancer cell lines: HBC4, Log_{10} GI_{50} (mol/L) = −4.22; BSY1, Log_{10} GI_{50} (mol/L) = −4.64; HBC5, Log_{10} GI_{50} (mol/L) = −4.56; MCF7, Log_{10} GI_{50} (mol/L) = −4.16; MDA-MB-231, Log_{10} GI_{50} (mol/L) = −4.51; U251, Log_{10} GI_{50} (mol/L) = −4.00; SF268, Log_{10} GI_{50} (mol/L) = −4.05; SF295, Log_{10} GI_{50} (mol/L) = −4.00; SF539, Log_{10} GI_{50} (mol/L) = −4.25; SNB75, Log_{10} GI_{50} (mol/L) = −4.21; SNB78, Log_{10} GI_{50} (mol/L) = −4.15; HCC2998, Log_{10} GI_{50} (mol/L) = −4.00; KM12, Log_{10} GI_{50} (mol/L) = −4.08; HT29, Log_{10} GI_{50} (mol/L) = −4.01; HCT15, Log_{10} GI_{50} (mol/L) = −4.14; HCT116, Log_{10} GI_{50} (mol/L) = −4.31; NCI-H23, Log_{10} GI_{50} (mol/L) = −4.00; NCI-H226, Log_{10} GI_{50} (mol/L) = −4.00; NCI-H522, Log_{10} GI_{50} (mol/L) = −4.81; NCI-H460, Log_{10} GI_{50} (mol/L) = −4.00; A549, Log_{10} GI_{50} (mol/L) = −4.00; DMS273, Log_{10} GI_{50} (mol/L) = −4.00;

DMS114, Log_{10} GI_{50} (mol/L) = −4.11; LOX-IMVI, Log_{10} GI_{50} (mol/L) = −4.24; OVCAR-3, Log_{10} GI_{50} (mol/L) = −4.61; OVCAR-4, Log_{10} GI_{50} (mol/L) = −4.18; OVCAR-5, Log_{10} GI_{50} (mol/L) = −4.13; OVCAR-8, Log_{10} GI_{50} (mol/L) = −4.00; SK-OV-3, Log_{10} GI_{50} (mol/L) = −4.00; RXF-631L, Log_{10} GI_{50} (mol/L) = −4.00; ACHN, Log_{10} GI_{50} (mol/L) = −4.19; St4, Log_{10} GI_{50} (mol/L) = −4.00; MKN1, Log_{10} GI_{50} (mol/L) = −4.52; MKN7, Log_{10} GI_{50} (mol/L) = −4.35; MKN28, Log_{10} GI_{50} (mol/L) = −4.22; MKN45, Log_{10} GI_{50} (mol/L) = −4.00; MKN74, Log_{10} GI_{50} (mol/L) = −4.34; DU145, Log_{10} GI_{50} (mol/L) = −4.00; PC3, Log_{10} GI_{50} (mol/L) = −4.41; MG-MID (mean value of Log_{10} GI_{50} (mol/L) for all cell lines tested) = −4.19; Δ (difference in Log_{10} GI_{50} (mol/L) value of most sensitive cell and MG-MID value) = 0.62; Range (difference in Log_{10} GI_{50} (mol/L) value of most sensitive cell and least sensitive cell) = 0.81 (effective value of selective cytotoxic activity: MG-MID <-5, Δ ≥ 0.5, and range ≥ 1.0). Ref: T. Yamada, et al, Tet. Lett., 2008, 49, 4192 | Y. Muroga, et al, J. Antibiot., 2008, 61, 615

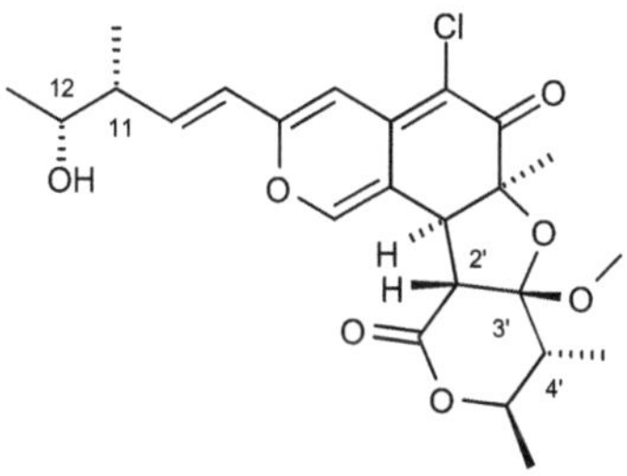

713 Chaetomugilin C

Type: Azaphilone 2-benzopyrans. $C_{23}H_{25}ClO_6$ Yellow oil, $[\alpha]_D$ = +103.3° (c = 0.04, EtOH). Source: Marine-derived fungus *Chaetomium globosum* OUPS-T106B-6 from fish *Mugil cephalus* (Katsuura Bay, Japan). Pharm: Cytotoxic (P_{388}, IC_{50} = 3.6 μmol/L, control 5-FU, IC_{50} = 1.7 μmol/L; HL60, IC_{50} = 2.7 μmol/L, 5-FU, IC_{50} = 2.7 μmol/L); cytotoxic selective (panel of 39 hmn cancer cell lines: HBC4, Log_{10} GI_{50} (mol/L) = −4.82; BSY1, Log_{10} GI_{50} (mol/L) = −5.54; HBC5, Log_{10} GI_{50} (mol/L) = −4.56; MCF7, Log_{10} GI_{50} (mol/L) = −5.00; MDA-MB-231, Log_{10} GI_{50} (mol/L) = −4.48; U251, Log_{10} GI_{50} (mol/L) = −4.78; SF268, Log_{10} GI_{50} (mol/L) = −4.63; SF295, Log_{10} GI_{50} (mol/L) = −4.89; SF539, Log_{10} GI_{50} (mol/L) = −5.37; SNB75, Log_{10} GI_{50} (mol/L) = −5.43; SNB78, Log_{10} GI_{50} (mol/L) = −4.70; HCC2998, Log_{10} GI_{50} (mol/L) = −4.78; KM12, Log_{10} GI_{50} (mol/L) = −4.68; HT29, Log_{10} GI_{50} (mol/L) = −4.51; HCT15, Log_{10} GI_{50} (mol/L) = −4.65; HCT116, Log_{10} GI_{50} (mol/L) = −5.33; NCI-H23, Log_{10} GI_{50} (mol/L) = −4.51; NCI-H226, Log_{10} GI_{50} (mol/L) = −4.51; NCI-H522, Log_{10} GI_{50} (mol/L) = −5.38; NCI-H460, Log_{10} GI_{50} (mol/L) = −4.68; A549, Log_{10} GI_{50} (mol/L) = −4.71; DMS273, Log_{10} GI_{50} (mol/L) = −5.30; DMS114, Log_{10} GI_{50} (mol/L) = −4.79; LOX-IMVI, Log_{10} GI_{50} (mol/L) = −4.81; OVCAR-3, Log_{10} GI_{50} (mol/L) = −4.96; OVCAR-4, Log_{10} GI_{50} (mol/L) = −4.36; OVCAR-5, Log_{10} GI_{50} (mol/L) = −4.40; OVCAR-8, Log_{10} GI_{50} (mol/L) = −4.71; SK-OV-3,

Log_{10} GI_{50} (mol/L) = –5.20; RXF-631L, Log_{10} GI_{50} (mol/L) = –4.47; ACHN, Log_{10} GI_{50} (mol/L) = –4.73; St4, Log_{10} GI_{50} (mol/L) = –5.03; MKN1, Log_{10} GI_{50} (mol/L) = –5.23; MKN7, Log_{10} GI_{50} (mol/L) = –4.78; MKN28, Log_{10} GI_{50} (mol/L) = –4.52; MKN45, Log_{10} GI_{50} (mol/L) = –5.00; MKN74, Log_{10} GI_{50} (mol/L) = –4.61; DU145, Log_{10} GI_{50} (mol/L) = –4.63; PC3, Log_{10} GI_{50} (mol/L) = –4.99; MG-MID (mean value of Log_{10} GI_{50} (mol/L) for all cell lines tested) = –4.83; Δ (difference in Log_{10} GI_{50} (mol/L) value of most sensitive cell and MG-MID value) = 0.71; Range (difference in Log_{10} GI_{50} (mol/L) value of most sensitive cell and least sensitive cell) = 1.19 (effective value of selective cytotoxic activity: MG-MID <-5, Δ ≥ 0.5, and range ≥ 1.0; COMPARE program suggested a novel mechanism of action). Ref: T. Yamada, et al, Tet. Lett., 2008, 49, 4192 | Y. Muroga, et al, J. Antibiot., 2008, 61, 615

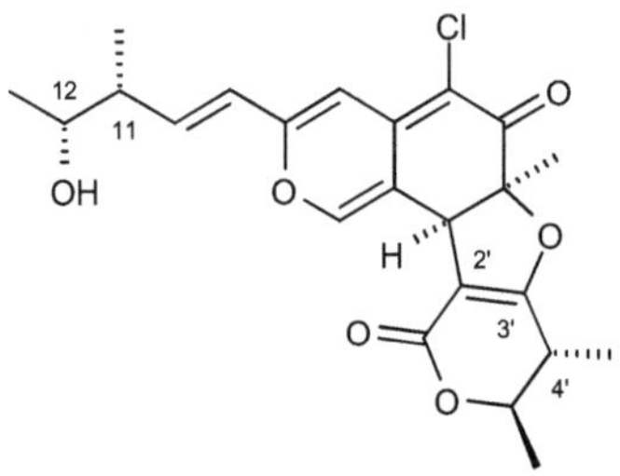

714 Chaetomugilin D

Type: Azaphilone 2-benzopyrans. $C_{23}H_{27}ClO_6$ Yellow gum or powder, mp 95–97 °C, $[\alpha]_D^{20}$ = –21° (c = 0.04, MeOH), $[\alpha]_D^{22}$ = –170.5° (c = 0.1, EtOH). Source: Marine-derived fungus *Chaetomium globosum* OUPS-T106B-6 from fish *Mugil cephalus* (Katsuura Bay, Japan), terrestrial fungus (*Chaetomium globosum*). Pharm: Cytotoxic (P_{388}, IC_{50} = 7.5 μmol/L, control 5-FU, IC_{50} = 1.7 μmol/L; HL60, IC_{50} = 6.8 μmol/L, 5-FU, IC_{50} = 2.7 μmol/L); cytotoxic selective (panel of 39 hmn cancer cell lines: HBC4, Log_{10} GI_{50} (mol/L) = –4.00; BSY1, Log_{10} GI_{50} (mol/L) = –4.55; HBC5, Log_{10} GI_{50} (mol/L) = –4.36; MCF7, Log_{10} GI_{50} (mol/L) = –4.08; MDA-MB-231, Log_{10} GI_{50} (mol/L) = –4.00; U251, Log_{10} GI_{50} (mol/L) = –4.00; SF268, Log_{10} GI_{50} (mol/L) = –4.00; SF295, Log_{10} GI_{50} (mol/L) = –4.00; SF539, Log_{10} GI_{50} (mol/L) = –5.00; SNB75, Log_{10} GI_{50} (mol/L) = –4.34; SNB78, Log_{10} GI_{50} (mol/L) = –4.14; HCC2998, Log_{10} GI_{50} (mol/L) = –4.00; KM12, Log_{10} GI_{50} (mol/L) = –4.00; HT29, Log_{10} GI_{50} (mol/L) = –4.00; HCT15, Log_{10} GI_{50} (mol/L) = –4.00; HCT116, Log_{10} GI_{50} (mol/L) = –4.00; NCI-H23, Log_{10} GI_{50} (mol/L) = –4.00; NCI-H226, Log_{10} GI_{50} (mol/L) = –4.00; NCI-H522, Log_{10} GI_{50} (mol/L) = –4.40; NCI-H460, Log_{10} GI_{50} (mol/L) = –4.68; A549, Log_{10} GI_{50} (mol/L) = –4.71; DMS273, Log_{10} GI_{50} (mol/L) = –5.30; DMS114, Log_{10} GI_{50} (mol/L) = –4.79; LOX-IMVI, Log_{10} GI_{50} (mol/L) = –4.81; OVCAR-3, Log_{10} GI_{50} (mol/L) = –4.96; OVCAR-4, Log_{10} GI_{50} (mol/L) = –4.36; OVCAR-5, Log_{10} GI_{50} (mol/L) = –4.40; OVCAR-8, Log_{10} GI_{50} (mol/L) = –4.71; SK-OV-3, Log_{10} GI_{50} (mol/L) = –5.20; RXF-631L, Log_{10} GI_{50} (mol/L) = –4.47; ACHN, Log_{10} GI_{50} (mol/L) = –4.73; St4,

Log_{10} GI_{50} (mol/L) = −5.03; MKN1, Log_{10} GI_{50} (mol/L) = −5.23; MKN7, Log_{10} GI_{50} (mol/L) = −4.78; MKN28, Log_{10} GI_{50} (mol/L) = −4.52; MKN45, Log_{10} GI_{50} (mol/L) = −5.00; MKN74, Log_{10} GI_{50} (mol/L) = −4.61; DU145, Log_{10} GI_{50} (mol/L) = −4.00; PC3, Log_{10} GI_{50} (mol/L) = −4.13; MG-MID (mean value of Log GI_{50} (mol/L) for all cell lines tested) = −4.05; Δ (difference in Log_{10} GI_{50} (mol/L) value of most sensitive cell and MG-MID value) = 0.50; Range (difference in Log_{10} GI_{50} (mol/L) value of most sensitive cell and least sensitive cell) = 0.55 (effective value of selective cytotoxic activity: MG-MID <-5, Δ ≥ 0.5, and range ≥ 1.0). Ref: Y. Muroga, et al, J. Antibiot., 2008, 61, 615 | J. -C. Qin, et al, BoMCL, 2009, 19, 1572

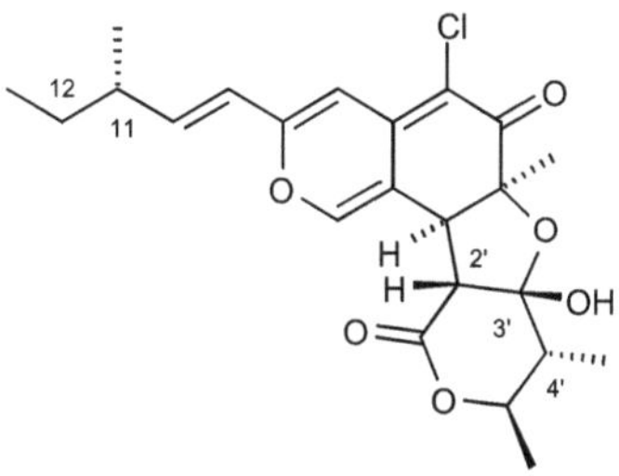

715 Chaetomugilin E

Type: Azaphilone 2-benzopyrans. $C_{24}H_{29}ClO_6$ Yellow powder, mp 95–97 °C, $[\alpha]_D$ = −155.4° (*c* = 0.11, EtOH). Source: Marine-derived fungus *Chaetomium globosum* OUPS-T106B-6 from fish *Mugil cephalus* (Katsuura Bay, Japan). Pharm: Cytotoxic (P_{388}, IC_{50} = 15.7 μmol/L, control 5-FU, IC_{50} = 1.7 μmol/L; HL60, IC_{50} = 13.2 μmol/L, 5-FU, IC_{50} = 2.7 μmol/L); cytotoxic selective (panel of 39 hmn cancer cell lines: HBC4, Log_{10} GI_{50} (mol/L) = −4.59; BSY1, Log_{10} GI_{50} (mol/L) = −4.75; HBC5, Log_{10} GI_{50} (mol/L) = −4.67; MCF7, Log_{10} GI_{50} (mol/L) = −4.44; MDA-MB-231, Log_{10} GI_{50} (mol/L) = −4.60; U251, Log_{10} GI_{50} (mol/L) = −4.55; SF268, Log_{10} GI_{50} (mol/L) = −4.59; SF295, Log_{10} GI_{50} (mol/L) = −4.51; SF539, Log_{10} GI_{50} (mol/L) = −4.72; SNB75, Log_{10} GI_{50} (mol/L) = −4.58; SNB78, Log_{10} GI_{50} (mol/L) = −4.68; HCC2998, Log_{10} GI_{50} (mol/L) = −4.63; KM12, Log_{10} GI_{50} (mol/L) = −4.64; HT29, Log_{10} GI_{50} (mol/L) = −4.64; HCT15, Log_{10} GI_{50} (mol/L) = −4.60; HCT116, Log_{10} GI_{50} (mol/L) = −4.72; NCI-H23, Log_{10} GI_{50} (mol/L) = −4.59; NCI-H226, Log_{10} GI_{50} (mol/L) = −4.64; NCI-H522, Log_{10} GI_{50} (mol/L) = −4.78; NCI-H460, Log_{10} GI_{50} (mol/L) = −4.44; A549, Log_{10} GI_{50} (mol/L) = −4.38; DMS273, Log_{10} GI_{50} (mol/L) = −4.66; DMS114, Log_{10} GI_{50} (mol/L) = −4.84; LOX-IMVI, Log_{10} GI_{50} (mol/L) = −4.71; OVCAR-3, Log_{10} GI_{50} (mol/L) = −4.67; OVCAR-4, Log_{10} GI_{50} (mol/L) = −4.72; OVCAR-5, Log_{10} GI_{50} (mol/L) = −4.61; OVCAR-8, Log_{10} GI_{50} (mol/L) = −4.54; SK-OV-3, Log_{10} GI_{50} (mol/L) = −4.38; RXF-631L, Log_{10} GI_{50} (mol/L) = −4.52; ACHN, Log_{10} GI_{50} (mol/L) = −4.67; St4, Log_{10} GI_{50} (mol/L) = −4.54; MKN1, Log_{10} GI_{50} (mol/L) = −4.74; MKN7, Log_{10} GI_{50} (mol/L) = −4.67; MKN28, Log_{10} GI_{50} (mol/L) = −4.47; MKN45, Log_{10} GI_{50} (mol/L) = −4.49; MKN74, Log_{10} GI_{50} (mol/L) = −4.60; DU145, Log_{10} GI_{50} (mol/L) = −4.52; PC3, Log_{10} GI_{50} (mol/L) = −4.68;

MG-MID (mean value of Log GI_{50} (mol/L) for all cell lines tested) = –4.61; Δ (difference in Log_{10} GI_{50} (mol/L) value of most sensitive cell and MG-MID value) = 0.23; Range (difference in Log_{10} GI_{50} (mol/L) value of most sensitive cell and least sensitive cell) = 0.46 (effective value of selective cytotoxic activity: MG-MID <-5, Δ ≥ 0.5, and range ≥ 1.0). Ref: Y. Muroga, et al, J. Antibiot., 2008, 61, 615 | Log_{10} GI50

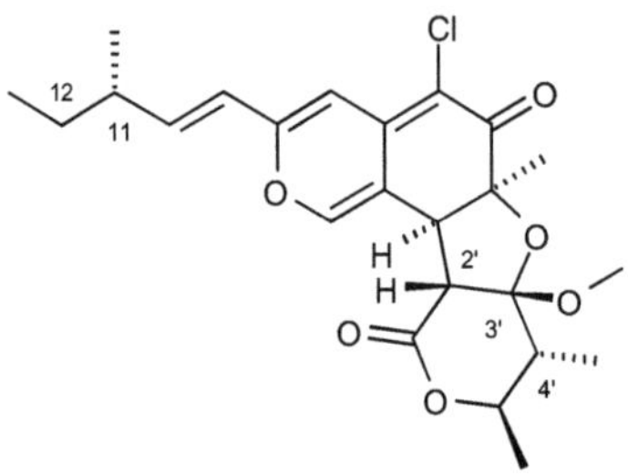

716 Chaetomugilin F

Type: Azaphilone 2-benzopyrans. $C_{23}H_{25}ClO_5$ Yellow powder, mp 92–94 °C, $[\alpha]_D^{22}$ = –185.1° (*c* = 0.13, MeOH). Source: Marine-derived fungus *Chaetomium globosum* OUPS-T106B-6 from fish *Mugil cephalus* (Katsuura Bay, Japan). Pharm: Cytotoxic (P_{388}, IC_{50} = 3.3 μmol/L, control 5-FU, IC_{50} = 1.7 μmol/L; HL60, IC_{50} = 1.3 μmol/L, 5-FU, IC_{50} = 2.7 μmol/L); cytotoxic selective (panel of 39 hmn cancer cell lines: HBC4, Log_{10} GI_{50} (mol/L) = –6.43; BSY1, Log_{10} GI_{50} (mol/L) = –5.63; HBC5, Log_{10} GI_{50} (mol/L) = –5.19; MCF7, Log_{10} GI_{50} (mol/L) = –6.14; MDA-MB-231, Log_{10} GI_{50} (mol/L) = –5.30; U251, Log_{10} GI_{50} (mol/L) = –5.50; SF268, Log_{10} GI_{50} (mol/L) = –5.37; SF295, Log_{10} GI_{50} (mol/L) = –5.85; SF539, Log_{10} GI_{50} (mol/L) = –6.66; SNB75, Log_{10} GI_{50} (mol/L) = –5.59; SNB78, Log_{10} GI_{50} (mol/L) = –5.26; HCC2998, Log_{10} GI_{50} (mol/L) = –4.83; KM12, Log_{10} GI_{50} (mol/L) = –6.21; HT29, Log_{10} GI_{50} (mol/L) = –5.02; HCT15, Log_{10} GI_{50} (mol/L) = –5.29; HCT116, Log_{10} GI_{50} (mol/L) = –5.61; NCI-H23, Log_{10} GI_{50} (mol/L) = –4.83; NCI-H226, Log_{10} GI_{50} (mol/L) = –5.26; NCI-H522, Log_{10} GI_{50} (mol/L) = –5.67; NCI-H460, Log_{10} GI_{50} (mol/L) = –5.23; A549, Log_{10} GI_{50} (mol/L) = –5.14; DMS273, Log_{10} GI_{50} (mol/L) = –6.35; DMS114, Log_{10} GI_{50} (mol/L) = –5.48; LOX-IMVI, Log_{10} GI_{50} (mol/L) = –5.34; OVCAR-3, Log_{10} GI_{50} (mol/L) = –5.51; OVCAR-4, Log_{10} GI_{50} (mol/L) = –4.69; OVCAR-5, Log_{10} GI_{50} (mol/L) = –5.39; OVCAR-8, Log_{10} GI_{50} (mol/L) = –5.11; SK-OV-3, Log_{10} GI_{50} (mol/L) = –5.47; RXF-631L, Log_{10} GI_{50} (mol/L) = –5.14; ACHN, Log_{10} GI_{50} (mol/L) = –5.33; St4, Log_{10} GI_{50} (mol/L) = –4.98; MKN1, Log_{10} GI_{50} (mol/L) = –5.55; MKN7, Log_{10} GI_{50} (mol/L) = –5.40; MKN28, Log_{10} GI_{50} (mol/L) = –5.32; MKN45, Log_{10} GI_{50} (mol/L) = –5.28; MKN74, Log_{10} GI_{50} (mol/L) = –5.49; DU145, Log_{10} GI_{50} (mol/L) = –5.42; PC3, Log_{10} GI_{50} (mol/L) = –5.27; MG-MID (mean value of Log GI_{50} (mol/L) for all cell lines tested) = –5.45; Δ (difference in Log_{10} GI_{50} (mol/L) value of most sensitive cell and MG-MID value) = 1.21; Range (difference in Log_{10} GI_{50} (mol/L) value of most sensitive cell and least sensitive cell) = 1.97 (effective value of selective cytotoxic activity: MG-MID <-5, Δ ≥ 0.5, and range ≥ 1.0; COMPARE program suggested a novel mechanism of action). Ref: Y. Muroga, et al, J. Antibiot., 2008, 61, 615

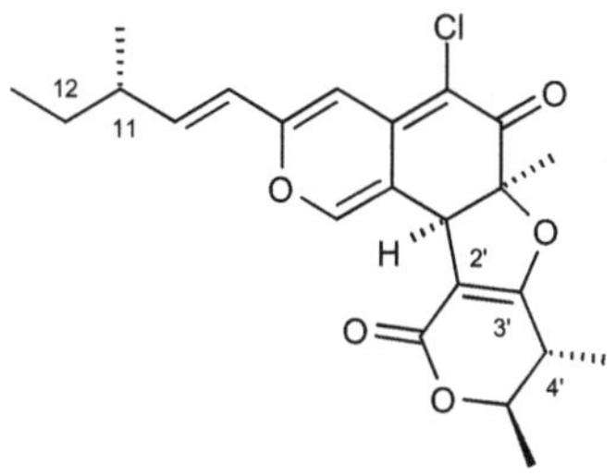

717 Chaetomugilin G

Type: Azaphilone 2-benzopyrans. $C_{24}H_{29}ClO_7$ Yellow powder, mp 165–167 °C, $[\alpha]_D^{22}$ = −80.1° (*c* = 0.13, EtOH). Source: Marine-derived fungus *Chaetomium globosum* OUPS-T106B-6 from fish *Mugil cephalus* (Katsuura Bay, Japan). Pharm: Cytotoxic (P_{388}, IC_{50} = 24.1 µmol/L, control 5-FU, IC_{50} = 1.7 µmol/L; HL60, IC_{50} = 19.8 µmol/L, 5-FU, IC_{50} = 2.7 µmol/L; L_{1210}, IC_{50} = 123.6 µmol/L; 5-FU, IC_{50} = 3.0 µmol/L; KB, IC_{50} = 137.8 µmol/L, 5-FU, IC_{50} = 6.0 µmol/L). Ref: T. Yamada, et al, J. Antibiot., 2009, 62, 353

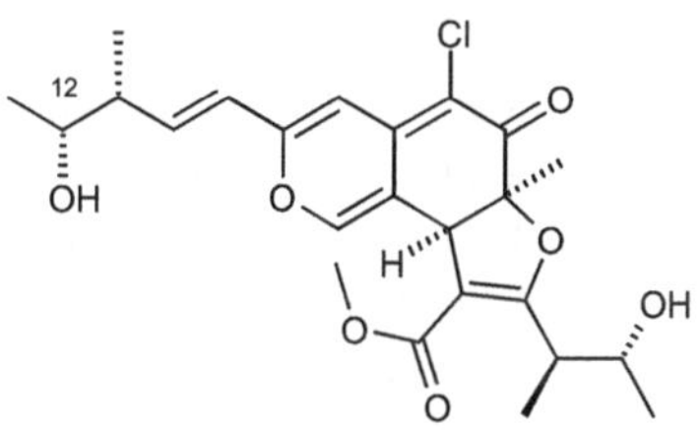

718 Chaetomugilin H

Type: Azaphilone 2-benzopyrans. $C_{24}H_{29}ClO_6$ Yellow powder, mp 119–121 °C, $[\alpha]_D^{22}$ = −175° (*c* = 0.09, EtOH). Source: Marine-derived fungus *Chaetomium globosum* OUPS-T106B-6. Pharm: Cytotoxic (P_{388}, IC_{50} = 12.3 µmol/L, control 5-FU, IC_{50} = 1.7 µmol/L; HL60, IC_{50} = 10.3 µmol/L, 5-FU, IC_{50} = 2.7 µmol/L; L_{1210}, IC_{50} = 93.3 µmol/L; 5-FU, IC_{50} = 3.0 µmol/L; KB, IC_{50} = 18.8 µmol/L, 5-FU, IC_{50} = 6.0 µmol/L). Ref: T. Yamada, et al, J. Antibiot., 2009, 62, 353

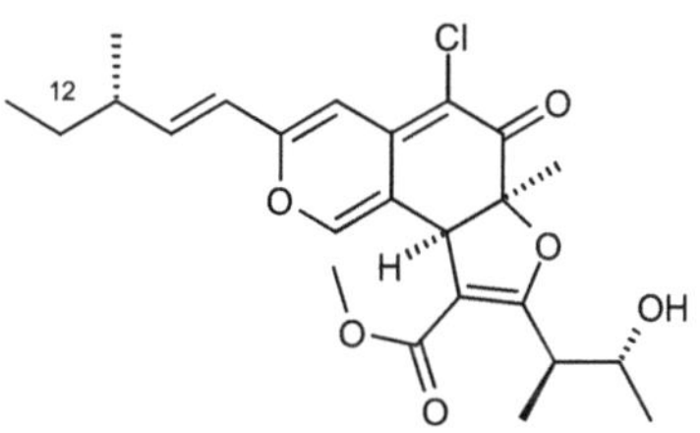

719 Chaetomugilin I

Type: Azaphilone 2-benzopyrans. $C_{22}H_{27}ClO_5$ Yellow powder ($CHCl_3$/MeOH), mp 98–100 °C, $[\alpha]_D$ = +210.2° (*c* = 0.04, EtOH). Source: Marine-derived fungus *Chaetomium globosum* OUPS-T106B-6. Pharm: Cytotoxic (P_{388}, IC_{50} = 1.1 μmol/L, control 5-FU, IC_{50} = 1.7 μmol/L; HL60, IC_{50} = 1.1 μmol/L, 5-FU, IC_{50} = 2.7 μmol/L; L_{1210}, IC_{50} = 1.9 μmol/L, 5-FU, IC_{50} = 1.1 μmol/L; KB, IC_{50} = 2.3 μmol/L, 5-FU, IC_{50} = 7.7 μmol/L) (in original paper, the unit is wrong as mM, corrected by book compiler to μmol/L); cytotoxic selective (panel of 39 hmn cancer cell lines: breast HBC4, Log_{10} GI_{50} (mol/L) = −5.07; BSY1, Log_{10} GI_{50} (mol/L) = −5.50; HBC5, Log_{10} GI_{50} (mol/L) = −5.25; MCF7, Log_{10} GI_{50} (mol/L) = −4.98; MDA-MB-231, Log_{10} GI_{50} (mol/L) = −5.38; U251, Log_{10} GI_{50} (mol/L) = −4.94; SF268, Log_{10} GI_{50} (mol/L) = −4.97; SF295, Log_{10} GI_{50} (mol/L) = −4.81; SF539, Log_{10} GI_{50} (mol/L) = −5.42; SNB75, Log_{10} GI_{50} (mol/L) = −5.42; SNB78, Log_{10} GI_{50} (mol/L) = −5.17; HCC2998, Log_{10} GI_{50} (mol/L) = −4.88; KM12, Log_{10} GI_{50} (mol/L) = −5.09; HT29, Log_{10} GI_{50} (mol/L) = −5.32; HCT15, Log_{10} GI_{50} (mol/L) = −5.32; HCT116, Log_{10} GI_{50} (mol/L) = −5.42; NCI-H23, Log_{10} GI_{50} (mol/L) = −4.87; NCI-H226, Log_{10} GI_{50} (mol/L) = −4.94; NCI-H522, Log_{10} GI_{50} (mol/L) = −5.78; NCI-H460, Log_{10} GI_{50} (mol/L) = −4.76; A549, Log_{10} GI_{50} (mol/L) = −4.68; DMS273, Log_{10} GI_{50} (mol/L) = −5.32; DMS114, Log_{10} GI_{50} (mol/L) = −5.32; LOX-IMVI, Log_{10} GI_{50} (mol/L) = −5.26; OVCAR-3, Log_{10} GI_{50} (mol/L) = −5.58; OVCAR-4, Log_{10} GI_{50} (mol/L) = −5.48; OVCAR-5, Log_{10} GI_{50} (mol/L) = −5.36; OVCAR-8, Log_{10} GI_{50} (mol/L) = −4.84; SK-OV-3, Log_{10} GI_{50} (mol/L) = −4.57; RXF-631L, Log_{10} GI_{50} (mol/L) = −4.63; ACHN, Log_{10} GI_{50} (mol/L) = −5.51; St4, Log_{10} GI_{50} (mol/L) = −4.85; MKN1, Log_{10} GI_{50} (mol/L) = −5.70; MKN7, Log_{10} GI_{50} (mol/L) = −5.44; MKN28, Log_{10} GI_{50} (mol/L) = −5.12; MKN45, Log_{10} GI_{50} (mol/L) = −4.71; MKN74, Log_{10} GI_{50} (mol/L) = −4.79; DU145, Log_{10} GI_{50} (mol/L) = −4.79; PC3, Log_{10} GI_{50} (mol/L) = −5.52; MG-MID (mean value of Log_{10} GI_{50} (mol/L) for all cell lines tested) = −5.15; Δ (difference in Log_{10} GI_{50} value of most sensitive cell and MG-MID value) = 0.63; Range (difference in Log_{10} GI_{50} value of most sensitive cell and least sensitive cell) = 1.21 (effective value of selective cytotoxic activity: MG-MID <-5, Δ ≥ 0.5, and range ≥ 1.0; COMPARE program suggested a novel mechanism of action). Ref: Y. Muroga, et al, Tetrahedron, 2009, 65, 7580 | T. Yamada, et al, BoMC, 2011, 19, 4106

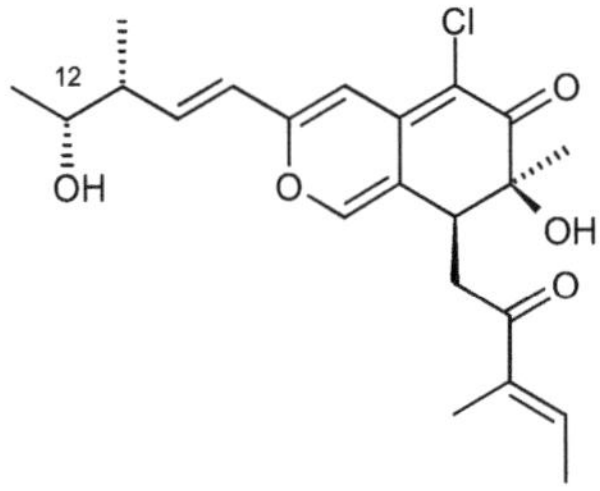

720 11-*epi*-Chaetomugilin I

Type: Azaphilone 2-benzopyrans. $C_{22}H_{27}ClO_5$ Yellow powder, mp 85–87 °C, ($CHCl_3$-MeOH),$[\alpha]_D^{22}$ = +77.5° (*c* = 0.12, EtOH). Source: Marine-derived fungus *Chaetomium globosum* from fish *Mugil cephalus* (Katsuura Bay, Japan). Pharm: Cytotoxic (P_{388}, IC_{50} = 0.7 pmol/L, control 5-FU, IC_{50} = 1.7 pmol/L; HL60, IC_{50} = 1.0 pmol/L, 5-FU, IC_{50} = 2.7 pmol/L; L_{1210}, IC_{50} = 1.6 pmol/L, 5-FU, IC_{50} = 1.1 pmol/L; KB, IC_{50} = 1.2 pmol/L, 5-FU, IC_{50} = 7.7 pmol/L). Ref: T. Yamada, et al, BoMC, 2011, 19, 4106

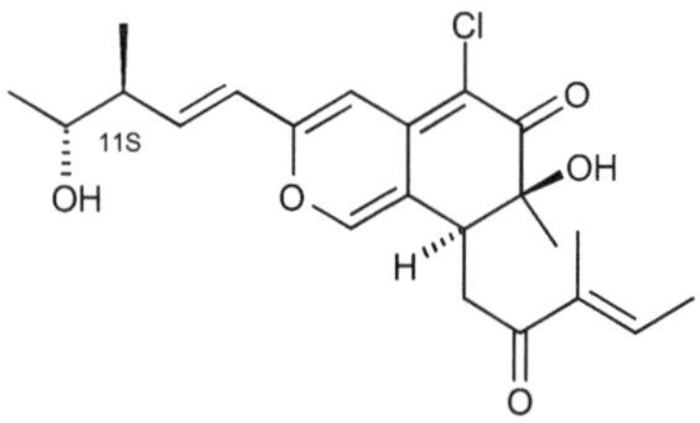

721 Chaetomugilin J

Type: Azaphilone 2-benzopyrans. $C_{22}H_{27}ClO_4$ Yellow powder ($CHCl_3$/MeOH), mp 99–101 °C, $[\alpha]_D$ = +243.8° (*c* = 0.1, EtOH). Source: Marine-derived fungus *Chaetomium globosum* OUPS-T106B-6. Pharm: Cytotoxic (P_{388}, IC_{50} = 12.6 μmol/L, control 5-FU, IC_{50} = 1.7 μmol/L; HL60, IC_{50} = 12.6 μmol/L, 5-FU, IC_{50} = 2.7 μmol/L; L_{1210}, IC_{50} = 2.8 μmol/L, 5-FU, IC_{50} = 1.1 μmol/L; KB, IC_{50} = 8.5 μmol/L, 5-FU, IC_{50} = 7.7 μmol/L) (in original paper, the unit is wrong as mM, corrected by book compiler to μmol/L). Ref: Y. Muroga, et al, Tetrahedron, 2009, 65, 7580

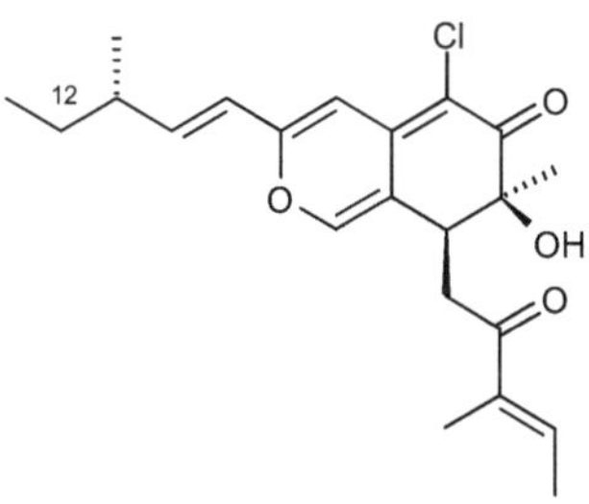

722 Chaetomugilin K

Type: Azaphilone 2-benzopyrans. $C_{23}H_{29}ClO_5$ Yellow powder ($CHCl_3$/MeOH), mp 222–224 °C, $[\alpha]_D$ = −65.7° (*c* = 0.14, EtOH). Source: Marine-derived fungus *Chaetomium globosum* OUPS-T106B-6. Pharm: Cytotoxic (P_{388}, IC_{50} = 8.2 μmol/L, control 5-FU, IC_{50} = 1.7 μmol/L; HL60, IC_{50} = 14.1 μmol/L, 5-FU, IC_{50} = 2.7 μmol/L; L1210, IC_{50} = 11.2 μmol/L, 5-FU, IC_{50} = 1.1 μmol/L; KB, IC_{50} = 18.7 μmol/L, 5-FU, IC_{50} = 7.7 μmol/L) (in original paper, the unit is wrong as mM, corrected by book compiler to μmol/L). Ref: Y. Muroga, et al, Tetrahedron, 2009, 65, 7580

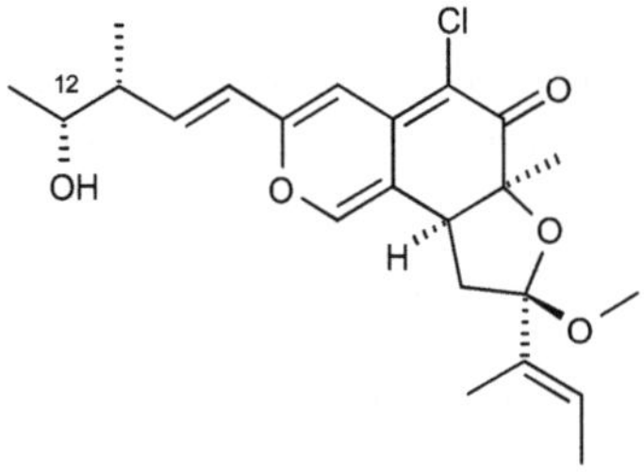

723 Chaetomugilin L

Type: Azaphilone 2-benzopyrans. $C_{23}H_{29}ClO_4$ Yellow powder ($CHCl_3$/MeOH), mp 152–154 °C, $[\alpha]_D$ = −319.3° (*c* = 0.09, EtOH). Source: Marine-derived fungus *Chaetomium globosum* OUPS-T106B-6. Pharm: Cytotoxic (P_{388}, IC_{50} = 10.9 μmol/L, control 5-FU, IC_{50} = 1.7 μmol/L; HL60, IC_{50} = 13.1 μmol/L, 5-FU, IC_{50} = 2.7 μmol/L; L_{1210}, IC_{50} = 15.6 μmol/L, 5-FU, IC_{50} = 1.1 μmol/L; KB, IC_{50} = 20.1 μmol/L, 5-FU, IC_{50} = 7.7 μmol/L) (in original paper, the unit is wrong as mM, corrected by book compiler to μmol/L). Ref: Y. Muroga, et al, Tetrahedron, 2009, 65, 7580

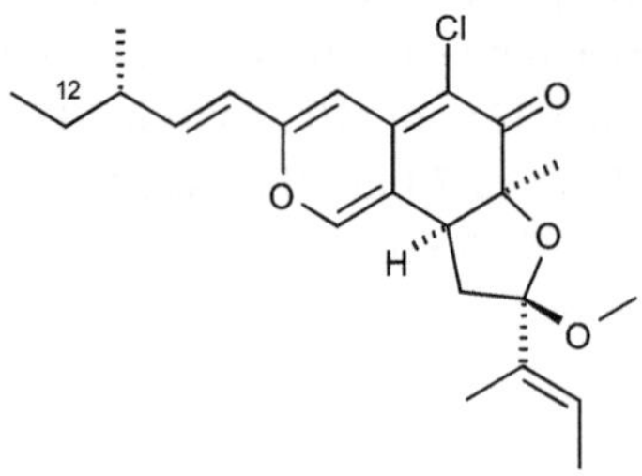

724 Chaetomugilin N

Type: Azaphilone 2-benzopyrans. $C_{23}H_{25}ClO_6$ Yellow powder ($CHCl_3$/MeOH), mp 127–129 °C, $[\alpha]_D$ = −71.8° (*c* = 0.12, EtOH). Source: Marine-derived fungus *Chaetomium globosum* OUPS-T106B-6. Pharm: Cytotoxic (P_{388}, IC_{50} = 2.3 μmol/L, control 5-FU, IC_{50} = 1.7 μmol/L; HL60, IC_{50} = 2.3 μmol/L, 5-FU, IC_{50} = 2.7 μmol/L; L_{1210}, IC_{50} = 10.6 μmol/L, 5-FU, IC_{50} = 1.1 μmol/L; KB, IC_{50} = 10.6 μmol/L, 5-FU, IC_{50} = 7.7 μmol/L) (in original paper, the unit is wrong as mM, corrected by book compiler to μmol/L). Ref: Y. Muroga, et al, Tetrahedron, 2009, 65, 7580

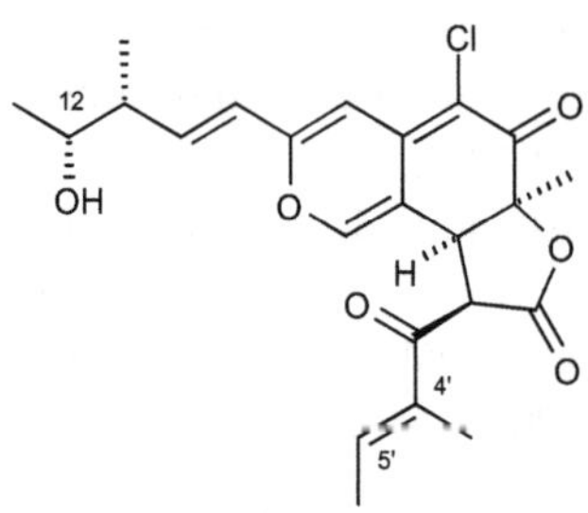

725 Chaetomugilin O

Type: Azaphilone 2-benzopyrans. $C_{23}H_{25}ClO_5$ Yellow powder ($CHCl_3$/MeOH), mp 95–97 °C, $[\alpha]_D$ = −116.1° (*c* = 0.14, EtOH). Source: Marine-derived fungus *Chaetomium globosum* OUPS-T106B-6. Pharm: Cytotoxic (P_{388}, IC_{50} = 11.1 μmol/L, control 5-FU, IC_{50} = 1.7 μmol/L; HL60, IC_{50} = 11.1 μmol/L, 5-FU, IC_{50} = 2.7 μmol/L; L_{1210}, IC_{50} = 10.1 μmol/L, 5-FU, IC_{50} = 1.1 μmol/L; KB, IC_{50} = 7.2 μmol/L, 5-FU, IC_{50} = 7.7 μmol/L) (in original paper, the unit is wrong as mM, corrected by book compiler to μmol/L). Ref: Y. Muroga, et al, Tetrahedron, 2009, 65, 7580

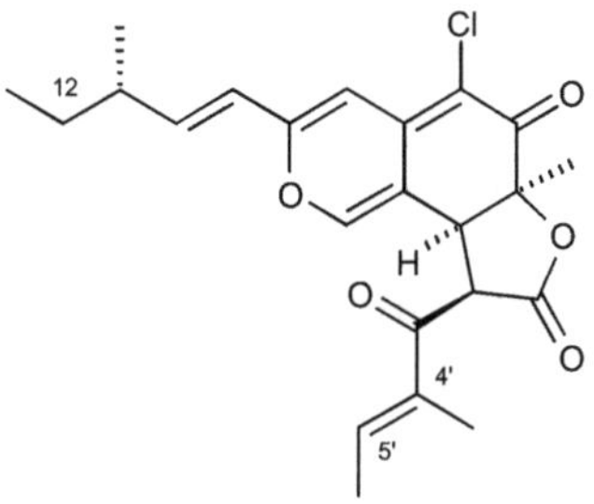

726 Chaetomugilin P

Type: Azaphilone 2-benzopyrans. $C_{22}H_{27}ClO_5$ Yellow powder, mp 103–105 °C, ($CHCl_3$-MeOH),$[\alpha]_D^{22}$ = −49.7° (*c* = 0.06, EtOH). Source: Marine-derived fungus *Chaetomium globosum* from fish *Mugil cephalus* (Katsuura Bay, Japan). Pharm: Cytotoxic (P_{388}, IC_{50} = 0.7 pmol/L, control 5-FU, IC_{50} = 1.7 pmol/L; HL60, IC_{50} = 1.2 pmol/L, 5-FU, IC_{50} = 2.7 pmol/L; L_{1210}, IC_{50} = 1.5 pmol/L, 5-FU, IC_{50} = 1.1 pmol/L; KB, IC_{50} = 1.8 pmol/L, 5-FU, IC_{50} = 7.7 pmol/L). Ref: T. Yamada, et al, BoMC, 2011, 19, 4106

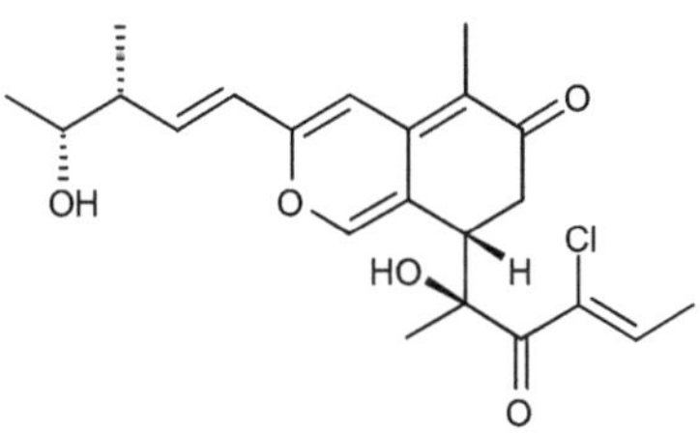

727 Chaetomugilin Q

Type: Azaphilone 2-benzopyrans. $C_{22}H_{29}ClO_6$ Yellow powder, mp 110–112 °C, ($CHCl_3$-MeOH),$[\alpha]_D^{22}$ = −18.3° (*c* = 0.14, EtOH). Source: Marine-derived fungus *Chaetomium globosum* from fish *Mugil cephalus* (Katsuura Bay, Japan). Pharm: Cytotoxic (P_{388}, IC_{50} = 49.5 pmol/L, control 5-FU, IC_{50} = 1.7 pmol/L; HL60, IC_{50} = 47.2 pmol/L, 5-FU, IC_{50} = 2.7 pmol/L; L_{1210}, IC_{50} = 80.2 pmol/L, 5-FU, IC_{50} = 1.1 pmol/L; KB, IC_{50} > 100 pmol/L, 5-FU, IC_{50} = 7.7 pmol/L). Ref: T. Yamada, et al, BoMC, 2011, 19, 4106

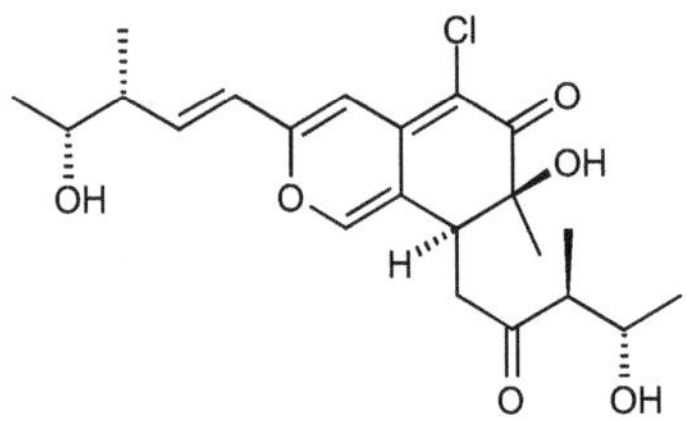

728 Chaetomugilin R

Type: Azaphilone 2-benzopyrans. $C_{17}H_{23}ClO_5$ Yellow powder, mp 104–106 °C, ($CHCl_3$-MeOH),$[\alpha]_D^{22}$ = −130.2° (*c* = 0.11, EtOH). Source: Marine-derived fungus *Chaetomium globosum* from fish *Mugil cephalus* (Katsuura Bay, Japan). Pharm: Cytotoxic (P_{388}, IC_{50} = 32.0 pmol/L, control 5-FU, IC_{50} = 1.7 pmol/L; HL60, IC_{50} = 51.8 pmol/L, 5-FU, IC_{50} = 2.7 pmol/L; L_{1210}, IC_{50} = 67.1 pmol/L, 5-FU, IC_{50} = 1.1 pmol/L; KB, IC_{50} = 67.1 pmol/L, 5-FU, IC_{50} = 7.7 pmol/L). Ref: T. Yamada, et al, BoMC, 2011, 19, 4106

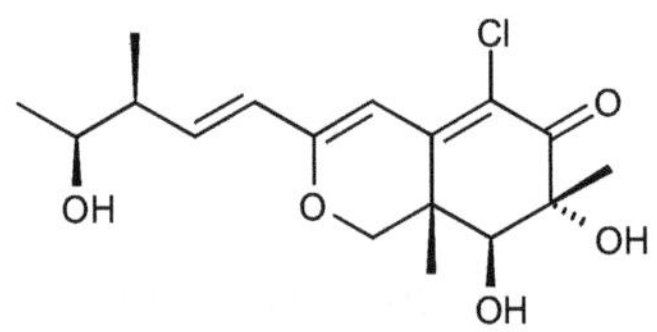

729 Chaetomugilin S

Type: Azaphilone 2-benzopyrans. $C_{23}H_{27}ClO_6$ Source: Marine-derived fungus *Chaetomium globosum* from fish *Mugil cephalus* (Katsuura Bay, Japan). Pharm: Cytotoxic (HTCLs and P_{388}, moderate, growth inhibitor). Ref: T. Yamada, et al, J. Antibiot., 2012, 65, 413

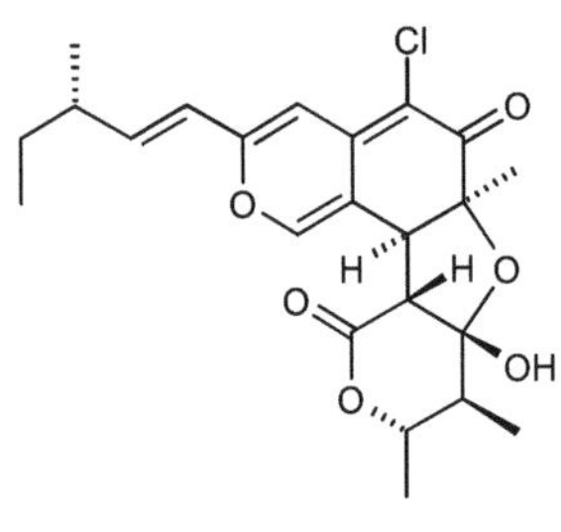

730 Comazaphilone A

Type: Azaphilone 2-benzopyrans. $C_{22}H_{26}O_7$ Yellow amorph. powder, $[\alpha]_D^{25}$ = +90.7° (*c* = 0.38, MeOH). Source: Marine-derived fungus *Penicillium commune* QSD-17 (sediment fermentation, southern China Sea). Pharm: Antibacterial (MRSA, MIC > 256 μg/mL, control Ampicillin, MIC = 8 μg/mL; *Pseudomonas aeruginosa*, MIC > 256 μg/mL, Ampicillin, MIC = 2 μg/mL; *Pseudomonas fluorescens*, MIC > 256 μg/mL, Ampicillin, MIC = 4 μg/mL; *Bacillus subtilis*, MIC = 256 μg/mL, Ampicillin,

MIC = 4 μg/mL); antifungal (*Candida albicans,* MIC > 256 μg/mL, control Nystatin, MIC = 2 μg/mL). Ref: S. -S. Gao, et al, JNP, 2011, 74, 256

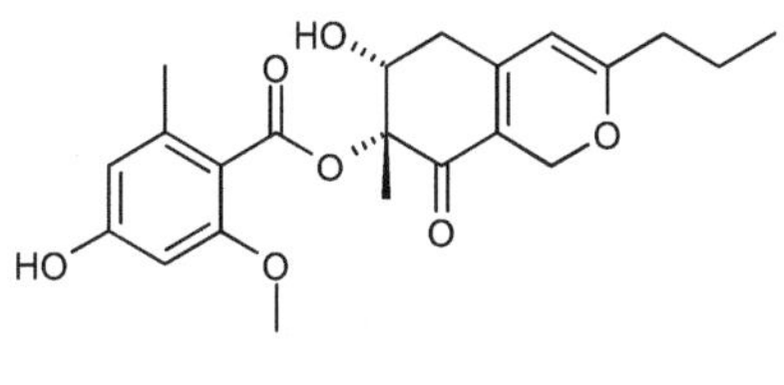

731 Comazaphilone B

Type: Azaphilone 2-benzopyrans. $C_{22}H_{26}O_8$ Yellow amorph. powder, $[\alpha]_D^{25}$ = +95.7° (*c* = 0.23, MeOH). Source: Marine-derived fungus *Penicillium commune* QSD-17 (sediment fermentation, southern China Sea). Pharm: Antibacterial (MRSA, MIC = 128 μg/mL, control Ampicillin, MIC = 8 μg/mL; *Pseudomonas aeruginosa,* MIC > 256 μg/mL, Ampicillin, MIC = 2 μg/mL; *Pseudomonas fluorescens,* MIC = 128 μg/mL, Ampicillin, MIC = 4 μg/mL; *Bacillus subtilis,* MIC = 64 μg/mL, Ampicillin, MIC = 4 μg/mL); antifungal (*Candida albicans,* MIC > 256 μg/mL, control Nystatin, MIC = 2 μg/mL). Ref: S. -S. Gao, et al, JNP, 2011, 74, 256

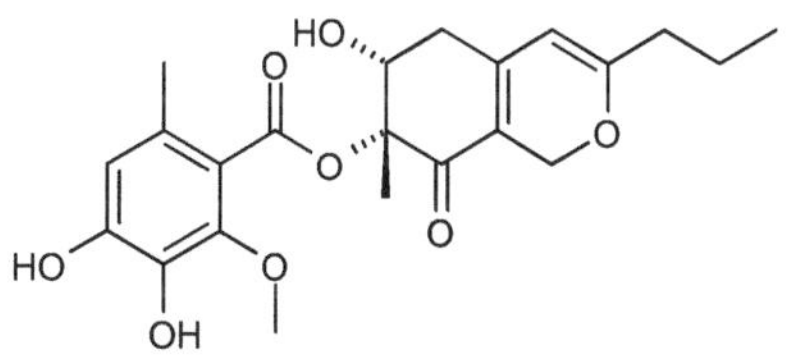

732 Comazaphilone C

Type: Azaphilone 2-benzopyrans. $C_{22}H_{24}O_8$ Yellow amorph. powder, $[\alpha]_D^{25}$ = +205° (*c* = 0.20, MeOH). Source: Marine-derived fungus *Penicillium commune* QSD-17 (sediment fermentation, southern China Sea). Pharm: Antibacterial (MRSA, MIC = 16 μg/mL, control Ampicillin, MIC = 8 μg/mL; *Pseudomonas aeruginosa,* MIC > 256 μg/mL, Ampicillin, MIC = 2 μg/mL; *Pseudomonas fluorescens,* MIC = 64 μg/mL, Ampicillin, MIC = 4 μg/mL;*Bacillus subtilis,* MIC = 32 μg/mL, Ampicillin, MIC = 4 μg/mL); antifungal (*Candida albicans,* MIC > 256 μg/mL, control Nystatin, MIC = 2 μg/mL). Ref: S. -S. Gao, et al, JNP, 2011, 74, 256

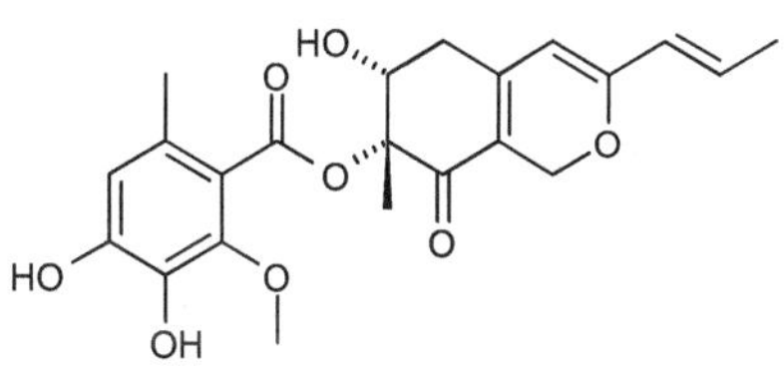

733 Comazaphilone D

Type: Azaphilone 2-benzopyrans. $C_{21}H_{22}O_7$ Yellow amorph. powder, $[\alpha]_D^{25}$ = +270° (c = 0.23, MeOH). Source: Marine-derived fungus *Penicillium commune* QSD-17 (sediment fermentation, southern China Sea). Pharm: Antibacterial (MRSA, MIC = 32 μg/mL, control Ampicillin, MIC = 8 μg/mL; *Pseudomonas aeruginosa,* MIC > 256 μg/mL, Ampicillin, MIC = 2 μg/mL; *Pseudomonas fluorescens,* MIC = 16 μg/mL, Ampicillin, MIC = 4 μg/mL; *Bacillus subtilis,* MIC > 256 μg/mL, Ampicillin, MIC = 4 μg/mL); antifungal (*Candida albicans,* MIC > 256 μg/mL, control Nystatin, MIC = 2 μg/mL); cytotoxic (SW1990, IC_{50} = 51 μmol/L, control Fluorouracil, IC_{50} = 120 μmol/L). Ref: S. -S. Gao, et al, JNP, 2011, 74, 256

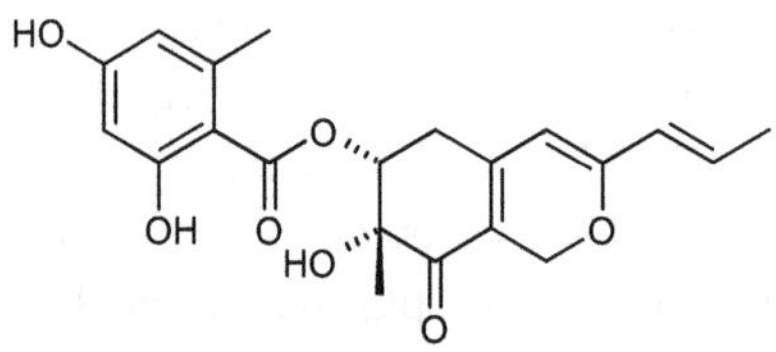

734 Comazaphilone E

Type: Azaphilone 2-benzopyrans. $C_{22}H_{24}O_8$ Yellow amorph. powder, $[\alpha]_D^{25}$ = +54.5° (c = 0.33, MeOH). Source: Marine-derived fungus *Penicillium commune* QSD-17 (sediment fermentation, southern China Sea). Pharm: Antibacterial (MRSA, MIC > 256 μg/mL, control Ampicillin, MIC = 8 μg/mL; *Pseudomonas aeruginosa,* MIC > 256 μg/mL, Ampicillin, MIC = 2 μg/mL; *Pseudomonas fluorescens,* MIC = 32 μg/mL, Ampicillin, MIC = 4 μg/mL; *Bacillus subtilis,* MIC = 16 μg/mL, Ampicillin, MIC = 4 μg/mL); antifungal (*Candida albicans,* MIC > 256 μg/mL, control Nystatin, MIC = 2 μg/mL); cytotoxic (SW1990, IC_{50} = 26 μmol/L, control Fluorouracil, IC_{50} = 120 μmol/L). Ref: S. -S. Gao, et al, JNP, 2011, 74, 256

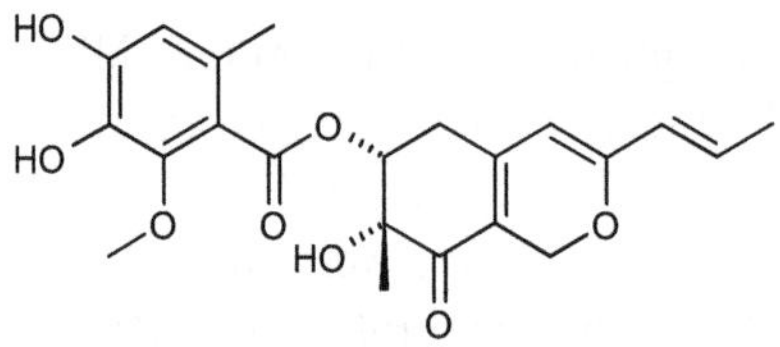

735 Comazaphilone F

Type: Azaphilone 2-benzopyrans. $C_{22}H_{26}O_8$ Yellow amorph. powder, $[\alpha]_D^{25}$ = +30.8° (c = 0.21, MeOH). Source: Marine-derived fungus *Penicillium commune* QSD-17 (sediment fermentation, southern China Sea). Pharm: Antibacterial (MRSA, MIC = 128 μg/mL, control Ampicillin, MIC = 8 μg/mL; *Pseudomonas aeruginosa,* MIC > 256 μg/mL, Ampicillin, MIC = 2 μg/mL; *Pseudomonas fluorescens,* MIC = 64 μg/mL, Ampicillin, MIC = 4 μg/mL; *Bacillus subtilis,* MIC = 128 μg/mL, Ampicillin, MIC = 4 μg/mL); antifungal (*Candida albicans,* MIC > 256 μg/mL, control Nystatin,

MIC = 2 μg/mL); cytotoxic (SW1990, IC_{50} = 53 μmol/L, control Fluorouracil, IC_{50} = 120 μmol/L). Ref: S. -S. Gao, et al, JNP, 2011, 74, 256

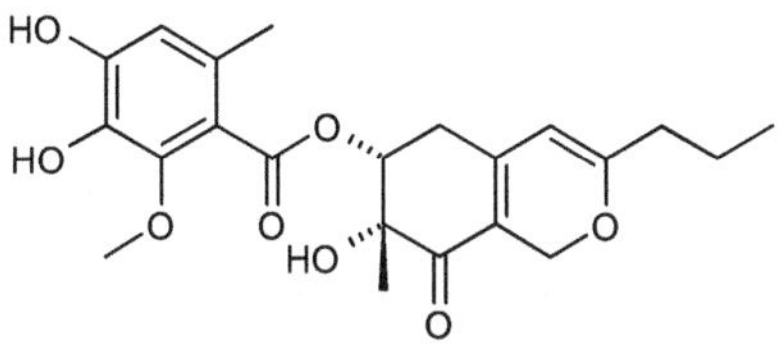

736 Purpurquinone B

Type: Azaphilone 2-benzopyrans. $C_{21}H_{20}O_{10}$ Source: Acid-tolerant fungus *Penicillium purpurogenum* JS03-21 (region of red soil by Lujiang River, Yunnan, China). Pharm: Antiviral (IFV virus, IC_{50} = 61.3 μmol/L). Ref: H. Wang, et al, JNP, 2011, 74, 2014

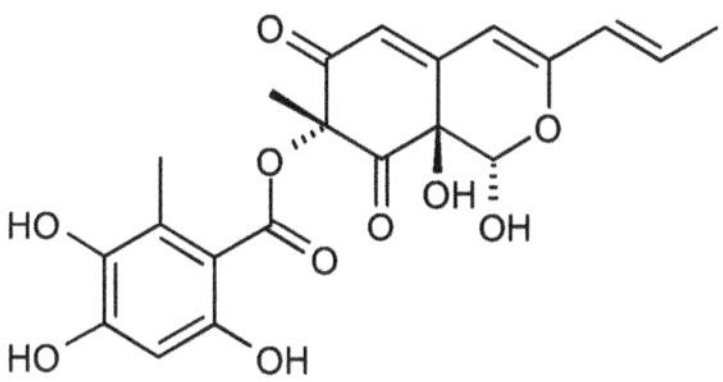

737 Purpurquinone C

Type: Azaphilone 2-benzopyrans. $C_{21}H_{20}O_8$ Source: Acid-tolerant fungus *Penicillium purpurogenum* JS03-21 (region of red soil by Lujiang River, Yunnan, China). Pharm: Antiviral (IFV virus, IC_{50} = 64.0 μmol/L). Ref: H. Wang, et al, JNP, 2011, 74, 2014

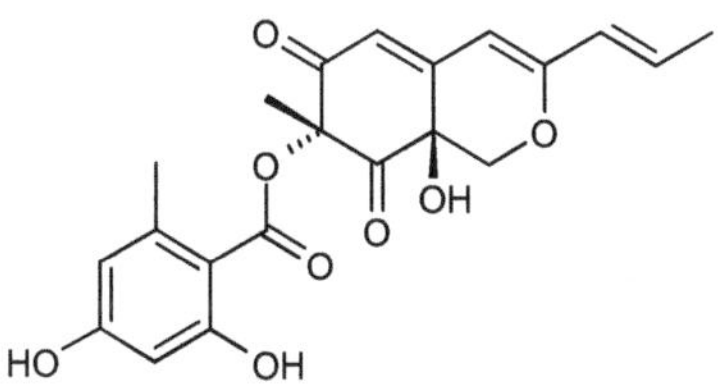

738 (+)-Sclerotiorin

Type: Azaphilone 2-benzopyrans. $C_{21}H_{23}ClO_5$ Orange-yellow needles, mp 206 °C, $[\alpha]_D^{20}$ = +500° (*c* = 1, $CHCl_3$). Source: Fungi *Penicillium sclerotiorum* and *Penicillium multicolor*. Pharm: Antifungal (yeast and filamentous fungi); anti-meiotic; inhibits cholesteryl ester transfer protein activity. Ref: J. -Y. Nam, et al, J. Microbiol. Biotechnol., 2000, 10, 544

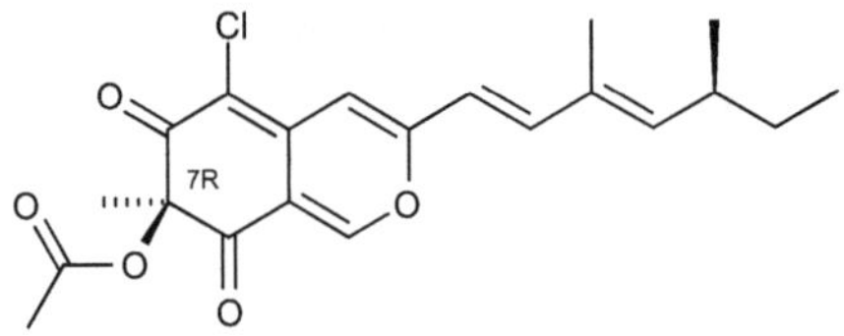

739 (–)-Sclerotiorin

Type: Azaphilone 2-benzopyrans. $C_{21}H_{23}ClO_5$ Yellow needles, mp 203–204 °C, $[\alpha]_D^{26}$ = –482° (*c* = 0.11, EtOH). Source: Unidentified marine fungus, fungi *Penicillium sclerotiorum* and *Penicillium multicolor*. Pharm: Antifungal (yeast and filamentous fungi); anti-meiotic; inhibits cholesteryl ester transfer protein activity. Ref: J. -Y. Nam, et al, J. Microbiol. Biotechnol., 2000, 10, 544 | L. Bao, et al, Nat. Prod. Commun., 2010, 5, 1789

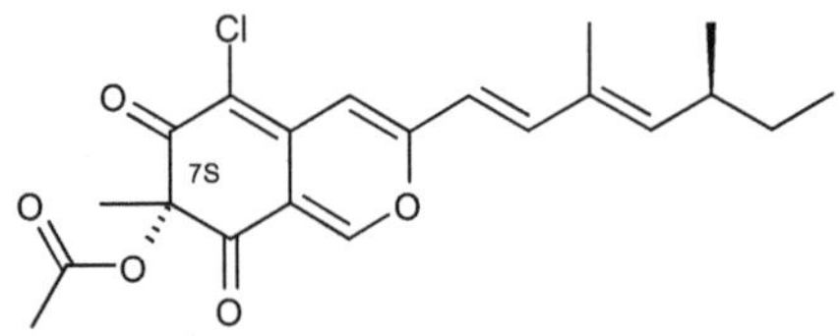

740 Secochaetomugilin A

Type: Azaphilone 2-benzopyrans. $C_{24}H_{31}ClO_8$ Pale yellow powder ($CHCl_3$/MeOH), mp 97–99 °C, $[\alpha]_D^{22}$ = +294° (*c* = 0.09, EtOH). Source: Marine-derived fungus *Chaetomium globosum* OUPS-T106B-6 from fish *Mugil cephalus*. Pharm: Cytotoxic (P_{388}, IC_{50} = 38.6 µmol/L; HL60, IC_{50} = 47.2 µmol/L; L_{1210}, IC_{50} = 53.6 µmol/L; KB, IC_{50} = 47.2 µmol/L). Ref: T. Yamada, et al, Mar. Drugs, 2009, 7, 249

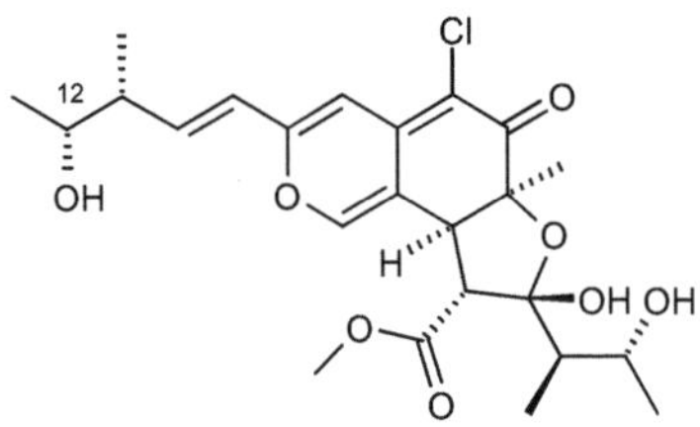

741 Secochaetomugilin D

Type: Azaphilone 2-benzopyrans. $C_{24}H_{31}ClO_7$ Pale yellow powder ($CHCl_3$/MeOH), mp 97–99 °C, $[\alpha]_D^{22}$ = +161.3° (*c* = 0.13, EtOH). Source: Marine-derived fungus *Chaetomium globosum* OUPS-T106B-6 from fish *Mugil cephalus*. Pharm: Cytotoxic (leukemia P_{388} cell line, IC_{50} = 38.6 µmol/L; hmn HL60 leukemia cell line, IC_{50} = 47.2 µmol/L; murine leukemia L_{1210} cell line, IC_{50} = 53.6 µmol/L; hmn epidermoid carcinoma KB cell line, IC_{50} = 47.2 µmol/L). Ref: T. Yamada, et al, Mar. Drugs, 2009, 7, 249

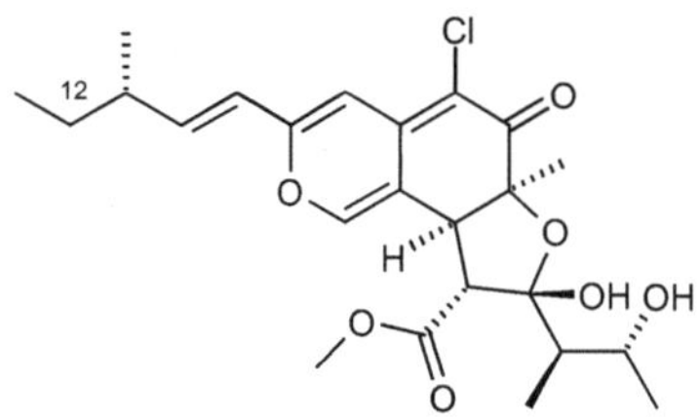

742 Aspergione B

Pseudodeflectusin Type: Furo-2-benzopyrans $C_{15}H_{16}O_4$ Needles (THF/hexane), mp 179–180 °C, $[\alpha]_D^{23}$ = +11° (c = 0.18, MeOH) (+1.9). Source: Marine-derived fungus *Aspergillus versicolor* from sponge *Xestospongia exigua*, marine-derived fungus *Aspergillus pseudodeflectus*. Pharm: Cytotoxic (NUGC-3, HeLa-S3, HL60, LD_{50} = 39 μmol/L). Ref: W. Lin, et al, JNP, 2003, 66, 57 | A. Ogawa, et al, BoMCL, 2004, 14, 3539

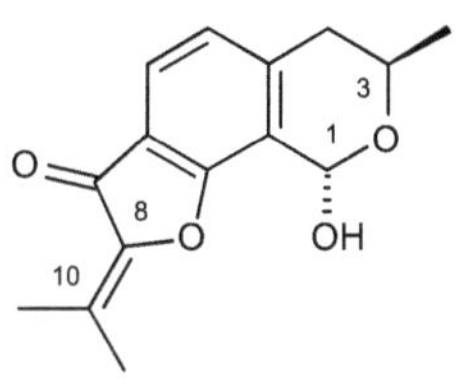

743 Pergillin

6,9-Dihydro-7-hydroxy-7-methyl-2-(1-methylethylidene)-7*H*-furo[3,2-*h*][2]benzopyran-3(2*H*)-one Type: Furo-2-benzopyrans $C_{15}H_{16}O_4$ Source: Mangrove-derived fungus *Aspergillus ustus* 094102 from mangrove *Bruguiera gymnorrhiza* (rhizosphere soil, China), terrestrial fungus (*Aspergillus ustus* from pea *Pisum sativum* seeds). Pharm: Plant growth inhibitor. Ref: H. G. Cutler, et al, J. Agric. Food Chem., 1980, 28, 989 | Z. Y. Lu, et al, JNP, 2009, 72, 1761

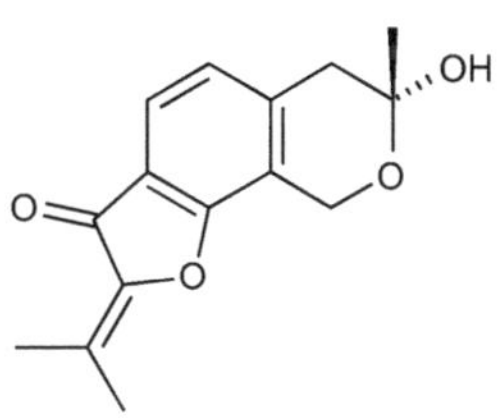

744 Amicoumacin A

Type: Isocomarines. $C_{20}H_{29}N_3O_7$ Powder (hydrochloride), mp 132–135 °C (dec) (hydrochloride), $[\alpha]_D^{23}$ = −97.2° (c = 1.0, MeOH) (hydrochloride). Source: Marine-derived bacteria *Bacillus subtilis* B1779 (sediment, Red Sea) and *Bacillus pumilus*. Pharm: Cytotoxic ((MTT assay, HeLa, IC_{50} = 4.32 μmol/L); antibacterial

(*Staphylococcus aureus, Bacillus subtilis, Loktanella hongkongensis*); anti-inflammatory; antiulcer; acaricide. Ref: J. Itoh, et al, Agric. Biol. Chem., 1982, 46, 1255 | Y. Li, et al, Mar. Drugs, 2012, 10, 319

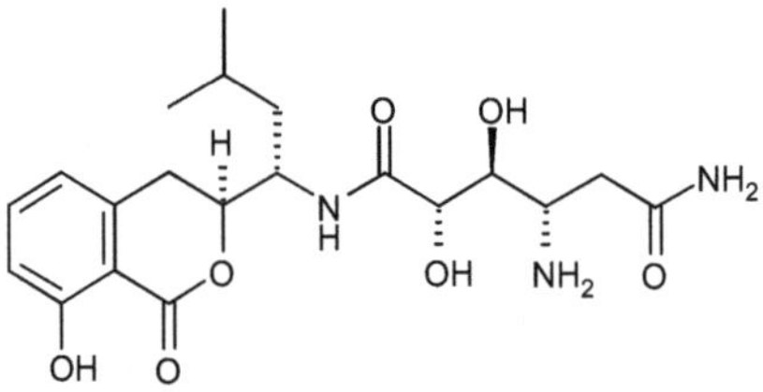

745 Amicoumacin B

Type: Isocomarines. $C_{20}H_{28}N_2O_8$ Cryst. $+H_2O$, mp 137–145 °C (dec), $[\alpha]_D^{23} = -106.1°$ (c = 0.5, MeOH). Source: Marine-derived bacterium *Bacillus subtilis* B1779 (sediment, Red Sea), marine-derived bacterium *Bacillus pumilus* from sponge *Dendrilla* sp., marine-derived fungus *Alternaria tenuis* Sg17-1. Pharm: Antibacterial (weak); antiulcer (weak). Ref: S. S. Afiyatullov, et al, Chem. Nat. Compd. (Engl. Transl.), 1991, 27, 765 | Y. Li, et al, Mar. Drugs, 2012, 10, 319

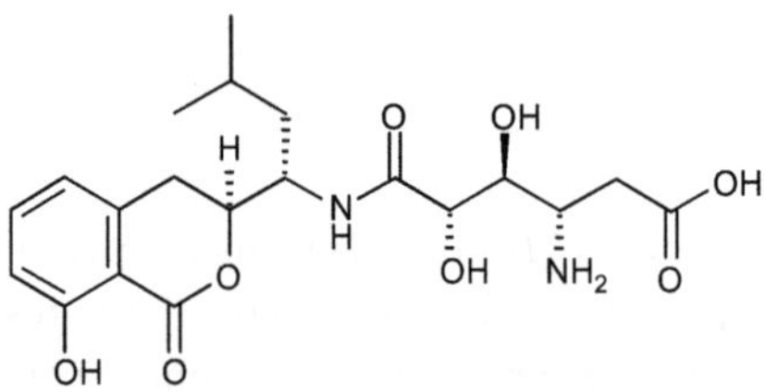

746 Amicoumacin C

Type: Isocomarines. $C_{20}H_{26}N_2O_7$ Powder, mp 131–133 °C (dec), $[\alpha]_D^{23} = -81.6°$ (c = 0.5, MeOH). Source: Marine-derived bacterium *Bacillus subtilis* B1779 (sediment, Red Sea). Pharm: Antibacterial; anti-inflammatory. Ref: J. Itoh, et al, Agric. Biol. Chem., 1982, 46, 1255; 2659 | Y. Li, et al, Mar. Drugs, 2012, 10, 319

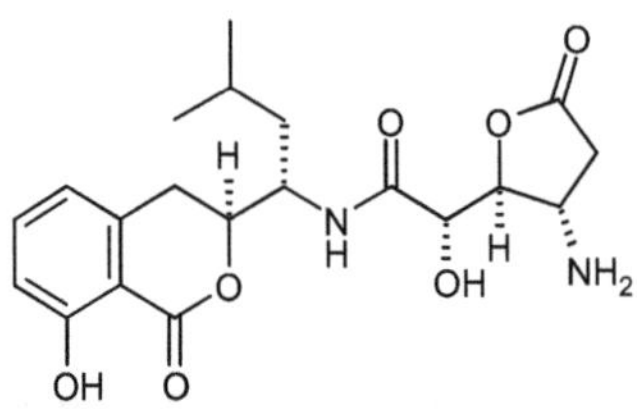

747 Antibiotic AI 77F

Type: Isocomarines. $C_{20}H_{23}NO_7$ Source: Marine-derived bacterium *Bacillus subtilis* B1779 (sediment, Red Sea), marine-derived bacterium *Bacillus pumilus* from sponge

Dendrilla sp., marine-derived fungus *Alternaria tenuis* Sg17-1. Pharm: Antibacterial; anti-inflammatory. Ref: Y. -F. Huang, et al, J. Antibiot., 2006, 59, 355 | Y. Li, et al, Mar. Drugs, 2012, 10, 319

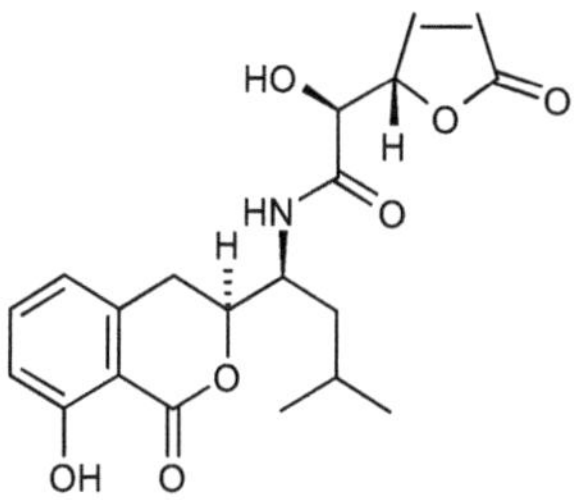

748 Antibiotic PM94128

Type: Isocomarines. $C_{22}H_{34}N_2O_6$ mp 172–173 °C, $[\alpha]_D^{25}$ = −88.9° (*c* = 2, $CHCl_3$). Source: Marine bacterium *Bacillus* sp. Pharm: Cytotoxic (P_{388}, IC_{50} = 0.05 µmol/L, A549, IC_{50} = 0.05 µmol/L, HT29, IC_{50} = 0.05 µmol/L, MEL28, IC_{50} = 0.05 µmol/L). Ref: L. M. Canedo, et al, J. Antibiot., 1997, 50, 175

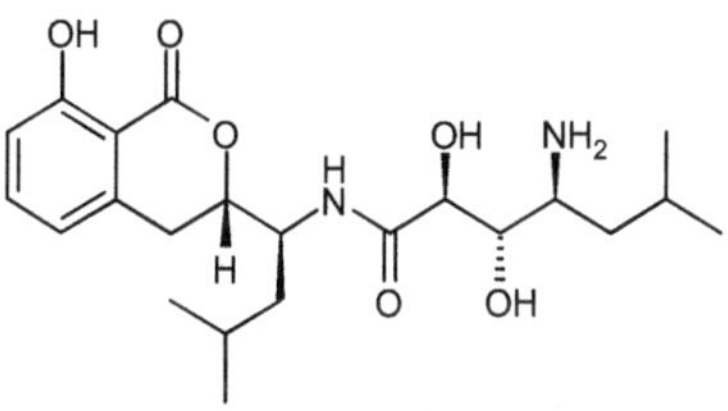

749 Antibiotic Sg17-1-4

Type: Isocomarines. $C_{24}H_{34}N_2O_9$ Amorph. powder (MeOH), mp 147–149 °C, $[\alpha]_D^{20}$ = −110° (*c* = 0.1, MeOH). Source: Marine-derived fungus *Alternaria tenuis* Sg17-1 from an unidentified alga (China waters). Pharm: Cytotoxic (A375-S2, weak; HeLa, moderate). Ref: Y. F. Huang, et al, J. Antibiot., 2006, 59, 355

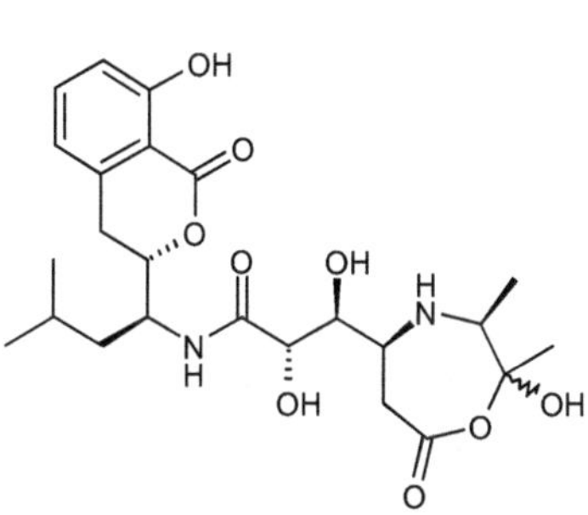

750 Aspergillumarin A

Type: Isocomarines. $C_{14}H_{16}O_4$ Source: Mangrove-derived fungus *Aspergillus* sp. from mangrove *Bruguiera gymnorrhiza* (leaf, South China Sea). Pharm: Antibacterial (50 µg/mL, *Staphylococcus aureus* and *Bacillus subtilis*, weak). Ref: S. Li, et al, Chem. Nat. Compd., 2012, 48, 371

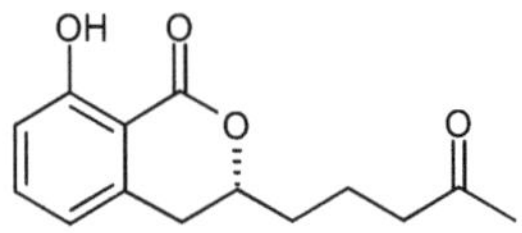

751 Aspergillumarin B

Type: Isocomarines. $C_{14}H_{18}O_4$ Source: Mangrove-derived fungus *Aspergillus* sp. from mangrove *Bruguiera gymnorrhiza* (leaf, South China Sea). Pharm: Antibacterial (50 µg/mL, *Staphylococcus aureus* and *Bacillus subtilis*, weak). Ref: S. Li, et al, Chem. Nat. Compd., 2012, 48, 371

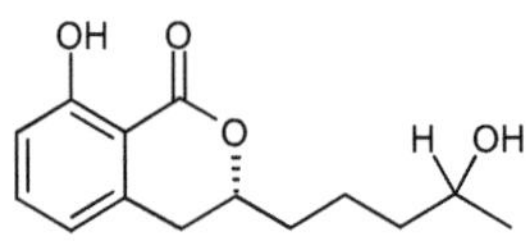

752 Bacilosarcin B

Type: Isocomarines. $C_{24}H_{35}N_3O_8$ Powder, $[\alpha]_D^{25} = -22.9°$ (c = 0.78, MeOH). Source: Marine-derived bacteria *Bacillus subtilis* B1779 (sediment, Red Sea) and *Bacillus subtilis* TP-B0611. Pharm: Cytotoxic (MTT assay, HeLa, IC_{50} = 33.60 µmol/L).antibacterial (strong). Ref: M. Azumi, et al, Tetrahedron, 2008, 64, 6420 | Y. Li, et al, Mar. Drugs, 2012, 10, 319

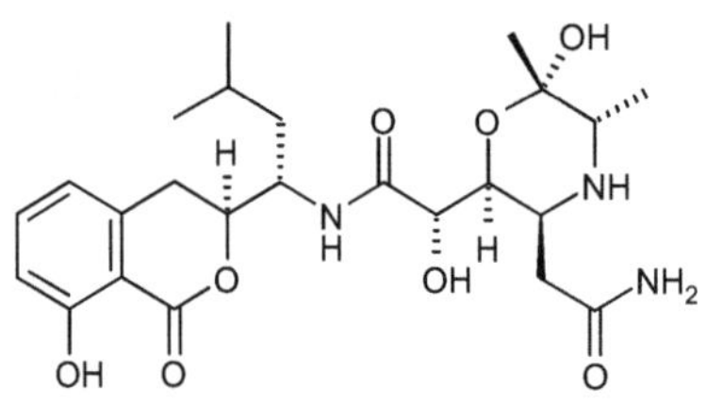

753 5-Carboxymellein

Type: Isocomarines. $C_{11}H_{10}O_5$ mp 250–252 °C, $[\alpha]_D^{25} = -195°$ (c = 0.14, EtOH). Source: Marine-derived fungus *Halorosellinia oceanica* BCC5149 (Thailand), marine-derived fungi *Hypoxylon* spp., unidentified mangrove-derived fungi Nos. 1839 and dz17 Pharm: Cytotoxic (KB, IC_{50} = 3 µg/mL; BC-1, IC_{50} = 3 µg/mL); antimalarial (*Plasmodium falciparum*, IC_{50} = 4 µg/mL). Ref: M. Chinworrungsee, et al, BoMCL, 2001, 11, 1965

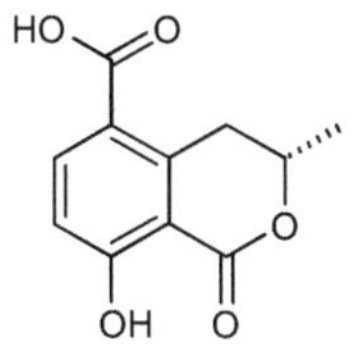

754 Decarboxyhydroxycitrinone

Type: Isocomarines. $C_{11}H_{10}O_5$ Source: Marine-derived fungus *Arthrinium sacchari* from an unidentified sponge (Atami-shi, Shizuoka, Japan). Antiangiogenic. Ref: M. Tsukada, et al, JNP, 2011, 74, 1645

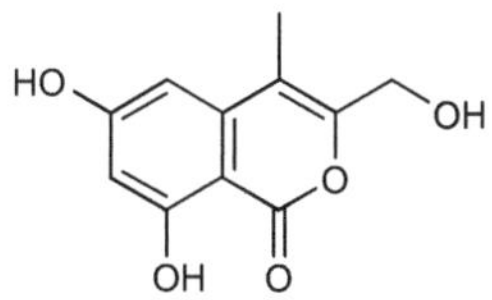

755 1-Deoxyrubralactone

Type: Isocomarines. $C_{14}H_{12}O_5$ Amorph. solid, $[\alpha]_D^{22} = -8.3°$ (c = 0.06, $CHCl_3$). Source: Marine-derived unidentified fungus from unidentified algae (Japan waters). Pharm: Eukaryotic DNA polymerases selective inhibitor (X and Y families, IC_{50} = 12–60 μmol/L, MMOA: specific inhibition of DNA polymerase β and κ); inhibitor inactive (other eukaryotic or prokaryotic polymerases, or other DNA metabolic enzymes). Ref: M. Naganuma, et al, BoMC, 2008, 16, 2939

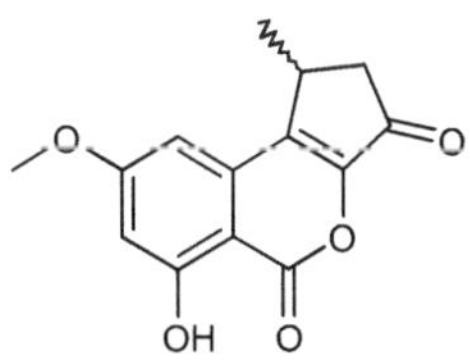

756 3,4-Dihydro-3,4,6,8-tetrahydroxy-3-methyl-1*H*-2-benzopyran-1-one

Type: Isocomarines. $C_{10}H_{10}O_6$ Source: Marine-derived fungus *Penicillium* sp. BM923 (from a marine sediment, offshore Japan). Pharm: Mammalian cell cycle inhibitor. Ref: C. -B. Cui, et al, J. Antibiotics, 1996, 49, 216

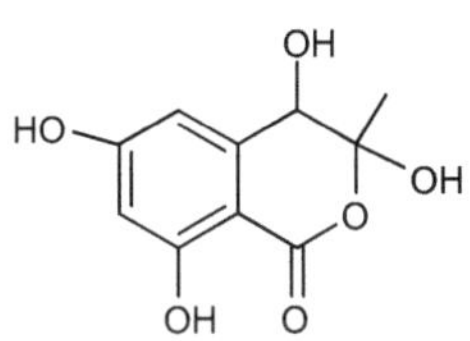

757 (3*R*,4*S*)-6,8-Dihydroxy-3,4,5-trimethyisochroman-1-one

Type: Isocomarines. $C_{12}H_{14}O_4$ Source: Deep-sea fungus *Aspergillus sydowi* YH11-2 (Guam, depth of 1000 m, E144°43′ N13°26′). Pharm: Cytotoxic (P_{388}, IC_{50} = 53.2 μmol/L, control CDDP, IC_{50} = 0.039 μmol/L). Ref: L. Tian, et al, Arch. Pharm. Res,. 2007, 30, 1051

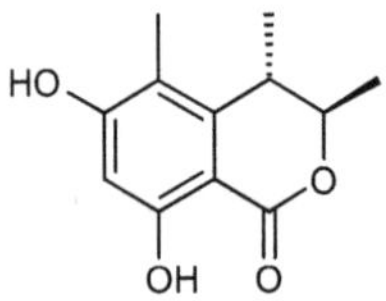

758 Hiburipyranone

Type: Isocomarines. $C_{12}H_{13}BrO_4$ $[\alpha]_D$ = −2.3° (*c* = 0.028, $CHCl_3$). Source: Sponge *Mycale adhaerens*. Pharm: Cytotoxic (P_{388}, IC_{50} = 0.19 μg/mL). Ref: N. Fusetani, et al, JOC, 1991, 56, 4971 | K. Uchida, et al, Tetrahedron, 1998, 54, 8975

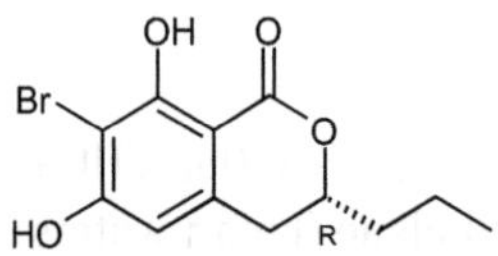

759 6-Hydroxymellein

Type: Isocomarines. $C_{10}H_{10}O_4$ Prisms (Me_2CO/petrol), mp 214–217 °C, $[\alpha]_D^{20}$ = −63° (*c* = 0.6, EtOH). Source: Marine-derived fungus *Aspergillus terreus* PT06-2 (Grown in high salt medium 10% salinity). Pharm: Phytotoxic. Ref: M. S. Islam,et al, Tetrahedron, 2007, 63, 1074 | Y. Wang, et al, Mar. Drugs, 2011, 9, 1368

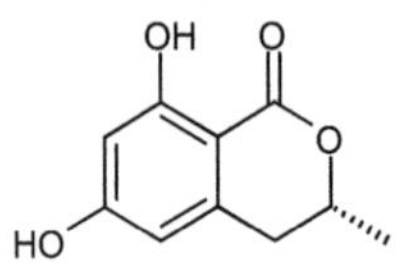

760 6-Hydroxy-5-methylramulosin

Type: Isocomarines. $C_{11}H_{16}O_4$ Oil, $[\alpha]_D^{25}$ = +30° (*c* = 0.24, EtOH). Source: Marine-derived fungus *Sterile mycelium* from green alga *Codium fragile* (Japan waters). Pharm: Cytotoxic (HeLa, 50 μg/mL, growth InRt = 65%, weak). Ref: A. A. El-Beih, et al, CPB, 2007, 55, 953

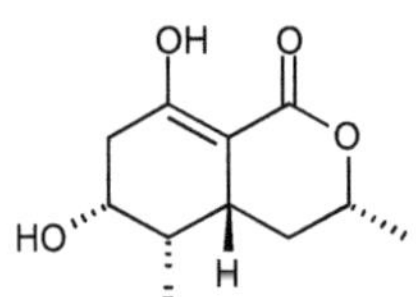

761 Irciniastatin A

Psymberin Type: Isocomarines. $C_{31}H_{47}NO_{11}$ Amorph. powder, $[\alpha]_D = +29°$ ($c = 0.02$, MeOH). Source: Sponges *Ircinia* cf. *ramosa* and *Psammocinia* aff. *bulbosa*. Pharm: Cytotoxic (murine and hmn cancer cell growth inhibitor, GI_{50} from 0.001 μg/mL to < 0.0001 μg/mL). Ref: G. R. Pettit, et al, JMC, 2004, 47, 1149 | R. H. Cichewitz, et al, Org. Lett., 2004, 6, 1951 | S. J. Robinson, et al, JNP, 2007, 70, 1002

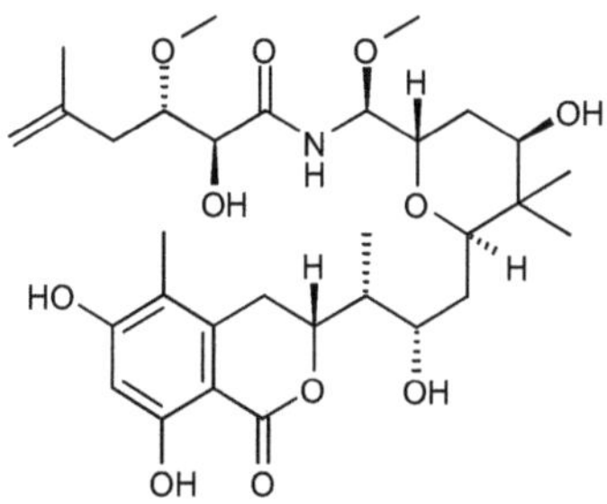

762 Irciniastatin B

Type: Isocomarines. $C_{31}H_{45}NO_{11}$ Amorph. powder, $[\alpha]_D = -4.7°$ ($c = 0.15$, MeOH). Source: Sponge *Ircinia* cf. *ramosa*. Pharm: Cytotoxic (murine and hmn cancer cell growth inhibitor, GI_{50} from 0.001 μg/mL to < 0.0001 μg/mL). Ref: G. R. Pettit, et al, JMC, 2004, 47, 1149

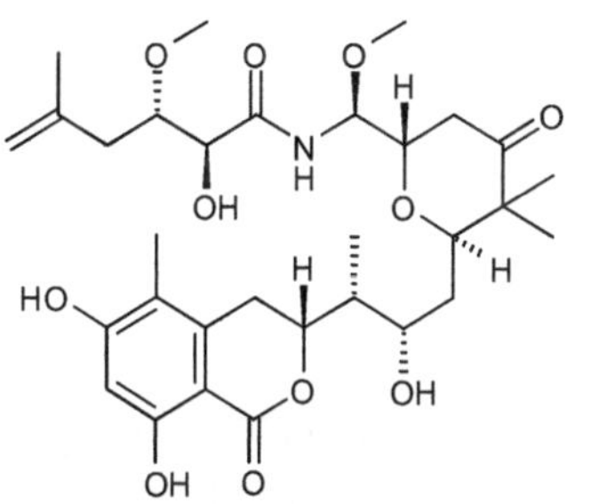

763 (–)-5-Methylmellein

Type: Isocomarines. $C_{11}H_{12}O_3$ Cryst., mp 126–127 °C, $[\alpha]_D^{15} = -105°$ ($c = 0.5$, $CHCl_3$). Source: Marine-derived fungus *Sterile mycelium* from green alga *Codium fragile* (Japan waters). Pharm: Antifungal (phytopathogenic fungi). Ref: A. Ballio, et al, Tet. Lett., 1966, 7, 3723 | A. A. El-Beih, et al, CPB, 2007, 55, 953

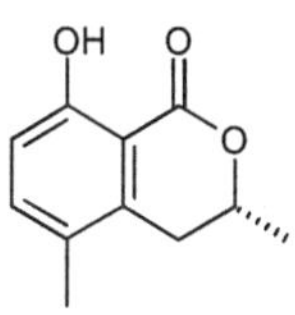

2.6 Flavonoids

764 Thalassiolin A

2″-*O*-Sulfoglucoluteolin Type: Flavones. $C_{21}H_{20}O_{14}S$. Source: Green alga *Thalassia testudinum*. Pharm: Inhibits growth of co-occuring thaustrochytrid *Schizochytrium aggregatum*. Ref: P. R. Jensen, et al, Appl. Envir. Microb., 1998, 64, 1490

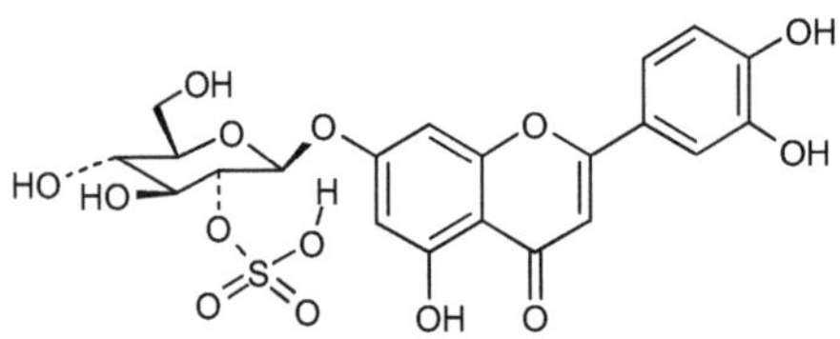

765 Actinoflavoside

Type: Dihydroflavones (Flavanones). $C_{27}H_{33}NO_9$ Non-cryst. solid, $[\alpha]_D$ = −110° (c = 1.3, MeOH). Source: Marine-derived streptomycete *Streptomyces* sp. (from estuarine sediment, New Zealand). Pharm: Antibacterial (gram-positive bacteria *Staphylococcus pneunoniae*, MIC = 64 μg/mL, *Staphylococcus aureus*, MIC = 64 μg/mL, *Micrococcus luteus*, MIC = 64 μg/mL). Ref: Z. -D. Jiang, et al, Tet. Lett., 1997, 38, 506

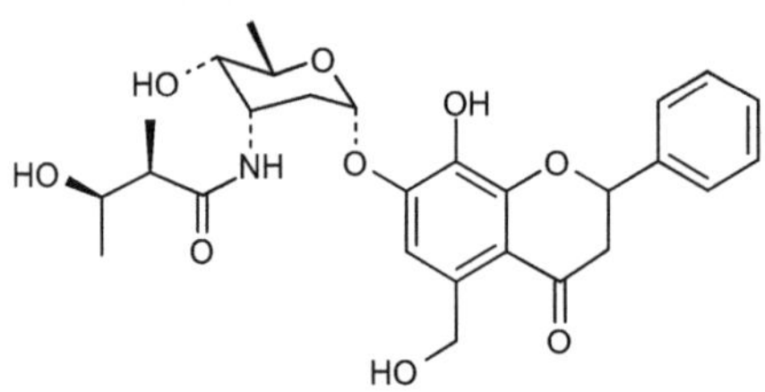

766 Populnin

Kaempferol 7-glucoside Type: Flavonols. $C_{21}H_{20}O_{11}$ Cryst. (H_2O), mp 250 °C. Source: Mangrove *Thespesia populnea*, and many other plants. Pharm: Anti-inflammatory. Ref: R. Slimestad, et al, Phytochemistry, 1995, 40, 1537

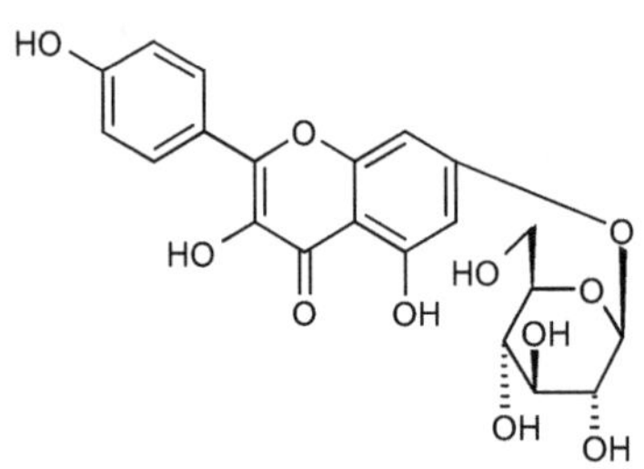

767 Genistein

4′,5,7-Trihydroxyisoflavone Type: Isoflavones. $C_{15}H_{10}O_5$ Prisms (EtOH aq), mp 301–302 °C (dec). Source: Marine-derived fungus *Ascochyta salicorniae* from green alga *Ulva* sp. (German coast of the North Sea), terrestrial plants (Leguminosae spp. and Papilionoideae spp.). Pharm: Cytotoxic (NCI-H460, GI_{50} = 1.5 μg/mL; KM20L2, GI_{50} = 3.7 μg/mL; DU145, GI_{50} = 1.9 μg/mL; BXPC3, GI_{50} = 2.5 μg/mL; MCF7, GI_{50} = 2.5 μg/mL; SF268, GI_{50} = 1.7 μg/mL) (Pettit, 2008); yrosine kinase inhibitor ($TKp56^{lck}$, 40 μg/mL, InRt = 45%, 200 μg/mL, InRt = 93%)); antiplasmodial (*Plasmodium falciparum* K1, IC_{50} = 9.272 μg/mL; *Plasmodium falciparum* NF 54, IC_{50} = 6.054 μg/mL); estrogen (weak); antioxidant. Ref: C. Osterhage, et al, JOC, 2000, 65, 6412 | G. R. Pettit, et al, JNP, 2008, 71, 438 | CRC Press, DNP on DVD, 2012, version 20.2

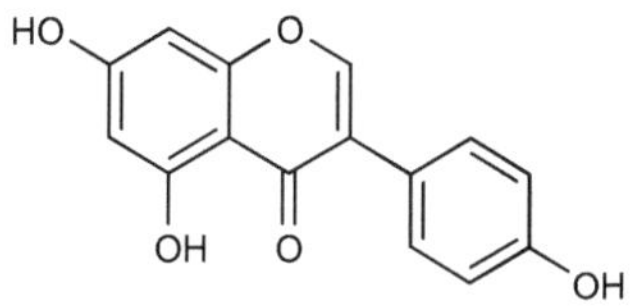

768 Karanjapin

Type: Chalcone flavonoids. $C_{18}H_{16}O_7$ Cryst. ($CHCl_3$/MeOH), mp 136 °C. Source: Semimangrove *Pongamia pinnata* (root bark). Pharm: Antioxidant Ref: A. Ghosh, et al, Nat. Prod. Commun., 2009, 4, 209

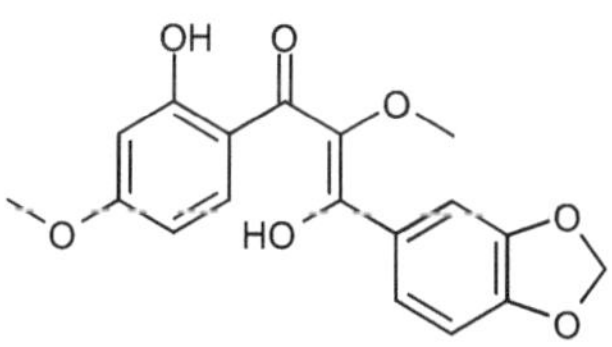

2.7 Tannin Glucose

769 Excoecariphenol D

Type: Tannin glucose. $C_{46}H_{36}O_{31}$ Source: Mangrove *Excoecaria agallocha* (China waters). Pharm: NS3-4A protease inhibitor; anti-hepatitis C virus (HCV). Ref: Y. Li, et al, BoMCL, 2012, 22, 1099

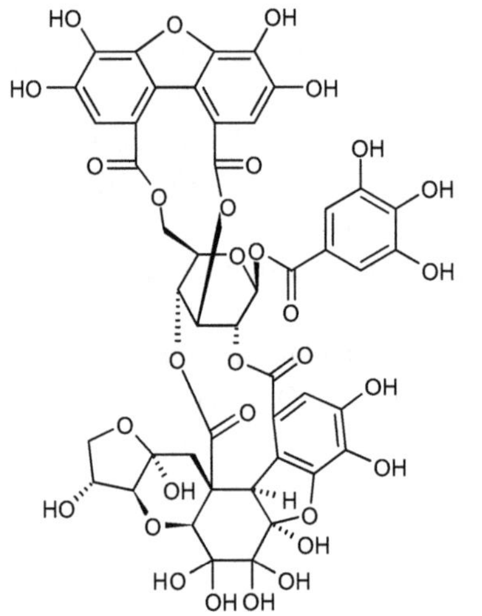

2.8 Neolignans

770 Kandelisesquilignan A

Type: Neolignans. $C_{31}H_{36}O_{11}$ Source: Mangrove *Kandelia obovata* (aerial parts, Ximen I., Zhejiang, China). Pharm: Antioxidant (DPPH scavenger, similar activity level with ascorbic acid). Ref: H. Nan, et al, Heterocycles, 2013, 87, 1093

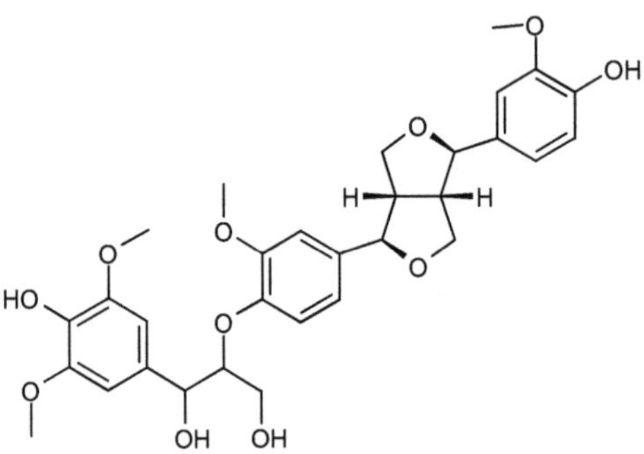

771 Kandelisesquilignan B

Type: Neolignans. $C_{32}H_{38}O_{12}$ Source: Mangrove *Kandelia obovata* (aerial parts, Ximen I., Zhejiang, China). Pharm: Antioxidant (DPPH scavenger, similar activity level with ascorbic acid). Ref: H. Nan, et al, Heterocycles, 2013, 87, 1093

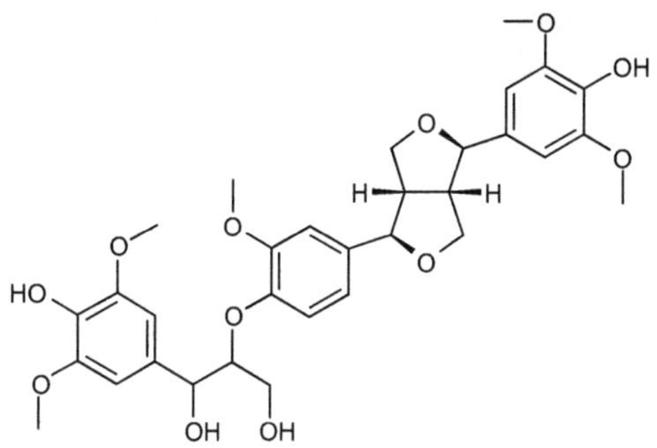

2.9 Naphthalene Derivatives

772 Accinal A

Type: Naphthalenes. $C_{12}H_{10}O_3$ Source: Marine fungus *Pestalotiopsis vaccinii.* Pharm: COX-2 inhibitor (potent). Ref: J. Wang, et al, Tetrahedron, 2014, 70, 9695

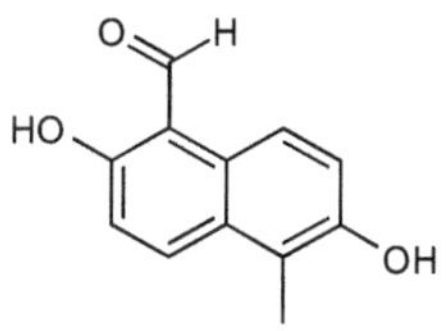

773 Balticol A

Type: Naphthalenes. $C_{14}H_{16}O_5$ Amorph. powder, $[\alpha]_D^{22} = -29.1°$ (c = 0.28, MeOH). Source: An unidentified marine-derived fungus (order Pleosporales, phylum Ascomycota) 222 from drift wood (Baltic coast, Germany). Pharm: Antiviral (influenza A virus/MDCK, no activity, control Amantadine sulfate, IC_{50} = 15.0 μg/mL; HSV-1 virus, IC_{50} = 1 μg/mL, control Aciclovir, IC_{50} = 0.1 μg/mL; tested at non-cytotoxic concentrations). Ref: M. A. M. Shushni, et al, Chem. Biodiversity, 2009, 6, 127

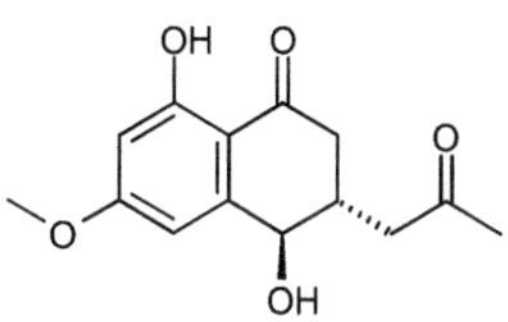

774 Balticol B

Type: Naphthalenes. $C_{12}H_{14}O_5$ Amorph. solid, $[\alpha]_D^{22} = -8.5°$ (c = 1.2, MeOH). Source: An unidentified marine-derived fungus (phylum Ascomycota, order Pleosporales, from drift wood, Baltic coast, Germany). Pharm: Antiviral (influenza A virus/MDCK, IC_{50} = 10 μg/mL, control Amantadine sulfate, IC_{50} = 15.0 μg/mL; HSV-1 virus,

IC_{50} = 1 μg/mL, control Aciclovir, IC_{50} = 0.1 μg/mL; tested at non-cytotoxic concentrations). Ref: M. A. M. Shushni, et al, Chem. Biodiversity, 2009, 6, 127

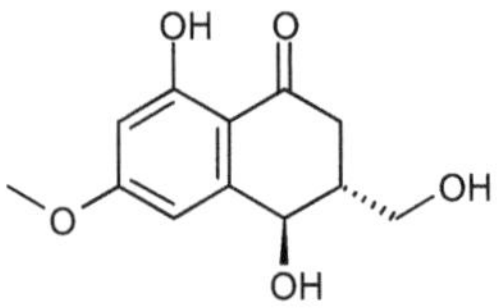

775 Balticol C

Type: Naphthalenes. $C_{12}H_{14}O_6$ Light brown oil, $[\alpha]_D^{22}$ = −18.7° (*c* = 0.45, MeOH). Source: An unidentified marine-derived fungus (phylum Ascomycota, order Pleosporales, from drift wood, Baltic coast, Germany). Pharm: Antiviral (influenza A virus/MDCK, IC_{50} = 1 μg/mL, control Amantadine sulfate, IC_{50} = 15.0 μg/mL; HSV-1 virus, IC_{50} = 1 μg/mL, control Aciclovir, IC_{50} = 0.1 μg/mL; tested at non-cytotoxic concentrations). Ref: M. A. M. Shushni, et al, Chem. Biodiversity, 2009, 6, 127

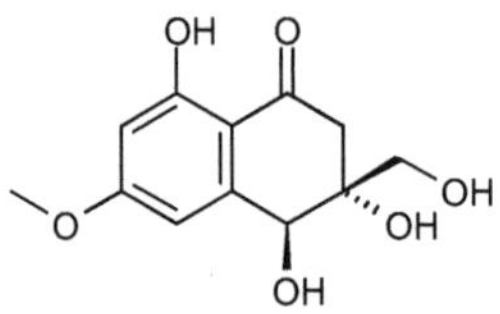

776 Balticol D

Type: Naphthalenes. $C_{14}H_{16}O_6$ Oil, $[\alpha]_D^{22}$ = −27.2° (*c* = 1, MeOH). Source: An unidentified marine-derived fungus (phylum Ascomycota, order Pleosporales, from drift wood, Baltic coast, Germany). Pharm: Antiviral (influenza A virus/MDCK, IC_{50} = 0.1 μg/mL, control Amantadine sulfate, IC_{50} = 15.0 μg/mL; HSV-1 virus, IC_{50} = 0.1 μg/mL, control Aciclovir, IC_{50} = 0.1 μg/mL; tested at non-cytotoxic concentrations). Ref: M. A. M. Shushni, et al, Chem. Biodiversity, 2009, 6, 127

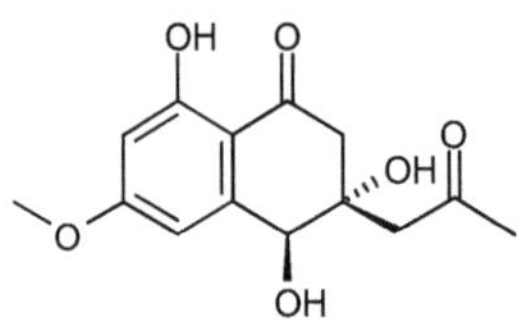

777 Balticol E

3,4-Dihydro-3,4,8-trihydroxy-3-(2-hydroxypropyl)-6-methoxy-1(2*H*)-naphthalenone
Type: Naphthalenes. $C_{14}H_{18}O_6$ Light brown oil, $[\alpha]_D^{22}$ = −13.8° (*c* = 0.24, MeOH). Source: An unidentified marine-derived fungus (phylum Ascomycota, order Pleosporales, from drift wood, Baltic coast, Germany). Pharm: Antiviral (influenza A virus/MDCK, no activity, control Amantadine sulfate, IC_{50} = 15.0 μg/mL; HSV-1

virus, IC_{50} = 0.01 µg/mL, control Aciclovir, IC_{50} = 0.1 µg/mL; tested at non-cytotoxic concentrations). Ref: M. A. M. Shushni, et al, Chem. Biodiversity, 2009, 6, 127

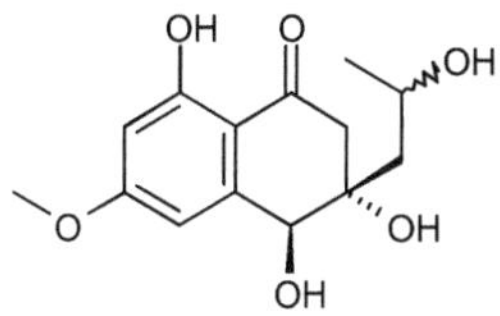

778 Balticol F

Type: Naphthalenes. $C_{15}H_{18}O_6$ Oil, $[\alpha]_D^{22}$ = +72.9° (*c* = 0.13, MeOH). Source: An unidentified marine-derived fungus (phylum Ascomycota, order Pleosporales, from drift wood, Baltic coast, Germany). Pharm: Antiviral (influenza A virus/MDCK, IC_{50} = 1 µg/mL, control Amantadine sulfate, IC_{50} = 15.0 µg/mL; HSV-1 virus, IC_{50} = 0.1 µg/mL, control Aciclovir, IC_{50} = 0.1 µg/mL; tested at non-cytotoxic concentrations). Ref: M. A. M. Shushni, et al, Chem. Biodiversity, 2009, 6, 127

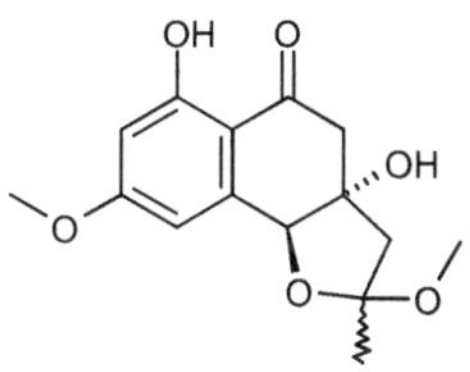

779 Coniolactone

Type: Naphthalenes. $C_{17}H_{16}O_5$ Source: Marine-derived fungus *Coniothyrium cereale* from green alga *Enteromorpha* sp. (Fehmarn I., Baltic Sea, Germany). Pharm: Antibacterial (*Mycobacterium phlei*). Ref: M. F. Elsebai, et al, Org. Biomol. Chem., 2011, 9, 802

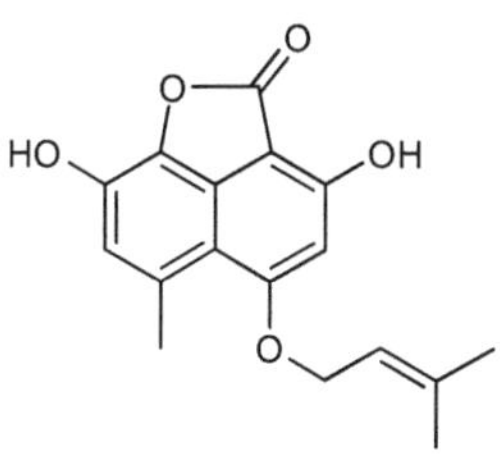

780 Keisstetralone

Type: Naphthalenes. $C_{19}H_{24}O_5$ Yellowish gum, $[\alpha]_D^{20}$ = −23.6° (*c* = 0.06, $CHCl_3$). Source: Marine-derived fungus *Keissleriella* sp. YS4108 (depth of 50 m, sediment sample, Yellow Sea near Sheyang Port, China). Pharm: Antifungal (yeast *Candida albicans*, MIC = 40 μg/mL; *Trichophyton rubrum*, MIC = 20 μg/mL; *Aspergillus niger*, MIC = 80 μg/mL). Ref: C. H. Liu, et al, PM, 2002, 68, 363

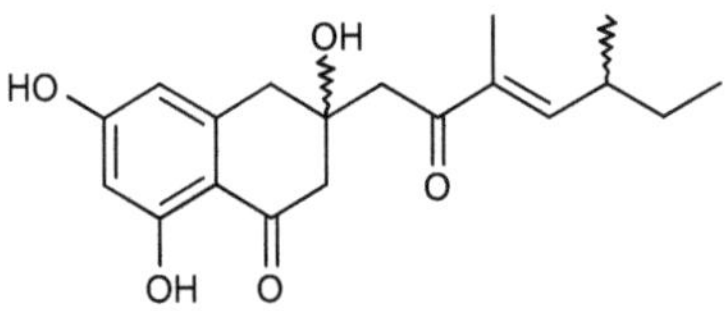

781 (4*R,5*R**,9*S**,10*R**,11*Z*)-4-Methoxy-9-((dimethylamino)-methyl)-12,15-epoxy-11(13)-en-decahydronaphthalen-16-ol**

Type: Naphthalenes. $C_{24}H_{43}NO_3$ Pale yellow oil, $[\alpha]_D^{24}$ = +72° (*c* = 0.67, MeOH). Source: Soft coral *Sinularia* sp. (Bowden Reef, Great Barrier Reef, Australia). Pharm: Cytotoxic (SF268, GI_{50} = 175 μmol/L, MCF7, GI_{50} = 70 μmol/L, H460, GI_{50} = 125 μmol/L). Ref: A. D. Wright, et al, Mar. Drugs, 2012, 10, 1619

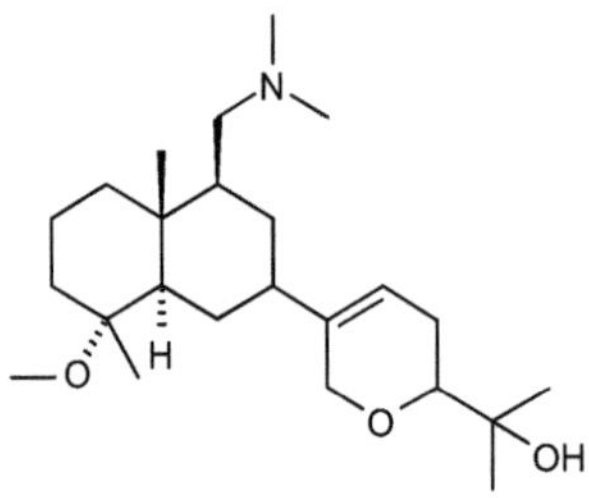

782 *N*-Phenyl-1-naphthylamin

1-Anilinonaphthalene; Antioxidant PAN Type: Naphthalenes. $C_{16}H_{13}N$ Prisms or needles (EtOH), leaflets (petrol), mp 62 °C, $bp_{528mmHg}$ 335 °C, bp_{8mmHg} 226 °C. Source: Marine-derived streptomycetes *Streptomyces* sp. B8335 Pharm: Antioxidant. Ref: K. A. Shaaban, et al, JNP, 2007, 70, 1545

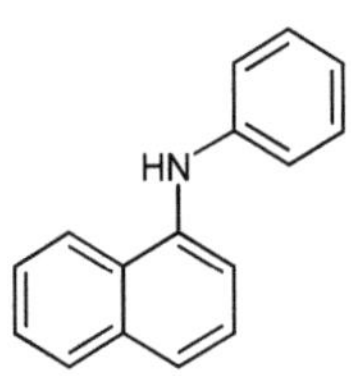

783 Phomopsidin

Antibiotic TUF 95F47 Type: Naphthalenes. $C_{21}H_{30}O_3$ $[\alpha]_D^{27}$ = +31° (*c* = 0.1, MeOH). Source: Mangrove-derived fungus *Phomopsis* sp. from an unidentified mangrove (submerged branch, Pohnpei I., Federated States of Micronesia, collected in tropical and sub-tropical coral reef environments). Pharm: Microtubule assembly inhibitor (IC_{50} = 5.7 μmol/L, control colchicines, IC_{50} = 10 μmol/L, rhizoxin, IC_{50} = 4 μmol/L);

antifungal; antibiotic. Ref: T. Yoshimoto, et al, Tennen Yuki Kagobutsu Toronkai Koen Yoshishu, 1997, 39, 637 | M. Namikoshi, et al, J. Antibiot., 1997, 50, 890 | M. Namikoshi, et al, CPB, 2000, 48, 1452 | M. Saleem, et al, NPR, 2007, 24, 1142 (rev)

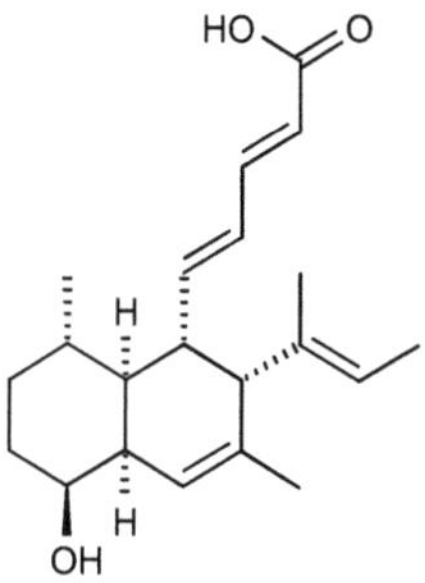

784 Tanzawaic acid A

Antibiotic GS 1302-3 Type: Naphthalenes. $C_{18}H_{26}O_2$ $[\alpha]_D^{28}$ = +147° (c = 0.04, $CHCl_3$). Source: Marine-derived fungus *Penicillium citrinum*. Pharm: Superoxide anion production inhibitor. Ref: M. Kuramoto, et al, Chem. Lett., 1997, 885

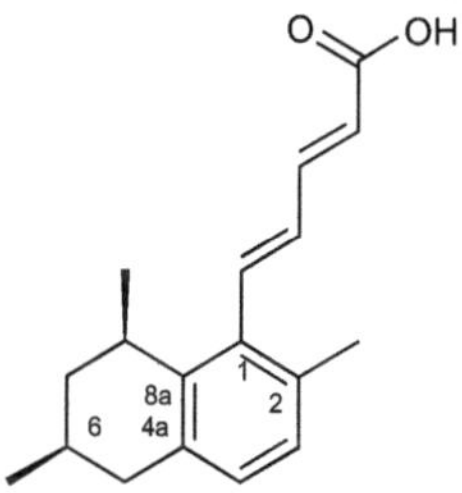

785 Tribromo-3-bromoamino-7-bromomethylnaphthalene

Type: Naphthalenes. $C_{12}H_7Br_5$ Colorless oil. Source: Red alga *Laurencia similis*. Pharm: Hypoglycemic (PTP1B inhibitor) (*in vitro* assay, IC_{50} = 102 μg/mL, control HD, IC_{50} = 0.80 μg/mL). Ref: J. Qin, et al, Bioorganic Med. Chem. Lett., 2010, *20*, 7152

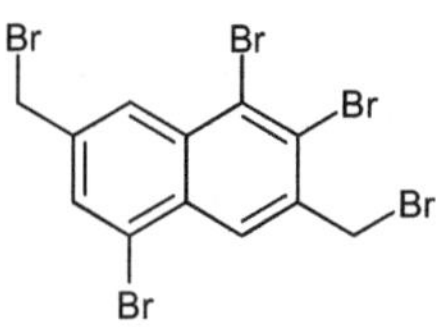

786 2,5,8-Tribromo-3-bromoamino-7-bromomethylnaphthalene

Type: Naphthalenes. $C_{12}H_7Br_5$ Colorless oil. Source: Red alga *Laurencia similis*. Pharm: Hypoglycemic (PTP1B inhibitor) (*in vitro* assay, IC_{50} = 65.30 μg/mL, control HD, IC_{50} = 0.80 μg/mL). Ref: J. Qin, et al, Bioorganic Med. Chem. Lett., 2010, *20*, 7152

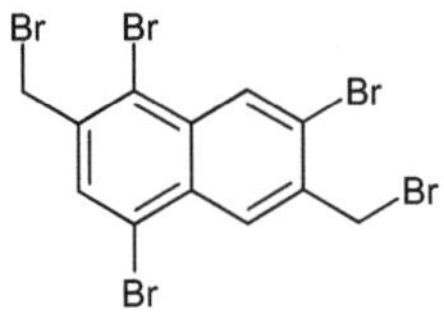

787 2,5,6-Tribromo-3-bromoamino-7-bromomethylnaphthalene

Type: Naphthalenes. $C_{12}H_7Br_5$ Colorless oil. Source: Red alga *Laurencia similis*. Pharm: Hypoglycemic (PTP1B inhibitor) (*in vitro* assay, IC_{50} = 69.80 μg/mL, control HD, IC_{50} = 0.80 μg/mL). Ref: J. Qin, et al, Bioorganic Med. Chem. Lett., 2010, *20*, 7152

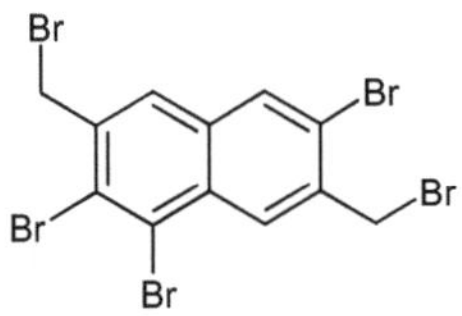

788 Trichoharzin

Type: Naphthalenes. $C_{25}H_{38}O_7$ Glassy solid, $[\alpha]_D$ = +38° (MeOH). Source: Marine-derived fungus *Trichoderma harzianum* from sponge *Mycale cecilia*. Pharm: Cytotoxic. Ref: M. Kobayashi, et al, Tet. Lett., 1993, 34, 7925

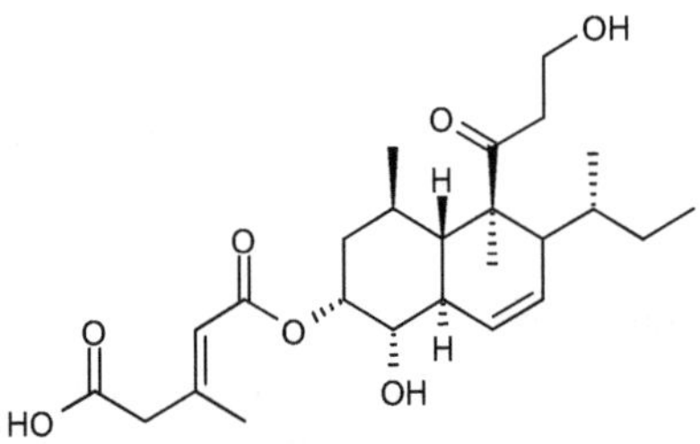

789 (–)-Cereoaldomine

Type: Furonaphthalenes. $C_{16}H_{15}NO_5$ Bright blood red crystalline solid, $[\alpha]_D^{23}$ = –320° (*c* = 0.30, MeOH). Source: Marine-derived fungus *Coniothyrium cereale* from green alga *Enteromorpha* sp. (Fehmarn I., Baltic Sea, Germany). Pharm: Hmn leukocyte elastase inhibitor (IC_{50} = 3.01 μmol/L, selective). Ref: M. F. Elsebai, et al, JNP, 2011, 74, 2282

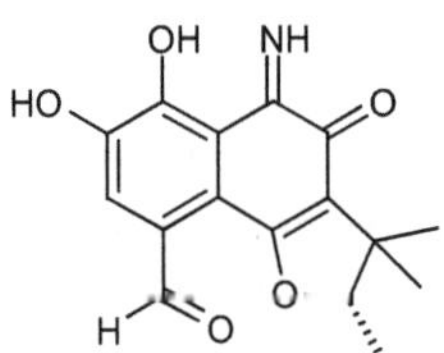

790 (–)-Cereolactam

Type: Furonaphthalenes. $C_{17}H_{17}NO_4$ Green amorph. solid, $[\alpha]_D^{23} = -70°$ (c = 0.10, MeOH). Source: Marine-derived fungus *Coniothyrium cereale* from green alga *Enteromorpha* sp. (Fehmarn I., Baltic Sea, Germany). Pharm: Hmn leukocyte elastase inhibitor (IC_{50} = 9.28 µmol/L, selective). Ref: M. F. Elsebai, et al, JNP, 2011, 74, 2282

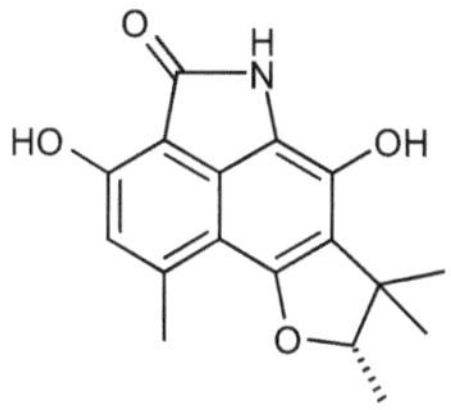

791 2,3-Dihydro-5-hydroxy-8-methoxy-2,4-dimethylnaphtho[1,2-*b*]furan-6,9-dione

Dihydroanhydrojavanicin Type: Furonaphthalenes. $C_{15}H_{14}O_5$ Red cryst. (EtOAc), mp 225–228 °C, $[\alpha]_D^{20} = -125°$ (c = 0.2, Me_2CO). Source: Marine-derived fungus *Fusarium* sp. PSU-F135. Pharm: Cytotoxic (KB, IC_{50} = 120 µmol/L, control doxorubicin, IC_{50} = 0.33 µmol/L; MCF7, IC_{50} = 101 µmol/L, control doxorubicin, IC_{50} = 2.18 µmol/L; Vero, inactive, control Ellipticine, IC_{50} = 4.47 µmol/L); antibacterial (*Staphylococcus aureus* and MRSA, MIC = 350 µmol/L). Ref: Y. Kimura, et al, Agric. Biol. Chem., 1988, 52, 1253 | J. H. Tatum, et al, Phytochemistry, 1989, 28, 283 | K. Trisuwan, et al, JNP, 2010, 73, 1507

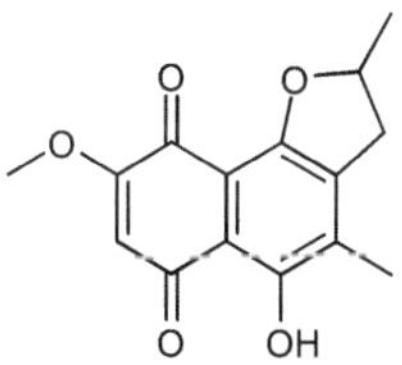

792 Marfuraquinocin A

Type: Furonaphthalenes. $C_{26}H_{32}O_5$ Source: Marine-derived streptomycete *Streptomyces niveus* (sediment, S. China Sea). Pharm: Cytotoxic (NCI–H460, growth inhibitor, moderate); antibacterial (*Staphylococcus aureus*, growth inhibitor, moderate). Ref: Y. Song, et al, JNP, 2013, 76, 2263

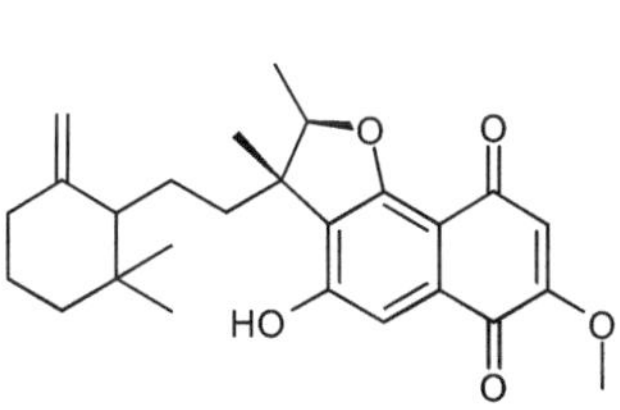

793 Marfuraquinocin C

Type: Furonaphthalenes. $C_{26}H_{32}O_6$ Source: Marine-derived streptomycete *Streptomyces niveus* (sediment, S. China Sea). Pharm: Cytotoxic (NCI–H460, growth inhibitor, moderate); antibacterial (*Staphylococcus aureus,* growth inhibitor, moderate); antibacterial (MRSE, inhibitor). Ref: Y. Song, et al, JNP, 2013, 76, 2263

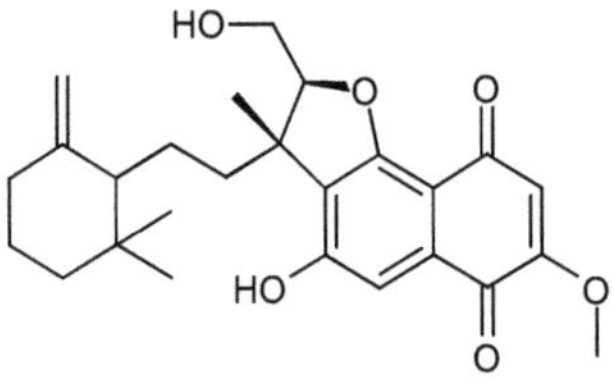

794 Marfuraquinocin D

Type: Furonaphthalenes. $C_{26}H_{32}O_6$ Source: Marine-derived streptomycete *Streptomyces niveus* (sediment, S. China Sea). Pharm: Antibacterial (*Staphylococcus aureus,* growth inhibitor, moderate); antibacterial (MRSE, inhibitor). Ref: Y. Song, et al, JNP, 2013, 76, 2263

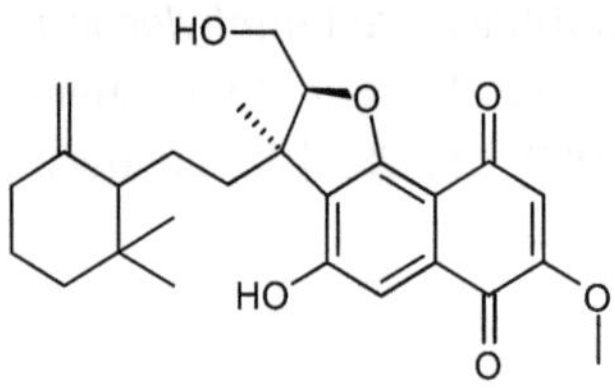

795 (−)-Trypethelone

Type: Furonaphthalenes. $C_{16}H_{16}O_4$ Dark violet-red crystalline solid, mp 250–253 °C, $[\alpha]_D^{23} = -355°$ (c = 0.10, MeOH). Source: Marine-derived fungus *Coniothyrium cereale* from green alga *Enteromorpha* sp. (Fehmarn I., Baltic Sea, Germany). Pharm: Antibacterial (*Mycobacterium phlei, Staphylococcus aureus* and *Escherichia coli*); cytotoxic (mouse broblast cells). Ref: M. F. Elsebai, et al, JNP, 2011, 74, 2282

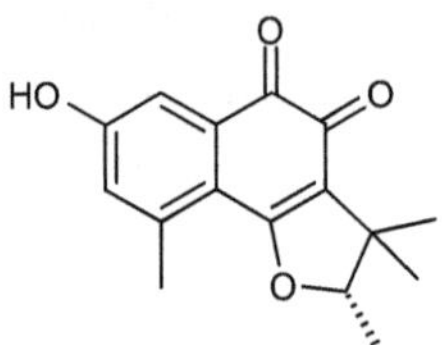

796 2-Benzyl-5-hydroxy-6,8-dimethoxy-4*H*-benzo[*g*]chromen-4-one

Type: Pyranonaphthalenes. $C_{22}H_{18}O_5$ Pale yellow needles, mp 223–225 °C. Source: Mangrove derived fungus *Phomopsis* sp. ZSU-H26 (endophytic) from mangrove *Excoecaria agallocha* (China waters). Pharm: Cytotoxic (Hep2 cell lines,

IC_{50} = 10 μg/mL; HepG2 cell lines, IC_{50} = 8 μg/mL, moderate). Ref: Z. -J. Huang, et al, Chem. Nat. Compd. (Engl. Transl.), 2010, 46, 15

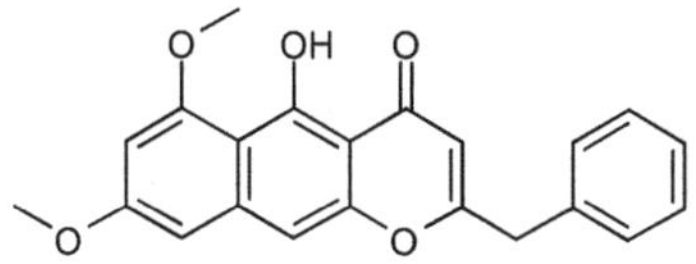

797 Fonsecin

Type: Pyranonaphthalenes. $C_{15}H_{14}O_6$ Bright yellow irregular prisms, mp 198 °C (dec). Source: Marine-derived fungus *Aspergillus carbonarius* WZ-4-11 (sediment, China). Pharm: Mycotoxin. Ref: B. Noureddine, et al, Nat. Prod. Res., 2005, 19, 653 | Y. P. Zhang, et al, Chem. Biodiversity, 2008, 5, 93

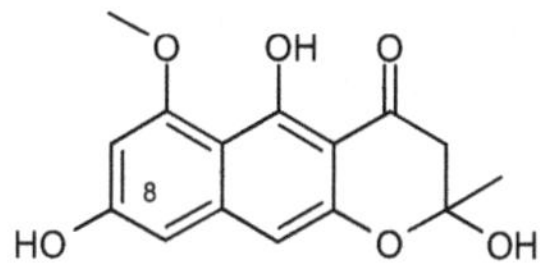

798 Nigerasperone A

Type: Pyranonaphthalenes. $C_{16}H_{14}O_6$ Yellow powder, mp 193–195 °C. Source: Marine-derived fungus *Aspergillus niger* EN-13. Pharm: Cytotoxic (hmn liver cancer SMMC-7721 and A549, IC_{50} > 10 μg/mL). Ref: Y. Zhang, et al, J. Antibiot., 2007, 60, 204

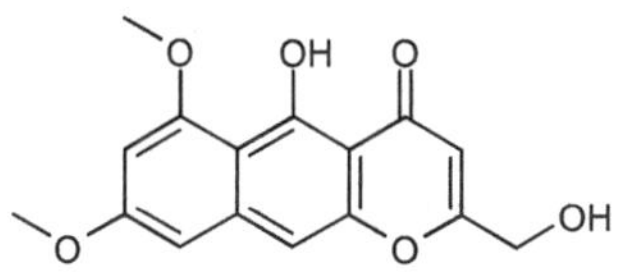

799 (−)-Scleroderolide

Type: Pyranonaphthalenes. $C_{18}H_{16}O_6$ Yellow cryst. (EtOH/petrol), mp 232–233 °C, $[\alpha]_D$ = −116° (*c* = 0.3, $CHCl_3$). Source: Marine-derived fungus *Coniothyrium cereal*, terrestrial fungus *Gremmeniella abietina*. Pharm: Antibacterial (gram-positive bacteria); antifungal (yeast); CETP inhibitor. Ref: H. Tomoda, et al, J. Antibiot., 1998, 51, 618 | M. F. Elsebai, et al, *Org. Biomol. Chem.*, 2011, *9*, 802

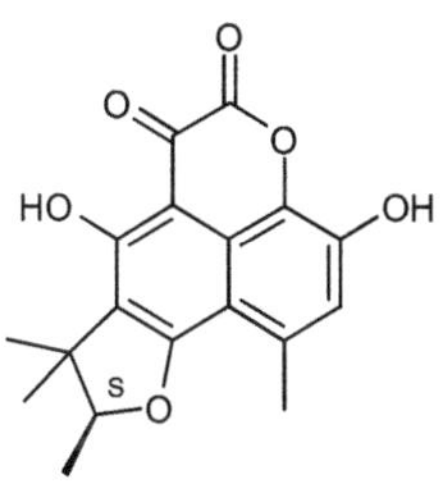

800 TMC 256A$_1$

6-*O*-Methylnorrubrofusarin Type: Pyranonaphthalenes. $C_{15}H_{12}O_5$ Yellow powder. Source: Marine-derived fungus *Aspergillus carbonarius* WZ-4-11 (sediment, China), marine-derived fungus *Aspergillus niger* TC 1629 from crinoid *Comantheria briareus*. Pharm: IL-4 signal transduction inhibitor. Ref: M. Sakurai, et al, J. Antibiot., 2002, 55, 685 | Y. P. Zhang, et al, Chem. Biodiversity, 2008, 5, 93

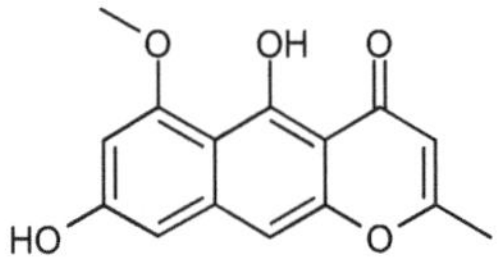

801 TMC 256A$_2$

Rubrofusarin B Type: Pyranonaphthalenes. $C_{16}H_{14}O_5$ Yellow cryst. ($CHCl_3/C_6H_6$), mp 213 °C, mp 178–180 °C. Source: Marine-derived fungus *Aspergillus carbonarius* WZ-4-11 (sediment, China). Pharm: Phytotoxic; mycotoxin. Ref: C .P. Gorst-Allman, et al, JCS Perkin I, 1980, 2474 | Y. P. Zhang, et al, Chem. Biodiversity, 2008, 5, 93

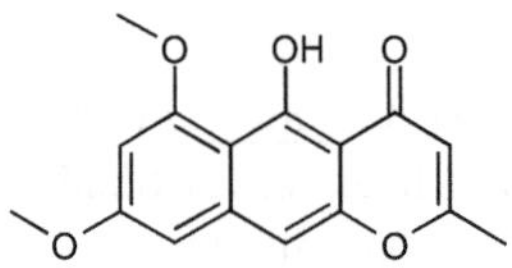

802 TMC 256C$_2$

Flavasperone Type: Pyranonaphthalenes. $C_{16}H_{14}O_5$ Yellow needles ($CHCl_3$/EtOH), mp 204 °C, mp 203–204 °C. Source: Marine-derived fungus *Aspergillus carbonarius* WZ-4-11 (sediment, China), terrestrial fungi (*Aspergillus* spp.), terrestrial *plants (Aspergillus niger* infected mango fruits and peanuts). Pharm: Mycotoxin. Ref: M. Sakurai, et al, J. Antibiot., 2002, 55, 685 | Y. P. Zhang, et al, Chem. Biodiversity, 2008, 5, 93

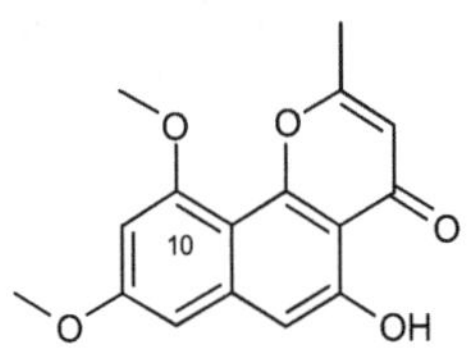

803 Antibiotic SC 2051

Type: Binaphthyls. $C_{32}H_{22}O_{12}$ Source: Marine-derived fungus *Hypocrea vinosa* AY380904. Pharm: Tyrosine kinase inhibitor (HUVEC lysate, tyrosine kinase assay, IC_{50} = 42.1 µmol/L; growth inhibition assay, IC_{50} = 17.4 µmol/L; migration assay, IC_{50} = 1.09 µmol/L; inhibitory effects on proliferation, migration, and tubule formation);

phosphodiesterase inhibitor. Ref: Japan. Pat., 1980, 80 54 897; CA, 93, 112329 | Y. Ohkawa, et al, JNP, 2010, 73, 579

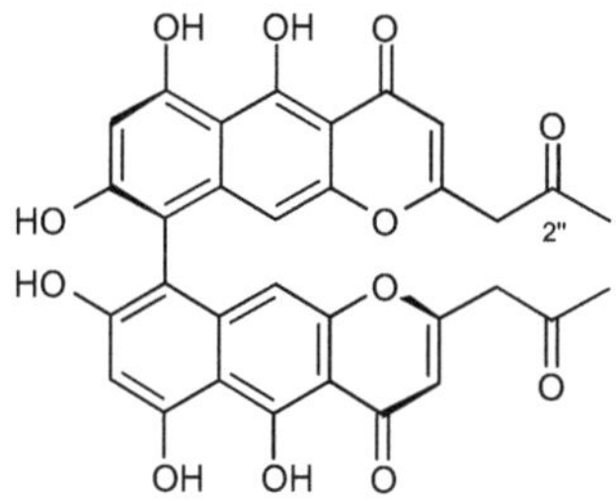

804 Asperpyrone A

Type: Binaphthyls. $C_{31}H_{24}O_{10}$ Yellow powder, $[\alpha]_D^{25} = +89°$ ($c = 0.13$, $CHCl_3$). Source: Marine-derived fungus *Aspergillus niger* EN-13, *Aspergillus niger* JV-33-48. Pharm: Cytotoxic (hmn liver cancer SMMC-7721 and A549, $IC_{50} > 10$ µg/mL); *Taq* DNA polymerase inhibitor. Ref: K. Akiyama, et al, JNP, 2003, 66, 136 | Y. Zhang, et al, J. Antibiot., 2007, 60, 204

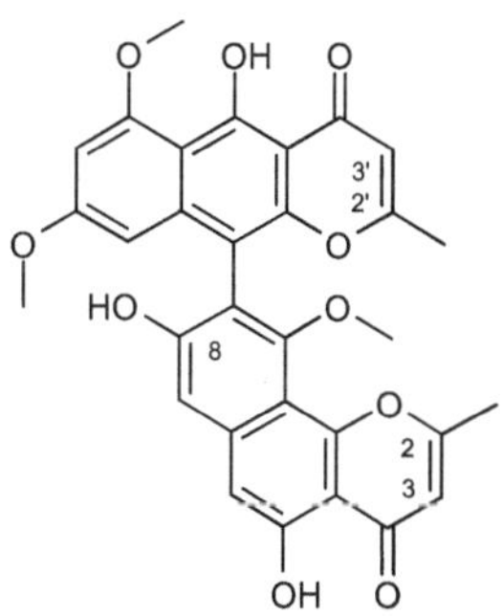

805 Asperpyrone C

Type: Binaphthyls. $C_{32}H_{26}O_{10}$ Yellow powder, $[\alpha]_D^{25} = +18°$ ($c = 0.01$, $CHCl_3$). Source: Marine-derived fungus *Aspergillus niger* EN-13, *Aspergillus niger* JV-33-48. Pharm: Cytotoxic (hmn liver cancer SMMC-7721 and A549, $IC_{50} > 10$ µg/mL); antifungal (*Candida albicans*, IZD = 10 mm, control Amphotericin B, IZD = 12 mm, weak). Ref: K. Akiyama, et al, JNP, 2003, 66, 136 | Y. Zhang, et al, J. Antibiot., 2007, 60, 204

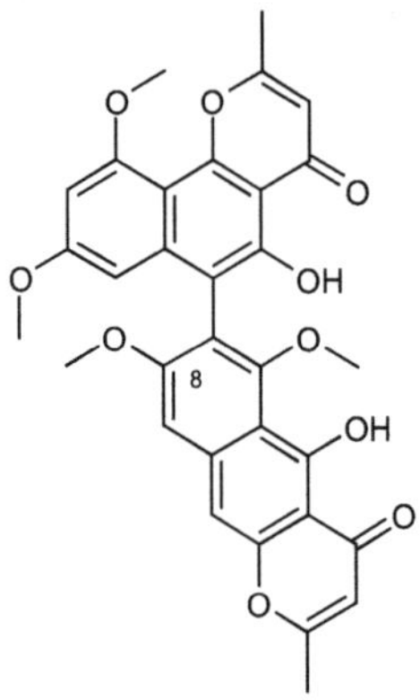

806 Aurasperone A

Type: Binaphthyls. $C_{32}H_{26}O_{10}$ Yellow plates ($CHCl_3$/propanol); needles (EtOH), mp 207 °C, mp 270 °C, mp 290–291 °C, $[\alpha]_D^{25} = -24.2°$ (c = 1.0, $CHCl_3$). Source: Marine-derived fungus *Aspergillus niger* EN-13, terrestrial fungi (*Aspergillus* spp.). Pharm: Cytotoxic (hmn liver cancer SMMC-7721 and A549, IC_{50} > 10 µg/mL); mycotoxin; insecticidal. Ref: H. Priestap, Tetrahedron, 1984, 40, 3617 | Y. Zhang, et al, J. Antibiot., 2007, 60, 204

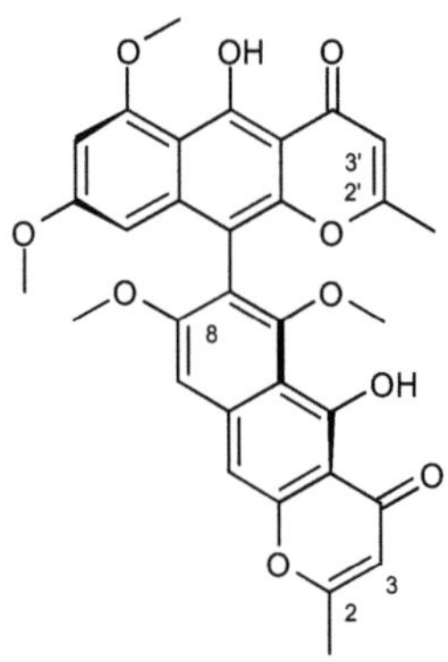

807 Aurasperone B

Type: Binaphthyls. $C_{32}H_{30}O_{12}$ Yellow, mp 186 °C (dec), $[\alpha]_D^{25} = +46.3°$ (c = 1.4, $CHCl_3$). Source: Marine-derived fungus *Aspergillus niger* EN-13, terrestrial fungi (*Aspergillus* spp.). Pharm: Cytotoxic (hmn liver cancer SMMC-7721 and A549, IC_{50} > 10 µg/mL) antifungal (*Candida albicans*, IZD = 10 mm, control Amphotericin B, IZD = 12 mm, weak); antioxidant (DPPH scavenger, 50 µg/mL, ScRt = 48.1% moderate, control BHT, ScRt = 80.4%); mycotoxin. Ref: H. Priestap, Tetrahedron, 1984, 40, 3617 | Y. Zhang, et al, J. Antibiot., 2007, 60, 204

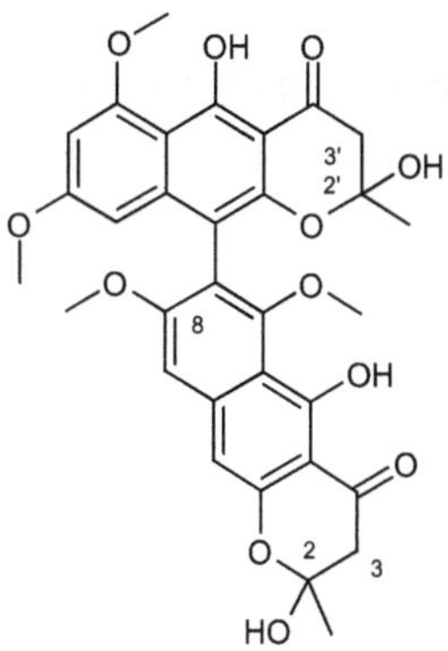

808 8′-*O*-Demethylisonigerone

Type: Binaphthyls. $C_{31}H_{24}O_{10}$ Yellow-brown powder, $[\alpha]_D^{20} = -90.8°$ (c = 0.15, MeOH). Source: Marine-derived fungus *Aspergillus carbonarius* WZ-4-11 (from sediment, China). Pharm: Antituberculosis (*Mycobacterium tuberculosis*, IC_{50} = 43.0 μmol/L). Ref: Y. P. Zhang, et al, Chem. Biodiversity, 2008, 5, 93

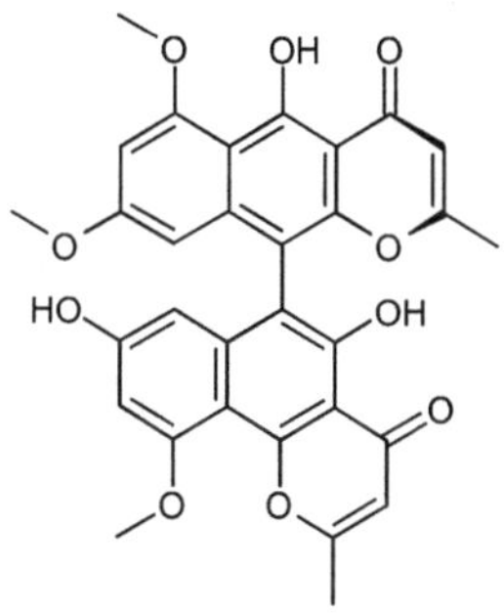

809 8-*O*-Demethylnigerone

8′-*O*-Demethylnigerone Type: Binaphthyls. $C_{31}H_{24}O_{10}$ Yellow–brown powder, $[\alpha]_D^{20}$ = −87.7° (c = 0.14, MeOH). Source: Marine-derived fungus *Aspergillus carbonarius* WZ-4-11 (sediment, China), marine-derived fungus *Exophiala* sp. Pharm: Antituberculosis (*Mycobacterium tuberculosis*, IC_{50} = 21.5 μmol/L). Ref: D. Zhang, et al, J. Antibiot. (Tokyo) 2008, 61, 40

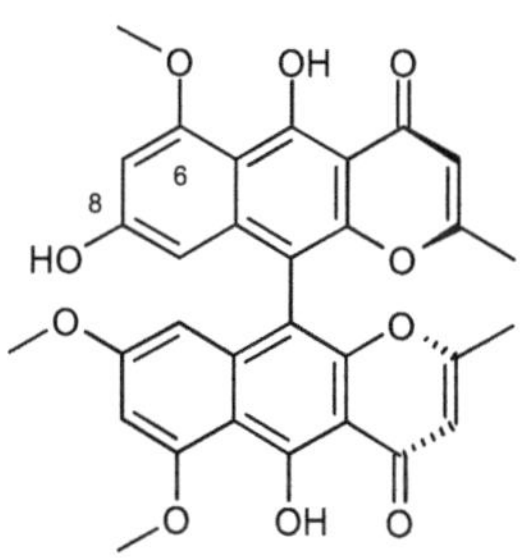

810 Dianhydroaurasperone C

Type: Binaphthyls. $C_{31}H_{24}O_{10}$ Source: Marine-derived fungus *Aspergillus niger* EN-13, terrestrial fungus (*Aspergillus niger*). Pharm: Cytotoxic (hmn liver cancer SMMC-7721 and A549, IC_{50} > 10 μg/mL). Ref: K. C. Ehrlich, et al, Appl. Environ. Microbiol., 1984, 48, 1 | Y. Zhang, et al, J. Antibiot., 2007, 60, 204

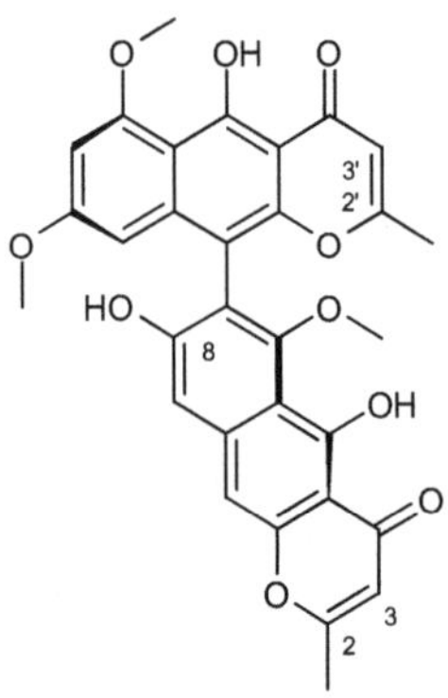

811 Fonsecinone A

Type: Binaphthyls. $C_{32}H_{26}O_{10}$ Yellow needles ($CHCl_3$/propanol), mp 280 °C. Source: Marine-derived fungus *Aspergillus niger* EN-13, terrestrial fungi (*Aspergillus fonsecaeus* and *Aspergillus carbonarius*). Pharm: Cytotoxic (hmn liver cancer SMMC-7721 and A549, IC_{50} > 10 μg/mL); antifungal (*Candida albicans*, IZD = 14 mm, control Amphotericin B, IZD = 12 mm, weak); mycotoxin; *Taq* DNA polymerase inhibitor. Ref: H. A. Priestap, Tetrahedron, 1984, 40, 3617 | K. Akiyama, et al, JNP, 2003, 66, 136 | F. R. Campos, et al, Magn. Reson. Chem., 2005, 43, 962 | Y. Zhang, et al, J. Antibiot., 2007, 60, 204 | S. de Lazaro, et al, Int. J. Quantum Chem., 2008, 108, 2408

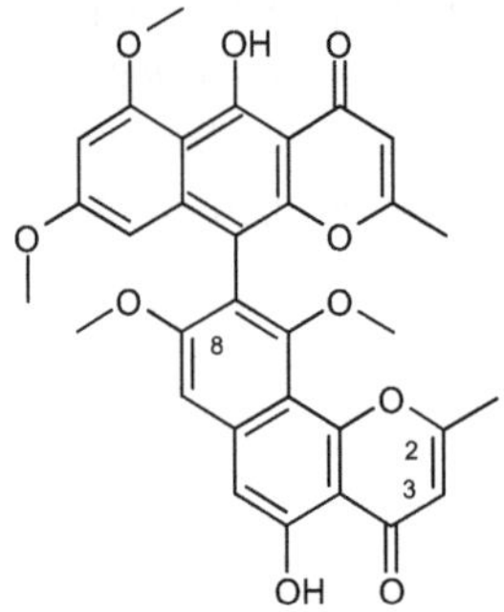

812 Fonsecinone B

Type: Binaphthyls. $C_{32}H_{28}O_{11}$ Amorph., mp 172–173 °C. Source: Marine-derived fungus *Aspergillus niger* EN-13, terrestrial fungus (*Aspergillus fonsecaeus*). Pharm:

Cytotoxic (hmn liver cancer SMMC-7721 and A549, IC_{50} > 10 µg/mL); antioxidant (DPPH scavenger, 50 µg/mL, ScRt = 13.2% moderate, control BHT, ScRt = 80.4%); mycotoxin. Ref: H. Priestap, Tetrahedron, 1984, 40, 3617 | Y. Zhang, et al, J. Antibiot., 2007, 60, 204

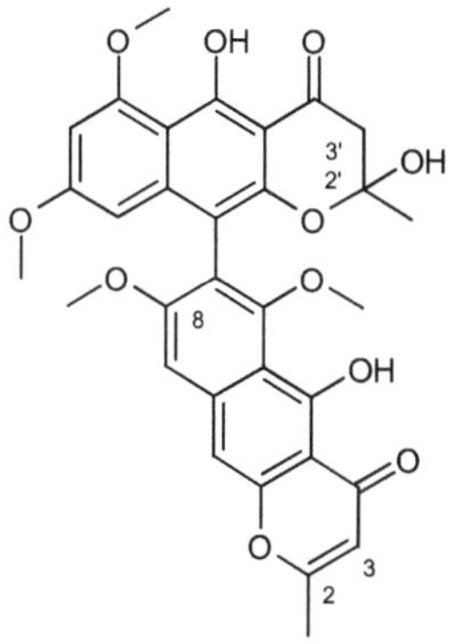

813 Fonsecinone C

Type: Binaphthyls. $C_{32}H_{28}O_{11}$ Amorph., mp 169–170 °C. Source: Marine-derived fungus *Aspergillus niger* EN-13, terrestrial fungus (*Aspergillus fonsecaeus*). Pharm: Cytotoxic (hmn liver cancer SMMC-7721 and A549, IC_{50} > 10 µg/mL); mycotoxin. Ref: H. Priestap, Tetrahedron, 1984, 40, 3617 | Y. Zhang, et al, J. Antibiot., 2007, 60, 204

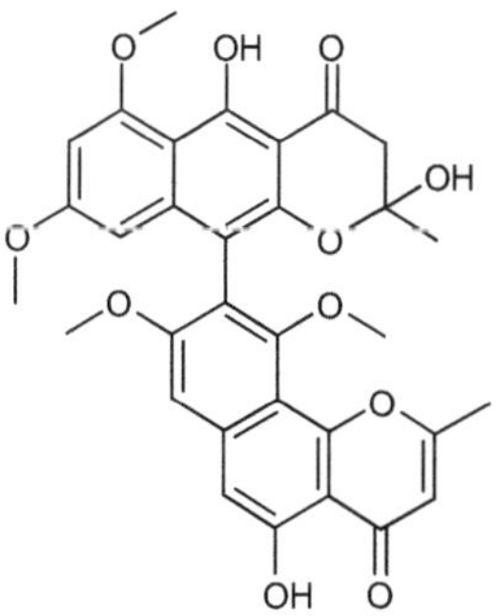

814 Fonsecinone D

Aurasperone E Type: Binaphthyls. $C_{32}H_{28}O_{11}$ Solid, mp 166–170 °C. Source: Marine-derived fungus *Aspergillus niger* EN-13, terrestrial fungi (*Aspergillus fonsecaeus* and *Aspergillus niger*). Pharm: Cytotoxic (hmn liver cancer SMMC-7721 and A549, IC_{50} > 10 µg/mL); antioxidant (DPPH scavenger, 50 µg/mL, ScRt = 37.5% moderate, control BHT, ScRt = 80.4%); mycotoxin. Ref: H. Priestap, Tetrahedron, 1984, 40, 3617 | Y. Zhang, et al, J. Antibiot., 2007, 60, 204

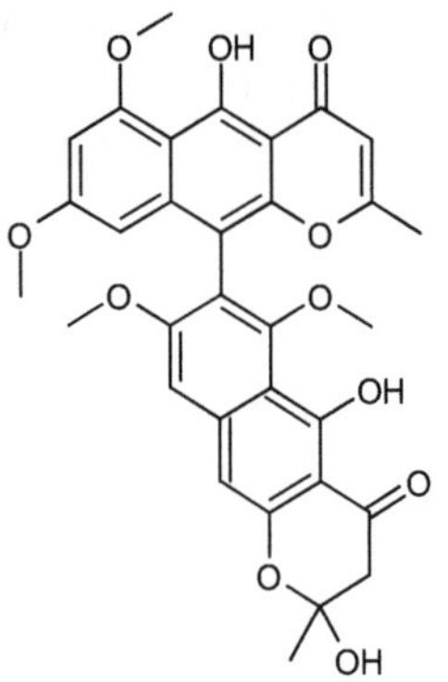

815 Hypochromin A

Type: Binaphthyls. $C_{32}H_{24}O_{12}$ Amorph. red solid, mp 259–261 °C (dec), $[\alpha]_D^{25}$ = +410° (c = 0.009, EtOH). Source: Marine-derived fungus *Hypocrea vinosa* AY380904 (sand, Okinawa). Pharm: Tyrosine kinase inhibitor (IC_{50} = 58.7 µmol/L); proliferation inhibitor (IC_{50} = 50.0 µmol/L); migration inhibitor (IC_{50} = 0.87 µmol/L); tubule formation inhibitor. Ref: Y. Ohkawa, et al, JNP, 2010, 73, 579

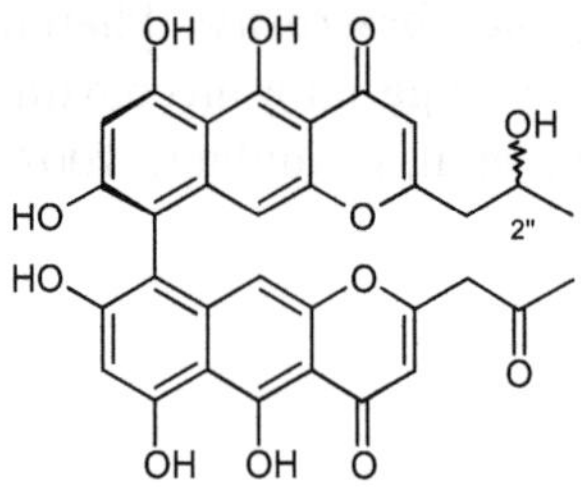

816 Hypochromin B

Type: Binaphthyls. $C_{30}H_{20}O_{11}$ Amorph. red solid, mp 238–240 °C (dec), $[\alpha]_D^{25}$ = +340° (c = 0.005, EtOH). Source: Marine-derived fungus *Hypocrea vinosa* AY380904 (sand, Okinawa). Pharm: Tyrosine kinase inhibitor (HUVEC lysate, tyrosine kinase assay, IC_{50} = 18.0 µmol/L; growth inhibition assay, IC_{50} = 13.1 µmol/L; migration assay, IC_{50} = 1.51 µmol/L; inhibitory effects on proliferation, migration, and tubule formation). Ref: Y. Ohkawa, et al, JNP, 2010, 73, 579

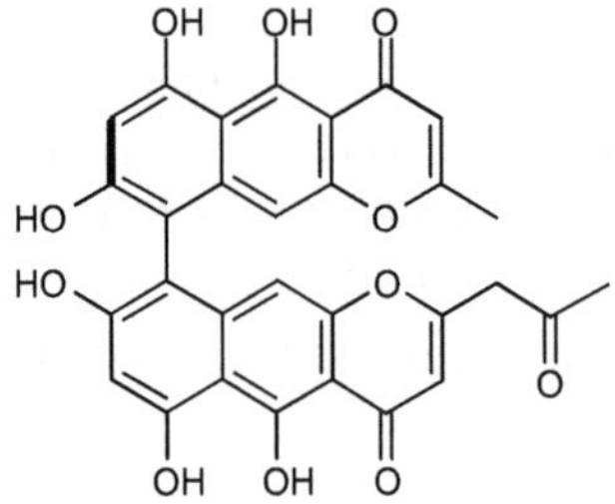

817 Isonigerone

Type: Binaphthyls. $C_{32}H_{26}O_{10}$ Orange solid ($CHCl_3$/MeOH), mp 330 °C, $[\alpha]_D^{20} = -275.7°$ (c = 0.15, $CHCl_3$), $[\alpha]_D^{20} = -93.1°$ (c = 0.74, $CHCl_3$). Source: Marine-derived fungus *Aspergillus carbonarius* WZ-4-11(sediment, China), terrestrial fungus (*Aspergillus niger*). Pharm: Mycotoxin. Ref: C. P. Gorst-Allman, et al, JCS Perkin I, 1980, 2474 | Y. P. Zhang, et al, Chem. Biodiversity, 2008, 5, 93

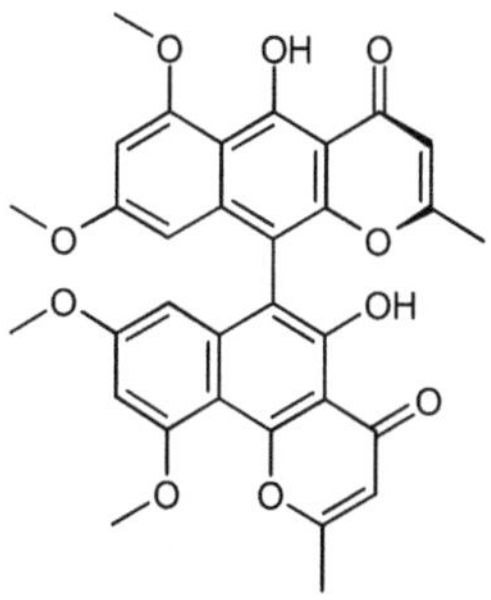

818 Nigerasperone B

Type: Binaphthyls. $C_{32}H_{30}O_{12}$ Yellow powder, mp 169–170 °C, $[\alpha]_D = -15.3°$ (c = 0.27, MeOH). Source: Marine-derived fungus *Aspergillus niger* EN-13. Pharm: Cytotoxic (hmn liver cancer SMMC-7721 and A549, IC_{50} > 10 μg/mL). Ref: Y. Zhang, et al, J. Antibiot., 2007, 60, 204

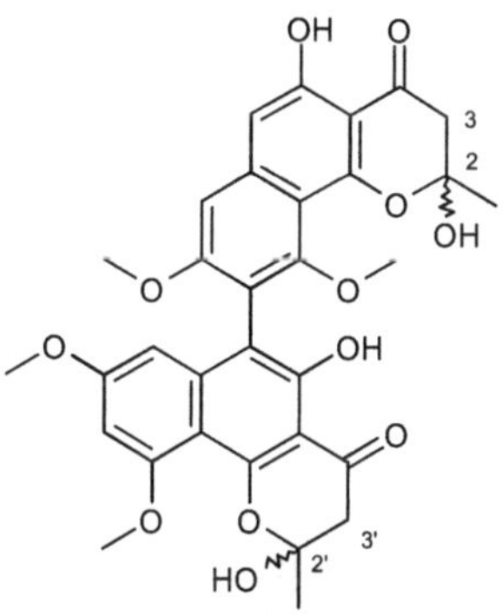

819 Nigerasperone C

Type: Binaphthyls. $C_{31}H_{26}O_{11}$ Yellow powder, mp 222–224 °C, $[\alpha]_D^{20} = -11.5°$ (c = 0.34, MeOH). Source: Marine-derived fungus *Aspergillus niger* EN-13. Pharm: Cytotoxic (hmn liver cancer SMMC-7721 and A549, IC_{50} > 10 μg/mL); antifungal (*Candida albicans*, IZD = 9 mm, control Amphotericin B, IZD = 12 mm, weak); antioxidant (DPPH scavenger, 50 μg/mL, ScRt = 41.6% moderate, control BHT, ScRt = 80.4%). Ref: Y. Zhang, et al, J. Antibiot., 2007, 60, 204

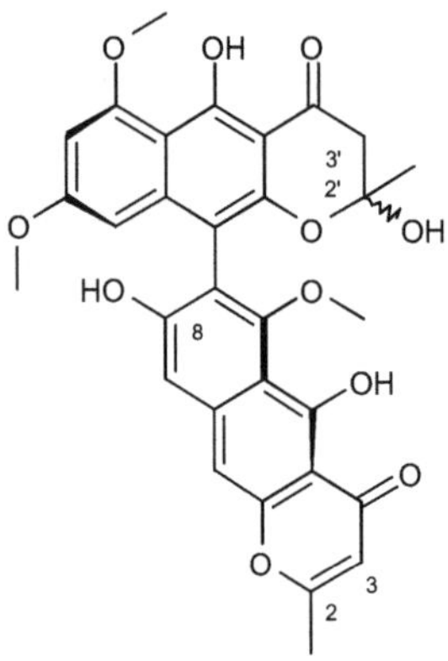

820 Nigerone

Type: Binaphthyls. $C_{32}H_{26}O_{10}$ Orange needles ($CHCl_3$/MeOH), mp 330 °C, $[\alpha]_D^{20}$ = −287.7° (*c* = 1, $CHCl_3$). Source: Marine-derived fungus *Aspergillus carbonarius* WZ-4-11 (sediment, China), terrestrial fungus (*Aspergillus niger*). Pharm: Mycotoxin. Ref: C. P. Gorst-Allman, et al, JCS Perkin I, 1980, 2474 | K. Koyama, et al, CPB, 1987, 35, 4049 | Y. P. Zhang, et al, Chem. Biodiversity, 2008, 5, 93

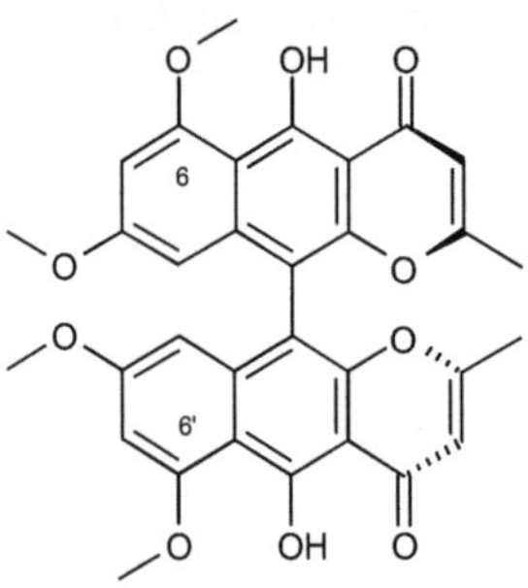

821 Xanalteric acid I

Type: Perylenes. $C_{20}H_{12}O_7$ Dark red powder, $[\alpha]_D^{20}$ = −120° (*c* = 0.03, MeOH). Source: Mangrove-derived fungus *Alternaria* sp. JCM9.2. Pharm: Antibacterial (MRSA, MIC = 125 µg/mL, weak); cytotoxic (L5178Y, 10 µg/mL, InRt = 45.0%, control Alternarian acid, InRt = 99.2%, marginal activity). Ref: J. Kjer, et al, JNP, 2009, 72, 2053

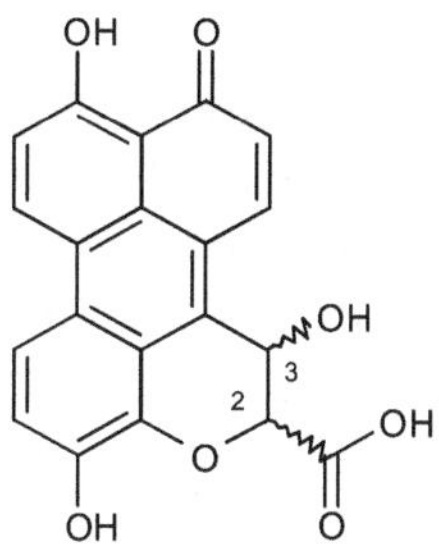

822 Xanalteric acid II

Type: Perylenes. $C_{20}H_{12}O_7$ Dark red powder, $[\alpha]_D^{20} = +40°$ ($c = 0.02$, MeOH). Source: Mangrove-derived fungus *Alternaria* sp. JCM9.2. Pharm: Antibacterial (MRSA, MIC = 250 μg/mL, weak); cytotoxic (L5178Y, 10 μg/mL, InRt = 87.5%, control Alternarian acid, InRt = 99.2%, marginal activity). Ref: J. Kjer, et al, JNP, 2009, 72, 2053

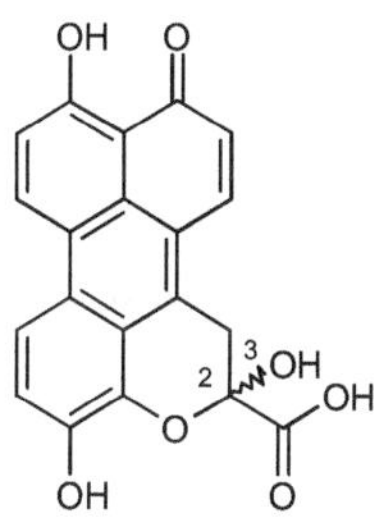

823 Bacillosporin A

Bacillisporin A Type: Duclauxin group. $C_{28}H_{20}O_{10}$ Pale yellow powder (Me_2CO/ $CHCl_3$), mp 282–285 °C (dcc), $[\alpha]_D^{24} = +484°$ ($c = 0.583$, Me_2CO). Source: Mangrove-derived fungus *Penicillium* sp. JP-1 from mangrove *Aegiceras corniculatum* (China waters), mangrove-derived fungus *Talaromyces bacillisporus* SBE-14. Pharm: Mycotoxin; antibacterial. Ref: M. Yamazaki, et al, CPB, 1980, 28, 3649 | T. Dethoup, et al, PM, 2006, 72, 957 | Z. J. Lin, et al, Phytochemistry, 2008, 69, 1273

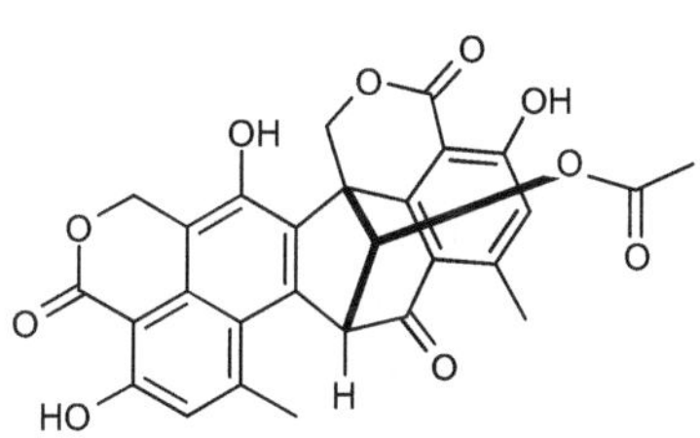

824 Bacillosporin C

Bacillisporin C Type: Duclauxin group. $C_{26}H_{18}O_{10}$ Pale yellow powder (THF/MeOH), mp > 280 °C, $[\alpha]_D^{27}$ = +490° (*c* = 0.2, MeOH). Source: Mangrove-derived fungus *Penicillium* sp. JP-1 from mangrove *Aegiceras corniculatum* (China waters), mangrove-derived fungus *Talaromyces bacillisporus* SBE-14. Pharm: Mycotoxin; antibacterial. Ref: M. Yamazaki, et al, CPB, 1980, 28, 3649 | T. Dethoup, et al, PM, 2006, 72, 957 | Z. Guo, et al, Magn. Reson. Chem., 2007, 45, 439 | Z. J. Lin, et al, Phytochemistry, 2008, 69, 1273

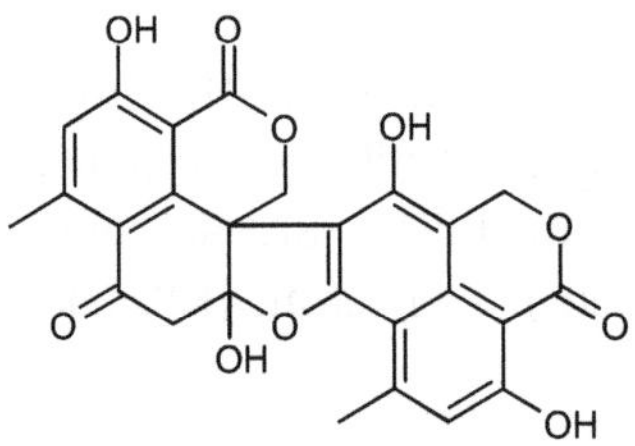

825 Bacillosporin D‡

Type: Duclauxin group. $C_{26}H_{16}O_{11}$ Pale yellow cryst., mp > 280 °C, $[\alpha]_D^{27}$ = +700° (*c* = 0.2, MeOH). Source: Mangrove-derived fungus *Talaromyces bacillisporus* SBE-14. Pharm: Mycotoxin; antibacterial. Ref: M. Yamazaki, et al, CPB, 1980, 28, 3649 | T. Dethoup, et al, PM, 2006, 72, 957 | Z. Guo, et al, Magn. Reson. Chem., 2007, 45, 439

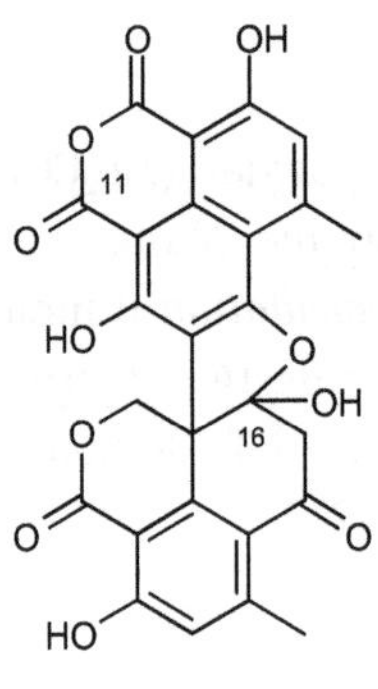

826 Annulin A

Type: Naphthoquinone derivatives. $C_{19}H_{20}O_7$ Orange cryst. (EtOH), mp 174–176 °C. Source: Hydroid *Garveia annulata* (order Anthoathecata, Northeastern Pacific). Pharm: Inhibits indoleamine 2,3-dioxygenase (IDO) (*in vitro, Ki* = 0.12–0.69 μmol/L); antibacterial. Ref: E. Fahy, et al, JOC, 1986, 51, 5145 | A. Pereira, et al, JNP, 2006, 69, 1496

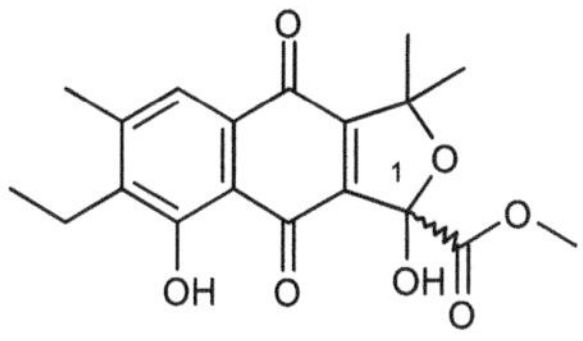

827 Annulin B

Type: Naphthoquinone derivatives. $C_{21}H_{22}O_7$ Orange oil, $[\alpha]_D$ = +8° (c = 0.2, $CHCl_3$). Source: Hydroid *Garveia annulata* (order Anthoathecata, Northeastern Pacific). Pharm: Inhibits indoleamine 2,3-dioxygenase (IDO) (*in vitro, Ki* = 0.12–0.69 μmol/L). Ref: E. Fahy, et al, JOC, 1986, 51, 5145 | A. Pereira, et al, JNP, 2006, 69, 1496

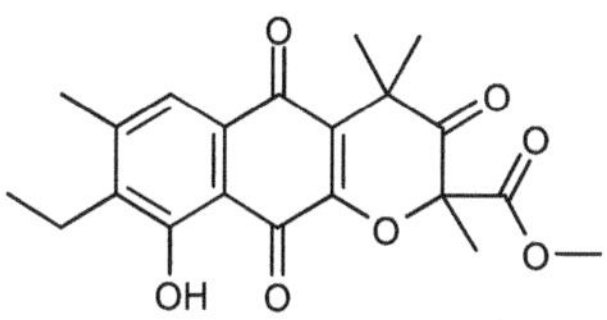

828 Annulin C

Type: Naphthoquinone derivatives. $C_{20}H_{22}O_7$ Yellow oil. Source: Hydroid *Garveia annulata* (order Anthoathecata, Northeastern Pacific). Pharm: Inhibits indoleamine 2,3-dioxygenase (IDO) (*in vitro, Ki* = 0.12–0.69 μmol/L). Ref: A. Pereira, et al, JNP, 2006, 69, 1496

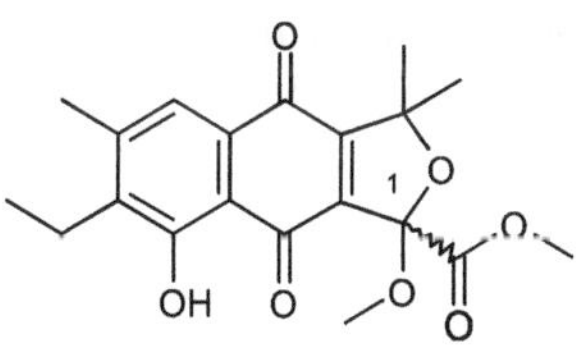

829 Debromomarinone

Type: Naphthoquinone derivatives. $C_{25}H_{28}O_5$ Source: An unidentified marine-derived actinomycete CNB-632 (shallow water sediment). Pharm: Antibacterial (*in vitro*, gram-positive bacteria MIC = 1–2 μg/mL; *Staphylococcus aureus, Staphylococcus epidermis, Staphylococcus pneumoniae* and *Staphylococcus pyrogenes*). Ref: Pathirana, et al, Tet. Lett., 1992, 33, 7663 | J. A. Kalaitzis, et al, Org. Lett., 2003, 5, 4449 (structure revised)

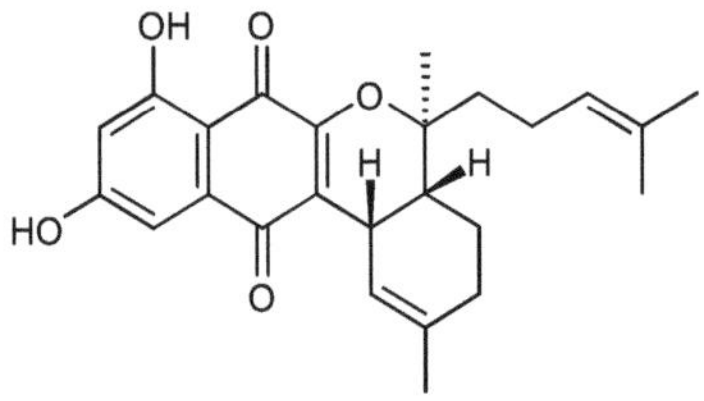

830 Debromomethoxymarinone

Methoxydebromomarinone Type: Naphthoquinone derivatives. $C_{26}H_{30}O_6$ $[\alpha]_D^{25}$ = +140° (*c* = 0.1, EtOH). Source: An unidentified marine-derived actinomycete CNH-099 (Batiquitos Lagoon, North of San Diego, CA). Pharm: Cytotoxic (colon carcinoma HCT116 cell lines, IC_{50} ≈ 8 μg/mL). Ref: I. H. Hardt,et al, Tet. Lett., 2000, 41, 2073

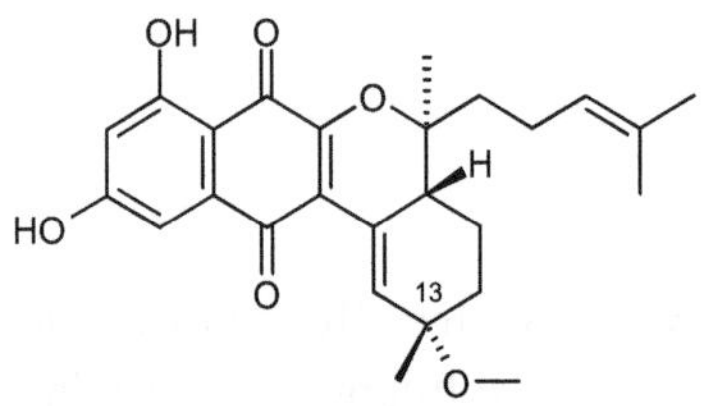

831 Deoxylapachol

2-Prenyl-1,4-naphthoquinone Type: Naphthoquinone derivatives. $C_{15}H_{14}O_2$ Bright-yellow prisms (petrol), mp 60–61 °C. Source: Brown alga *Landsburgia quercifolia*. Pharm: Cytotoxic; skin irritant. Ref: N. B. Perry, et al, JNP, 1991, 54, 978

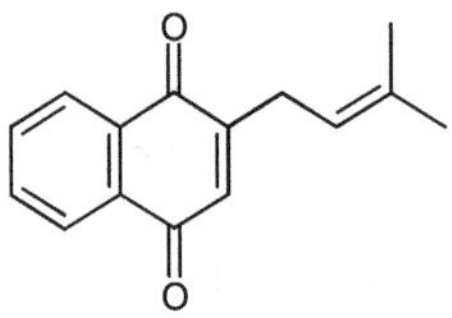

832 Fusarnaphthoquinone A

Type: Naphthoquinone derivatives. $C_{15}H_{18}O_7$ Source: Marine-derived fungi *Fusarium* sp. from gorgonian sea fan *Annella* sp. (Koh Hin Ran Pet, Surat Thani Province, Thailand). Pharm: Cytotoxic (KB, IC_{50} = 130 μmol/L, control doxorubicin, IC_{50} = 0.33 μmol/L; MCF7, IC_{50} = 22 μmol/L, control doxorubicin, IC_{50} = 2.18 μmol/L; Vero, inactive, control Ellipticine, IC_{50} = 4.47 μmol/L). Ref: K. Trisuwan, et al, JNP, 2010, 73, 1507

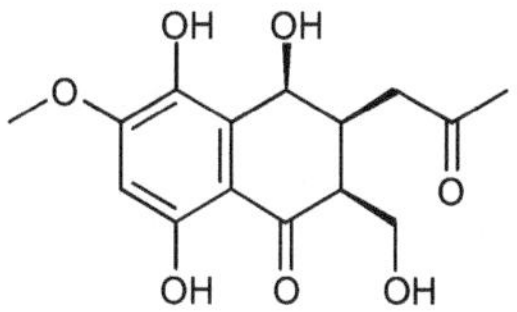

833 Fusarnaphthoquinone B

Type: Naphthoquinone derivatives. $C_{15}H_{16}O_5$ Source: Marine-derived fungi *Fusarium* sp. from gorgonian sea fan *Annella* sp. (Koh Hin Ran Pet, Surat Thani Province, Thailand). Pharm: Antifungal (*Cryptococcus neoformans* and *Microsporum gypseum*, MIC = 200 μmol/L). Ref: K. Trisuwan, et al, JNP, 2010, 73, 1507

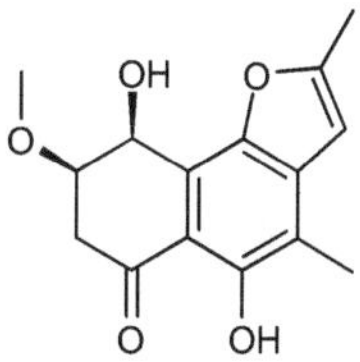

834 Griseusin C

Type: Naphthoquinone derivatives. $C_{20}H_{16}O_9$ Orange powder. Source: Mangrove-derived fungus *Penicillium* sp. GT200261505 from mangrove *Kandelia candel* (China waters). Pharm: 3*α*-Hydroxysteroid dehydrogenase inhibitor (mild). Ref: X. Li, et al, Arch. Pharmacal Res., 2006, 29, 942

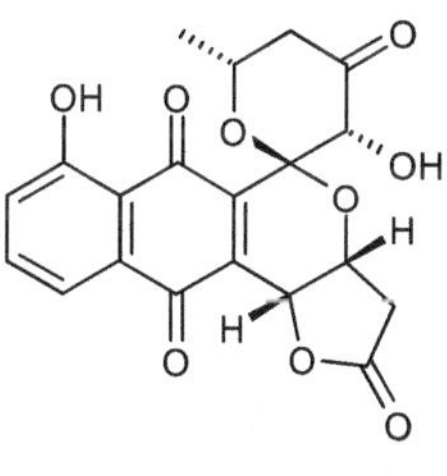

835 Hydroxydebromomarinone

Type: Naphthoquinone derivatives. $C_{25}H_{28}O_6$ $[\alpha]_D^{25}$ = +280° (*c* = 0.2, EtOH). Source: An unidentified marine-derived actinomycete CNH-099 (Batiquitos Lagoon, North of San Diego, CA). Pharm: Cytotoxic (colon carcinoma HCT116 cell lines, IC_{50} ≈ 8 μg/mL). Ref: I. H. Hardt, et al, Tet. Lett., 2000, 41, 2073

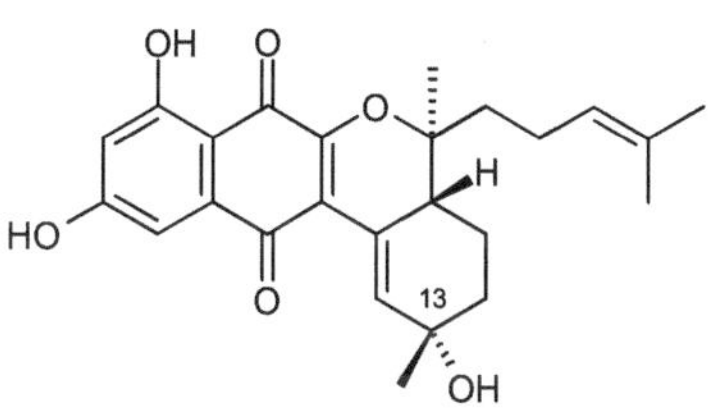

836 Isomarinone

Type: Naphthoquinone derivatives. $C_{25}H_{27}BrO_5$ $[\alpha]_D^{25}$ = −120° (*c* = 0.2, MeOH). Source: An unidentified marine-derived actinomycete CNH-099 (Batiquitos Lagoon, North of San Diego, CA). Pharm: Cytotoxic (colon carcinoma HCT116 cell lines, $IC_{50} \approx 8$ μg/mL). Ref: I. H. Hardt, et al, Tet. Lett., 2000, 41, 2073

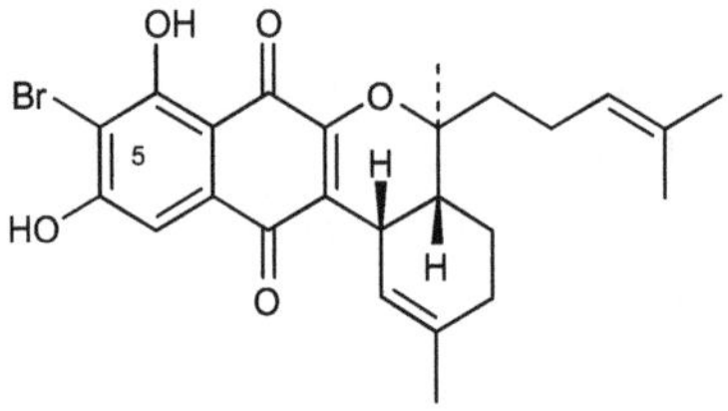

837 Javanicin tautomeric with 5,8-quinone structure

Type: Naphthoquinone derivatives. $C_{15}H_{14}O_6$ Source: Marine-derived fungus *Fusarium* sp. PSU-F135 Pharm: Antiplasmodial (*in vitro Plasmodium falciparum* K1, IC_{50} = 12 μmol/L, control Dihydroartemisinin, IC_{50} = 0.004 μmol/L); cytotoxic (KB, IC_{50} = 5.7 μmol/L, control doxorubicin, IC_{50} = 0.33 μmol/L; MCF7, IC_{50} = 13 μmol/L, control doxorubicin, IC_{50} = 2.18 μmol/L; Vero, IC_{50} = 170 μmol/L, control Ellipticine, IC_{50} = 4.47 μmol/L). Ref: K. Trisuwan, et al, JNP, 2010, 73, 1507

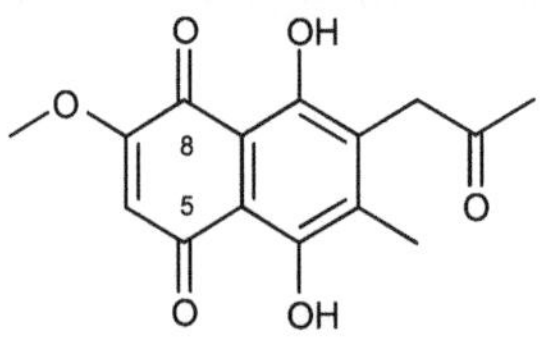

838 Marinone

Type: Naphthoquinone derivatives. $C_{25}H_{27}BrO_5$ $[\alpha]_D^{25}$ = −170° (*c* = 0.15, MeOH). Source: An unidentified marine-derived actinomycete CNB-632 (shallow water sediment), an unidentified marine-derived actinomycete CNH-099 (Batiquitos Lagoon, North of San Diego, CA). Pharm: Antibacterial (*in vitro*, gram-positive bacteria MIC = 1.0 μg/mL; *Bacillus subtilis*). Ref: C. Pathirana, et al, Tet. Lett., 1992, 33, 7663 | I. H. Hardt, et al, Tet. Lett., 2000, 41, 2073

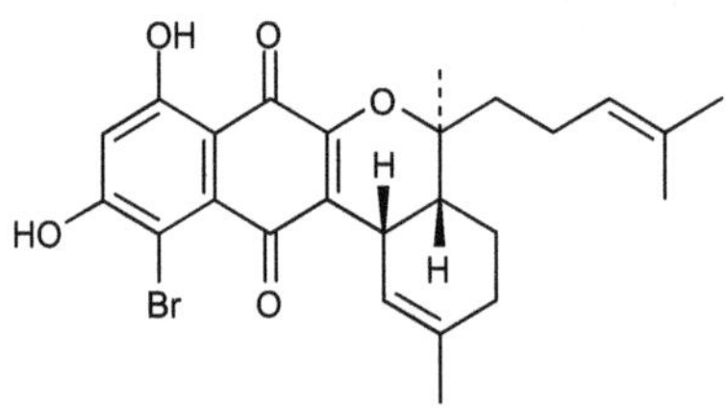

839 Neomarinone

Type: Naphthoquinone derivatives. $C_{26}H_{32}O_5$ $[\alpha]_D^{25}$ = +86° (c = 0.5, MeOH). Source: An unidentified marine-derived actinomycete CNH-099 (Batiquitos Lagoon, North of San Diego, CA). Pharm: Cytotoxic (colon carcinoma HCT116 cell lines, IC_{50} ≈ 8 µg/mL); cytotoxic (NCI's 60 cancer cell line panel, mean IC_{50} = 10 µg/mL). Ref: I. H. Hardt, et al, Tet. Lett., 2000, 41, 2073 | J. A. Kalaitzis, et al, Org. Lett., 2003, 5, 4449 | M. Peña-López, et al, Chem. Eur. J., 2009, 15, 910

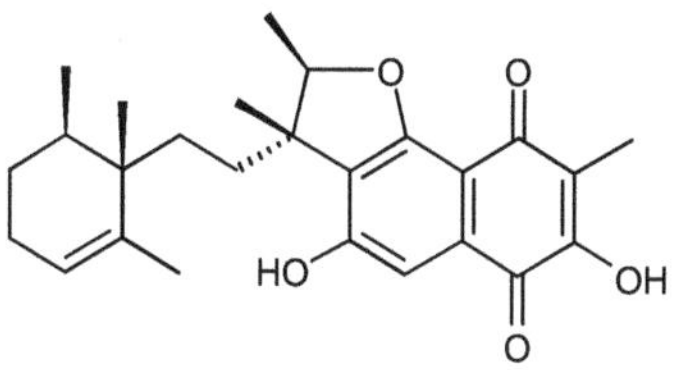

840 Seriniquinone

Type: Naphthoquinone derivatives. $C_{20}H_8O_4S$ Source: Marine bacterium *Serinicoccus* sp. Pharm: Cytotoxic (selective). Ref: L. Trzoss, et al, Proc. Natl. Acad. Sci. USA, 2014, 111, 14687

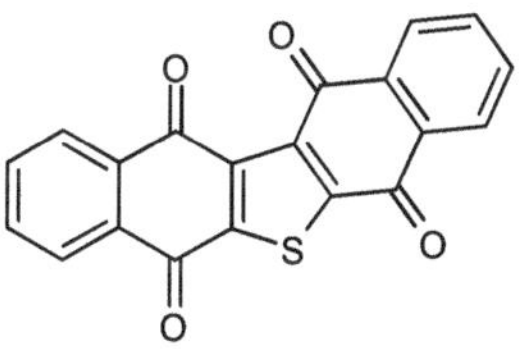

841 Antibiotic A 80915A

Type: Benzochromanquinones. $C_{26}H_{31}Cl_3O_5$ Pale yellow powder, $[\alpha]_D^{25}$ = −89.7° (c = 1, MeOH). Source: An unidentified marine-derived actinomycete CNQ-525. Pharm: Antibacterial (including MRSA); cytotoxic. Ref: D. S. Fukuda, et al, J. Antibiot., 1990, 43, 623 | I. E. Soria-Mercado, et al, JNP, 2005, 68, 904 | N. M. Haste, et al, Mar. Drugs, 2011, 9, 680

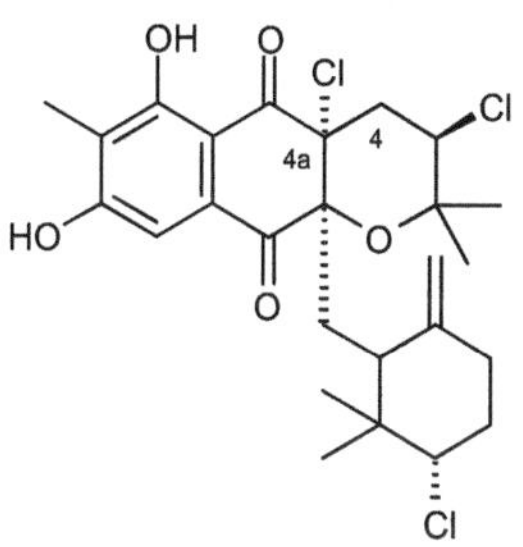

842 Antibiotic A 80915C

Type: Benzochromanquinones. $C_{26}H_{33}Cl_3O_6$ Yellow prisms, $[\alpha]_D^{25} = -190°$ (c = 0.03, $CHCl_3$), $[\alpha]_D^{25} = -115.4°$ (c = 0.56, MeOH). Source: An unidentified marine-derived actinomycete CNQ-525. Pharm: Antibacterial; cytotoxic. Ref: D. S. Fukuda, et al, J. Antibiot., 1990, 43, 623 | Soria-Mercado, et al, JNP, 2005, 68, 904

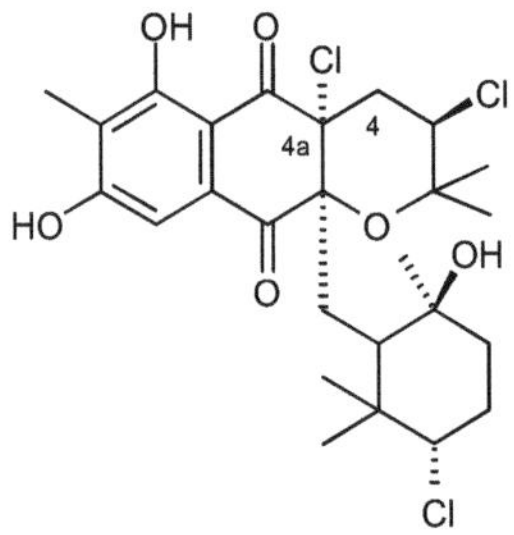

843 4*a*-Dechloroantibiotic A 80915C

Type: Benzochromanquinones. $C_{26}H_{34}Cl_2O_6$ Pale yellow cryst., mp 197–198 °C, $[\alpha]_D^{25} = -24.3°$ (c = 0.14, MeOH). Source: An unidentified marine-derived actinomycete CNQ-525. Pharm: Cytotoxic. Ref: I. E. Soria-Mercado, et al, JNP, 2005, 68, 904

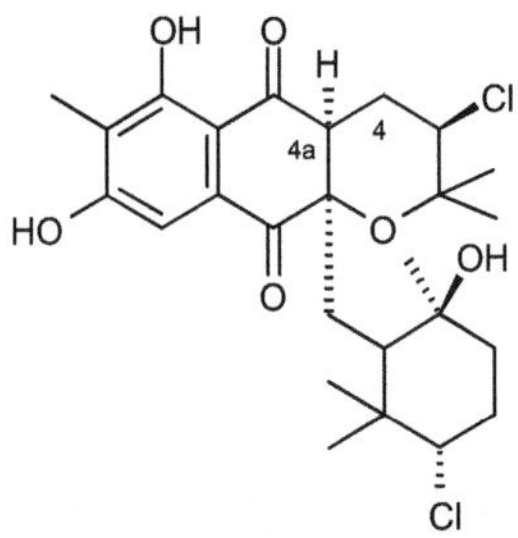

844 4*a*-Dechloro-4,4*a*-didehydro-antibiotic A 80915A

Type: Benzochromanquinones. $C_{26}H_{30}Cl_2O_5$ Pale yellow cryst., mp 179–181 °C, $[\alpha]_D^{25} = -7.7°$ (c = 0.4, MeOH). Source: An unidentified marine-derived actinomycete CNQ-525. Pharm: Cytotoxic. Ref: N. M. Haste, et al, Mar. Drugs, 2011, 9, 680

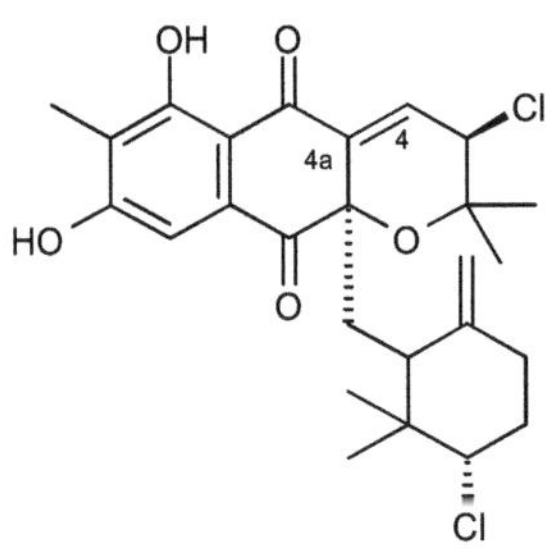

845 Napyradiomycin A

Type: Benzochromanquinones. $C_{25}H_{27}ClO_6$ Source: Marine-derived streptomycete *Streptomyces* sp. CNQ-329 (sediment, San Diego, California, USA). Pharm: Cytotoxic (HCT116, moderate); antibacterial (MRSA, moderate). Ref: J. W. Blunt, et al, NPR, 2015, 32, 116 (rev)

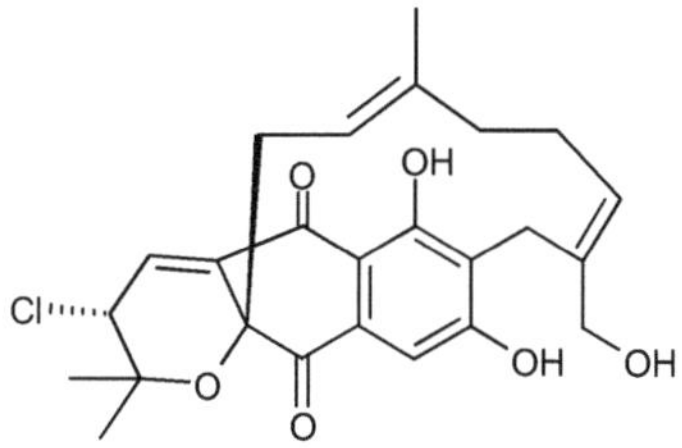

846 Napyradiomycin B

Type: Benzochromanquinones. $C_{25}H_{28}O_7$ Source: Marine-derived streptomycete *Streptomyces* sp. CNQ-329 (sediment, San Diego, California, USA). Pharm: Antibacterial (MRSA, moderate). Ref: J. W. Blunt, et al, NPR, 2015, 32, 116 (rev)

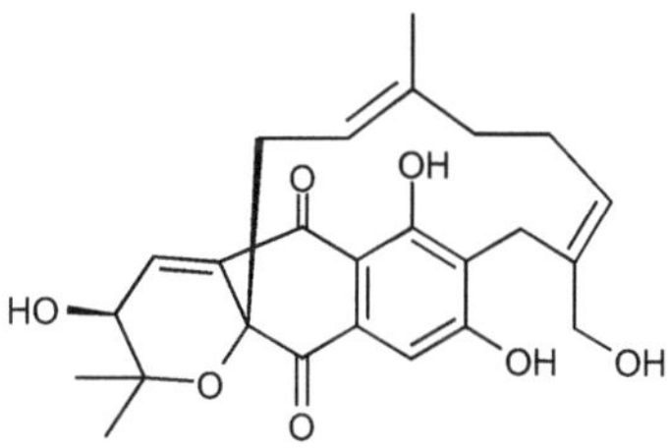

847 Napyradiomycin D

Type: Benzochromanquinones. $C_{25}H_{26}Cl_2O_5$ Source: Marine-derived streptomycete *Streptomyces* sp. CNQ-329 (sediment, San Diego, California, USA). Pharm: Cytotoxic (HCT116, moderate). Ref: J. W. Blunt, et al, NPR, 2015, 32, 116 (rev)

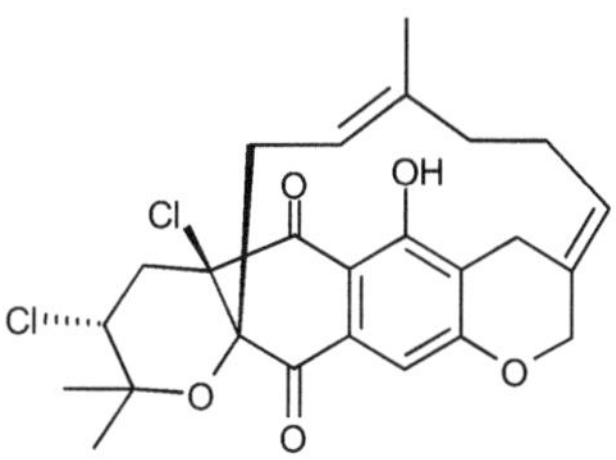

848 Napyradiomycin E

Type: Benzochromanquinones. $C_{25}H_{28}BrClO_5$ Source: Marine-derived streptomycete *Streptomyces* sp. CNQ-329 (sediment, San Diego, California, USA). Pharm: Cytotoxic (HCT116, moderate). Ref: J. W. Blunt, et al, NPR, 2015, 32, 116 (rev)

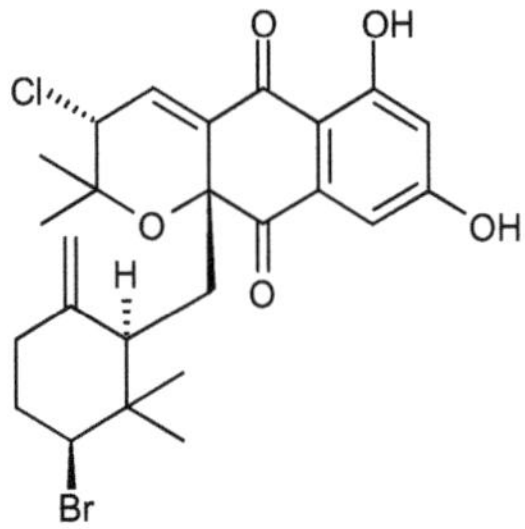

849 Napyradiomycin F

Type: Benzochromanquinones. $C_{25}H_{32}Cl_2O_6$ Source: Marine-derived streptomycete *Streptomyces* sp. CNH-070 (sediment, San Elijo Lagoon, Encinitas, California, USA). Pharm: Cytotoxic (HCT116, moderate). Ref: J. W. Blunt, et al, NPR, 2015, 32, 116 (rev)

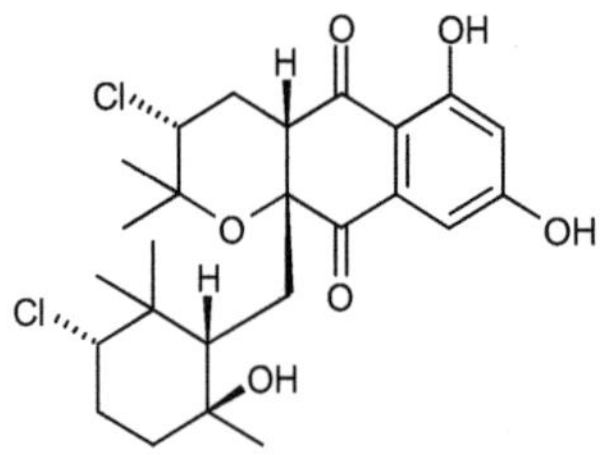

850 Anhydroexfoliamycin

Type: Benzoisochromanquinones. $C_{22}H_{24}O_8$ Dark red solid, mp 167 °C, $[\alpha]_D^{20}$ = +633° (*c* = 0.14, MeOH). Source: Marine-derived streptomycete *Streptomyces exfoliatus* from marine soil. Pharm: Antibacterial (gram-positive bacteria); anti-AD preclinical trial (Target: GSK3*β* mediated by the JNK pathway. Animal Model: 3xTg-AD mice. Effect: GSK3*β* inhibition τ, phosphorylation reduction) (Russo, 2016). Ref: O. Potterat, et al, J. Antibiot., 1993, 46, 346 | C. Volkmann, et al, J. Antibiot., 1995, 48, 431 | P. Russo, et al, Mar. Drugs, 2016, 14, 5 (review)

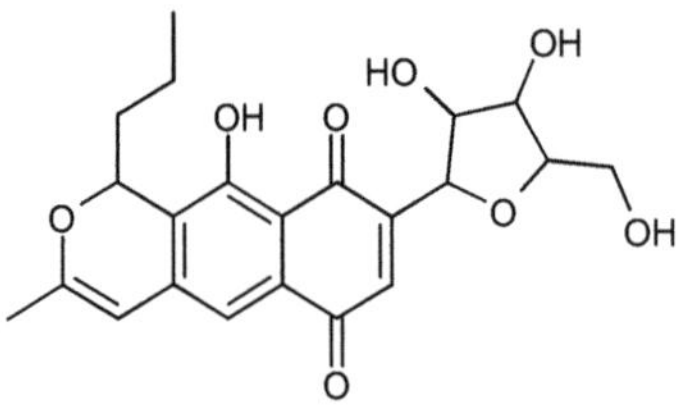

851 Anhydrofusarubin

Type: Benzoisochromanquinones. $C_{15}H_{12}O_6$ Violet-black needles (MeOH), mp 193–201 °C. Source: Marine-derived fungi *Fusarium* sp. from gorgonian sea fan *Annella* sp. (Koh Hin Ran Pet, Surat Thani Province, Thailand), marine-derived fungus *Fusarium* sp. PSU-F135. Pharm: Antiplasmodial (*in vitro Plasmodium falciparum* K1, IC_{50} = 14 μmol/L, control Dihydroartemisinin, IC_{50} = 0.004 μmol/L); antituberculosis (*Mycobacterium tuberculosis*, MIC = 87 μmol/L, control Isoniazid, MIC = 0.17–0.34 μmol/L); cytotoxic (KB, IC_{50} = 2.0 μmol/L, control doxorubicin, IC_{50} = 0.33 μmol/L; MCF7, IC_{50} = 0.9 μmol/L, control doxorubicin, IC_{50} = 2.18 μmol/L; Vero, IC_{50} = 58 μmol/L, control Ellipticine, IC_{50} = 4.47 μmol/L); antibacterial (*Staphylococcus aureus* and MRSA, MIC = 350 μmol/L). Ref: B. E. Cross, et al, JCS(C), 1970, 930 | N. Claydon, et al, J. Invertebr. Pathol., 1977, 30, 216 | K. Trisuwan, et al, JNP, 2010, 73, 1507

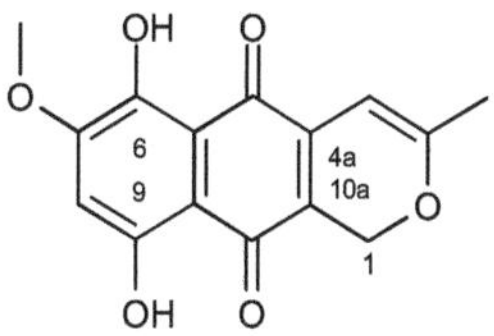

852 Fusarubin

Oxyjavanicin; Hydroxyjavanicin Type: Benzoisochromanquinones. $C_{15}H_{14}O_7$ Red prisms (C_6H_6), mp 218 °C. Source: Marine-derived fungi *Fusarium* sp. PSU-F135. Pharm: Cytotoxic (KB, IC_{50} = 14 μmol/L, control doxorubicin, IC_{50} = 0.33 μmol/L; MCF7, IC_{50} = 9.8 μmol/L, control doxorubicin, IC_{50} = 2.18 μmol/L; Vero, IC_{50} = 79 μmol/L, control Ellipticine, IC_{50} = 4.47 μmol/L); antibacterial (*Staphylococcus aureus* and MRSA, MIC = 350 μmol/L); antifungal (*Cryptococcus neoformans* and *Microsporum gypseum*, MIC = 200 μmol/L). Ref: K. Trisuwan, et al, JNP, 2010, 73, 1507

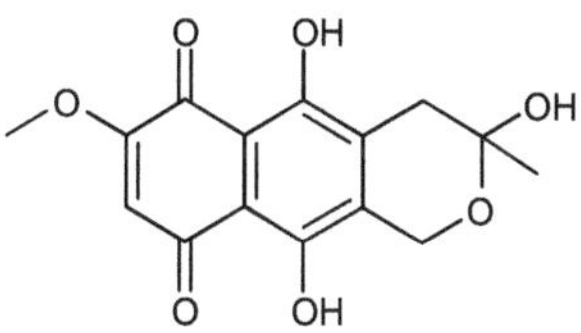

853 Halawanone A

Type: Benzoisochromanquinones. $C_{23}H_{22}O_9$ Source: Marine-derived streptomycete *Streptomyces* sp. BD-18T (shallow water sediment on Oahu). Pharm: Antibacterial (*Bacillus subtilis, Staphylococcus aureus*, gram-positive bacteria). Ref: P. W. Ford, et al, JNP, 1998, 61, 1232

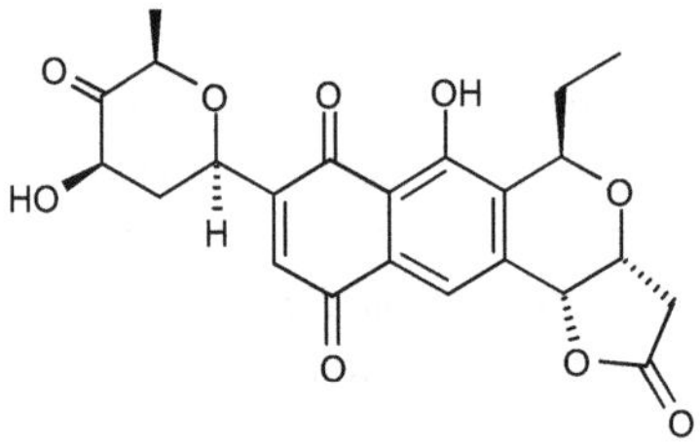

854 Halawanone B

Type: Benzoisochromanquinones. $C_{22}H_{20}O_9$ Source: Marine-derived streptomycete *Streptomyces* sp. BD-18T (shallow water sediment on Oahu). Pharm: Antibacterial (*Bacillus subtilis, Staphylococcus aureus*, gram-positive bacteria). Ref: P. W. Ford, et al, JNP, 1998, 61, 1232

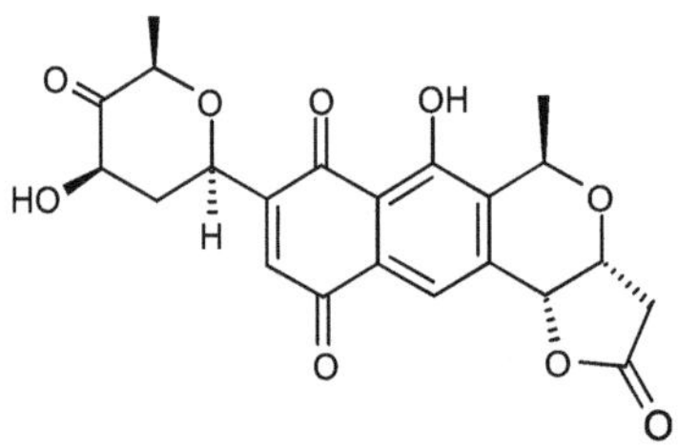

855 3-*O*-Methylfusarubin

Type: Benzoisochromanquinones. $C_{16}H_{16}O_7$ Reddish–orange needles ($CHCl_3$/EtOH), mp 157–158 °C, mp 188–189 °C. Source: Marine-derived fungi *Fusarium* sp. PSU-F135. Pharm: Antibacterial (gram-positive bacteria), antifungal; cytotoxic (mouse leukaemia). Ref: K. Trisuwan, et al, JNP, 2010, 73, 1507

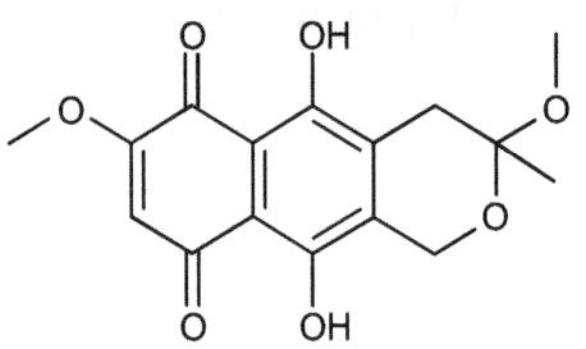

856 Obionin A

Type: Benzoisochromanquinones. $C_{21}H_{24}O_5$ Brown-red solid, mp 168–169 °C, $[\alpha]_D$ = +28.5° (*c* = 0.01, $CHCl_3$). Source: Marine-derived fungus *Leptosphaeria obiones*. Pharm: Inhibits binding of dopamine D-1 selective ligand to bovine corpus striatum membrane (IC_{50} = 2.5 μg/mL). Ref: G. K. Poch, et al, Tet. Lett., 1989, 30, 3483 | T. S. Bugni, et al, NPR, 2004, 21, 143 (rev)

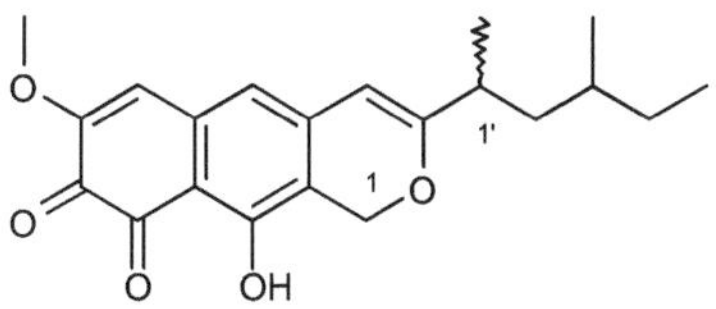

857 Obioninene

Leptosphaerodione Type: Benzoisochromanquinones. $C_{21}H_{22}O_5$ Deep red solid, $[\alpha]_D^{20} = -53.3°$ (c = 0.02, MeOH). Source: Marine-derived fungus *Leptosphaeria oraemaris, Stagonospora* spp. Pharm: Phytotoxin. Ref: A. Guerriero, et al, Helv. Chim. Acta, 1991, 74, 1445

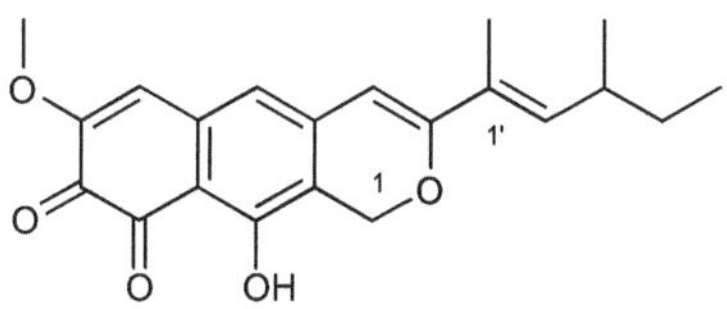

858 5-Bromo-4,7-dihydroxy-1-indanone

Type: Indenes and indanes. $C_9H_7BrO_3$ Source: An unidentified sponge. Pharm: Antineoplastic. Ref: Japan. Pat., 1996, 96 198 798; CA, 125, 217489a

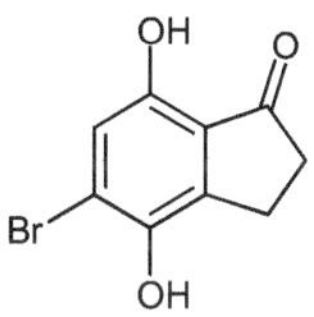

859 2,3-Dihydro-1-methoxy-6-methyl-3-oxo-1*H*-indene-4-carboxaldehyde

7-Formyl-3-methoxy-5-methyl-1-indanone Type: Indenes and indanes. $C_{12}H_{12}O_3$ Clear oil, $[\alpha]_D = +1.3°$ (c = 1.1, MeOH). Source: Cyanobacterium *Lyngbya majuscula* (Guam). Pharm: Inhibits hypoxia-induced activation of VEGF gene promotor (Hep3B). Ref: D. G. Nagle, et al, JNP, 2000, 63, 1431

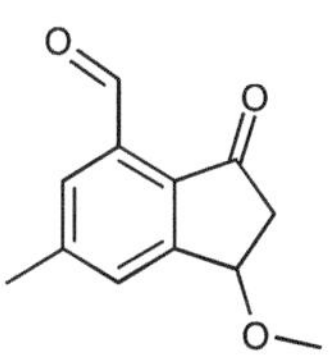

860 Penostatin E

Type: Indenes and indanes. $C_{22}H_{32}O_3$ Oil, $[\alpha]_D = +48.5°$ (c = 0.16, $CHCl_3$). Source: Marine-derived fungus *Penicillium* sp. OUPS-79 from green alga *Enteromorpha intestinalis*. Pharm: Cytotoxic (P_{388}, ED_{50} = 0.9 μg/mL). Ref: C. Takahashi, et al, Tet. Lett., 1996, 37, 655 | C. Iwamoto, et al, Tetrahedron, 1999, 55, 14353

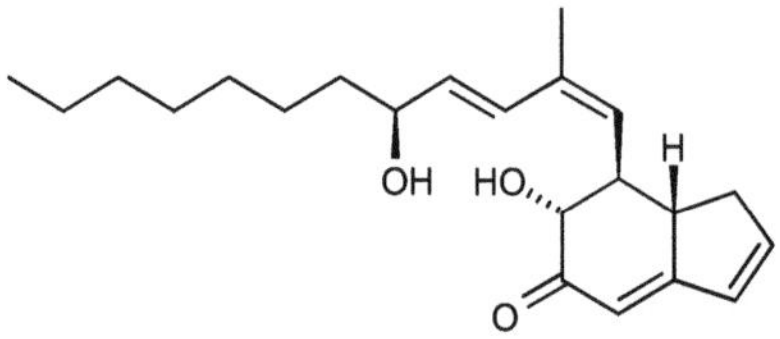

2.10 Anthracene Derivatives

861 Abietinarin A

Type: Anthracenes. $C_{19}H_{22}O_4$ Yellow amorph. solid. Source: Hydroid *Abietinaria* sp. Pharm: Cytotoxic (L_{1210}, ED_{50} < 10 μg/mL). Ref: C.Pathirana, et al, Can. J. Chem., 1990, 68, 394

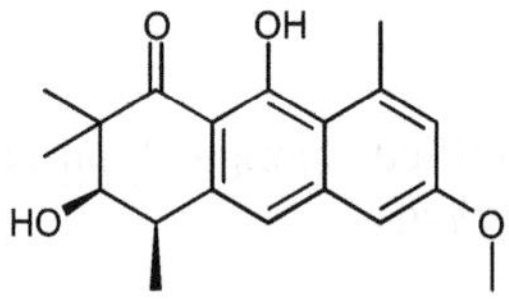

862 Abietinarin B

Type: Anthracenes. $C_{19}H_{22}O_4$ Yellow amorph. solid. Source: Hydroid *Abietinaria* sp. Pharm: Cytotoxic. Ref: C. Pathirana, et al, Can. J. Chem., 1990, 68, 394

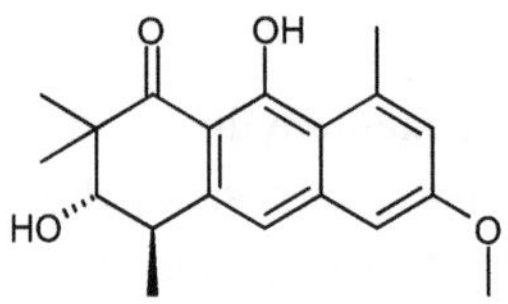

863 Anthracene derivative 100

Type: Anthracenes. $C_{20}H_{18}O_3$ Source: Marine-derived streptomycete *Streptomyces* sp. (sediment, Kiaochow Bay, Qingdao, China). Pharm: Cytotoxic (A549 cell line). Ref: H. Zhang, et al, Mar. Drugs, 2011, 9, 1502

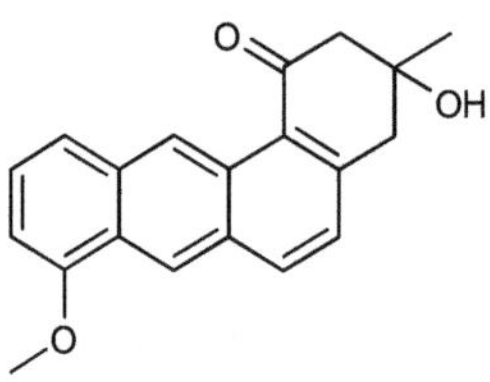

864 Asperflavin 6-*O*-α-*D*-Ribofuranoside

Type: Anthracenes. $C_{21}H_{24}O_9$ Yellow oil, $[\alpha]_D^{20}$ = +23.8° (*c* = 0.2, MeOH). Source: Marine-derived fungus *Microsporum* sp. from red alga *Lomentaria catenata* (surface, Korea waters). Pharm: Antioxidant (DPPH scavenger, IC_{50} = 14.2 μmol/L, control Ascorbic acid, IC_{50} = 20 μmol/L); antibacterial (MRSA and MDRSA, MIC = 50 μg/mL). Ref: Y. Li, et al, CPB, 2006, 54, 882

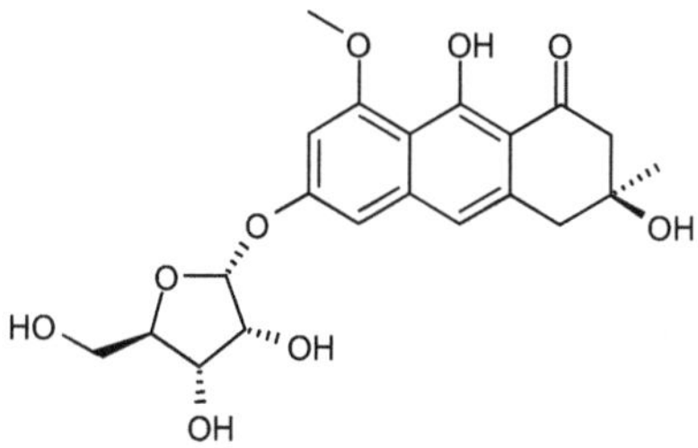

865 Eurorubrin

5,5′-Methylenebisasperflavin Type: Anthracenes. $C_{33}H_{32}O_{10}$ Amorph. brown powder, $[\alpha]_D^{25}$ = +21.1° (*c* = 0.3, MeOH). Source: Mangrove-derived fungus *Eurotium rubrum*, from semimangrove *Hibiscus tiliaceus* (stem, Hainan, I., China). Pharm: Antioxidant (DPPH scavenger, strong exceeding control BHT). Ref: D. -L. Li, et al, J. Microbiol. Biotechnol., 2009, 19, 675

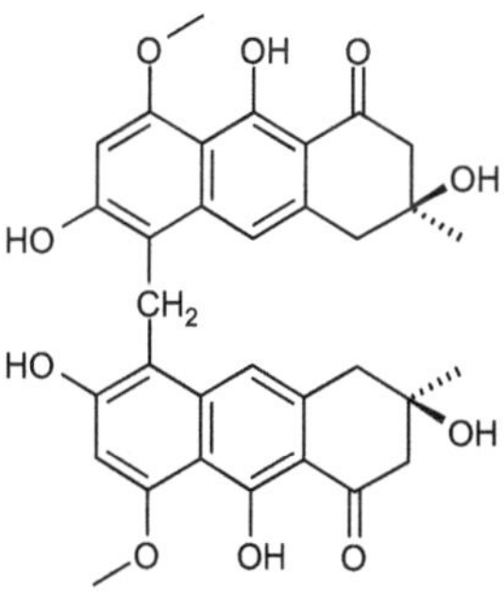

866 Fusaquinone B

Tetrahydrobostrycin Type: Anthracenes. $C_{16}H_{20}O_8$ Needles, $[\alpha]_D^{20}$ = −93.2° (*c* = 0.07, DMSO) (*Fusarium* sp.), $[\alpha]_D^{18}$ = −116.6° (*c* = 0.8, MeOH) (*Aspergillus* sp.). Source: Mangrove-derived fungus *Fusarium* sp. ZH-210 from an unidentified mangrove (sediment), marine-derived fungus *Aspergillus* sp. 05F16. Pharm: Antibacterial (100 μg/disc, *Staphylococcus aureus*, IZD = 15 mm; *Escherichia coli*, IZD = 9.2 mm). Ref: J. Xu, et al, J. Antibiot., 2009, 61, 415 (Aspergillus) | Y. Chen, et al, Magn. Reson. Chem., 2009, 47, 362 | K. Trisuwan, et al, JNP, 2010, 73, 1507

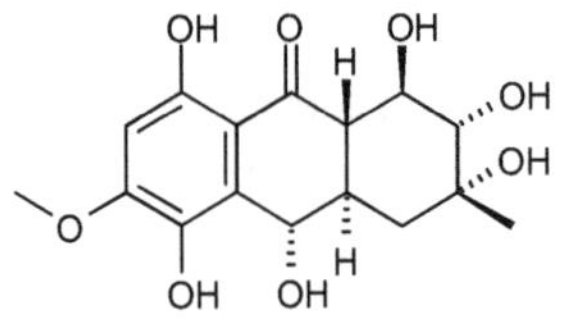

867 Fusaquinone C

1-Deoxytetrahydrobostrycin Type: Anthracenes. $C_{16}H_{20}O_7$ Red cryst., mp 234–235 °C, $[\alpha]_D^{20} = -87.8°$ (c = 0.11, DMSO) (*Fusarium* sp.), $[\alpha]_D^{18} = -69.3°$ (c = 0.4, MeOH) (*Aspergillus* sp.). Source: Marine-derived fungus *Aspergillus* sp. from an unidentified alga (Indonesia), mangrove-derived fungus *Fusarium* sp. ZH-210 from an unidentified mangrove (sediment). Pharm: Antibacterial (100 μg/disc, *Staphylococcus aureus*, IZD = 12 mm). Ref: J. Z. Xu, et al, J. Antibiot., 2008, 61, 415 | Y. Chen, et al, Magn. Reson. Chem., 2009, 47, 362 | K. Trisuwan, et al, JNP, 2010, 73, 1507

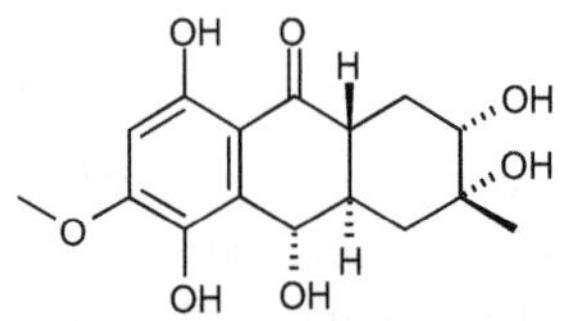

868 Fusaranthraquinone

Type: Anthracenes. $C_{16}H_{20}O_7$ Source: Marine-derived fungi *Fusarium* sp. PSU-F14. Pharm: antifungal (*Cryptococcus neoformans* and *Microsporum gypseum*, MIC = 200 μmol/L). Ref: K. Trisuwan, et al, JNP, 2010, 73, 1507

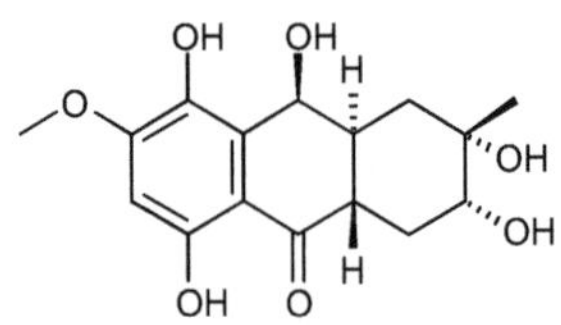

869 9α-Hydroxydihydrodesoxybostrycin

Type: Anthracenes. $C_{16}H_{20}O_7$ Source: Marine-derived fungi *Fusarium* sp. PSU-F14. Pharm: Cytotoxic (KB, IC_{50} = 19 μmol/L, control doxorubicin, IC_{50} = 0.33 μmol/L; MCF7, IC_{50} = 15 μmol/L, control doxorubicin, IC_{50} = 2.18 μmol/L; Vero, IC_{50} = 57 μmol/L, control Ellipticine, IC_{50} = 4.47 μmol/L); antibacterial (*Staphylococcus aureus* and MRSA, MIC = 350 μmol/L). Ref: K. Trisuwan, et al, JNP, 2010, 73, 1507

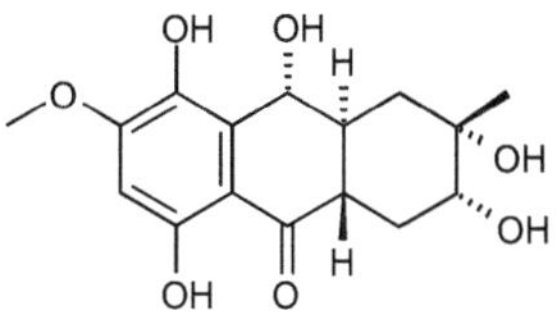

870 9α-Hydroxyhalorosellinia A

Type: Anthracenes. $C_{16}H_{20}O_8$ Source: Marine-derived fungi *Fusarium* sp. PSU-F14. Pharm: Antiplasmodial (*in vitro Plasmodium falciparum* K1, IC_{50} = 25 μmol/L, control Dihydroartemisinin, IC_{50} = 0.004 μmol/L); antituberculosis (*Mycobacterium tuberculosis*, MIC = 39 μmol/L, control Isoniazid, MIC = 0.17–0.34 μmol/L); cytotoxic (KB, IC_{50} = 49 μmol/L, control doxorubicin, IC_{50} = 0.33 μmol/L; MCF7, IC_{50} = 62 μmol/L, control doxorubicin, IC_{50} = 2.18 μmol/L; Vero, IC_{50} = 54 μmol/L, control Ellipticine, IC_{50} = 4.47 μmol/L). Ref: K. Trisuwan, et al, JNP, 2010, 73, 1507

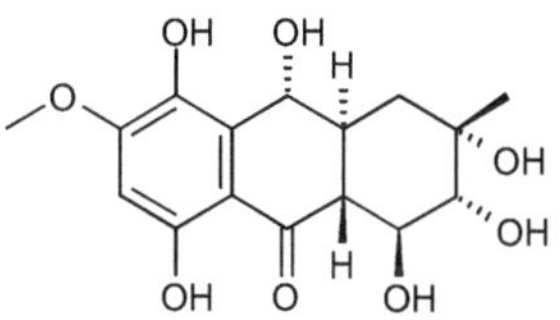

871 Physcion-10,10′-*cis*–bianthrone

Type: Anthracenes. $C_{32}H_{26}O_8$ Source: Marine-derived fungus *Aspergillus glaucus*. Pharm: Cytotoxic (MTT assay, HL60, IC_{50} = 44.0 μmol/L; SRB method, A549, IC_{50} = 14.2 μmol/L). Ref: L. Du, et al, JNP, 2008, 71, 1837

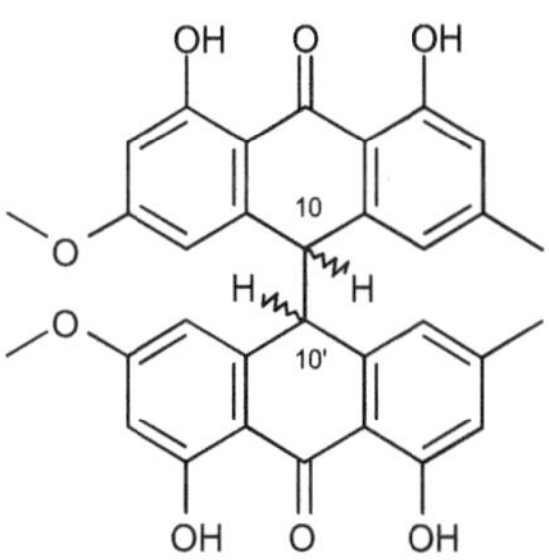

H-10, H-10′ *cis*

872 Physcion-10,10′-*trans*–bianthrone

Type: Anthracenes. $C_{32}H_{26}O_8$ Source: Marine-derived fungus *Aspergillus glaucus*. Pharm: Cytotoxic (MTT assay, HL60, IC_{50} = 7.8 μmol/L; SRB method, A549, IC_{50} = 9.2 μmol/L). Ref: L. Du, et al, JNP, 2008, 71, 1837

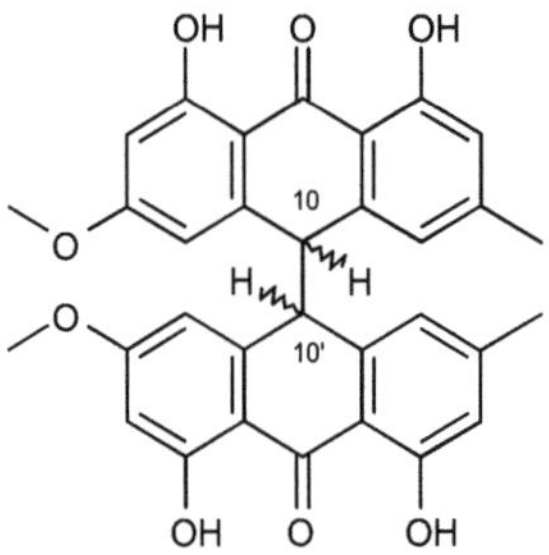

H-10, H10' *trans*

873 Trioxacarcin A

Antibiotic DC 45A Type: Anthracenes. $C_{42}H_{52}O_{20}$ Yellow powder +$2H_2O$, mp 177–183 °C (dec), $[\alpha]_D^{25} = -15.3°$ (c = 1, EtOH). Source: Marine-derived streptomycete *Streptomyces* sp. B8652, terrestrial streptomycetes (*Streptomyces ochraceus* and *Streptomyces bottropensis*). Pharm: Antibacterial (gram-positive and -negative bacteria); LD_{50} (mus, ipr) = 1 mg/kg. Ref: F. J. Tomita, J. Antibiot., 1981, 34, 1519; 1525 | R. P. Maskey, et al, J. Antibiot., 2004, 57, 771

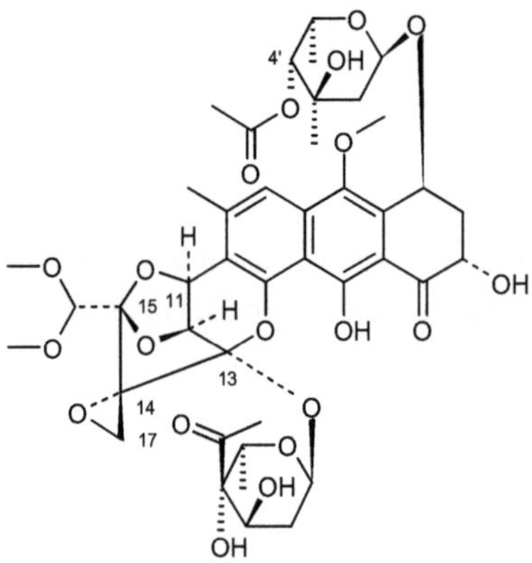

874 Trioxacarcin B

Antibiotic DC $45B_1$ Type: Anthracenes. $C_{42}H_{54}O_{21}$ Yellow powder +$2H_2O$, mp 193–194 °C (dec), $[\alpha]_D^{25} = -122.7°$ (c = 1, $CHCl_3$). Source: Marine-derived streptomycete *Streptomyces* sp. B8652, terrestrial streptomycetes (*Streptomyces ochraceus* and *Streptomyces bottropensis*). Pharm: Antibacterial (*Bacillus subtilis, Streptomyces viridochromogenes, Staphylococcus aureus* and *Escherichia coli*, MIC = 0.15–2.5 μg/mL); cytotoxic; LD_{50} (mus, ipr) = 100 mg/kg. Ref: F. J. Tomita, J. Antibiot., 1981, 34, 1519; 1525 | R. P. Maskey, et al, J. Antibiot., 2004, 57, 771

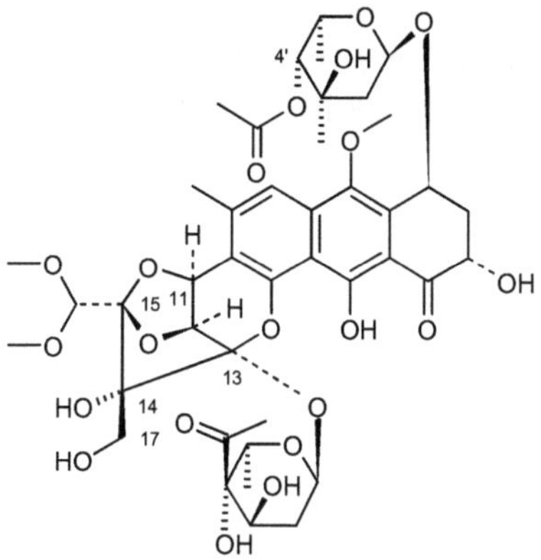

875 Trioxacarcin C

Antibiotic DC $45B_2$ Type: Anthracenes. $C_{42}H_{54}O_{20}$ Yellow powder $+2H_2O$, mp 181–182 °C (dec), $[\alpha]_D^{25}$ = −10° (*c* = 0.2, EtOH). Source: Marine-derived streptomycete *Streptomyces* sp. B8652, terrestrial streptomycetes (*Streptomyces ochraceus* and *Streptomyces bottropensis*). Pharm: Antibacterial (gram-positive and -negative bacteria); cytotoxic; LD_{50} (mus, ipr) = 2 mg/kg, LD_{50} (mus, ivn) = 1 mg/kg. Ref: F. J. Tomita, J. Antibiot., 1981, 34, 1519; 1525 | R. P. Maskey, et al, J. Antibiot., 2004, 57, 771

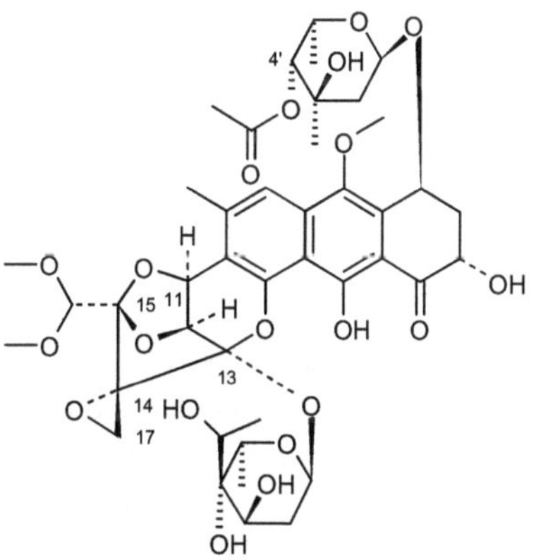

876 Austrocortirubin

Type: 1,4-Anthraquinones. $C_{16}H_{16}O_7$ Dark red needles (C_6H_6/petrol), mp 193–195 °C, $[\alpha]_D^{20}$ = +34° (*c* = 0.543, $CHCl_3$), $[\alpha]_D^{20}$ = +109° (*c* = 0.824, EtOH). Source: Marine-derived fungus *Fusarium* sp. from gorgonian sea fan *Annella* sp. (Koh Hin Ran Pet, Surat Thani Province, Thailand), marine-derived fungus *Fusarium* sp. PSU-F14, terrestrial fungi (toadstools *Cortinarius toadstools* and *Dermocybe splendid*). Pharm: Cytotoxic (KB, inactive, control doxorubicin, IC_{50} = 0.33 μmol/L; MCF7, IC_{50} = 6.3 μmol/L, control doxorubicin, IC_{50} = 2.18 μmol/L; Vero, inactive, control

Ellipticine, IC_{50} = 4.47 μmol/L); antibacterial (*Staphylococcus aureus* and MRSA, MIC = 350 μmol/L). Ref: M. A. Archard, et al, Phytochemistry, 1985, 24,2755 | M. Gill, et al, Tet. Lett., 1985, 26, 2593 | K. Trisuwan, et al, JNP, 2010, 73, 1507

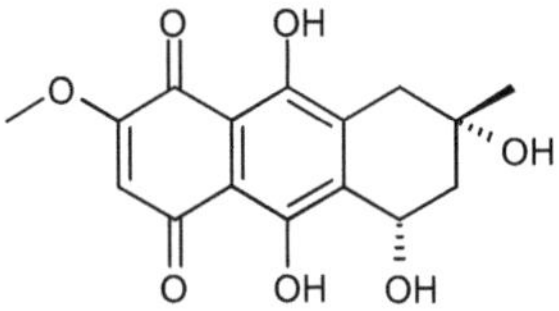

877 Bostrycin

Rhodosporin Type: 1,4-Anthraquinones. $C_{16}H_{16}O_8$ Red cryst. (Py aq), mp 222–224 °C. Source: Marine-derived fungi *Aspergillus* sp. from an unidentified alga (Indonesia) and *Fusarium* sp. PSU-F14 and 05F13. Pharm: Antiplasmodial (*in vitro Plasmodium falciparum* K1, IC_{50} = 98 μmol/L, control Dihydroartemisinin, IC_{50} = 0.004 μmol/L); cytotoxic (KB, IC_{50} = 0.9 μmol/L, control doxorubicin, IC_{50} = 0.33 μmol/L; MCF7, IC_{50} = 2.7 μmol/L, control doxorubicin, IC_{50} = 2.18 μmol/L; Vero, IC_{50} = 4.2 μmol/L, control Ellipticine, IC_{50} = 4.47 μmol/L); inducer of mitochondria-mediated apoptosis (yeast *Saccharomyces cerevisiae*); LD_{50} (mus, ipr) = 200 mg/kg. Ref: B. Beagley, et al, Chem. Comm., 1989, 17 | J. Xu, et al, Yakugaku Zasshi, 2006, 126, 234 | J. Z. Xu, et al, J. Antibiot., 2008, 61, 415 | K. Trisuwan, et al, JNP, 2010, 73, 1507

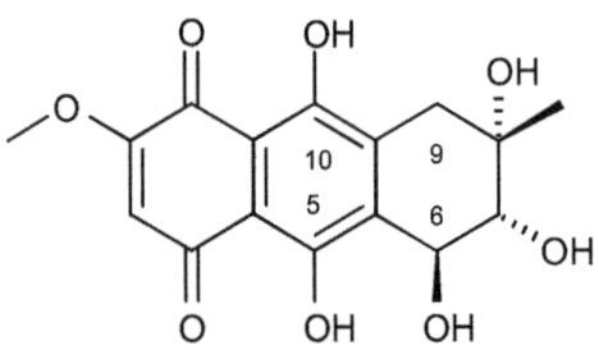

878 5,7-Dideoxybostrycin

Type: 1,4-Anthraquinones. $C_{16}H_{16}O_6$ Source: Marine-derived fungus *Nigrospora* sp. 1403 from an unidentified sea anemone. Pharm: Antibacterial (*Bacillus subtilis,* MIC = 0.625 μmol/L, *Bacillus cereus,* MIC = 10.0 μmol/L, *Micrococcus luteus,* MIC = 20.0 μmol/L, *Staphylococcus albus,* MIC = 5.00 μmol/L, *Staphylococcus aureus,* MIC = 2.50 μmol/L, *Micrococcus tetragenus,* MIC = 1.25 μmol/L, *Escherichia coli,* MIC = 2.50 μmol/L, *Vibrio anguillarum,* MIC = 2.50 μmol/L, *Vibrio parahaemolyticus,* MIC = 1.25 μmol/L). Ref: K. L. Yang, et al, *JNP,* 2012, *75,* 935

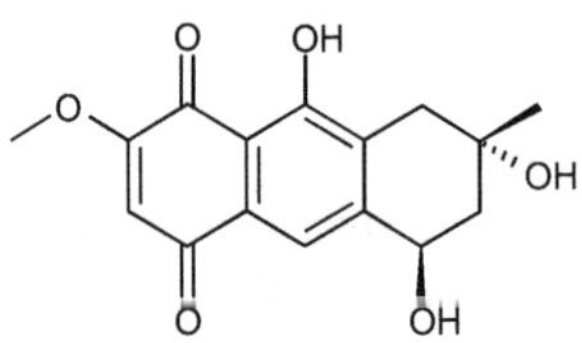

879 5-Deoxybostrycin

Type: 1,4-Anthraquinones. $C_{16}H_{16}O_7$ Source: Marine-derived fungus *Nigrospora* sp. from an unidentified sea anemone (Weizhou I., Guangxi, China). Pharm: Antibacterial (*Bacillus cereus,* strong); cytotoxic (A549, strong). Ref: K. -L. Yang, et al, JNP, 2012, 75, 935

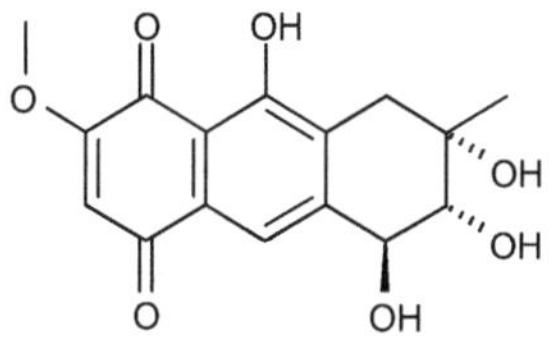

880 Nigrosporin B

Type: 1,4-Anthraquinones. $C_{16}H_{16}O_6$ Yellow oil, $[\alpha]_D^{28}$ = −16.7° (*c* = 0.02, $CHCl_3$). Source: Marine-derived fungi *Fusarium* sp. PSU-F14. Pharm: Antiplasmodial (*in vitro Plasmodium falciparum* K1, IC_{50} = 13 μmol/L, control Dihydroartemisinin, IC_{50} = 0.004 μmol/L); antituberculosis (*Mycobacterium tuberculosis*, MIC = 41 μmol/L, control Isoniazid, MIC = 0.17–0.34 μmol/L); cytotoxic (KB, IC_{50} = 88 μmol/L, control doxorubicin, IC_{50} = 0.33 μmol/L; MCF7, IC_{50} = 5.4 μmol/L, control doxorubicin, IC_{50} = 2.18 μmol/L; Vero, IC_{50} = 29 μmol/L, control Ellipticine, IC_{50} = 4.47 μmol/L). Ref: K. Trisuwan, et al, JNP, 2010, 73, 1507

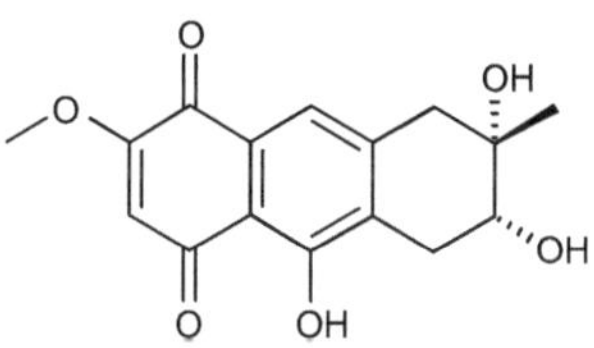

881 Albopunctatone

Type: 9,10-Anthraquinones. $C_{28}H_{16}O_8$ Source: Ascidian *Didemnum albopunctatum* (Swain Reefs, Great Barrier Reef). Pharm: Antiplasmodial (CSPF and CRPF). Ref: A. R. Carroll, et al, JNP, 2012, 75, 1206

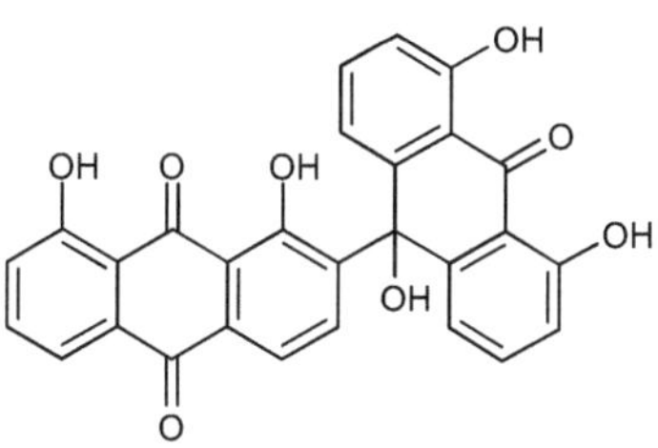

882 Alterporriol K

Type: 9,10-Anthraquinones. $C_{32}H_{26}O_{11}$ Red powder, $[\alpha]_D^{25}$ = +690° (*c* = 1.0, MeOH). Source: Mangrove-derived fungus *Alternaria* sp. from mangrove *Aegiceras corniculatum* (Guangdong, China). Pharm: Cytotoxic (MTT assay, MDA-MB-435, IC_{50} = 26.97 μmol/L; MCF7, IC_{50} = 29.11 μmol/L). Ref: C. -H. Huang, et al, Mar. Drugs, 2011, 9, 832

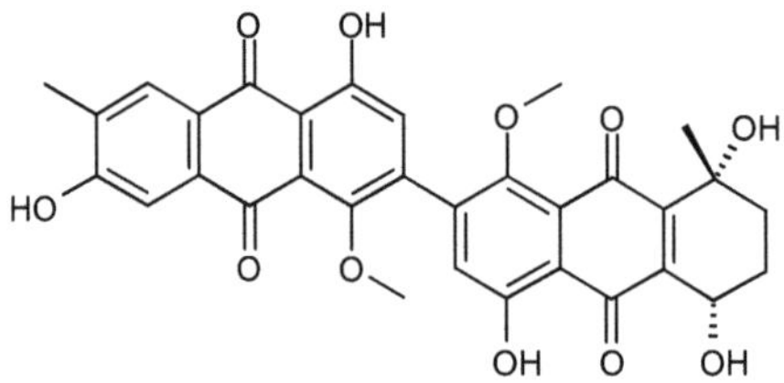

883 Alterporriol L

Type: 9,10-Anthraquinones. $C_{32}H_{26}O_{12}$ Red powder, $[\alpha]_D^{25}$ = +30° (*c* = 1.0, MeOH). Source: Mangrove-derived fungus *Alternaria* sp. from mangrove *Aegiceras corniculatum* (Guangdong, China). Pharm: Cytotoxic (MTT assay, MDA-MB-435, IC_{50} = 13.11 μmol/L; MCF7, IC_{50} = 20.04 μmol/L). Ref: C. -H. Huang, et al, Mar. Drugs, 2011, 9, 832

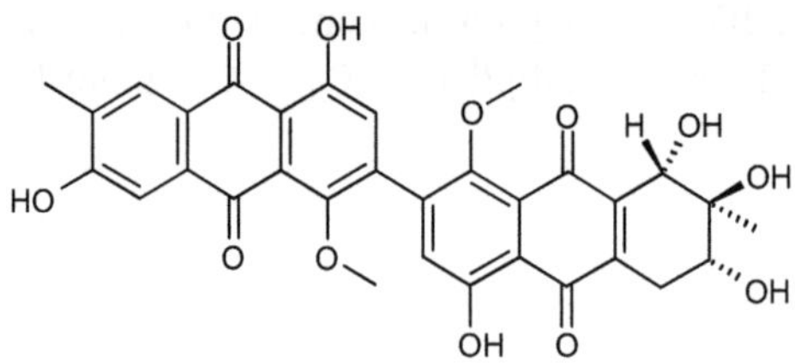

884 Alterporriol P

Type: 9,10-Anthraquinones. $C_{32}H_{26}O_{12}$ Source: Marine-derived fungus *Alternaria* sp. from soft coral *Sarcophyton* sp. (Weizhou coral reef, Giuangxi, China). Pharm: Cytotoxic (HTCLs cells). Ref: C. -J. Zheng, et al, JNP, 2012, 75, 189

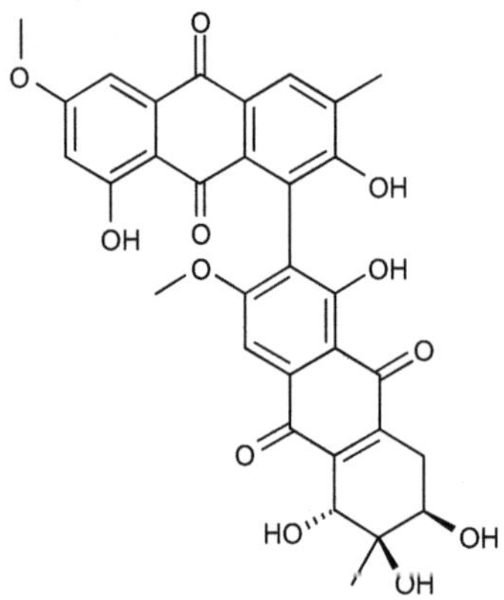

885 Alterporriol Q

Type: 9,10-Anthraquinones. $C_{32}H_{22}O_{10}$ Source: Marine-derived fungus *Alternaria* sp. from soft coral *Sarcophyton* sp. ZJ-2008003 (Weizhou coral reef, Giuangxi, China). Pharm: Antiviral (porcine reproductive and respiratory syndrome (PRRS) virus, IC_{50} = 39 μmol/L). Ref: C. -J. Zheng, et al, JNP, 2012, 75, 189 | S. Z. Moghadamtousi, et al, Mar. Drugs, 2015, 13, 4520 (rev)

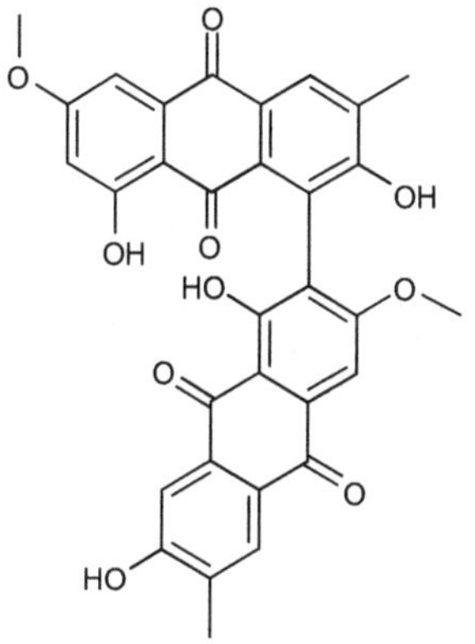

886 Antibiotic BE 43472A

Type: 9,10-Anthraquinones. $C_{32}H_{24}O_{10}$ Amorph. yellow solid, $[\alpha]_D^{23}$ = +538° (c = 0.23, $CHCl_3$). Source: Marine-derived streptomycete *Streptomyces* sp. N1-78-1. Pharm: Antibacterial; cytotoxic. Ref: A. M. Socha, et al, BoMC, 2006, 14, 8446 | A. M. Socha, et al, JNP, 2006, 69, 1070 | K. C. Nicolaou, et al, JACS, 2009, 131, 14812

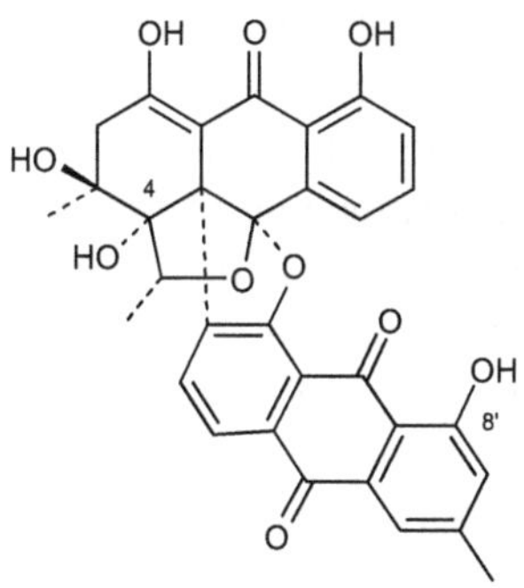

887 7,7′-Bihelminthosporin

2,2′-Bihelminthosporin; Antibiotic 2240A Type: 9,10-Anthraquinones. $C_{30}H_{18}O_{10}$ Orange-red solid, mp 300 °C, $[\alpha]_D^{20}$ = +62.5° (c = 0.08, dioxin). Source: An unidentified mangrove-derived fungus No. 2240 (endophytic). Pharm: hTopo I isomerase inhibitor. Ref: N. Tan, et al, J. Asian Nat. Prod. Res., 2008, 10, 607 | N. Tan, et al, Yingyong Huaxue, 2009, 26, 277

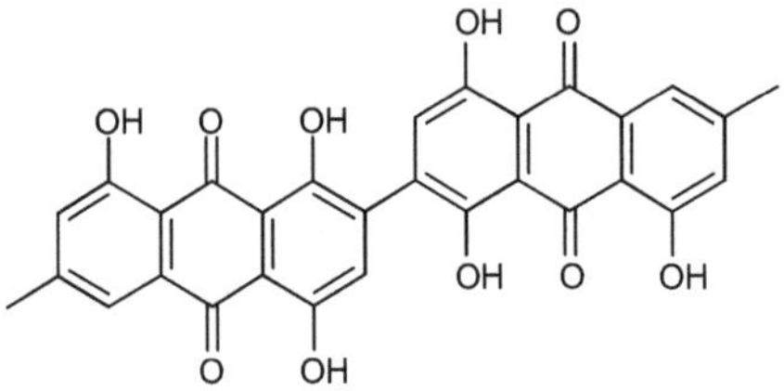

888 Chrysophanol

1,8-Dihydroxy-3-methylanthraquinone Type: 9,10-Anthraquinones. $C_{15}H_{10}O_4$ Golden-yellow plates (C_6H_6), mp 200–201 °C, mp 196 °C. Source: Marine-derived fungus *Monodictys* sp. from urchin *Anthocidaris crassispina* (Japan waters), annelid (spoon worm) *Urechis unicinctus*, mangrove-derived endophytic fungi *Halorosellinia* sp. 1403 and *Guignardia* sp. 4382, widely distributed in plants. Pharm: Antimicrobial; cathartic; cytotoxic (drug-sensitive parental KB, IC_{50} = 174.87 μmol/L, control Adriamycin, IC_{50} = 0.034 μmol/L; KBV200, IC_{50} = 331.97 μmol/L, Adriamycin, IC_{50} = 1.894 μmol/L). Ref: A. A. El-Beih, et al, CPB, 2007, 55, 1097 | J. Y. Zhang, et al, Mar. Drugs, 2010, 8, 1469

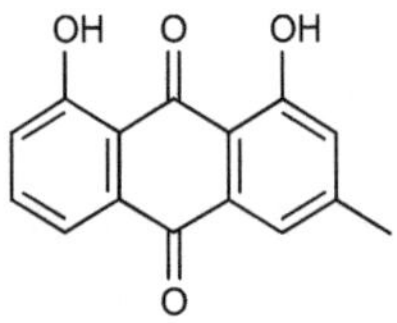

889 1,8-Dihydroxyanthraquinone

Dantron Type: 9,10-Anthraquinones. $C_{14}H_8O_4$ Square orange plates (Me_2CO), mp 193 °C. Source: Marine-derived fungus *Beauveria bassiana* from an unidentified sponge (Iriomote I., Okinawa, Japan), bryozoans *Dakaria subovoidea* (Japan waters) and *Watersipora subtorquata*, mangrove-derived endophytic fungi *Halorosellinia* sp. 1403 and *Guignardia* sp. 4382, terrestrial plants (*Rheum palmatum, Xyris semifuscata, Cinchona ledgeriana* and *Pyrrhalta luteola*). Pharm: Antioxidant; cathartic; cytotoxic (drug-sensitive parental KB, IC_{50} = 56.56 μmol/L, control Adriamycin, IC_{50} = 0.034 μmol/L; KBV200, IC_{50} = 109.15 μmol/L, Adriamycin, IC_{50} = 1.894 μmol/L). Ref: M. P. Kuntsmann, et al, JOC, 1966, 31, 2920 | T. Shindo, et al, Experientia, 1993, 49, 177 | J. Y. Zhang, et al, Mar. Drugs, 2010, 8, 1469 | H. Yamazaki, et al, Mar. Drugs, 2012, 10, 2691 | CRC Press, DNP on DVD, 2012, version 20.2

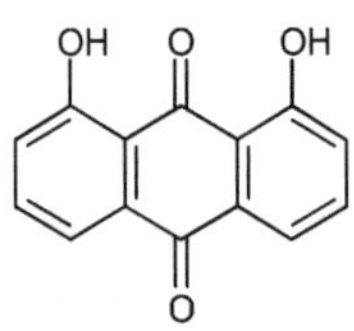

890 1,7-Dihydroxy-2,4-dimethoxy-6-methylanthraquinone

Type: 9,10-Anthraquinones. $C_{17}H_{14}O_6$ Source: Mangrove-derived endophytic fungi *Halorosellinia* sp. 1403 and *Guignardia* sp. 4382. Pharm: Cytotoxic (drug-sensitive parental KB, IC_{50} > 500 μmol/L, control Adriamycin, IC_{50} = 0.034 μmol/L; KBV200, IC_{50} = 301.47 μmol/L, Adriamycin, IC_{50} = 1.894 μmol/L). Ref: J. Y. Zhang, et al, Mar. Drugs, 2010, 8, 1469

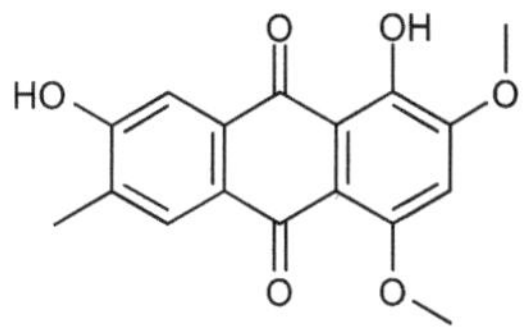

891 1,4-Dihydroxy-2-methoxy-7-methylanthraquinone

Austrocortinin Type: 9,10-Anthraquinones. $C_{16}H_{12}O_5$ Red cryst. ($CHCl_3$/petrol), mp 237–240 °C, mp 208–209 °C. Source: Marine-derived fungus *Halorosellinia* sp. 1403 (decayed wood in Hong Kong and a salt lake in Bahamas), mangrove-derived endophytic fungi *Halorosellinia* sp. 1403 and *Guignardia* sp. 4382, terrestrial fungi (*Cortinarius* spp.). Pharm: Cytotoxic (drug-sensitive parental KB, IC_{50} = 305.14 μmol/L, control Adriamycin, IC_{50} = 0.034 μmol/L; KBV200, IC_{50} > 500 μmol/L, Adriamycin, IC_{50} = 1.894 μmol/L). Ref: M. Archard, et al, Phytochemistry, 1985, 24, 2755 | Z. -G. She, et al, Acta Cryst. E, 2006, 62, 3737 | X. -K. Xia, et al, Magn. Reson. Chem., 2007, 45, 1006 | J. Y. Zhang, et al, Mar. Drugs, 2010, 8, 1469

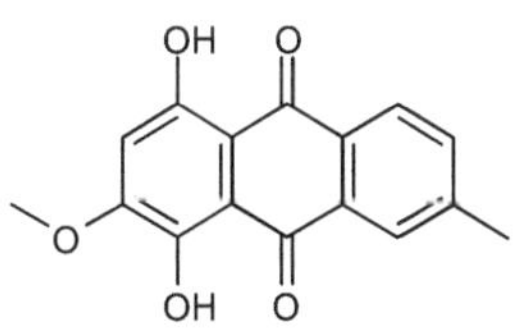

892 1,3-Dihydroxy-6-methoxy-8-methylanthraquinone

Type: 9,10-Anthraquinones. $C_{16}H_{12}O_5$ Source: Mangrove-derived endophytic fungi *Halorosellinia* sp. 1403 and *Guignardia* sp. 4382. Pharm: Cytotoxic (drug-sensitive parental KB, IC_{50} = 38.05 μmol/L, control Adriamycin, IC_{50} = 0.034 μmol/L; KBV200, IC_{50} = 34.64 μmol/L, Adriamycin, IC_{50} = 1.894 μmol/L). Ref: J. Y. Zhang, et al, Mar. Drugs, 2010, 8, 1469

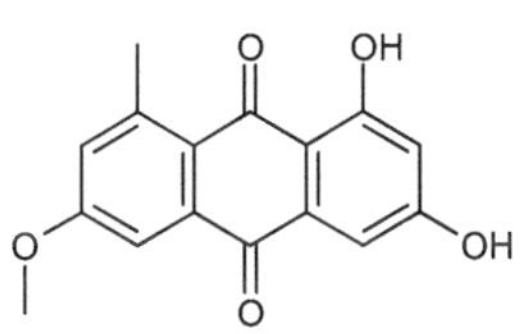

893 1,4-Dihydroxy-6-methylanthraquinone

6-Methylquinizarin Type: 9,10-Anthraquinones. $C_{15}H_{10}O_4$ Orange cryst. (AcOH), red needles (EtOH), mp 175–177 °C, mp 167–168 °C. Source: Marine-derived fungus *Halorosellinia* sp. 1403 (decayed wood in Hong Kong and a salt lake in Bahamas), terrestrial plant (*Rubia cordifolia*, root), mangrove-derived endophytic fungi *Halorosellinia* sp. 1403 and *Guignardia* sp. 4382. Pharm: Cytotoxic (drug-sensitive parental KB, IC_{50} = 114.09 μmol/L, control Adriamycin, IC_{50} = 0.034 μmol/L; KBV200, IC_{50} = 86.45 μmol/L, Adriamycin, IC_{50} = 1.894 μmol/L). Ref: A. M. Tessier, et al, PM, 1981, 41, 337 | X. -K. Xia, et al, Magn. Reson. Chem., 2007, 45, 1006 | J. Y. Zhang, et al, Mar. Drugs, 2010, 8, 1469

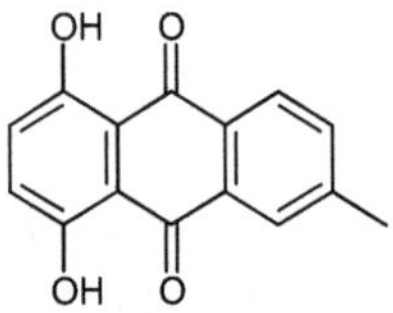

894 1,3-Dihydroxy-6-methyl-8-methoxyanthraquinone

Type: 9,10-Anthraquinones. $C_{16}H_{12}O_5$ Source: Mangrove-derived endophytic fungi *Halorosellinia* sp. 1403 and *Guignardia* sp. 4382. Pharm: Cytotoxic (drug-sensitive parental KB, IC_{50} > 500 μmol/L, control Adriamycin, IC_{50} = 0.034 μmol/L; KBV200, IC_{50} = 109.81 μmol/L, Adriamycin, IC_{50} = 1.894 μmol/L). Ref: J. Y. Zhang, et al, Mar. Drugs, 2010, 8, 1469

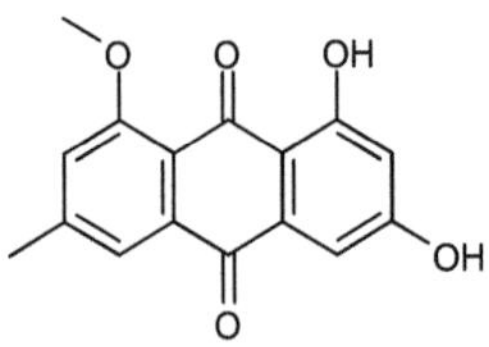

895 1,3-Dimethoxy-6-methylanthraquinone

Type: 9,10-Anthraquinones. $C_{17}H_{14}O_4$ Source: Mangrove-derived endophytic fungi *Halorosellinia* sp. 1403 and *Guignardia* sp. 4382. Pharm: Cytotoxic (drug-sensitive parental KB, IC_{50} = 57.32 μmol/L, control Adriamycin, IC_{50} = 0.034 μmol/L; KBV200, IC_{50} = 90.86 μmol/L, Adriamycin, IC_{50} = 1.894 μmol/L). Ref: J. Y. Zhang, et al, Mar. Drugs, 2010, 8, 1469

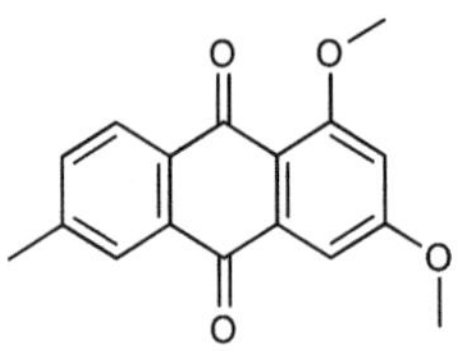

896 6,8-Di-*O*-methylaverantin

Type: 9,10-Anthraquinones. $C_{22}H_{24}O_7$ Source: Marine-derived fungus *Aspergillus versicolor* EN-7from brown alga *Sargassum thunbergii* (Qingdao, Shandong, China). Pharm: Antibacterial (*Escherichia coli*, 20 μg/disk, IZ = 7 mm). Ref: Y. Zhang, et al, Biosci., Biotechnol., Biochem., 2012, 76, 1774

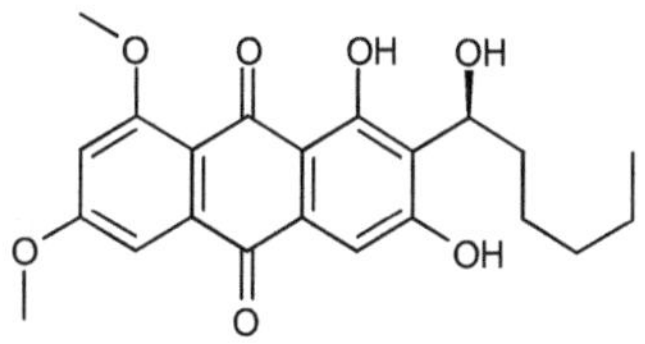

897 Dioxamycin

Type: 9,10-Anthraquinones. $C_{38}H_{40}O_{15}$ Orange amorph. powder, mp 176–178 °C (dec), $[\alpha]_D^{23}$ = +43.8° (*c* = 0.05, MeOH), $[\alpha]_D^{24}$ = +49° (*c* = 0.2, MeOH). Source: Marine-derived streptomycete *Streptomyces fradiae* PTZ00025 (sediment, source unspecified). Pharm: Antibacterial (gram-positive bacteria); cytotoxic. Ref: R. Sawa, et al, J. Antibiot., 1991, 44, 396 | W. Xin, et al, Mar. Drugs, 2012, 10, 2388

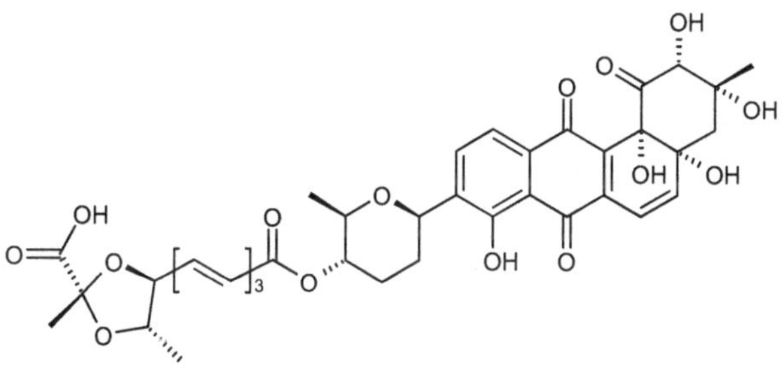

898 Emodin

1,3,8-Trihydroxy-6-methylanthraquinone Type: 9,10-Anthraquinones. $C_{15}H_{10}O_5$ Orange or yellow-brown needles (Py aq or MeOH), mp 266–268 °C, mp 254–256 °C. Source: Marine-derived fungus *Monodictys* sp. from urchin *Anthocidaris crassispina* (Japan waters), mangrove-derived endophytic fungi *Halorosellinia* sp. 1403 and *Guignardia* sp. 4382 and occurs in terrestrial plants. Pharm: Antimicrobial; antineoplastic; cathartic; monoamine oxidase inhibitor; cytotoxic (drug-sensitive parental KB, IC_{50} > 500 μmol/L, control Adriamycin, IC_{50} = 0.034 μmol/L; KBV200, IC_{50} > 500 μmol/L, Adriamycin, IC_{50} = 1.894 μmol/L). Ref: A. A. El-Beih, et al, CPB, 2007, 55, 1097 | J. Y. Zhang, et al, Mar. Drugs, 2010, 8, 1469

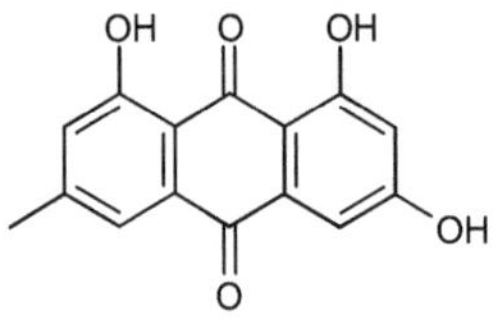

899 Evariquinone

Type: 9,10-Anthraquinones. $C_{16}H_{12}O_6$ Orange needles, mp 238–242 °C (subl). Source: Marine-derived fungus *Emericella variecolor* from sponge *Haliclona valliculata*. Pharm: Antiproliferative (KB and NCI-H460, 3.16 μg/mL effective). Ref: G. Bringmann, et al, Phytochemistry, 2003, 63, 437 | M. Saleem, et al, NPR, 2007, 24, 1142 (rev)

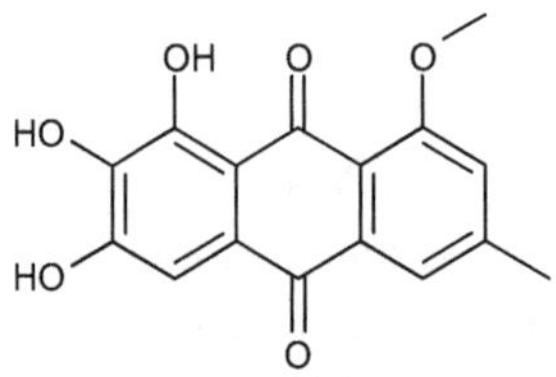

900 Fradimycin A

Type: 9,10-Anthraquinones. $C_{39}H_{42}O_{15}$ Orange amorph. powder, $[\alpha]_D^{23}$ = +39.5° (*c* = 0.05, MeOH). Source: Marine-derived streptomycete *Streptomyces fradiae* PTZ00025 (sediment, source unspecified). Pharm: Antibacterial (*Staphylococcus aureus*, MIC = 2.0–6.0 μg/mL); cytotoxic (hmn colon HCT15, IC_{50} = (0.52 ± 0.11)μmol/L; hmn colon SW620, IC_{50} = (6.46 ± 1.44)μmol/L; Rat glioma C6, IC_{50} = (1.28 ± 0.37)μmol/L). Ref: W. Xin, et al, Mar. Drugs, 2012, 10, 2388

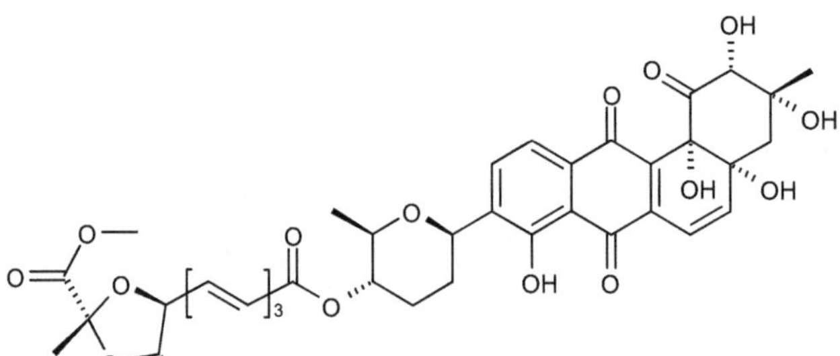

901 Fradimycin B

Type: 9,10-Anthraquinones. $C_{38}H_{38}O_{14}$ Orange amorph. powder, $[\alpha]_D^{23}$ = +36.3° (*c* = 0.05, MeOH). Source: Marine-derived streptomycete *Streptomyces fradiae* PTZ00025 (sediment, source unspecified). Pharm: Antibacterial (*Staphylococcus aureus*, MIC = 2.0–6.0 μg/mL); cytotoxic (Hmn colon HCT15, IC_{50} = (0.13 ± 0.04)μmol/L;

hmn colon SW620, IC_{50} = (4.33 ± 1.56)μmol/L; rat glioma C6, IC_{50} = (0.47 ± 0.09)μmol/L). Ref: W. Xin, et al, Mar. Drugs, 2012, 10, 2388

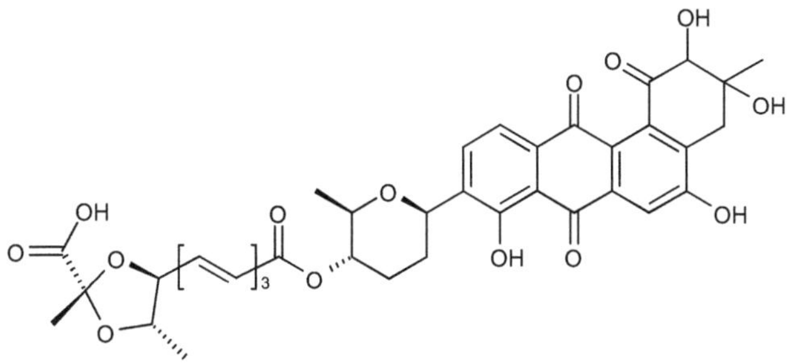

902 Galvaquinone B

Type: 9,10-Anthraquinones. $C_{21}H_{20}O_6$ Source: Marine-derived streptomycete *Streptomyces spinoverrucosus* (sediment, Trinity Bay, Galveston, Texas). Pharm: Epigenetic modulatory activity; cytotoxic (NSCLC, Calu3 and H2887, moderate). Ref: Y. Hu, et al, JNP, 2012, 75, 1759

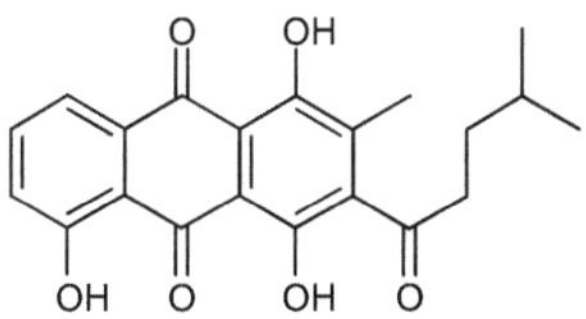

903 Halawanone C

Type: 9,10-Anthraquinones. $C_{21}H_{20}O_7$ Source: Marine-derived streptomycete *Streptomyces* sp. BD-18T (shallow water sediment on Oahu). Pharm: Antibacterial (*Bacillus subtilis, Staphylococcus aureus*, gram-positive bacteria). Ref: P. W. Ford, et al, JNP, 1998, 61, 1232

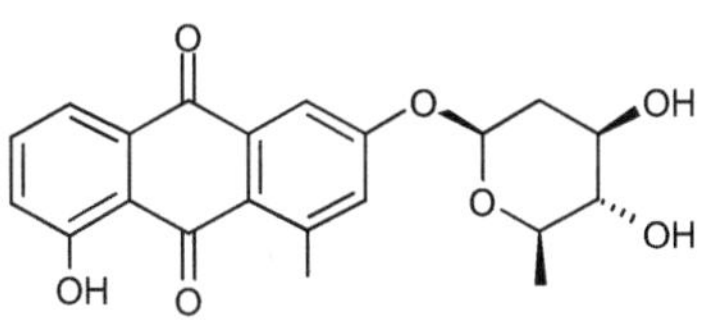

904 Halawanone D

Type: 9,10-Anthraquinones. $C_{22}H_{22}O_7$ Source: Marine-derived streptomycete *Streptomyces* sp. BD-18T (shallow water sediment on Oahu). Pharm: Antibacterial (*Bacillus subtilis, Staphylococcus aureus*, gram-positive bacteria). Ref: P. W. Ford, et al, JNP, 1998, 61, 1232

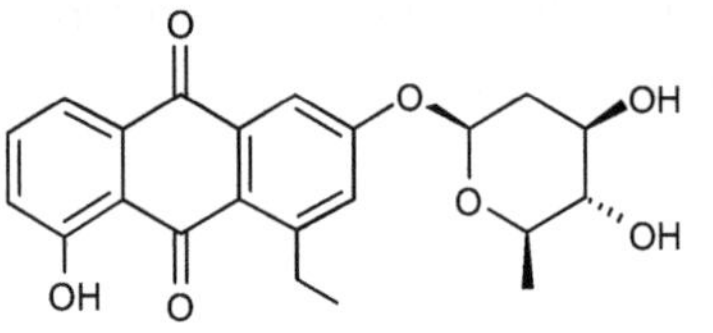

905 1-Hydroxy-2,4-dimethoxy-7-methylanthraquinone

Type: 9,10-Anthraquinones. $C_{17}H_{14}O_5$ Source: Mangrove-derived endophytic fungi *Halorosellinia* sp. 1403 and *Guignardia* sp. 4382. Pharm: Cytotoxic (drug-sensitive parental KB, IC_{50} = 68.39 μmol/L, control Adriamycin, IC_{50} = 0.034 μmol/L; KBV200, IC_{50} = 243.69 μmol/L, Adriamycin, IC_{50} = 1.894 μmol/L). Ref: J. Y. Zhang, et al, Mar. Drugs 2010, 8, 1469

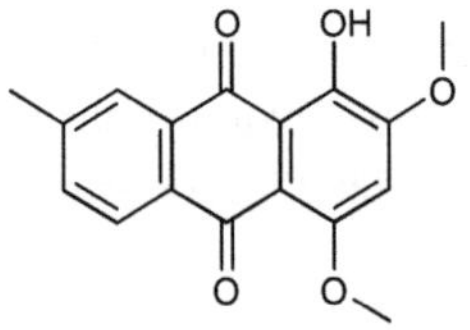

906 Isorhodoptilometrin-1-methyl ether

Type: 9,10-Anthraquinones. $C_{18}H_{16}O_6$ Source: Marine-derived fungus *Aspergillus versicolor* from green alga *Halimeda opuntia* (Rass Mohamed, south Sinai, Egypt). Pharm: Antibacterial (3 gram-positive bacteria, 50 μg/disk: *Bacillus cereus*, IZ = 2 mm, *Bacillus subtilis*, IZ = 3 mm, *Staphylococcus aureus*, IZ = 5 mm). Ref: U. W. Hawas, et al, Arch. Pharmacal Res., 2012, 35, 1749

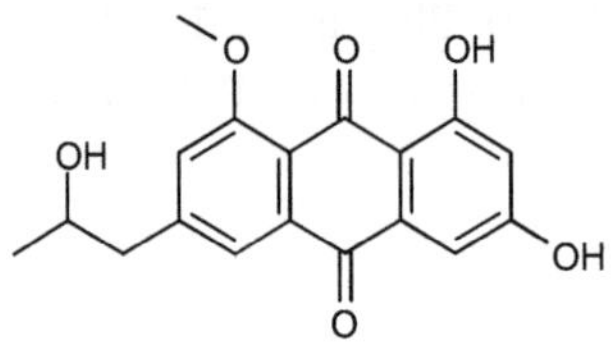

907 1-Methoxy-3-methyl-8-hydroxyanthraquinone

Type: 9,10-Anthraquinones. $C_{16}H_{12}O_4$ Source: Mangrove-derived endophytic fungi *Halorosellinia* sp. 1403 and *Guignardia* sp. 4382. Pharm: Cytotoxic (drug-sensitive parental KB, IC_{50} > 500 μmol/L, control Adriamycin, IC_{50} = 0.034 μmol/L; KBV200, IC_{50} = 185.68 μmol/L, Adriamycin, IC_{50} = 1.894 μmol/L). Ref: J. Y. Zhang, et al, Mar. Drugs, 2010, 8, 1469

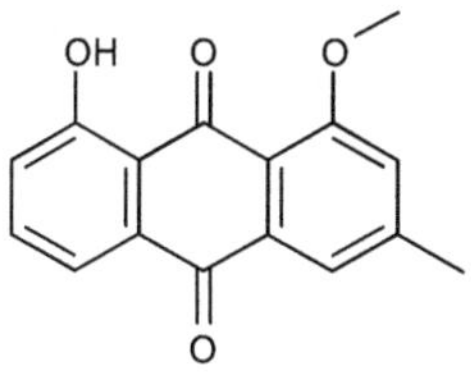

908 MK844-mF10

Type: 9,10-Anthraquinones. $C_{38}H_{38}O_{13}$ Orange amorph. powder, $[\alpha]_D^{23} = +39.6°$ (c = 0.01, MeOH). Source: Marine-derived streptomycete *Streptomyces fradiae* PTZ00025 (sediment, source unspecified). Pharm: Antibacterial (*Staphylococcus aureus*, MIC = 2.0–6.0 μg/mL); cytotoxic (Hmn colon HCT15, IC_{50} = (0.30 ± 0.07)μmol/L; hmn colon SW620, IC_{50} = (4.39 ± 0.93)μmol/L; rat glioma C6, IC_{50} = (1.31 ± 0.32)μmol/L). Ref: W. Xin, et al, Mar. Drugs, 2012, 10, 2388

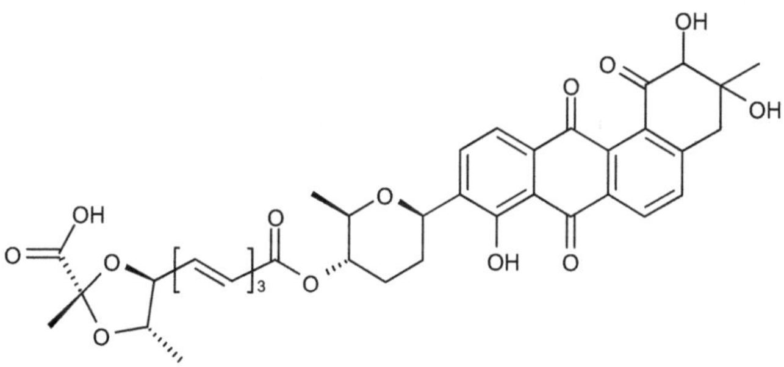

909 Monodictyquinone

1,8-Dihydroxy-2-methoxy-6-methylanthraquinone Type: 9,10-Anthraquinones. $C_{16}H_{12}O_5$ Yellow solid. Source: Marine-derived fungus *Monodictys* sp. from urchin *Anthocidaris crassispina* (Japan waters). Pharm: Antibacterial (Paper disks diameter 6 mm, impregnated with compound 2.5 μg, were incubated on agar plates containing Microorganisms: *Bacillus subtilis*, IZD = 7 mm; *Escherichia coli*, IZD = 8 mm); antifungal (yeast *Candida albicans*, IZD = 7 mm). Ref: A. A. El-Beih, et al, CPB, 2007, 55, 1097

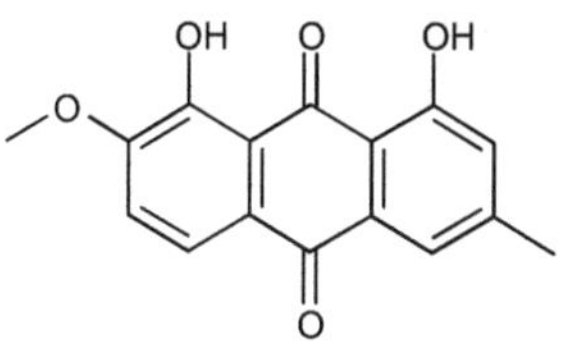

910 Pachybasin

1-Hydroxy-3-methylanthraquinone Type: 9,10-Anthraquinones. $C_{15}H_{10}O_3$ Yellow needles, mp 175–176 °C, mp 234–237 °C. Source: Marine-derived fungus *Halorosellinia* sp. 1403 (decayed wood in Hong Kong and a salt lake in Bahamas), mangrove-derived endophytic fungi *Halorosellinia* sp. 1403 and *Guignardia* sp. 4382. Pharm: Cytotoxic (KB, IC_{50} = 1.40 μg/mL; KBV200, IC_{50} = 2.58 μg/mL); cytotoxic (drug-sensitive parental KB, IC_{50} = 3.17 μmol/L, control Adriamycin, IC_{50} = 0.034 μmol/L; KBV200, IC_{50} = 3.21 μmol/L, Adriamycin, IC_{50} = 1.894 μmol/L, might induce apoptosis *via* a mitochondrial pathway). Ref: X. -K. Xia, et al, Magn. Reson. Chem., 2007, 45, 1006 | A. A. El-Beih, et al, CPB, 2007, 55, 1097 | J. Y. Zhang, et al, Mar. Drugs 2010, 8, 1469

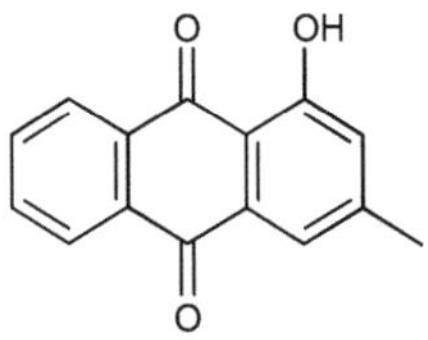

911 1,2,3,6,8-Pentahydroxy-7-(1*R*-methoxyethyl)anthraquinone

Type: 9,10-Anthraquinones. $C_{17}H_{14}O_8$ Orange-yellow powder, $[\alpha]_D$ = −380.2° (*c* = 0.1, EtOH). Source: Marine-derived fungus *Microsphaeropsis* sp. from sponge *Aplysina aerophoba* (Mediterranean Sea). Pharm: Kinases inhibitor (PKC-ε, IC_{50} = 54.0 μmol/L; CDK4/cyclin D1, IC_{50} = 37.5 μmol/L; and EGFR, IC_{50} = 41.0 μmol/L). Ref: G. Brauers, et al, JNP, 2000, 63, 739

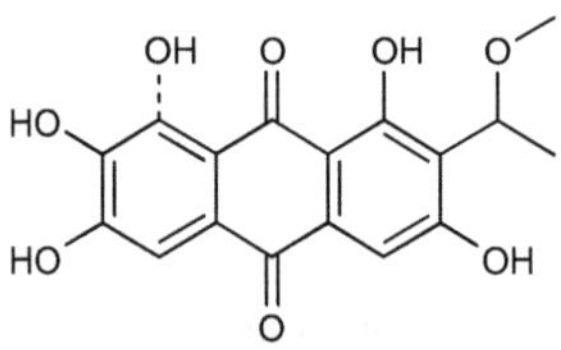

912 Physcion

Type: 9,10-Anthraquinones. $C_{16}H_{12}O_5$ Orange needles (EtOAc/petrol), mp 209–210 °C. Source: Mangrove-derived fungus *Alternaria* sp. from mangrove *Aegiceras corniculatum* (Guangdong, China), an unidentified mangrove-derived fungus 2526 (South China Sea), and widely distributed in lichens, e.g. *Parmelia* spp., and higher plants. Pharm: Antimicrobial; possesses cathartic properties; LD_{50} (mus, ipr) = 10 mg/kg. Ref: M. Saleem, et al, NPR, 2007, 24, 1142 (rev) | C. -H. Huang, et al, Mar. Drugs, 2011, 9, 832

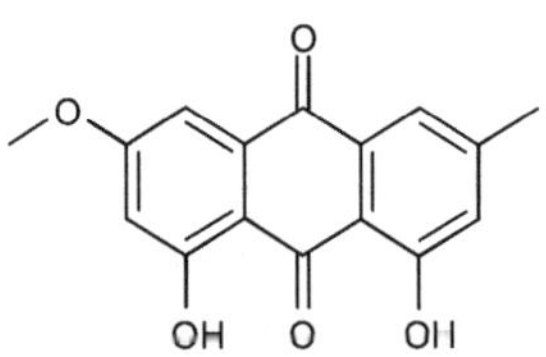

913 Questin 3-α-D-ribofuranoside

3-*O*-(*a*-*D*-Ribofuranosyl)questin Type: 9,10-Anthraquinones. $C_{21}H_{20}O_9$ Amorph. orange powder, $[\alpha]_D^{25}$ = +89.4° (*c* = 0.02, MeOH). Source: Mangrove-derived fungus *Eurotium rubrum* from semimangrove *Hibiscus tiliaceus* (China waters). Pharm: Antioxidant (DPPH scavenger, moderate). Ref: D. -L. Li, et al, J. Microbiol. Biotechnol., 2009, 19, 675

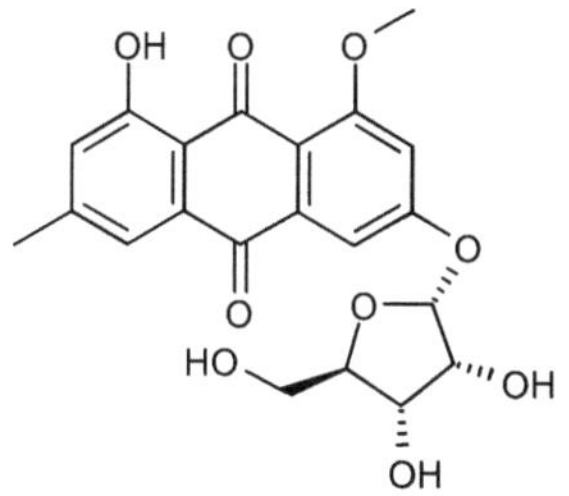

914 Tetrahydroaltersolanol C

Type: 9,10-Anthraquinones. $C_{16}H_{20}O_6$ Source: Marine-derived fungus *Alternaria* sp. ZJ-2008003 from soft coral *Sarcophyton* sp. (Weizhou coral reef, Giuangxi, China). Pharm: Antiviral (porcine reproductive and respiratory syndrome (PRRS) virus, IC_{50} = 65 μmol/L). Ref: C. -J. Zheng, et al, JNP, 2012, 75, 189 | S. Z. Moghadamtousi, et al, *Mar. Drugs*, 2015, *13*, 4520 (rev)

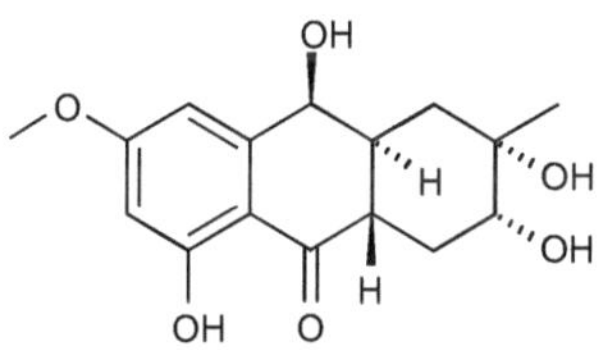

915 1,3,6,8-Tetrahydroxy-2-(1-hydroxyethyl)anthraquinone

Type: 9,10-Anthraquinones. $C_{16}H_{12}O_7$ Orange-yellow powder, $[\alpha]_D$ = +151.5° (*c* = 0.1, EtOH). Source: Marine-derived fungus *Microsphaeropsis* sp. from sponge *Aplysina aerophoba* (Mediterranean Sea). Pharm: Kinases inhibitors (PKC-ε, IC_{50} = 18.5 μmol/L; CDK4/cyclin D1, IC_{50} = 43.5 μmol/L; and EGFR, IC_{50} = 37.5 μmol/L). Ref: G. Brauers, et al, JNP, 2000, 63, 739

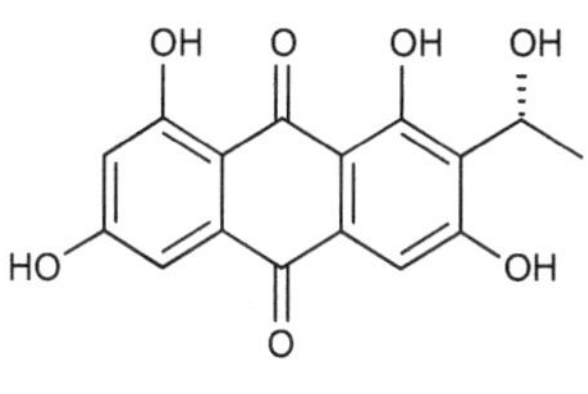

916 1,3,6,8-Tetrahydroxy-2-(1-methoxyethyl)anthraquinone

Type: 9,10-Anthraquinones. $C_{17}H_{14}O_7$ Orange-yellow powder, $[\alpha]_D = -193.8°$ ($c = 0.1$, EtOH). Source: Marine-derived fungus *Microsphaeropsis* sp. from sponge *Aplysina aerophoba* (Mediterranean Sea). Pharm: Kinases inhibitors (PKC-ε, IC_{50} = 27.0 μmol/L; CDK4/cyclin D1, IC_{50} = 22.5 μmol/L; and EGFR, IC_{50} = 27.5 μmol/L). Ref: G. Brauers, et al, JNP, 2000, 63, 739

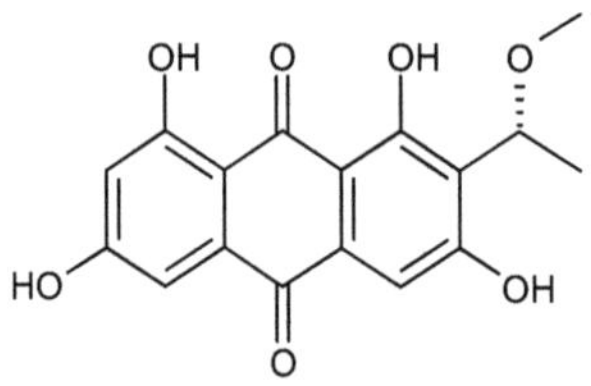

917 Trichodermaquinone

Type: 9,10-Anthraquinones. $C_{15}H_{14}O_6$ Source: Marine-derived fungus *Trichoderma aureoviride* PSU-F95 from gorgonian sea fan *Annella* sp. (Similan Is., Thailand). Pharm: Antibacterial (MRSA, MIC = 8 μg/mL). Ref: N. Khamthong, et al, Arch. Pharmacal Res., 2012, 35, 461

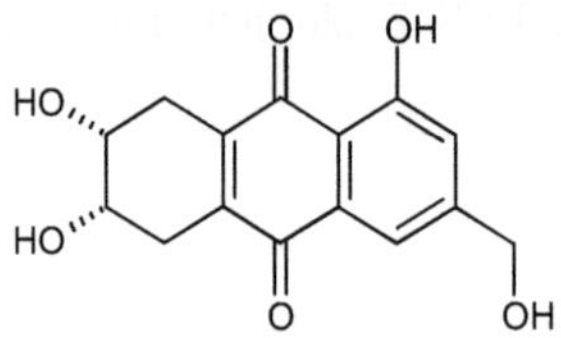

918 1,3,8-Trihydroxyanthraquinone

Type: 9,10-Anthraquinones. $C_{14}H_8O_5$ Source: Mangrove-derived endophytic fungi *Halorosellinia* sp. 1403 and *Guignardia* sp. 4382. Pharm: Cytotoxic (drug-sensitive parental KB, IC_{50} > 500 μmol/L, control Adriamycin, IC_{50} = 0.034 μmol/L; KBV200, IC_{50} = 72.60 μmol/L, Adriamycin, IC_{50} = 1.894 μmol/L). Ref: J. Y. Zhang, et al, Mar. Drugs 2010, 8, 1469

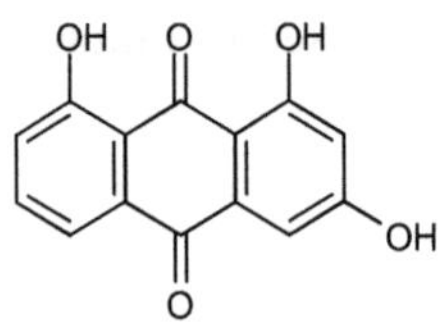

919 1,4,7-Trihydroxy-2-methoxy-6-methylanthraquinone

Type: 9,10-Anthraquinones. $C_{16}H_{12}O_6$ Source: Mangrove-derived endophytic fungi *Halorosellinia* sp. 1403 and *Guignardia* sp. 4382. Pharm: Cytotoxic (drug-sensitive parental KB, IC_{50} > 500 μmol/L, control Adriamycin, IC_{50} = 0.034 μmol/L; KBV200,

IC_{50} > 500 μmol/L, Adriamycin, IC_{50} = 1.894 μmol/L). Ref: J. Y. Zhang, et al, Mar. Drugs 2010, 8, 1469

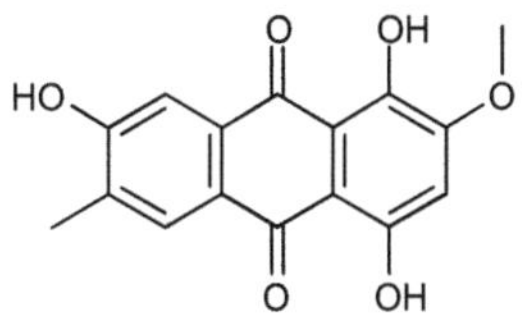

920 γ-Indomycinone

Type: Pyrano[*b*]anthraquinones. $C_{22}H_{18}O_6$ Yellow powder. Source: Marine-derived streptomycete *Streptomyces* sp. (psychrophilic, cold water, depth of 4680 m sediment core, culture broth). Pharm: Cytotoxic (hmn colon cancer line HCT116); inhibits DNA synthesis; pluramycin antibiotics. Ref: R. W. Schumacher, et al, JNP, 1995, 58, 613 | M. D. Lebar, et al, NPR, 2007, 24, 774 (rev)

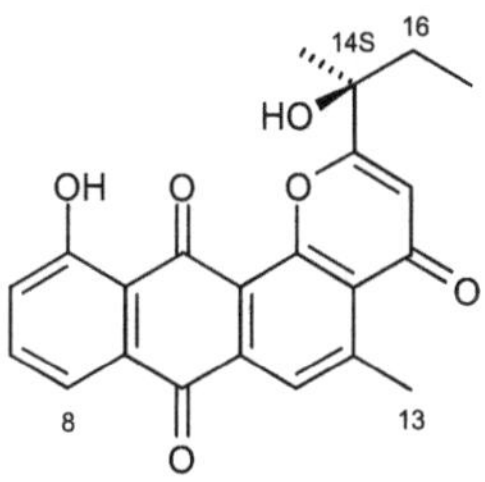

921 δ-Indomycinone

Type: Pyrano[*b*]anthraquinones. $C_{24}H_{22}O_7$ Yellow solid, mp 181 °C. Source: Marine-derived streptomycete *Streptomyces* sp. B8300 (Lagoon, Gulf of Mexico), marine-derived streptomycete *Streptomyces* sp. (psychrophilic, cold water, depth of 4680 m sediment core, culture broth). Pharm: Antioxidant.antibiotic (pluramycin-class); Ref: R. W. Schumacher, et al, JNP, 1995, 58, 613 | M. A. F. Biabani, et al, J. Antibiot., 1997, 50, 874 | L. F. Tietze, et al, Chem. Eur. J., 2007, 13, 9939 | M. D. Lebar, et al, NPR, 2007, 24, 774 (rev)

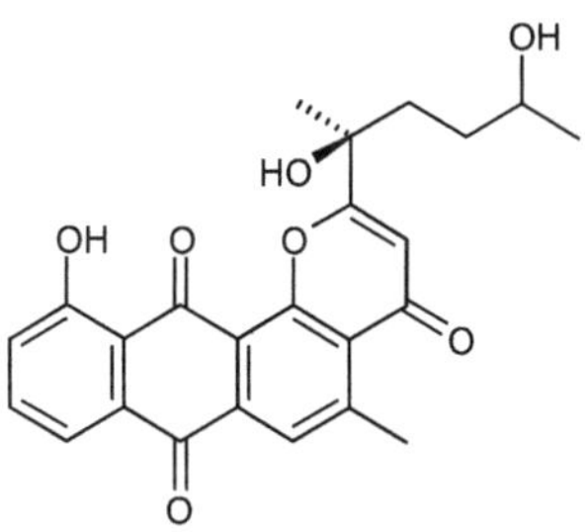

922 Saliniquinone A

Type: Pyrano[*b*]anthraquinones. $C_{23}H_{16}O_7$ Yellow solid, $[\alpha]_D$ = +440° (*c* = 0.2, $CHCl_3$). Source: Marine-derived bacterium *Salinispora arenicola* CNS-325 (sediment, Palau, Oceania). Pharm: Cytotoxic (HCT116, potent); antibacterial (MRSA, weak). Ref: B. T. Murphy, et al, Aust. J. Chem., 2010, 63, 929

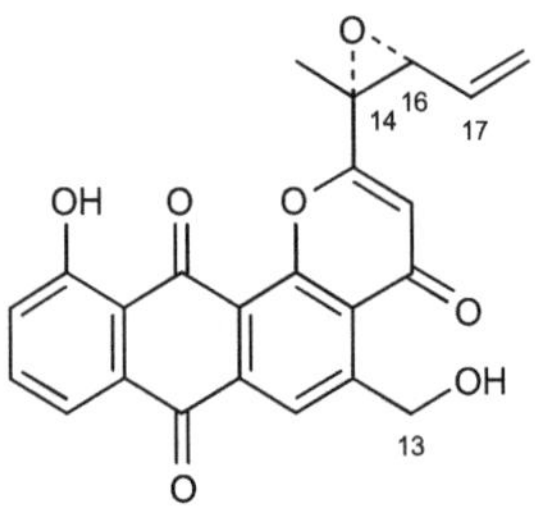

923 Aranciamycin

Antibiotic SM 173A Type: Anthracyclinones. $C_{27}H_{28}O_{12}$ Orange-yellow cryst. (MeOH), mp 240 °C (dec), $[\alpha]_D$ = +149.5° (*c* = 0.5, MeOH). Source: Marine-derived streptomycete *Streptomyces* sp. from sponge *Haliclona* sp. (Tateyama City, Chiba, Japan), terrestrial streptomycetes (*Streptomyces echinatus* and *Streptomyces chromofuscus*). Pharm: Cytotoxic (HeLa, IC_{50} = 2.7 μmol/L; HL60, IC_{50} = 4.1 μmol/L); antibacterial (gram-positive bacteria). Ref: K. Schmidt-Bäse, et al, Acta Cryst. C, 1993, 49, 250 | K. Motohashi, et al, JNP, 2010, 73, 755

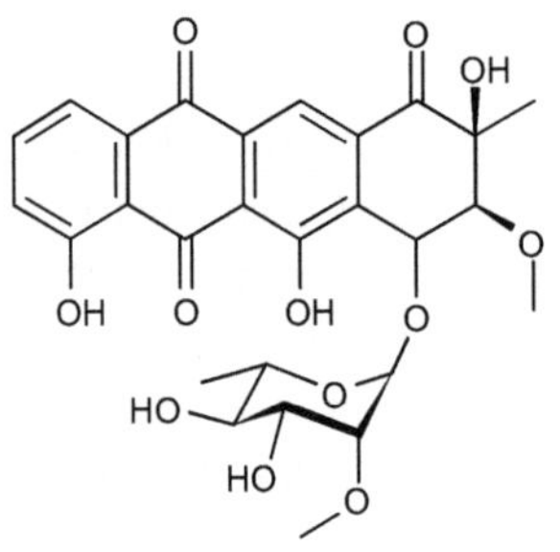

924 Tetracenoquinocin

Type: Anthracyclinones. $C_{26}H_{24}O_9$ Red oil, $[\alpha]_D^{25}$ = −69° (*c* = 0.1, $CHCl_3$). Source: Marine-derived streptomycete *Streptomyces* sp. from sponge *Haliclona* sp. (Tateyama City, Chiba, Japan). Pharm: Cytotoxic (HeLa, IC_{50} = 120 μmol/L; HL60, IC_{50} = 210 μmol/L). Ref: K. Motohashi, et al, JNP, 2010, 73, 755

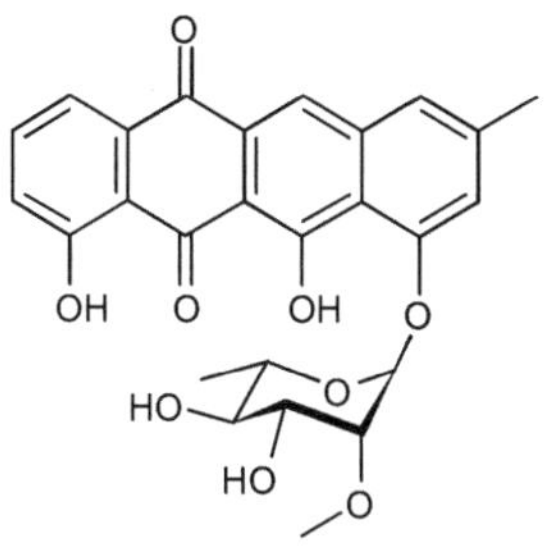

2.11 Extended Quinones

925 Aspergiolide A

Type: Extended quinones. $C_{25}H_{16}O_9$ Red needles ($CHCl_3$/MeOH). Source: Mangrove-derived fungus *Aspergillus glaucus* HB1-119 (cultured from sediment around mangrove roots, China). Pharm: Cytotoxic (MTT method: HL60, IC_{50} = 0.28 µmol/L; P_{388}, IC_{50} = 35.0 µmol/L; SRB method: A549, IC_{50} = 0.13 µmol/L; Bel7402, IC_{50} = 7.5 µmol/L). Ref: L. Du, et al, Tetrahedron, 2007, 63, 1085; 2008, 64, 4657 (corrigendum)

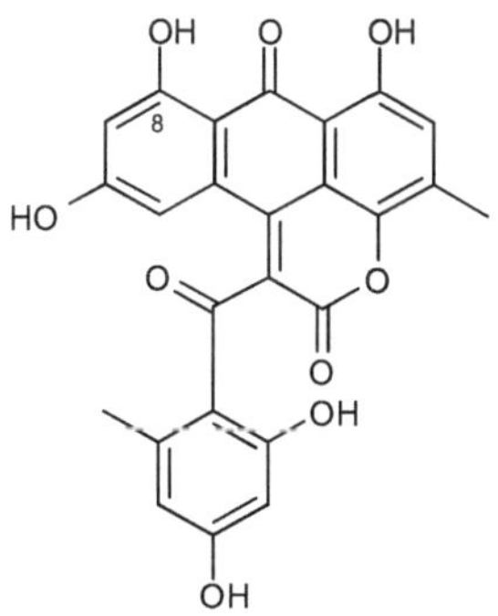

926 Aspergiolide B

Type: Extended quinones. $C_{26}H_{18}O_9$ Red solid (MeOH). Source: Mangrove-derived fungus *Aspergillus glaucus* HB1-119 (cultured from sediment around mangrove roots, China). Pharm: Cytotoxic (MTT assay, HL60, IC_{50} = 0.51 µmol/L; SRB method, A549, IC_{50} = 0.24 µmol/L). Ref: L. Du, et al, JNP, 2008, 71, 1837

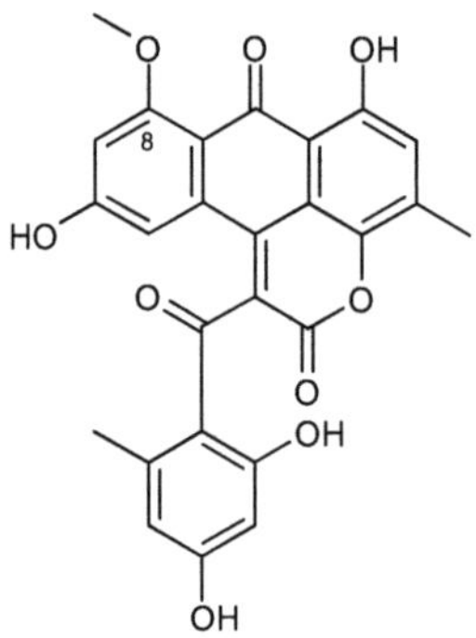

2.12 Phenalenes

927 Conioscleroderolide

Type: Phenalenes. $C_{18}H_{16}O_6$ Source: Marine-derived fungus *Coniothyrium cereale* from green alga *Enteromorpha* sp. (Fehmarn I., Baltic Sea, Germany). Pharm: Antibacterial (*Staphylococcus aureus*); hmn leukocyte elastase (HLE) inhibitor. Ref: M. F. Elsebai, et al, Org. Biomol. Chem., 2011, 9, 802

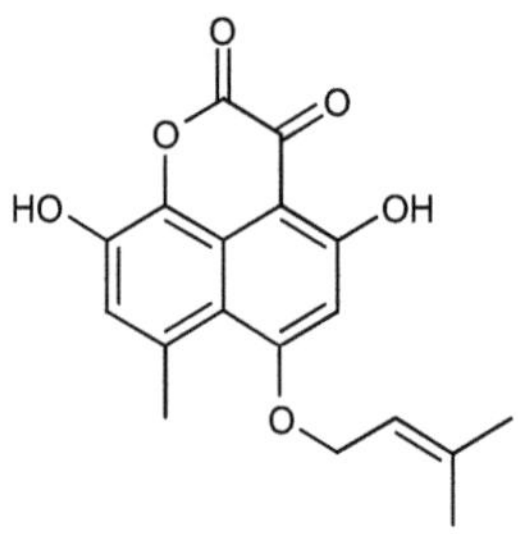

928 (*E*)-Coniosclerodinol

Type: Phenalenes. $C_{18}H_{16}O_7$ Source: Marine-derived fungus *Coniothyrium cereale* from green alga *Enteromorpha* sp. (Fehmarn I., Baltic Sea, Germany). Pharm: Antibacterial (*Mycobacterium phlei*). Ref: M. F. Elsebai, et al, Org. Biomol. Chem., 2011, 9, 802

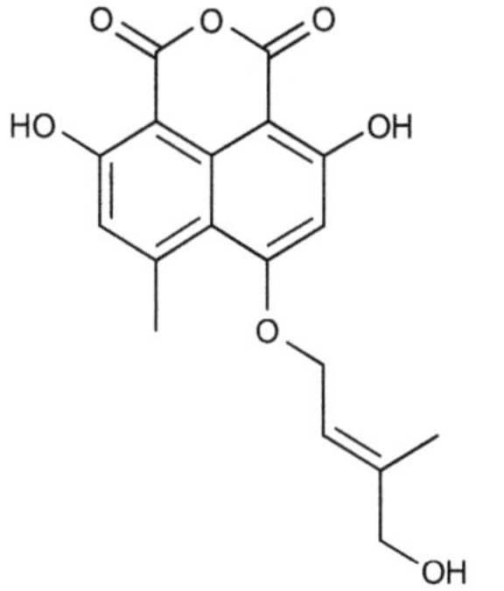

929 (Z)-Coniosclerodinol

Type: Phenalenes. $C_{18}H_{16}O_7$ Source: Marine-derived fungus *Coniothyrium cereale* from green alga *Enteromorpha* sp. (Fehmarn I., Baltic Sea, Germany). Pharm: Antibacterial (*Mycobacterium phlei*, (20 µg/disk, IZ = 16 mm). Ref: M. F. Elsebai, et al, Org. Biomol. Chem., 2011, 9, 802

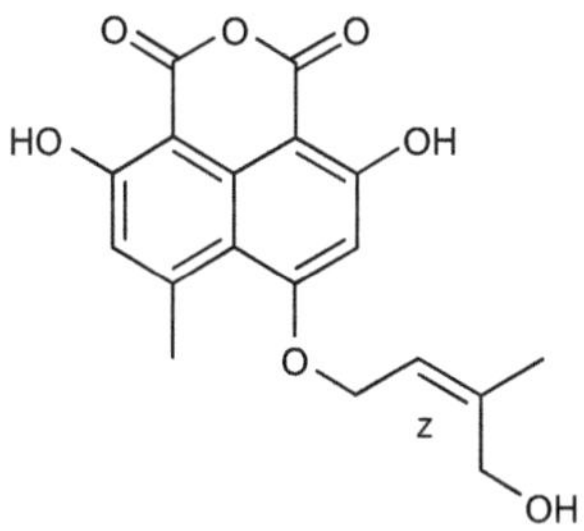

930 (15S,17S)-(−)-Sclerodinol

Type: Phenalenes. $C_{18}H_{16}O_7$ Source: Marine-derived fungus *Coniothyrium cereale*. Pharm: Antibacterial (*Mycobacterium phlei*, 20 µg/disk, IZ = 20 mm). Ref: M. F. Elsebai, et al, Org. Biomol. Chem., 2011, 9, 802

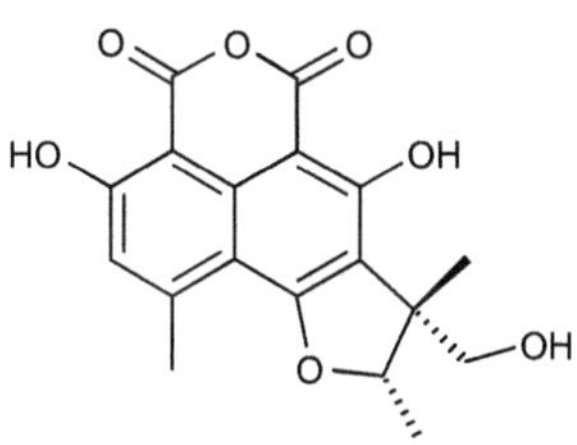

931 Sculezonone A

Type: Phenalenes. $C_{20}H_{20}O_8$ Amorph. yellow solid, $[\alpha]_D^{23}$ = +45° (c = 0.2, MeOH). Source: Marine-derived fungus *Penicillium* sp. (cultured broth) from mussel *Mytilus coruscus* (Okinawa). Pharm: DNA polymerases inhibitor (bovine DNA polymerase *a* and *c*; moderate activity for bovine DNA polymerase). Ref: K. Komatsu, et al, JNP, 2000, 63, 408 | M. Perpelescu, et al, Biochemistry, 2002, 41, 7610 | M. Saleem, et al, NPR, 2007, 24, 1142 (rev)

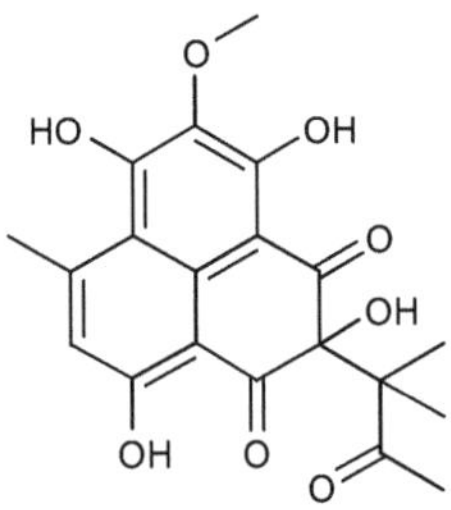

932 Sculezonone B

Type: Phenalenes. $C_{20}H_{20}O_9$ Amorph. yellow solid, $[\alpha]_D^{23}$ = +130° (*c* = 0.2, MeOH). Source: Marine-derived fungus *Penicillium* sp. (cultured broth) from mussel *Mytilus coruscus* (Okinawa). Pharm: DNA polymerases inhibitor (bovine DNA polymerase *a* and *c*; moderate activity for bovine DNA polymerase). Ref: K. Komatsu, et al, JNP, 2000, 63, 408 | M. Perpelescu, et al, Biochemistry, 2002, 41, 7610 | M. Saleem, et al, NPR, 2007, 24, 1142 (rev)

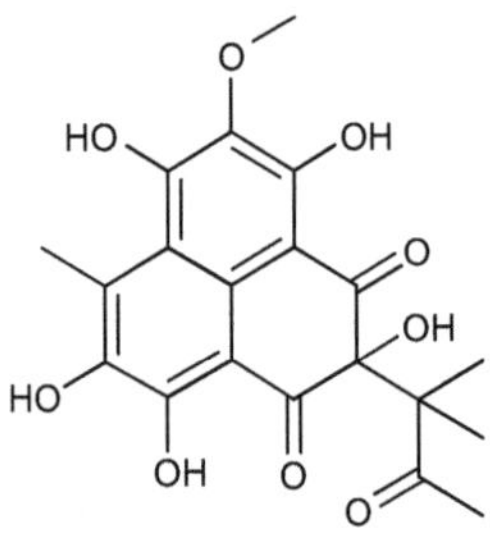

2.13 Acenaphthalenes

933 Coniosclerodione

Type: Acenaphthalenes. $C_{18}H_{16}O_5$ Source: Marine-derived fungus *Coniothyrium cereale* from green alga *Enteromorpha* sp. (Fehmarn I., Baltic Sea, Germany). Pharm: Antibacterial (20 μg/disk: *Micrococcus luteus*, IZ = 10 mm, *Mycobacterium phlei*, IZ = 12 mm; *Staphylococcus aureus*, MIC = 65.7 μmol/L). Ref: M. F. Elsebai, et al, Org. Biomol. Chem., 2011, 9, 802

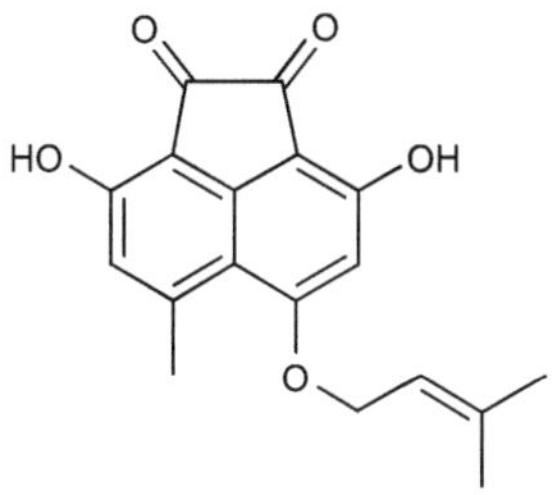

934 (–)-Sclerodione

Type: Acenaphthalenes. $C_{18}H_{16}O_5$ Bright-red or red-brown cryst. (CH_2Cl_2/hexane), mp 210–212 °C, mp 204–207 °C, $[\alpha]_D = -115.3°$ (c = 0.17, $CHCl_3$). Source: Marine-derived fungus *Coniothyrium cereal*, terrestrial fungi *Gremmeniella abietina* and *Roesleria hypogeal*. Pharm: Antibacterial (gram-positive bacteria); antifungal (yeast); Phytotoxin. Ref: W. A. Ayer, et al, Can. J. Chem., 1986, 64, 1585 | M. F. Elsebai, et al, Org. Biomol. Chem., 2011, *9*, 802

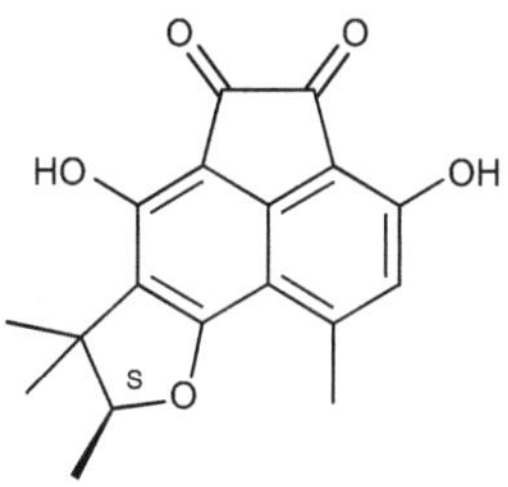

2.14 Miscellaneous Polycyclic Aromatics

935 Arenjimycin

Type: Miscellaneous polycyclic aromatics. $C_{33}H_{33}NO_{14}$ Source: Marine-derived bacterium *Salinispora arenicola* from ascidian *Ecteinascidia turbinata* (Sweetings Cay, Grand Bahama I., Bahamas). Pharm: Cytotoxic (HCT116, potent); antibacterial (DRS, other gram-positive pathogens, *Mycobacterium bacille*). Ref: R. N. Asolkar, et al, J. Antibiot., 2010, 63, 37

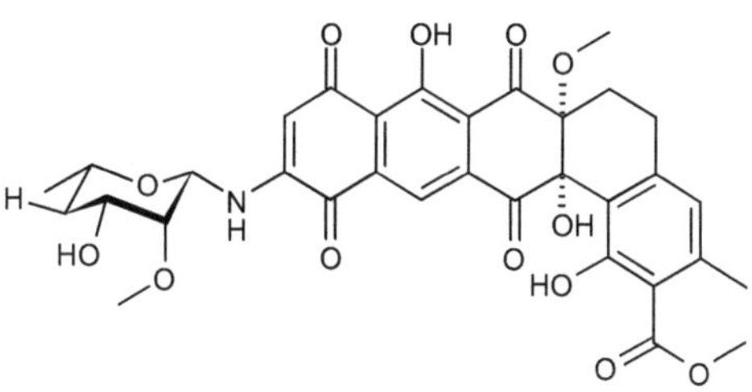

936 Bryoanthrathiophene

5,7-Dihydroxy-1-methyl-6*H*-anthra[1,9-*bc*]thiophen-6-one Type: Miscellaneous polycyclic aromatics. $C_{16}H_{10}O_3S$ Yellow powder. Source: Bryozoan *Watersipora subtorquata*. Pharm: Antiangiogenic. Ref: S. -J. Jeong, et al, JNP, 2002, 65, 1344

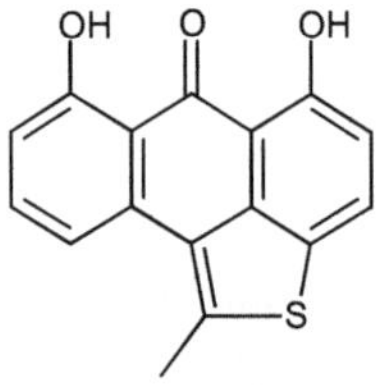

937 Citreaglycon A

Type: Miscellaneous polycyclic aromatics. $C_{27}H_{20}O_{11}$ Source: Marine-derived streptomycete *Streptomyces caelestis* (seawater, Jiddah, Saudi Arabia). Pharm: Antibacterial (*Staphylococcus haemolyticus* UST950701-004, MIC = 8.0 μg/mL, control Penicillin G, MIC = 0.13 μg/mL, Streptomycin, MIC = 8.0 μg/mL; *Staphylococcus aureus* UST950701-005, MIC = 16.0 μg/mL, Penicillin G, MIC = 0.25 μg/mL, Streptomycin, MIC = 8.0 μg/mL; *Bacillus subtilis* 769, MIC = 8.0 μg/mL, Penicillin G, MIC = 0.13 μg/mL, Streptomycin, MIC = 8.0 μg/mL; MRSA ATCC43300, MIC = 8.0 μg/mL, Penicillin G, MIC > 50 μg/mL, Streptomycin, MIC > 50 μg/mL); cytotoxic (HeLa, IC_{50} > 40 μg/mL, control Cisplatin, IC_{50} = 18.14 μg/mL). Ref: L. -L. Liu, et al, Mar. Drugs, 2012, 10, 2571

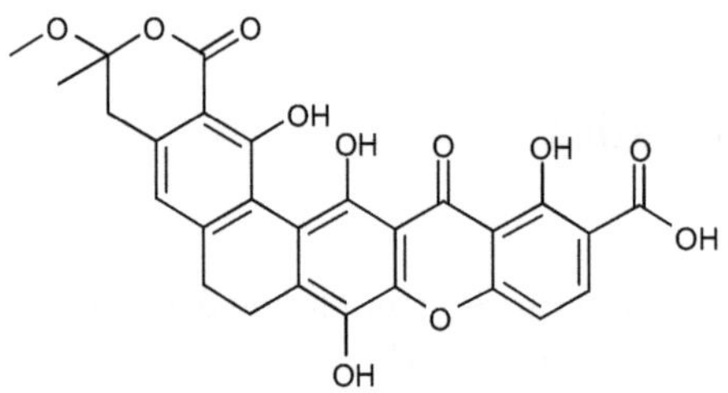

938 Citreamicin θA

Type: Miscellaneous polycyclic aromatics. $C_{30}H_{25}NO_{11}$ Source: Marine-derived streptomycete *Streptomyces caelestis* (seawater, Jiddah, Saudi Arabia). Pharm: Antibacterial (*Staphylococcus haemolyticus* UST950701-004, MIC = 0.5 μg/mL, control Penicillin G, MIC = 0.13 μg/mL, Streptomycin, MIC = 8.0 μg/mL; *Staphylococcus aureus* UST950701-005, MIC = 1.0 μg/mL, Penicillin G, MIC = 0.25 μg/mL, Streptomycin, MIC = 8.0 μg/mL; *Bacillus subtilis* 769, MIC = 0.25 μg/mL, Penicillin G, MIC = 0.13 μg/mL, Streptomycin, MIC = 8.0 μg/mL; *Staphylococcus aureus* (MRSA) ATCC43300, MIC = 0.25 μg/mL, Penicillin G, MIC > 50 μg/mL, Streptomycin,

MIC > 50 µg/mL); cytotoxic (HeLa, IC_{50} = 0.055 µg/mL, control Cisplatin, IC_{50} = 18.14 µg/mL). Ref: L. -L. Liu, et al, Mar. Drugs, 2012, 10, 2571

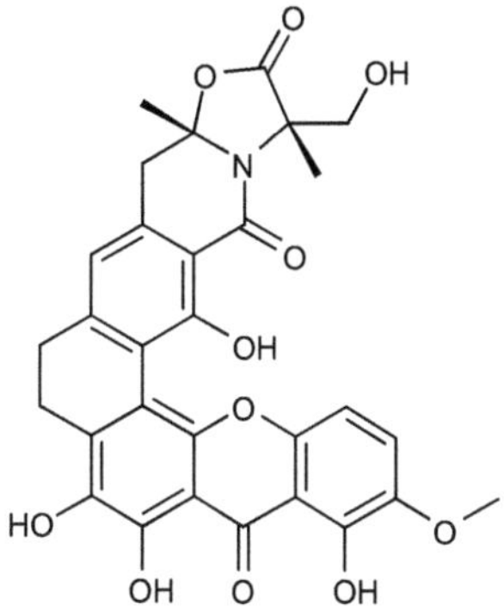

939 Citreamicin θB

Type: Miscellaneous polycyclic aromatics. $C_{30}H_{25}NO_{11}$ Source: Marine-derived streptomycete *Streptomyces caelestis* (seawater, Jiddah, Saudi Arabia). Pharm: Antibacterial (*Staphylococcus haemolyticus* UST950701-004, MIC = 0.5 µg/mL, control Penicillin G, MIC = 0.13 µg/mL, Streptomycin, MIC = 8.0 µg/mL; *Staphylococcus aureus* UST950701-005, MIC = 1.0 µg/mL, Penicillin G, MIC = 0.25 µg/mL, Streptomycin, MIC = 8.0 µg/mL; *Bacillus subtilis* 769, MIC = 0.25 µg/mL, Penicillin G, MIC = 0.13 µg/mL, Streptomycin, MIC = 8.0 µg/mL; *Staphylococcus aureus* (MRSA) ATCC43300, MIC = 0.25 µg/mL, Penicillin G, MIC > 50 µg/mL, Streptomycin, MIC > 50 µg/mL); cytotoxic (HeLa, IC_{50} = 0.072 µg/mL, control Cisplatin, IC_{50} = 18.14 µg/mL). Ref: L. -L. Liu, et al, Mar. Drugs, 2012, 10, 2571

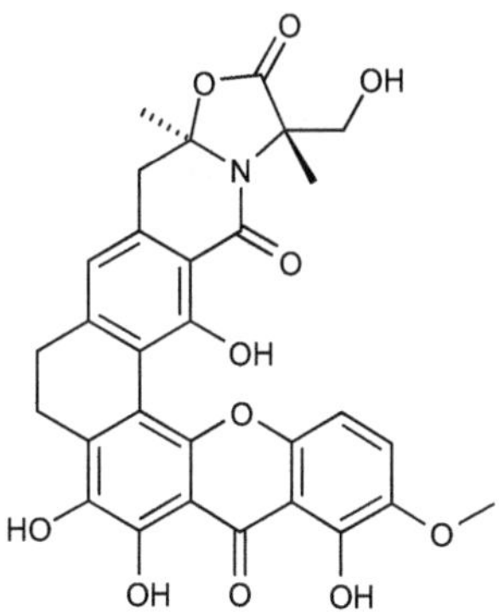

940 Dehydrocitreaglycon A

Type: Miscellaneous polycyclic aromatics. $C_{27}H_{20}O_{10}$ Source: Marine-derived streptomycete *Streptomyces caelestis* (seawater, Jiddah, Saudi Arabia). Pharm: Antibacterial (*Staphylococcus haemolyticus* UST950701-004, MIC = 8.0 µg/mL, control Penicillin G,

MIC = 0.13 μg/mL, Streptomycin, MIC = 8.0 μg/mL; *Staphylococcus aureus* UST950701-005, MIC = 16.0 μg/mL, Penicillin G, MIC = 0.25 μg/mL, Streptomycin, MIC = 8.0 μg/mL; *Bacillus subtilis* 769, MIC = 8.0 μg/mL, Penicillin G, MIC = 0.13 μg/mL, Streptomycin, MIC = 8.0 μg/mL; *Staphylococcus aureus* (MRSA) ATCC43300, MIC > 50 μg/mL, Penicillin G, MIC > 50 μg/mL, Streptomycin, MIC > 50 μg/mL); cytotoxic (HeLa, IC_{50} > 40 μg/mL, control Cisplatin, IC_{50} = 18.14 μg/mL). Ref: L. -L. Liu, et al, Mar. Drugs, 2012, 10, 2571

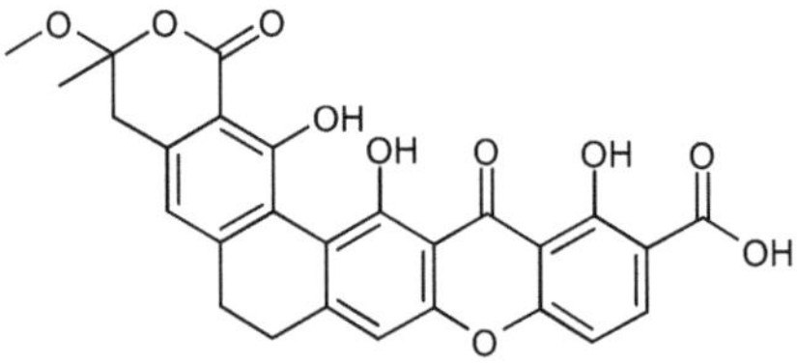

941 5,7-Dihydroxy-1-(hydroxymethyl)-6*H*-anthra[1,9-*bc*]thiophen-6-one

Type: Miscellaneous polycyclic aromatics. $C_{16}H_{10}O_4S$ mp > 260 °C. Source: Bryozoan *Dakaria subovoidea* (Japan waters). Pharm: Antioxidant; hypolipemic. Ref: T. Shindo, et al, Experientia, 1993, 49, 177

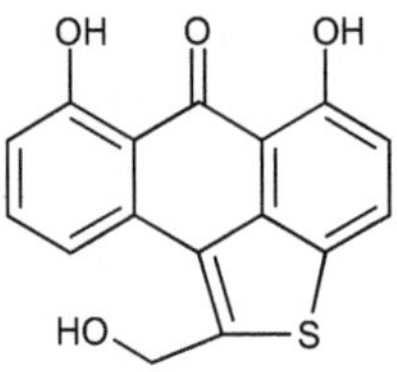

942 5,7-Dihydroxy-6-oxo-6*H*-anthra[1,9-*bc*]thiophene-1-carboxylic acid, methyl ester

Type: Miscellaneous polycyclic aromatics. $C_{17}H_{10}O_5S$ mp 230–235 °C. Source: Bryozoan *Dakaria subovoidea* (Japan waters). Pharm: Antioxidant; hypolipemic. Ref: T. Shindo, et al, Experientia, 1993, 49, 177 | T. R. Kelly, et al, Org. Lett., 2000, 2, 2351-2352

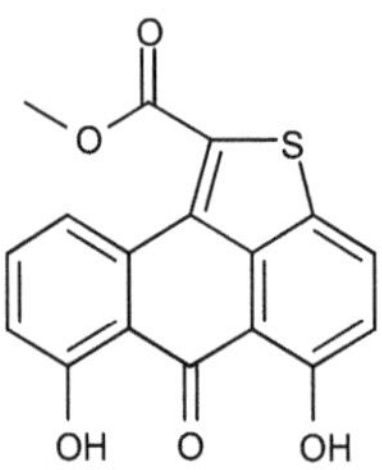

943 1-Hydroxy-1-norresistomycin

1,3,5,7,10-Pentahydroxy-1,9-dimethyl-2*H*-benzo[*cd*]pyrene-2,6(1*H*)-dione Type: Miscellaneous polycyclic aromatics. $C_{21}H_{14}O_7$ Orange solid. Source: Marine-derived streptomycete *Streptomyces chibaensis* AUBN1/7. Pharm: Cytotoxic (gastric adenocarcinoma HM02 and hmn liver cancer HepG2 cells); antibacterial (gram-positive and -negative bacteria). Ref: A. Gorajana, et al, J. Antibiot., 2005, 58, 526 | I. Kock, et al, J. Antibiot., 2005, 58, 530

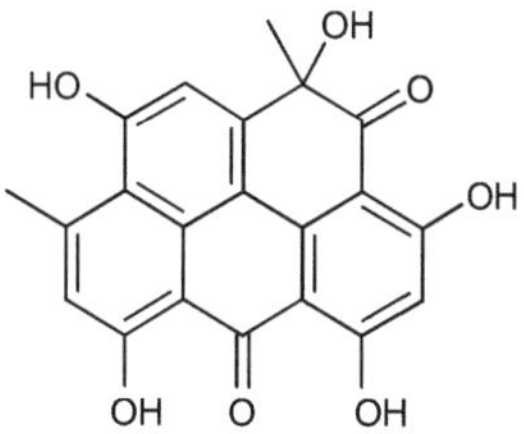

944 Kiamycin

Type: Miscellaneous polycyclic aromatics. $C_{20}H_{18}O_4$ Source: Marine-derived streptomycete *Streptomyces* sp. M268 (sediment, Qingdao, China). Pharm: Cytotoxic (100 μmol/L: HL60, InRt = 68.8%, control Adriamycin, InRt = 70.0%; Bel7402, InRt = 31.7%, Adriamycin, InRt = 79.3%; A549, InRt = 55.9%, Adriamycin, InRt = 80.8%). Ref: Z. Xie, et al, Mar. Drugs, 2012, 10, 551

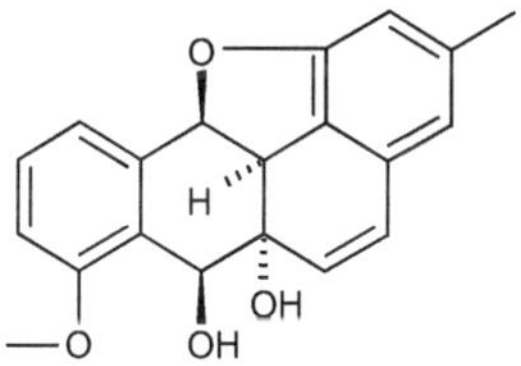

945 (–)-Lomaiviticin A

Type: Miscellaneous polycyclic aromatics. $C_{68}H_{80}N_6O_{24}$ Amorph. red powder, $[\alpha]_D$ = –101.4° (*c* = 0.07, MeOH). Source: Marine-derived bacterium *Micromonospora lomaivitiensis* LL-37I366. Pharm: Cytotoxic (biochemical induction assay, potent DNA damaging agent with minimum induction concentration ≤ 0.1 ng/spot); cytotoxic (24 cancer cell line panel, cytotoxicity profile was compared to known DNA-damaging anticancer drugs adriamycin and mitomicin C, suggesting different mechanism of interaction with DNA molecules); cytotoxic (potent: K562, IC_{50} = 11 nmol/L, LNCaP, IC_{50} = 2 nmol/L, HCT116, IC_{50} = 2 nmol/L, HeLa, IC_{50} = 7 nmol/L);cytotoxic (number of cancer cell lines, IC_{50} = 0.01–98 ng/mL); antibacterial (plate assay, gram-positive bacteria, MIC = 6–25 ng/spot). Ref: H. He, et al, JACS, 2001, 123, 5362 | C. M. Woo, et al, JACS, 2012, 134, 15285

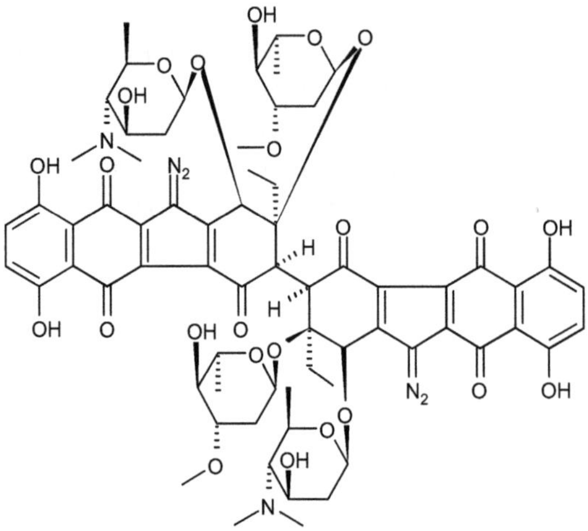

946 Lomaiviticin B

Type: Miscellaneous polycyclic aromatics. $C_{54}H_{56}N_6O_{18}$ Amorph. red powder, $[\alpha]_D$ = −71.4° (*c* = 0.07, MeOH). Source: Marine-derived bacterium *Micromonospora lomaivitiensis* LL-371366. Pharm: Cytotoxic (biochemical induction assay, potent DNA damaging agent with minimum induction concentration ≤ 0.1 ng/spot); cytotoxic (24 cancer cell line panel, cytotoxicity profile was compared to known DNA-damaging anticancer drugs adriamycin and mitomicin C, suggesting different mechanism of interaction with DNA molecules); antibacterial (plate assay, gram-positive bacteria, MIC = 6–25 ng/spot). Ref: H. He, et al, JACS, 2001, 123, 5362 | C. M. Woo, et al, JACS, 2012, 134, 15285

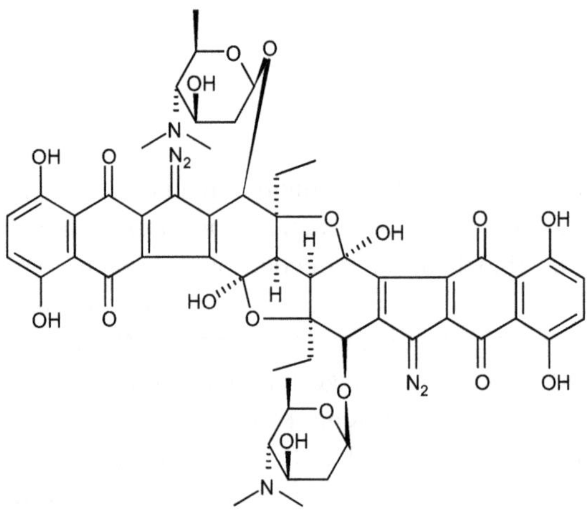

947 (–)-Lomaiviticin C

Type: Miscellaneous polycyclic aromatics. $C_{68}H_{82}N_4O_{24}$ Source: Marine-derived bacterium *Salinispora pacifica* (USDA Agricultural Research Service). Pharm: Cytotoxic (K562, IC_{50} = 472 nmol/L, LNCaP, IC_{50} = 332 nmol/L, HCT116, IC_{50} = 223 nmol/L, HeLa, IC_{50} = 589 nmol/L). Ref: C. M. Woo, et al, JACS, 2012, 134, 15285

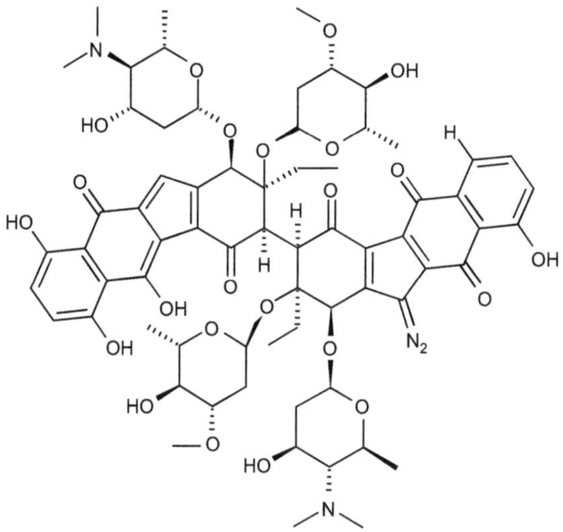

948 (–)-Lomaiviticin D

Type: Miscellaneous polycyclic aromatics. $C_{69}H_{84}N_4O_{24}$ Source: Marine-derived bacterium *Salinispora pacifica* (USDA Agricultural Research Service). Pharm: Cytotoxic (K562, IC_{50} = 197 nmol/L, LNCaP, IC_{50} = 196 nmol/L, HCT116, IC_{50} = 167 nmol/L, HeLa, IC_{50} = 161 nmol/L). Ref: C. M. Woo, et al, JACS, 2012, 134, 15285

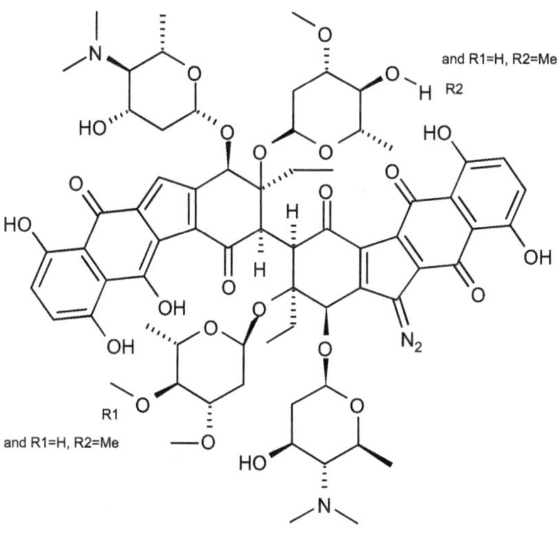

949 (–)-Lomaiviticin E

Type: Miscellaneous polycyclic aromatics. $C_{70}H_{86}N_4O_{24}$ Source: Marine-derived bacterium *Salinispora pacifica* (USDA Agricultural Research Service). Pharm: Cytotoxic (K562, IC_{50} = 496 nmol/L, LNCaP, IC_{50} = 964 nmol/L, HCT116, IC_{50} = 255 nmol/L, HeLa, IC_{50} = 292 nmol/L). Ref: C. M. Woo, et al, JACS, 2012, 134, 15285

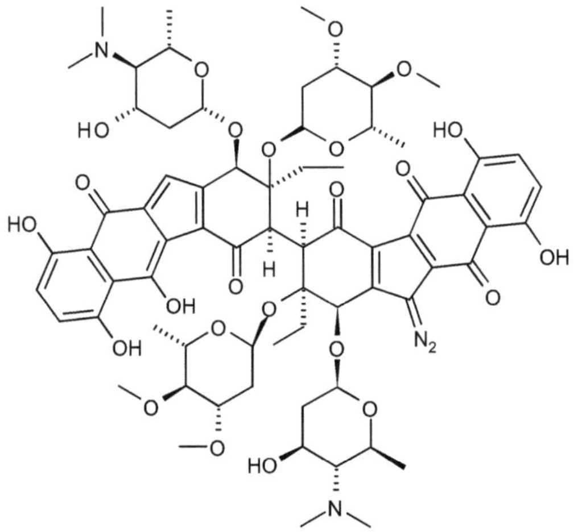

950 Talaromycesone A

Type: Miscellaneous polycyclic aromatics. $C_{28}H_{24}O_{11}$ Source: Marine fungus *Talaromyces* sp. LF458. Pharm: Antibacterial (hmn pathogenic *Staphylococcus epidermidis*, IC_{50} = 3.70 µmol/L; MRSA, IC_{50} = 5.48 µmol/L); acetylcholinesterase inhibitor. Ref: B. Wu, et al, Mar. Biotechnol., 2015, 17, 110

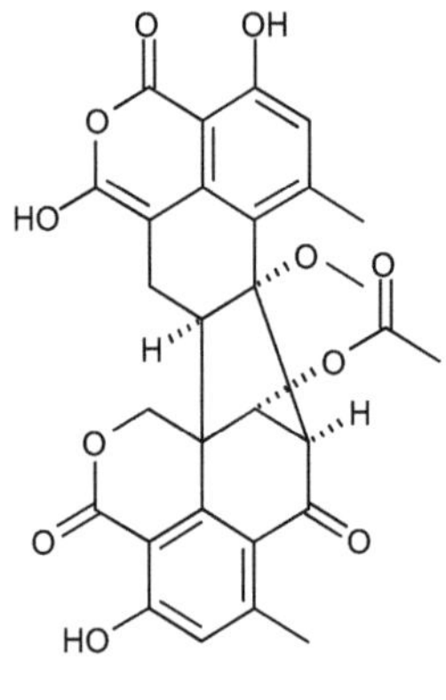

951 Talaromycesone B

Type: Miscellaneous polycyclic aromatics. $C_{27}H_{18}O_{10}$ Source: Marine fungus *Talaromyces* sp. LF458. Pharm: Antibacterial (hmn pathogenic *Staphylococcus epidermidis*, IC_{50} = 17.36 μmol/L; MRSA, IC_{50} = 19.50 μmol/L); acetylcholinesterase inhibitor. Ref: B. Wu, et al, Mar. Biotechnol., 2015, 17, 110

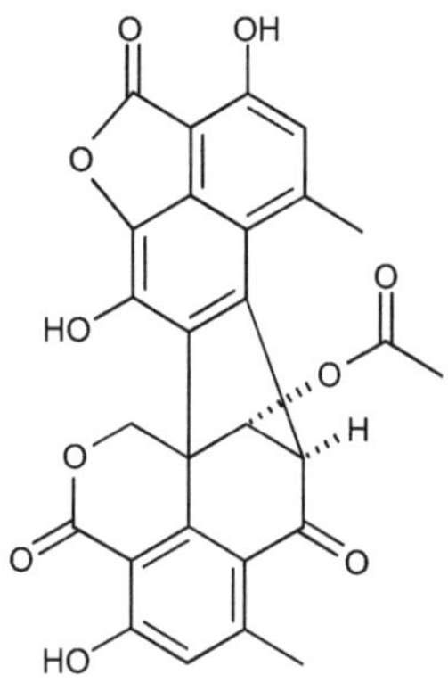

Index 1 Compound Name and Synonym Index

This index lists in alphabetical order all active compound's 1,084 entry names including both 951 key names and 133 synonym names contained in the bodies of compound entries. A equal sign (=) and compound code number (from 1 to 951) follow the name immediately for locating the compound in the "*Handbook of Active Marine Natural Products Volume 7*" book. Following symbols are ineffective in ordering: *D-*, *L-*, *dl*, *R-*, *S-*, *E-*, *Z-*, *O-*, *N-*, *C-*, *H-*, *cis-*, *trans-*, *ent-*, *epi-*, *meso-*, *erythro-*, *threo-*, *sec-*, *seco-*[§], *m-*, *o-*, *p-*, *n-*, *α-*, *β-*, *γ-*, *δ-*, *ε-*, *κ-*, *ξ-*, *ψ-*, *ω-*, (+), (–), (±) etc., and: 0, 1, 2, 3, 4, 5, 6, 7, 8, 9, {,}, [,], (,),.,:,;, *, ', ", "', →, etc. ([§]note: In the books regular "seco-" is effective in ordering as "nor-".)

https://doi.org/10.1515/9783110655810-003

I

J

Index 2 Compound Molecular Formula Index

The Molecular Formula Index of Volume 7 lists the molecular formulae of all 951 active isolated compounds from marine organisms given in the HAMNP Volume 7 in Hill convention order. Under a bold formula, all related compound names following code numbers are listed in the code number order too.

Index 3 Compound Organism Source Index

This index lists in alphabetical order all 594 marine organism in Latin names in **HAMNP Volume 7**, following a code sequence of related active compounds. When one hopes to know the English type name of any marine organism, please see an entry of a related compound in the code sequence. For example, if one hopes to know the English common type name of "*Haliclona* sp.", from entry 923 of this index, one will know that the *Haliclona* sp. is a sponge.

Index 4 Compound Sampling Geographic Locality Index

In this index, all geographic locations in **HAMNP Volume 7** have been devided as 17 large areas:

1 **CHINA**
2 **JAPAN**
3 **RUSSIA**
4 **KOREA WATERS**
5 **R. O. KOREA**
6 **ASIA**
7 **AUSTRALIA**
8 **OCEANIA**
9 **EUROPE**
10 **AFRICA**
11 **USA**
12 **NORTH AMERICA**
13 **CARIBBEAN SEA**
14 **SOUTH AMERICA**
15 **PACIFIC OCEAN**
16 **ATLANTIC OCEAN**
17 **ANTARCTIC/ARCTIC**

For all 147 compound sampling geographic locations, each of them has put into one large area, and then within the area, all related geographic places are listed in alphabetical order with the detail information in the texts of the "Handbook of Active Marine Natural Products" books and a number code sequence of the related compounds follows the detail information immediately. There are 555 related compounds with geographic information in HAMNP Volume 7.

1 CHINA

2 JAPAN

3 RUSSIA

4 KOREA WATERS

5 R. O. KOREA

6 ASIA

7 AUSTRALIA

8 OCEANIA

9 EUROPE

10 AFRICA

11 USA

12 NORTH AMERICA

13 CARIBBEAN SEA

14 SOUTH AMERICA

15 PACIFIC OCEAN

16 ATLANTIC OCEAN

17 ANTARCTIC/ARCTIC

Index 5 Compound Pharmacological Activity Index

In this index, a set of very formatted pharmacological activity codes have been used, specially for all types of cancer cells, please see **"List of Cancer Cells Codes"**. A special note is that the word "Cytotoxic" means ***in vitro*** anticancer activities, while the word "Antineoplastic" means ***in vivo*** anticancer activities.

(The End of HAMNP Volume 7)

Bei Fragen zur Produktsicherheit wenden Sie sich bitte an:
If you have any questions regarding product safety,
please contact:

Walter de Gruyter GmbH
Genthiner Straße 13
10785 Berlin
productsafety@degruyterbrill.com